Carbonate Petroleum Reservoirs

Edited by
Perry O. Roehl and Philip W. Choquette

With 386 Illustrations

Springer-Verlag
New York Berlin Heidelberg Tokyo

Editors

Perry O. Roehl, Department of Geology, Trinity University, San Antonio, TX 78282, USA

Philip W. Choquette, Marathon Oil Company, P.O. Box 296, Littleton, CO 80160, USA

Series Editor

Robert N. Ginsburg, University of Miami, School of Marine and Atmospheric Science, Fisher Island Station, Miami, FL 33139, USA

On the front cover: Photo of Cerro Azul No. 4, reprinted from *Mexican Petroleum,* copyright 1922 by Pan-American Petroleum and Transport Co., New York.

On the front endpaper: Tectonic map of the world redrafted from a map that appeared in *Plate Tectonics and Crustal Evolution,* Second Edition, by Kent C. Condie, copyright 1982 by Pergamon Press Inc., Oxford. Reprinted by permission.

Library of Congress Cataloging in Publication Data
Main entry under title:
Carbonate petroleum reservoirs.
   (Casebooks in earth sciences)
   Bibliography: p.
   Includes index.
   1. Petroleum—Geology—Case studies. 2. Rocks,
Carbonate—Case studies. I. Roehl, Perry O.
II. Choquette, Philip W. III. Series.
TN870.5.C34 1985   553.2′8   85-4725

Media conversion by Ampersand Publisher Services, Inc., Rutland, Vermont.
Printed and bound by Halliday Lithographers, West Hanover, Massachusetts.
Printed in the United States of America.

9 8 7 6 5 4 3 2 1

ISBN 0-387-96012-0 Springer-Verlag New York Berlin Heidelberg Tokyo
ISBN 3-540-96012-0 Springer-Verlag Berlin Heidelberg New York Tokyo

Series Preface

The case history approach has an impressive record of success in a variety of disciplines. Collections of case histories, casebooks, are now widely used in all sorts of specialties other than in their familiar application to law and medicine. The case method had its formal beginning at Harvard in 1871 when Christopher Lagdell developed it as a means of teaching. It was so successful in teaching law that it was soon adopted in medical education, and the collection of cases provided the raw material for research on various diseases. Subsequently, the case history approach spread to such varied fields as business, psychology, management, and economics, and there are over 100 books in print that use this approach.

The idea for a series of *Casebooks in Earth Sciences* grew from my experience in organizing and editing a collection of examples of one variety of sedimentary deposits. The project began as an effort to bring some order to a large number of descriptions of these deposits that were so varied in presentation and terminology that even specialists found them difficult to compare and analyze. Thus, from the beginning, it was evident that something more than a simple collection of papers was needed. Accordingly, the nearly fifty contributors worked together with George de Vries Klein and me to establish a standard format for presenting the case histories. We clarified the terminology and some basic concepts, and when the drafts of the cases were completed we met to discuss and review them. When the collection was ready to submit to the publisher, and I was searching for an appropriate subtitle, a perceptive colleague, R. Michael Lloyd pointed out that it was a collection of case histories comparable in principle to the familiar casebooks of law and medicine. After this casebook [*Tidal Deposits* (1975)] was published and accorded a warm reception, I realized that the same approach could well be applied to many other subjects in earth science.

It is the aim of this new series, *Casebooks in Earth Sciences,* to apply the discipline of compiling and organizing truly representative case histories to accomplish various objectives: establish a collection of case histories for both reference and teaching; clarify terminology and basic concepts; stimulate and facilitate synthesis and classification; and encourage the identification of new questions and new approaches. There are no restrictions on the subject matter for the casebook series save that they concern earth science. However, it is clear that the most appropriate subjects are those that are largely descriptive. Just as there are no fixed boundaries on subject matter, so is the format and approach of individual volumes open to the discretion of the editors working with their contributors. Most casebooks will of necessity be communal efforts with one or more editors working with a group of contributors. However, it is also likely that a collection of case histories could be assembled by one person drawing on a combination of personal experience and the literature.

Clearly the case history approach has been successful in a wide range of disciplines. The systematic application of this proven method to earth science subjects holds the promise of producing valuable new resources for teaching and research.

Miami, Florida
February, 1985

Robert N. Ginsburg
Series Editor

Preface

The success enjoyed in recent years by modern investigators of carbonate reservoirs has produced a vast new literature the volume and sophistication of which have challenged efforts at synthesis. In this book we have confined ourselves to a selection of representative, well-documented accounts of carbonate reservoirs, in the form of 35 case studies. Collectively, these accounts illustrate the variations of form and subtlety of development that typify so many fields productive from carbonate reservoir traps. The case studies also demonstrate the application and need for integration of modern concepts of carbonate depositional settings and resulting facies, diagenesis and the resulting petrophysical modification of these facies, and the reservoir properties that result from all these interrelated factors. Finally, the case studies reveal the strategic importance of a regional tectono-sedimentary framework, both for setting the depositional conditions that govern the initial accumulation of carbonates, and for establishing the complex of climatic, burial-diagenetic, and tectonic effects that finally determine whether a body of limestone or dolomite will actually become a reservoir of petroleum.

For the purposes of managerial and engineering requirements in the development and efficient production of carbonate reservoirs, the complexities seen in virtually all of these examples—in lateral and vertical geometry and continuity and in all measurable petrophysical properties—call attention once again to an urgent need. This is the need for close interaction between petroleum engineers on the one hand, concerned with efficiently producing oil and gas, and geologists on the other hand, with modern sedimentological approaches, skilled at defining and characterizing the reservoirs for these fluids.

In all of these case studies there is a strong emphasis on modern concepts and methods in carbonate sedimentology as applied to geologic studies of reservoirs. However, we have tried to make these studies both accessible and intelligible to readers trained in other disciplines. Each reservoir example is summarized in some detail in a table at the front of the case study, with characterization of the geologic setting; nature of reservoir and trap; inferences about hydrocarbon source, porosity, permeability, and other reservoir properties and parameters; general production and reserves data; and other information. The geologic classifications and many of the more esoteric terms used in the case studies, familiar to most carbonate sedimentologists but likely to be generally unfamiliar to their reservoir–engineer colleagues, are summarized in a Glossary and an illustrated outline of Classifications toward the end of the book. We have attempted to employ geographic names in accord with principal usage by the international petroleum community. It is not the purpose of this publication to validate or establish precedence for the recognition of any geopolitical or geographic entity. The standard of authority for geographic names used in this publication is *The International Petroleum Encyclopedia for 1983* (The Petroleum Publishing Co., Tulsa, OK). Finally, we have extracted generalizations about the collection of reservoirs sampled in this book, and present these together with a brief perspective in the Introduction. We hope these features of the book will encourage engineers as well as geologists to read further.

Our attempt to compile a sampling of carbonate reservoir studies began with a call for papers and solicitation for support from segments of the worldwide petroleum industry and a few associated governmental entities. This appeal brought an enthusiastic response in the form of titles and abstracts from some 50 authors, as well as the generous support of 27 sponsors, separately acknowledged herein.

Bolstered by this response, we developed a set of guidelines and presentation format that we hoped would be acceptable to the authors and to a wide readership. A workshop for the authors who had submitted initial manuscripts was convened at Vail, Colorado, in June 1980. Each prospective author presented his paper for review by the others. For this conference we were fortunate in being able to produce a volume of preprints, which allowed each contributor to review and evaluate his contributions in the context of all the submitted manuscripts. Committees of authors then worked to develop recommendations for content, methodology, format, illustration styles, terminology, documentation, and other aspects germane to a collection of case studies. Although a variety of emphases and approaches in the case studies made adherence to rigid format impractical, many of the recommendations made at Vail were adopted and are very much in evidence in this book.

Following the Vail workshop, a few additional case studies were solicited in an effort to fill in gaps in geologic and/or geographic coverage. Readers will recognize that comprehensive sampling is difficult to achieve in a manageable number of adequately documented case studies, as we attempt to address in the Introduction that follows. But we have tried.

The workshop/preprint approach would have been impossible without the generous help and financial support of many people and organizations. Robert N. Ginsburg, University of Miami, gave helpful advice in planning and implementing the Vail workshop. Members of the staff of Marathon Oil Company's Denver Research Center compiled and printed the preprint volume, which was distributed to all the authors and sponsors of the workshop.

The book manuscript went through numerous drafts, which were prepared by DeAnne Hite, Linda Millarke, and Connie Pedde of Marathon Oil Company and Linda McCabe of Trinity University. Many of the line drawings submitted by the authors were either modified or redrafted by Susan Hartline and Arzell Thompson. Final proofreading was done by Sally Andrews of Marathon and Lynn Travis and Randall Walters of Trinity. Jane C. Olson did the bulk of the subject index and Mary Roehl prepared the author index. We extend particular thanks to the management of Marathon's Research Center and the administration of Trinity University for ongoing support of this project, and to our wives, Mary Roehl and Jean Choquette, for steadfast moral support during the book's long gestation.

We thank all of the contributors to this casebook for their patience and cooperation, over an extended period of time, in working with us as we attempted to weld a wide diversity of case studies into a cohesive format. Some of our colleagues, notably Dexter H. Craig of Marathon and Robert N. Ginsburg, served as much-needed sounding boards, gave council, and reviewed some of our efforts, and we gratefully acknowledge their support. Having said all this, however, we bow to the inevitable and accept the principal responsibility for the collective outcome, which we hope will be useful and instructive.

San Antonio, Texas   *Perry O. Roehl*
Littleton, Colorado   *Philip W. Choquette*
February, 1985

Project Sponsors

We are pleased to acknowledge the substantial and timely monetary support and encouragement of the following industrial and governmental organizations:

Abu Dhabi Marine Operating Company, United Arab Emirates

AGIP S.p.A., Milan, Italy

Anschutz Corporation, Denver, Colorado

Aquitaine Company of Canada, Ltd., Calgary, Alberta, Canada

ARCO Oil and Gas Company, Dallas, Texas

Conoco, Inc., Houston, Texas

Exxon Company, USA, Houston, Texas

Gulf Canada Resources, Inc., Calgary, Alberta, Canada

Michel T. Halbouty, Houston, Texas

Home Oil Company, Ltd., Calgary, Alberta, Canada

Hudson's Bay Oil and Gas Company, Ltd., Calgary, Alberta, Canada

Japan Petroleum Exploration Company, Ltd., Tokyo, Japan

Koninklijke/Shell Exploratie en Produktie Laboratorium, The Hague, The Netherlands

Marathon Oil Company, Littleton, Colorado

Northwest Exploration Company, Denver, Colorado

Occidental Petroleum Company, Bakersfield, California

Petrobras Brasileiro, S.A., Rio de Janeiro, Brazil

Petro-Lewis Corporation, Denver, Colorado

Phillips Petroleum Company, Bartlesville, Oklahoma

Placid Oil Company, Dallas, Texas

Shell Oil Company, Houston, Texas

Sohio Petroleum Company, San Francisco, California
Southland Royalty Company, Fort Worth, Texas
Springer-Verlag New York, Inc., New York, New York
Superior Oil Company, Houston, Texas
Tenneco Oil Company, Houston, Texas
Yacimientos Petroliferos Fiscales, Buenos Aires, Argentina

Abbreviations and Conversion Factors

bbl	barrels(s)	cp	centipoise(s)
BCPD	barrels condensate per day	G	pore geometrical factor (see Roehl, this volume)
BGC	barrels gas condensate	GOR	gas-oil ratio
BLG	barrels liquid gas	IP	initial production
BO	barrels oil	md	millidarcys
BOPD	barrels oil per day	NA	not available and/or not known
BPH	barrels per hour		
BW	barrels water	P_D	extrapolated displacement capillary pressure (see Roehl, this volume)
BWPD	barrels water per day		
$BV_{P\infty}$	bulk volume occupied, extrapolated to infinite capillary pressure (see Roehl, this volume)	psi	pounds per square inch
		SCFG	standard cubic feet of gas
		S_o	oil saturation
CFG	cubic feet gas	S_w	water saturation
CFGPD	cubic feet gas per day	URE	ultimate recovery efficiency

Non-Metric Unit	Conversion Factor		Metric Unit
inches	× 2.54	=	centimeters
feet	× 0.3048	=	meters
meters	× 3.281	=	feet
miles	× 1.609	=	kilometers
acre	× 0.004047	=	square kilometers
cubic feet (standard)	× 0.02817	=	cubic meters
barrels (42 US gallons)	× 0.15891	=	cubic meters

Contents

Mesozoic Reservoirs

Cenozoic Reservoirs

Contributors

Charles W. Achauer
ARCO Oil and Gas Company, Dallas, TX 75221, USA

James H. Anderson
Exxon Production Research Company, Houston, TX 77252, USA

Koichi Aoyagi
Japan National Oil Corporation, Tokyo 100, Japan

George B. Asquith
Pioneer Production Corporation, Amarillo, TX 79189, USA

Lawrence R. Baria
Consulting Geologist, Box 369, Jackson, MS 39205, USA

Pedro Bartok
Sohio Petroleum Company, Houston, TX 77210, USA

Alfred E. Budwill
Mobil Oil Canada Ltd., Calgary, Alberta, Canada

Albert V. Carozzi
Department of Geology, University of Illinois at Urbana-Champaign, Urbana, IL 61801, USA

Philip W. Choquette
Marathon Oil Company, Denver Research Center, Littleton, CO 80160, USA

Stewart Chuber
Consulting Geologist, Schulenburg, TX 78956, USA

James H. Clement
Shell Oil Company, Rocky Mountain Division Corporation, Houston, TX 77001, USA

Thomas C. Connally, Jr.
Amoco International, Houston, TX 77210, USA

Paul D. Crevello
Marathon Oil Company, Littleton, CO 80160, USA

John M. Cys
Consulting Geologist, Midland, TX 79701, USA

S. Depowski
Instytut Geologiczny, Warsaw, Poland

James R. Derby
Derby and Associates, Inc., Tulsa, OK 74135, USA

John F. Drake
Mendota Mining Company, Canadian, TX 79014, USA

Yehezkeel Druckman
Geological Survey of Israel, Oil Research Division, 95501 Jerusalem, Israel

William J. Ebanks, Jr.
ARCO Oil and Gas Company, Dallas, TX 75221, USA

Paul Enos
Department of Geology, University of Kansas, Lawrence, KS 66044, USA

Frank U. H. Falkenhein
Petroleo Brasileiro S. A. (PETROBRAS), Exploration Department, DEPEX-DIRNOE, Rio de Janeiro, R. J., 20035, Brazil

Charles T. Feazel
Phillips Petroleum Company, Phillips Research Center, Bartlesville, OK 74004, USA

Gerald M. Friedman
Department of Geology, Rensselaer Polytechnic Institute, Troy, NY 12181, USA

Lee C. Gerhard
Department of Geology, Colorado School of Mines, Golden, CO 80401, USA

Dan Gill
Geological Survey of Israel, 95501 Jerusalem, Israel

Robert B. Halley
U. S. Geological Survey, Denver, CO 80225, USA

Paul M. Harris
Chevron Oil Field Research Company, La Habra, CA 90631, USA

Clifton F. Jordan, Jr.
Mobil Research and Development Corporation, Farmers Branch, TX 75234, USA

John Keany
Slawson Oil Company, Amarillo, TX 79105, USA

Christopher G. St. C. Kendall
Department of Geology, University of South Carolina, Columbia, SC 29208, USA

John T. Kilpatrick
Consolidated Oil & Gas Corporation, Denver, CO 80295, USA

Robert F. Lindsay
Gulf Oil Exploration & Production Company, Gulf Exploration Technology Center, Houston, TX 77236, USA

Mark W. Longman
Consulting Geologist, Lakewood, CO 80214, USA

Robert G. Loucks
ARCO Oil and Gas Company, Dallas, TX 75221, USA

Morad Malek-Aslani
Tenneco Oil Company, Houston, TX 77001, USA

Salvatore J. Mazzullo
Petroleum Geological Consultant, Midland, TX 79701, USA

Ian A. McIlreath
Petro-Canada Ltd., Calgary, Alberta, Canada

Harry McQuillan
Consulting Geologist, Upper Moutere, Nelson, New Zealand

James A. Miller
Union Oil Company of California, Science & Technology Division, Brea, CA 92621,USA

Clyde H. Moore, Jr.
Applied Carbonate Research Program, Department of Geology, Louisiana State University, Baton Rouge, LA 70803, USA

William A. Morgan
Conoco Exploration Research, Ponca City, OK 74603, USA

Tadeuscc M. Peryt
Instytut Geologiczny, Warsaw, Poland

R. Michael Peterson
Keplinger & Associates, Inc., Tulsa, OK 74112, USA

Bruce H. Purser
Laboratoire de Petrologie, Sedimentologie, & Paleontologie, Faculte des Sciences, Universite de Paris XI, 91405 Orsay, France

Walter C. Pusey
Director of Geology, International, Conoco, Inc., Houston, TX 77046, USA

Thomas J. A. Reijers
Shell Internationale Petroleum, The Hague, The Netherlands

Perry O. Roehl
Department of Geology, Trinity University, San Antonio, TX 78282, USA

Kenneth Ruzyla
Exxon Production Research Company, Houston, TX 77001, USA

Volkmar Schmidt
Petro-Canada Ltd., Calgary, Alberta, Canada

Randolph P. Steinen
Department of Geology and Geophysics, University of Connecticut, Storrs, CT 06268, USA

David L. Stoudt
Mosbacher Production Company, Houston, TX 77002, USA

Harry A. Vest
Conoco, Inc., Houston, TX 77046, USA

Willard L. Watney
Kansas Geological Survey, Lawrence, KS 66044, USA

Richard M. Weinbrandt
Consultant, Jackson, WY 83001, USA

Augustus O. Wilson
Saker Geological Services, Houston TX 77018, USA

Introduction

Perry O. Roehl and Philip W. Choquette

Although carbonate reservoirs have been important contributors to world oil and gas production for several decades, their importance has increased dramatically as a result of sharp changes in world demand in conjunction with restricted geopolitical locations of many of the truly giant carbonate fields. In a survey of 266 giant oil and gas fields with recoverable reserves of 500 million barrels equivalent or more discovered through 1967, Halbouty *et al.* (1970) cite 116 fields, or 44 percent, that produce substantially or entirely from carbonate reservoirs. A review of the statistics suggests that these carbonate reservoirs contain about 61 percent of the recoverable oil in giant fields.

It is uncertain how relevant these values may be in assessing the totality of in-place and recoverable reserves in *all* known fields of all sizes. However, if giant fields contain some 85 percent of the world's recoverable petroleum (Fitzgerald, 1980), statistics for the giants should heavily influence those for all petroleum fields. If one assumes a direct relationship for the entire range of volumetric productivity, then carbonate reservoirs should contain about 60 percent, or 486 billion, of the 810 billion barrels of recoverable oil estimated for the total recoverable world supply (Oil and Gas Journal, December 29, 1983). Even these numbers may have to be revised upward as detailed studies of large reservoirs reveal hitherto unexpected carbonates of diagenetic origin as contributors to the world's population of large reservoirs. Some Miocene reservoirs in the circum-Pacific region—on Honshu, Japan, and in parts of onshore and offshore California—are examples of this type (see the case studies by Aoyagi and by Roehl and Weinbrandt).

Aims of this Book

Some relevant statistics for petroleum production and reserves are reported, but such was not our primary intent when we set out to assemble this collection of case studies. Our main objective was to focus on the geology of a fairly significant number of carbonate reservoirs and to present each example as a condensed, yet well-integrated geologic case study. In addition, we have included as much petrophysical and reservoir–volumetric information as could be supplied by the authors. We have departed from the scheme of most other "resource" evaluations by focusing on estimates of both ultimate recoverable reserves and total initial oil in place, feeling that the size of a field as measured by volume of original oil is significantly related to the geologic details of reservoir formation. By taking this approach, we hope to diminish somewhat the geologic importance attributed to recovery efficiencies, which often are biased by the age of discovery of individual fields and the exploitation methods, or lack thereof, employed over the life of such fields.

Some recent compilations (*e.g.*, Mazzullo, 1980) highlight the increasing tempo of geologic efforts being focused on studies of carbonate reservoirs for exploration and reservoir development purposes. Until recent years, however, published case studies documenting the role of depositional facies and their diagenesis in localizing carbonate reservoirs have been relatively few.[1] With very few exceptions in the literature— notably the work of Cussey and other French sedimentologists in Elf Aquitaine, translated with additions by Reeckman and Friedman, 1982— case studies of carbonate oil and gas reservoirs emphasizing sedimentological controls on reservoir localization have not been assembled between the covers of a single volume. Even the French synthesis includes little information about the petrophysics, production data, and other reservoir characteristics of specific fields.

This book attempts to fill part of that void. In it we have assembled 35 carbonate reservoir case studies of some 41 oil and gas fields distributed geographically as shown in Figure I-1 and Table I-1. We believe these represent a significant part of the spectrum of geologic provinces, tectonic settings, and geologic ages of reservoirs, varieties of pore systems, reservoirs facies, and diagenetic histories involved in creating reservoirs in carbonates. Most of these case studies utilize a variety of approaches involving, in various mixtures, the disciplines of sedimentology (both depositional and diagenetic), petrophysics, logging and seismic geophysics, geochemistry, rock mechanics, and classical stratigraphy. Common to all of the case studies, however, is a strong emphasis on the importance (or, in fractured reservoirs, the limited relevance) of sedimentary facies and their diagenesis in the making of the reservoirs.

This collection of case studies is designed to serve as both a review and a synthesis of the progress made in recent years in geological analysis and interpretation of carbonate reservoirs. The wealth of knowledge and understanding that can now be applied in integrated exploration for new carbonate reservoirs and in production studies of existing carbonate reservoirs has encompassed a broad front of science and technology. It has come from basic studies of Recent carbonates; from new techniques for interpreting organic matter maturation; from arrays of new geophysical wireline logs; from the new "seismic stratigraphy"; from plate-tectonic and paleogeographic reconstructions; from new approaches in biostratigraphy and lithostratigraphy; and from a variety of methods for characterizing the petrophysics of porous strata. Geologists trained and experienced in modern concepts and methods for interpreting sedimentary carbonates can be vastly more successful than their predecessors in exploring for and helping optimize the productivity of reservoirs in carbonates. The emergence of carbonate geologists so armed is timely, because exploring for and developing reservoirs in stratigraphic traps usually require much higher levels of sophistication in interpreting sedimentary facies than are needed for purely structural traps.

In 1961, one of us wrote to his corporate employer: "In order to broaden his purview, it is incumbent on the geologist and engineer to undertake the study of petroleum accumulations by contemplating each of three major aspects of the reservoir, *viz*. (1) the generation, evolution, and present physiochemical characteristics of the rocks, (2) the geometry, surface properties, and volumetrics of included void space, and (3) the history and interrelated status of contained fluids. Intuitively, as well as by physical measurement, we recognize the successive controls operative in this sequence." Most of the case studies in this book have a degree of balance that was, with rare exception, only a perception two decades ago. The reader will find that each of the three related aspects pointed out above has been addressed to considerable extent in these examples, and with considerable sophistication.

Some Roots of Modern Carbonate Reservoir Geology

How did our advances in understanding limestone and dolomite reservoirs come about? Perhaps the critical step was the study of modern sedimentary facies and processes. Starting in earnest in the 1950s, geologists searched Recent carbonate environments for clues to the fundamentals of deposition and early burial history in order to establish analogs for ancient reservoirs in Recent

---

[1] Early examples are those of Barnetche and Illing (1956), Edie (1958), Andrichuk (1959), Thomas and Glaister (1960), Choquette and Traut (1963), Klovan (1964), Roehl (1967), and Hemphill *et al.* (1970).

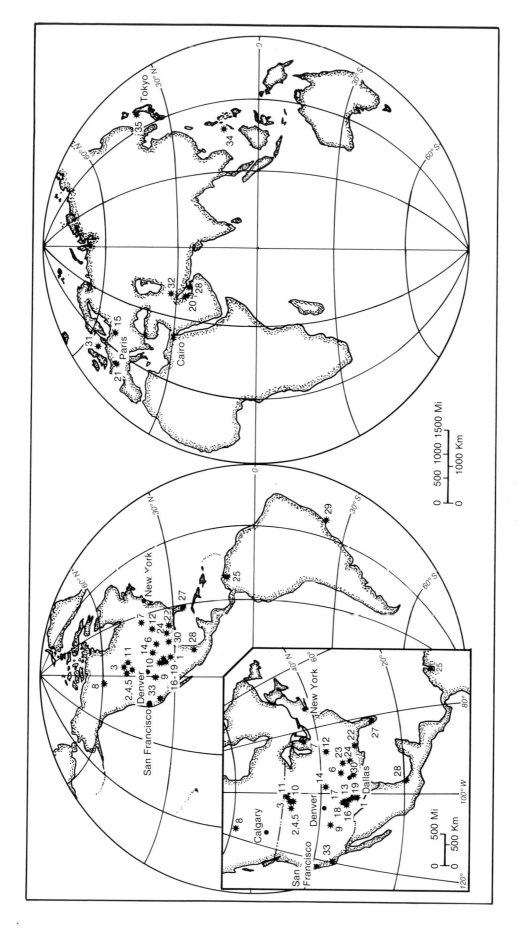

Fig. I-1. Geographic distribution of oil and gas fields in this collection. The maps are Lambert projections.

Table I.1. A Classification of Carbonate Reservoir Types Discussed in this Volume (Based in part on Wilson, 1980a)

Reservoir Rock Type	Dominant Porosity		Age and Region Example
	Origins	Characteristics	
Sub-unconformity dolomite and limestone (*subaerial diagenetic terranes*)	Near-surface dolomitization and/or solution and fracture-brecciation	Varied: moldic, vug, channel, cavern, solution-enlarged fracture and breccia. Matrix porosity often low and generally coarse	Paleozoic in Anadarko, Williston, Paradox, and Midland basins, USA; Mesozoic in Persian Gulf region, Paris Basin, and south Florida
Dolomite, subtidal to supratidal, in cyclic sequences	Dolomitization and attendant solution	Intercrystal and associated interparticle, moldic, other porosity moderate to very light	Paleozoic in Williston and Midland basins, USA
Carbonate sands on shelves and ramps; some associated dolomite	Original; some solution effects	Interparticle, other primary types; local moldic. Porosity moderate to high, may be very high in dolomites	Paleozoic in Anadarko, Illinois, and Permian basins, USA; Mesozoic in USA Gulf Coast, Paris Basin, Persian Gulf, and offshore Brazil
Biogenic reefs and reef-mounds A. Shallow shelf/ platforms B. Shelf/platform margin (incl. atoll) C. Outer/deep shelf/ ramp ("pinnacle reefs")	Primary (biogenic and other), fracturing, solution (variable), and solution enlargement	Varied: interparticle, growth-framework, shelter, etc.; moldic, various solution-enlarged types; vugs, etc. Porosity very low to high, variable, coarse	Paleozoic of Delaware, Midland and Basins; Mesozoic of USA Gulf Coast. Pinnacle reefs: Silurian in Michigan Basin, Devonian in Western Canada Basin, Tertiary in Philippines Basin
Debris deposits	Original; may be modified by solution, fracturing	Interparticle; some others. Porosity very low to high, variable, often coarse	Cretaceous, intraplatform basin, NE Mexico
Chalks, dominantly pelagic and micrograined	Original; often fractured	Inter- and intraparticle; often enhanced. Matrix porosity moderate to very high, but micro-sized	Cretaceous/Tertiary in North Sea Basin
Fractured "basinal" and tight shelf sands	Fracturing, local brecciation	Fracture; breccia. Matrix porosity low to nil	Tertiary in circum-Pacific basins (Japan, California) and NE Persian Gulf region fold belt
Other	E.g., probable burial-diagenetic solution (mesogenetic)	Solution-interparticle, moldic. Porosity moderate to high	Jurassic, USA Gulf Coast

sedimentary models. Especially noteworthy were early studies of modern biogenic settings conducted by such workers as Black (1933), Newell *et al.* (1951), Ginsburg (1956), Lowenstam and Epstein (1957), Goreau (1959), Lowenstam (1963), and Ball (1967), among others. For modern evaporative deposits, including dolomites, such early pioneering studies as those by Illing (1954), Alderman and Skinner (1957), Adams and Rhodes (1960), Shearman (1963) and his students, Illing *et al.* (1965), and Shinn *et al.* (1965) were important forerunners to a rapidly growing number of detailed investigations. Perhaps the latest new studies of importance are of the modern deep-marine deposits. Chalks, cherts, and syngenetic deep-water dolomites have been and continue to be the object of task force investigation under the aegis of the Deep Sea Drilling Project and in other groups, and are targets of large-scale marine exploration drilling by major oil companies exploring for petroleum. Examples of these more recent investigations include those by Beall and Fischer (1969), Schlanger and Douglas (1974), Scholle (1977), Schlager and James (1978), and Kelts and McKenzie (1982), among others. The latest interpretations of deepwater dolomite are only now being applied to important reservoir problems (*e.g.*, Roehl and Weinbrandt, this volume).

Somewhat later, major advances began to be made in understanding of the diagenetic changes that are taking place at many stages, beginning with early burial of a sedimentary sequence. Among the key early studies bearing on diagenesis and pore-space modification were those of Bathurst (*e.g.*, 1958, 1966), Folk (*e.g.*, 1959), Ginsburg (1957), Dunham (1969), Land (1967), Purdy (1968), Shinn (*e.g.*, 1969), Choquette and Pray (1970), Füchtbauer (1970), and Matthews and his students (*e.g.*, Matthews, 1974; Steinen and Matthews, 1973), to name only a few. Excellent reviews on diagenetic changes in limestone that affect pore systems have appeared in recent years, for example those by Bathurst (1971, 1980), Longman (1980), and James and Choquette (1983).

Our understanding of dolomitization and its effects on porosity has been substantially advanced by a number of modern studies beginning with those of Weyl (1960) and Murray (1960), Lucia (1962), Shinn *et al.* (1965), Illing *et al.* (1965), and Hsü and his associates (*e.g.*, Hsü and Siegenthaler, 1969), followed shortly by studies

of Badiozamani, (1973) Land (1973a and b), and many others. Useful reviews on dolomites and dolomitization have been prepared by Bathurst (1971), Zenger (1972), and most recently Morrow (1982a and b). Two symposia of particular importance have been compiled under the leadership of Pray and Murray (1965) and Zenger *et al.* (1980).

During the early post-World War II period, pore geometry and petrophysics in general also came under close investigation (for example, Purcell, 1949; Archie, 1952; Thomeer, 1960; Murray, 1960; Stout, 1964). Progress continues with modeling techniques and estimates of recovery efficiency such as those by Wardlaw (1976, 1980) and others. Down-hole logging has made enormous progress, and quantitative measures of petrophysical properties in subsurface reservoirs using multiple porosity and fluid-saturation tools are now commonplace.

The origins and modifications of pore systems in limestones and dolomites, as products of particular depositional environments and later diagenetic histories, were addressed in a few early studies (*e.g.*, Murray, 1960; Lucia, 1962; Pray and Choquette, 1966 Choquette and Pray, 1970), and more recently have been evaluated by many investigators (*e.g.*, Wilson, 1980a and b; Cussey and his coworkers in Reeckmann and Friedman, 1982; and Feazel and Schatzinger, in press). It is now more widely recognized that carbonate strata in which significant amounts of porosity actually are preserved in the record are relatively uncommon, and owe their pore space to combinations of rather special circumstances (Wilson, 1980a; Feazel and Schatzinger, in press). For more extended reviews of the depositional and diagenetic factors that create and modify porosity in carbonates, we recommend the recent papers by Wilson (1980a and b).

Some Principles Restated

From the body of work achieved over the past two decades or so, it is abundantly clear that the nature of newly deposited carbonate sediments, their depositional environments, and their early diagenetic pathways are key determinants of the likelihood and extent of potential reservoir rocks. Sedimentary facies provide the initial templates for the occurrence of porosity and the extent of a prospective reservoir. Necessarily, all subsequent modifications to these sediments—such as those due to percolation of meteoric waters, exposure to hypersaline brines, compaction, tectonic and non-tectonic fracturing—depend on the initial sedimentologic content and setting.

The vast majority of shallow marine carbonate sediments are comprised in varying combinations of biogenic skeletal debris derived from marine organisms that dwell chiefly on stable shelves or in epeiric environments, along with ooids, fecal pellets, and other non-skeletal grains produced in these environments. On many shelf and ramp settings and most isolated carbonate platforms, dilution by non-carbonate sediments is minimal.

Except for usually brief geologic episodes of tectonic instability and interludes of marine transgression, shoaling-upward sequences and/or seaward progradation are most usual. These base-leveling episodes lead to the post-depositional exposure of sediments and the vagaries of climate. Early diagenesis ensues as a consequence of the percolation of meteoric waters and, in arid regions, the transformation of exposure surfaces into sabkhas and diagenetic terranes whose extremes lead to dolomitization and facies succession by evaporites. Our collective experience has shown that many carbonate reservoir rocks have vestiges of primary porosity, but most have undergone substantial enhancement of pore space, or wholesale occlusion, because of these environmental dynamics.

An important exception to the predominance of shallow, clear-water marine environments in the deposition of most carbonates has been encountered recently along active plate margins where deep-water carbonate and silica sediments of microfloral and microfaunal origin occur interstratified with fine, petroleum-bearing noncarbonate clastics and volcanics. Much of this sediment, carbonate and noncarbonate, has been altered syngenetically to dolomite, which now produces large quantities of oil following a post-depositional history of natural hydraulic fracturing. Miocene rocks of such origin girdle the Pacific and suggest new carbonate targets for future exploration worldwide.

but depending on what attributes one chooses to emphasize, a variety of schemes can be proposed. A practical classification for both exploration and production geology purposes is one emphasizing the geology of the reservoir facies and the associations in which they occur. The classification in Table I-1 uses this approach. It is based in part on reservoir categories proposed by Wilson (1980a and b), but differs in some aspects and includes additional categories as well. In our experience, the majority of carbonate reservoir facies fall into seven main groupings. An eighth category, in which porosity originated by dissolution during deep/late burial diagenesis, seems likely in principle but has yet to be reported in an "end-member" situation unaccompanied by primary or near-surface solution porosity. The Jurassic Smackover grainstone reservoir in the Mt. Vernon Field discussed by Druckman and Moore in this volume is partly of this type, and in the future additional reservoirs of this type are likely to be recognized.

Although the classification in Table I-1 does not emphasize porosity characteristics, these do differ from one reservoir group to another, not only qualitatively as noted in the table, but often quantitatively. These pore-space types and attributes can in fact be expressed in terms of porosity–permeability relationships, fluid saturations, capillary pressure characteristics, and other petrophysical attributes. Reservoir data in many of the case studies illustrate the petrophysical differences between reservoir types. Many carbonate reservoirs do not fall simply in any one of the groups in Table I-1, but instead comprise two or more reservoir facies that may have comparable economic importance. This is particularly true of many shoal-water carbonate reservoirs in which, for example, ooid or skeletal grainstone facies may be productive along with various kinds of skeletal buildups and tidal-flat wackestone and mudstone units. A listing of such multiple-facies reservoirs would be very long indeed and might well include most of the carbonate giants and supergiants.

Principal Carbonate Reservoir Types

It is difficult to organize carbonate reservoirs into simple categories because of their diversity,

Overview of Reservoirs in this Volume

Though it seemed unlikely to us that the small number of reservoirs cited in this collection could

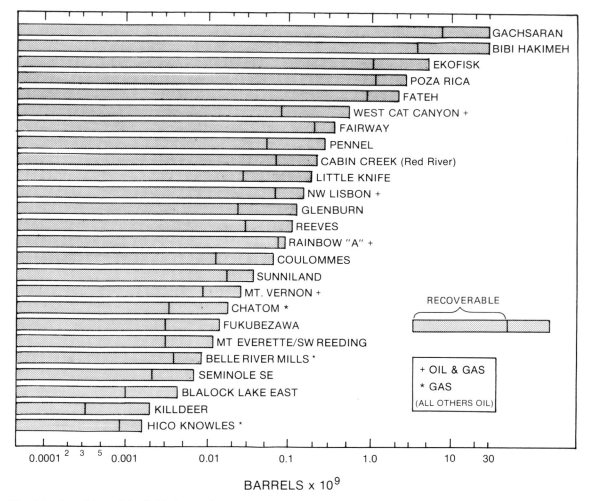

Fig. I-2. A ranking of the fields by total equivalent oil-in-place reserves. Includes 18 oil, 3 gas, and 3 oil and gas fields; gas-to-oil volumetric conversion factor 6000 SCFB/BO. Summary ages and some geologic attributes of these reservoirs are provided in Table I-2.

provide any sort of reasonable representation of the known carbonate reservoirs of the world, it appears that we have been fortunate enough to acquire a sampling of fields that range from the very small, with less than a million equivalent barrels of initial oil-in-place reserves (IOIP), to supergiants with over 30 billion barrels. The IOIP reserves for those fields for which statistics are available show a nearly smooth logarithmic distribution, seen in Figure I-2. We are not in a position to judge whether or not this distribution is typical, but clearly it spans most of the size spectrum of fields in terms of reserves. Recoverable reserves, shown in Figure I-3, tend to be grouped more in step-like distribution, but nonetheless span much of the spectrum for carbonates—from

around 100,000 barrels to some 9 billion barrels equivalent.

A problem we encountered was how best to arrange and group the case studies. Our initial inclination was to organize the fields on the basis of the depositional settings inferred for the reservoir rocks. This would have resulted in assemblages based on the nature of reservoir facies, as in Table I-1, and the principal factors responsible for localization of each field, as in Table I-2. We found, however, that this approach had limitations because many of the fields have highly complex developmental geologic histories, and commonly more than one set of factors has been responsible for localizing a given petroleum accumulation.

On the other hand, there are distinct advan-

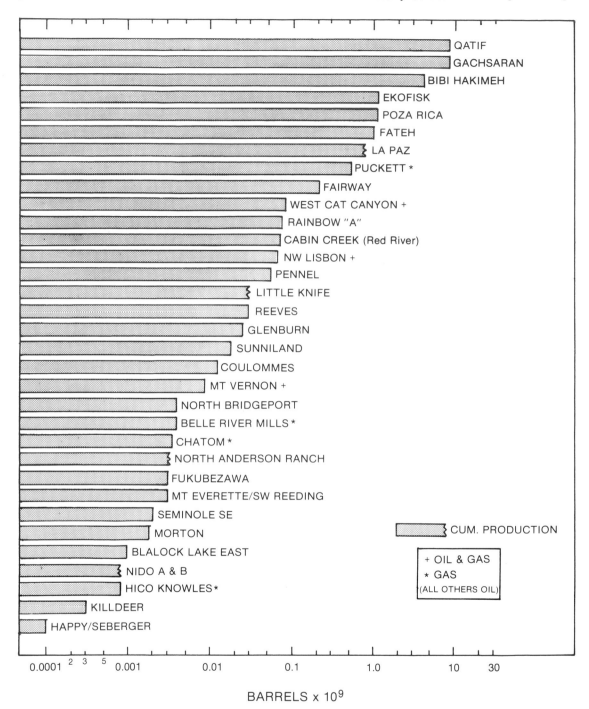

Fig. I-3. A ranking of the fields by equivalent recoverable oil reserves. Includes 25 oil, 4 gas, and 4 oil and gas fields; gas-to-oil volumetric conversion factor 6000 SCFB/BO. Note that the ranking is distorted somewhat by the availability for a few fields of only cumulative production data. In most cases (exceptions noted in reservoir summary tables accompanying individual case studies), oil and equivalent oil reserves are volume recoverable by primary methods. Summary ages and some geologic attributes of these reservoirs are provided in Table I-2.

Table I-2. Summary of fields discussed in this volume by age, localizing factors, and origins of porosity.

Localization of fields
1. Unconformities (incl. subaerial diag. terrane)
2. Facies (depos. and/or diag.)
3. Constructional (reefs)
4. Uplift or tilting
5. Salt bodies (subjacent)
6. Mobile belt/plate margin bathymetry & tectonics
7. Migration or hydrodynamics

Nature of reservoir rocks
1. Sub-unconformity carbonates (subaerial diag. terrane)
2. Peritidal/supratidal dolomites
3. Shelf/ramp sands
4. Reef-mounds
   A. Shelf
   B. Shelf/platform margin & atoll
   C. Outer/deeper shelf/ramp ("pinnacle")
5. Debris deposits
6. Chalks & other "basinal" facies
7. Fractured thin-bedded "basinal" & tight shelf carbonates
8. Other

Origin of porosity
1. Original porosity
2. Dissolution, near surface
3. Dolomitization
4. Fracturing
5. Other

Paper	Authors	Field(s)	Age
35	Aoyagi	Fukubezawa	Miocene
34	Longman	Nido B	
33	Roehl & Weinbrandt	West Cat Canyon	
32	McQuillan	Bibi Hakimeh / Gachsaran	Olio-Miocene
31	Feazel et al.	Ekofisk	Paleocene-Cret
30	Achauer	Fairway	Cretaceous
29	Carozzi & Falkenhein	Garoupa-Pampa	
28	Enos	Poza Rica	
27	Halley	Sunniland	
26	Jordan et al.	Fateh	
25	Reijers & Bartok	La Paz	
24	Crevello et al.	Hico Knowles	Jurassic
23	Druckman & Moore	Mt. Vernon	
22	Feazel	Chatom	
21	Purser	Coulommes	
20	Wilson	Qatif	
19	Asquith & Drake	Blalock Lake East	Permian
18	Chuber & Pusey	Reeves	
17	Cys & Mazzullo	Morton	
16	Malek-Aslani	North Anderson Ranch	
15	Depowski & Peryt	Tarchaly, Rybaki & Sulecin	
14	Ebanks & Watney	Happy, Seberger	Pennsylvanian
13	Mazzullo	Seminole SW	
12	Choquette & Steinen	North Bridgeport	Mississippian
11	Gerhard	Glenburn	
10	Lindsay & Kendall	Little Knife	
9	Miller	NW Lisbon	
8	Schmidt et al.	Rainbow	Devonian
7	Gill	Belle River Mills	Silurian
6	Morgan	Mt. Everette & SW / Reeding	
5	Roehl	Cabin Creek	
4	Clement	Pennel	Ordovician
3	Derby & Kilpatrick	Killdeer	
2	Ruzyla & Friedman	Cabin Creek	
1	Loucks & Anderson	Puckett	Cambro-Ordovician

tages in grouping reservoirs chronologically as in Table I-2, solely on the basis of their geologic age. One sees in this way broad trends through time of such things as the origin of carbonate grain types; the biotic makeup, geometry, and abundance of reefs and other skeletal-carbonate buildups; the type of plate-margin, cratonic, or epeiric depositional facies; and variations in local subaerial exposure histories. Presumably a much larger sampling would reveal such trends more clearly. The balance in age representation among fields in this collection can also be seen in Table I-2. The only large parts of Phanerozoic time unrepresented are the Cambrian and Triassic periods.

Over the range of time represented by these examples, we see a perceptible shift in predominant carbonate reservoir types from peritidal and subaerially exposed facies in Paleozoic examples, through a long-term sequence of shallow-shelf sands and reefs, to relatively deep-marine facies in the Cretaceous and Tertiary ranging from periplatform debris flows to deep-shelf pelagic chalks and deep-basinal mixed lithofacies that have been diagenetically altered, mainly to dolomite. On the other hand, we see no "trends" in the occurrence of pinnacle or other reefs through the Phanerozoic.

Substantial modifications to this picture would appear in a larger sampling of carbonate reservoirs. For example, to our chagrin an important group of reservoirs not included in the classification of Table I-1—the "late" and commonly tectonically or karst-localized, coarsely crystalline dolomite reservoirs—is not represented in this collection. Among reservoirs of this type, in our experience, are those in such major fields as Albion-Scipio (Ordovician) in the Michigan Basin, Indian Basin (Pennsylvanian) in southeastern New Mexico, and Samah (Cretaceous) in the Sirte Basin of Libya. Moreover, the fields in this particular collection are unduly clustered because of three representatives each of Red River reservoirs (Ordovician) from the Williston Basin and Smackover reservoirs (Jurassic) from the U.S. Gulf Coast. Nevertheless, this sampling seems to us somewhat representative of the kinds of long-term trends through Phanerozoic time that appear in this and earlier compilations (*e.g.*, Halbouty *et al.*, 1970; Mazzullo, 1980).

It is common practice to characterize petroleum reservoir traps simply as stratigraphic, or structural, or some combination. From informa-

tion supplied by the authors, we have tried to characterize more explicitly in Table I-2 the main factors believed to have been responsible for localizing the traps in this sampling of reservoirs. Elaboration of these factors, along with the more conventional descriptions of traps, may be found in the case studies and their accompanying reservoir summary tabulations.

If we turn again to the bar plots of Figures I-2 and I-3 in the light of Tables I-1 and I-2, some interesting generalizations emerge, in spite of the fact that not all of the 35 case studies include useful reservoir fluid data. Many of the largest fields occur in Tertiary mobile belts and other regions affected by Tertiary tectonics, principally in carbonates of Cretaceous to Miocene age. Furthermore, at least nine of the largest fields owe their reservoir productivity largely to fracturing, and at least ten smaller fields involve fracturing to some extent.

On the other hand, the great number and diversity of reservoirs in carbonates of shallow-epeiric, platform, peritidal, sabkha origins are also evident and include a wide range of fields with between 70 and 700 million barrels IOIP. That carbonate facies commonly develop into porous reservoirs where warm, shallow phototropic waters existed is evident for fields resulting from a combination of shallow reef-mound, atoll, or pinnacle reef growth over structures, particularly subjacent salt diapirs. Ekofisk, Fateh, Garoupa, and Fairway fields, though very large reservoirs and each composed of carbonates of special origin, were nonetheless all dependent for their development and localization on salt tectonics (halokinesis). Non-salt-supported shallow shelves and reefs have smaller fields of less than 200 million barrels IOIP.

Tables I-3 and I-4 represent a further attempt at a synthesis of information in the 35 case studies. In Table I-3 we outline various generalizations about the reservoirs, and in Table I-4 focus on some genetic and descriptive aspects of the pore systems in these reservoirs. Several generalizations beyond those pointed out earlier can be seen from these tables:

1. Of the eight giant fields, all but one (Puckett) are reservoired in geologically young carbonates; the proportion is surprisingly like the preponderance of Mesozoic and Cenozoic giant fields seen in much larger samplings (*e.g.* Halbouty *et al.*, 1970).

2. Limestones are the principal reservoirs in seven of the eight giants and in something over half of the 35 fields surveyed in Table I-3. Perhaps not surprisingly, a review of the reservoir summary tables shows a preponderance of limestone reservoirs among Mesozoic fields and of dolomite reservoirs among Paleozoic fields.

3. The shallow shelf, ramp, epeiric-sea, and peritidal environments are by far the most prominent sites for carbonate reservoir facies, as noted earlier, but a number of reservoirs, some of them very large, occur in facies of deep-shelf to deep "basinal" origin.

4. Porosity and permeability values range widely, the highest values comparing favorably in our experience with those for the best of the sandstone reservoirs.

5. Ultimate recovery efficiences show a very wide range; these must surely reflect in many cases both the inherent variations in reservoir productivities and, perhaps as importantly, the reservoir engineering state of the art and vagaries of government regulations at the time of field development.

6. In propensity for textural types, the limestone and dolomite reservoirs differ in significant ways (Table I-4A); this is due in part to the strong tendency for limestones of reefal and high-energy carbonate-sand origins to show retention or enhancement of original porosity, and also the tendency for dolomite to develop preferentially in lime-mud-rich carbonates.

7. Pore systems of "secondary," post-depositional origin are strongly predominant in these reservoirs, not only among the dolomites as expected, but also among the limestone reservoirs (Table I-4B); however, partial retention of original porosity is very common, despite modifications by solution and dolomitization.

8. The large majority of these carbonate reservoirs acquired their favorable pore-system characteristics during diagenesis at or near the earth's surface, either by meteoric-water alteration or by alteration in more saline waters (*e.g.*, sabkha-related); burial-diagenetic changes affecting pore systems are dominated by cementation and compaction (Table I-4C), although solution and various carbonate–mineral transformations (such as the formation of burial-diagenetic dolomite) seem more common than has generally been believed.

Table I.3. A Synthesis of Carbonate Reservoir Attributes Represented in This Collection

Number of Fields Surveyed	41
Reservoir Rocks	
Original Depositional Setting	
Peritidal	7
Shallow Shelf	16
Deep Shelf	2
Deep, Bathyal	2
Pinnacle	3
Ramp	5
Lithology	
Limestone	15
Dolomite	11
Combinations	6
Limestone Predominance	3
Petrophysical Properties, Range	
Porosity, %	1–45
Permeability, md	0.3–9,500
Fractures Dominant Contributor, Fields[1]	9
Oil gravity, degrees API	12–54
Field Volumetrics	
Initial oil in place (million BO equiv)	1.7–32,000
Recoverable oil (million BO equiv)	0.1–9,000
Ultimate recovery efficiency, %	15–88
Giant Fields[2]	8
Age	
Oligo-Miocene	2
Cretaceous (& Paleocene[3])	4
Jurassic	1
Paleozoic (Ordovician)	1
Lithology	
Limestone	5
Dominantly Limestone	2
Dolomite	1

[1]Some contribution of fractures to productivity reported for many other fields.
[2]Giant fields defined as exceeding 500 million equivalent barrels recoverable oil; gas-to-oil conversion factor 6000 SCFB/BO.
[3]One dual reservoir age (Ekofisk)

Epilogue

What can we learn from this information? On at least two fronts there is a clear need to increase our level of exploration understanding. The first is the extent, the likely diversity, and the recognition of reservoirs in deep-marine carbonates. The second is the apparent importance of fractures which, though not widespread in all reservoirs, may nevertheless be present in reservoirs of any

Table I.4. A Summation of Pore Origins, General Pore Types, and Sedimentary-texture Types Associated with Reservoirs Discussed in This Volume

A. Dominant Sedimentary Textural Type (see Appendix)

	Limestone	*Dolomite*
Reef/Boundstone	9	3
Grainstone	17	3
Oolite or Pisolite	8	1
Packstone	14	10
Wackestone	7	11
Mudstone	—	6
Not determined[1]	—	3

B. General Pore Types (see Appendix)

	Limestone	*Dolomite*
Primary (BP, WP, GF, etc.)[2]	6	2
Solution-enlarged primary	14	2
Solution (MO, VG, CV, CH)	14	12
Intercrystal	2	17
Fracture and/or Breccia	4	7

C. Pore System Genesis

Primary[2]	8
Surface/Near-Surface Diagenesis	
Limestone solution	21
Dolomitization	11
Deeper-burial Diagenesis[3]	
Solution	6
Cementation	18
Mineral transformation	3

[1]Includes some sucrosic dolomite with obscure original texture.
[2]Little modified except by cement.
[3]Generally includes mechanical and solution compaction.

age and any location. Fractures apparently have several modes of development. Their size, orientation, and frequency have so far provided major obstacles to direct observation and analysis. Significant progress is nevertheless being made in down-hole logging measurements and on theoretical fronts.

An important, obvious implication is that in carbonate reservoir development and exploitation, any one of a wide variety of geologic and engineering "reservoir models" may be applicable. Little imagination is needed to visualize the effects on productivity of misinterpreting or failing to establish the geometry, distribution, and petrophysics of a given reservoir.

Perhaps a major area of concern for future workers on carbonate reservoirs is that of supplemental recovery. Table I-3 seems to confirm what most workers may already know: that there is a very wide range in recovery efficiencies based on primary production drive. Hence, economics permitting, every known carbonate reservoir is a future candidate for further detailed geologic analysis. Information from such analysis must be obtained by careful post-mortem studies long after initial field discovery, and such studies must be designed to provide highly applicable information for the use of the drilling and reservoir engineers. Geologists have legitimate cause for carefully designing thorough and ongoing programs for the acquisition, retrieval, and study of both primary reservoir-rock material and secondary, indirect information in the form of wireline logs, seismic data, and the like. A very large information base is often required to elucidate the important characteristics of carbonate reservoirs. Such an information base can be crucial for the design of adequate supplemental recovery programs.

References

ADAMS, J.E., and M.L. RHODES, 1960, Dolomitization by seepage refluxion: Amer. Assoc. Petroleum Geologists Bull., v. 44, p. 1912–20.

ALDERMAN, A.R., and H.C.W. SKINNER, 1957, Dolomite sedimentation in the southwest of Australia: Amer. Jour. Science, v. 255, p. 561-567.

ANDRICHUK, J.M., 1959, Ordovician and Silurian stratigraphy and sedimentation in southern Manitoba, Canada: Amer. Assoc. Petroleum Geologists Bull., v. 43, no. 10, p. 2333–2398.

ARCHIE, G.E., 1952, Classification of carbonate reservoir rocks and petrophysical considerations: Amer. Assoc. Petroleum Geologists Bull. v. 36, no. 2, p. 278–298.

BADIOZAMANI, K., 1973, The *dorag* dolomitization model—application to the Middle Ordovician of Wisconsin: Jour. Sedimentary Petrology, v. 43, p. 965–984.

BALL, M.M., 1967, Carbonate sand bodies of Florida and the Bahamas: Jour. Sedimentary Petrology, v. 37, p. 556–591.

BARNETCHE, A. and L.V. ILLING, 1956, The Tamabra Limestone of the Poza Rica oil field, Veracruz, Mexico: 20th Internat. Geol. Cong. Proc., Mexico City, 38 p.

BATHURST, R.G.C., 1958, Diagenetic fabrics in some British limestones: Liverpool and Manchester Geol. Jour., v. 2, p. 11–36.

BATHURST, R.G.C., 1966, Boring algae, micrite envelopes and lithification of molluscan biosparites: Geol. Jour., v. 5, p. 15–32.

BATHURST, R.G.C., 1971, Carbonate Sediments and their Diagenesis: Developments in Sedimentology 12, Elsevier Scientific Publ. Co., Amsterdam, 620 p.

BATHURST, R.G.C., 1980, Lithification of carbonate sediments: Sci. Prog. Oxf., v. 66, p. 451–471.

BEALL, A.O., and A.G. FISCHER, 1969, Sedimentology: *in* Worzel, J.L. *et al.*, eds., Deep Sea Drilling Project, v. 10, U.S. Govt. Printing Office, Washington, D.C., p. 521–593.

BLACK, M., 1933, The algal sedimentation of Andros Island, Bahamas: Royal Soc. London, Philosoph. Trans., ser. B, v. 222, p. 165–192.

CHOQUETTE, P.W., and L.C. PRAY, 1970, Geological nomenclature and classification of porosity in sedimentary carbonates: Amer. Assoc. Petroleum Geologists Bull., v. 54, p. 207–250.

CHOQUETTE, P.W., and J.D. TRAUT, 1963, Pennsylvanian carbonate reservoirs, Ismay Field, Utah and Colorado: *in* Bass, R.O. and S.L. Sharp, eds., Shelf Carbonates of the Paradox Basin—a symposium, Four Corners Geol. Soc., 4th Field Conf., p. 157–184.

DUNHAM, R.J., 1969, Early vadose silt in Townsend mound (reef), New Mexico: *in* Friedman, G.M., ed., Depositional Environments in Carbonate Rocks—a symposium, Soc. Econ. Paleontologists and Mineralogists, Spec. Publ. 14, p. 139–181.

EDIE, R.W., 1958, Mississippian sedimentation and oil fields in southeastern Saskatchewan: Amer. Assoc. Petroleum Geologists Bull., v. 42, p. 94–126.

FEAZEL, C.T., and SCHATZINGER, R.A., Prevention of carbonate cementation in petroleum reservoirs: *in* Schneidermann, N. and P.M. Harris, eds., Carbonate Cements Revisited, Soc. Econ. Paleontologists and Mineralogists, Spec. Publ. 37, 1984 (in press).

FITZGERALD, T.A., 1980, Giant field discoveries 1968–78—an overview: *in* Halbouty, M.T., ed., Giant Oil and Gas Fields of the Decade 1968–1978, Amer. Assoc. Petroleum Geologists Mem. 30, p. 1–5.

FOLK, R.L., 1959, Practical petrographic classification of limestones: Amer. Assoc. Petroleum Geologists Bull., v. 43, p. 1–38.

FÜCHTBAUER, H., 1970, Karbonatgesteine: *in* Füchtbauer, H. and G. Müller, eds., Sedimente und Sedimentgesteine, Schweizerbart, Stüttgart, p. 275–417.

GINSBURG, R.N., 1956, Environmental relationships of grain size, Florida carbonate sediments: Amer. Assoc. Petroleum Geologists Bull., v. 40, p. 2384–2427.

GINSBURG, R.N., 1957, Early diagenesis and lithification of shallow-water carbonate sediments in south Florida: *in* LeBlanc, R.J. and J.G. Breeding, eds., Regional aspects of carbonate deposition, Soc. Econ. Paleontologists and Mineralogists, Spec. Publ. 5, p. 80–99.

GOREAU, T.F., 1959, The ecology of Jamaican coral reefs: Ecology, v. 40, p. 79–90.

HALBOUTY, M.T., A.A. MEYERHOFF, R.E. KING, R.H. DOTT, SR., H.D. KLEMME, and T. SHABAD, 1970, World's giant oil and gas fields, geologic factors affecting their formation, and basin classification—Part I, Giant oil and gas fields: *in* Halbouty, M.T., ed., Geology of Giant Petroleum Fields: Amer. Assoc. Petroleum Geologists Mem. 14, p. 502–528.

HEMPHILL, C.R., R.I. SMITH, and F. SZABO, 1970, Geology of Beaverhill Lake reefs, Swan Hills area, Alberta: *in* Halbouty, M.T., ed., Geology of Giant Petroleum Fields: Amer. Assoc. Petroleum Geologists Mem. 14, p. 50–90.

HSÜ, K.J., and SIEGENTHALER, C., 1969, Preliminary experiments on hydrodynamic movements induced by evaporation and their bearing on the dolomite problem: Sedimentology, v. 12, p. 11–25.

ILLING, L.V., 1954, Bahaman calcareous sands: Amer. Assoc. Petroleum Geologists Bull. v. 38, p. 1–95.

ILLING, L.V., A.J. WELLS, and J.C.M. TAYLOR, 1965, Penecontemporaneous dolomite in the Persian Gulf: *in* Pray, L.C. and R.C. Murray, eds., Dolomitization and Limestone Diagenesis—a symposium, Soc. Econ. Paleontologists and Mineralogists, Spec. Publ. 13, p. 89–111.

JAMES, N.P., and P.W. CHOQUETTE, 1983, Diagenesis 6. Limestones—the sea floor diagenetic environment: Geoscience Canada, v. 10, p. 162–179.

KELTS, K., and J.A. MCKENZIE, 1982, Diagenetic dolomite formation in Quaternary anoxic diatomaceous muds of Deep Sea Drilling Project, Leg 64, Gulf of California: *in* Curray, J.R., D.G. Moore *et al.*, eds., Initial Reports, Deep Sea Drilling Project, v. 64, U.S. Govt. Printing Office, Washington, D.C., p. 553–569.

KLOVAN, J.E., 1964, Facies analysis of the Redwater reef complex, Alberta, Canada: Canadian Soc. Petroleum Geologists Bull., v. 12, p. 1–100.

LAND, L.S., 1967, Diagenesis of skeletal carbonates: Jour. Sedimentary Petrology, v. 37, p. 914–930.

LAND, L.S., 1973a, Contemporaneous dolomitization of Middle Pleistocene reefs by meteoric water, North Jamaica: Bull. Marine Sciences Gulf Caribbean, v. 23, p. 64–92.

LAND, L.S., 1973b, Holocene meteoric dolomitization of Pleistocene limestones, north Jamaica: Sedimentology, v. 20, p. 411–424.

LONGMAN, M.W., 1980, Carbonate diagenetic textures from nearshore diagenetic environments: Amer. Assoc. Petroleum Geologists Bull., v. 64, p. 461–487.

LUCIA, F.J., 1962, Diagenesis of a crinoidal sediment: Jour. Sedimentary Petrology, v. 38, p. 845–858.

LOWENSTAM, H.A., and S. EPSTEIN, 1957, The origin of the sedimentary aragonite needle muds of the Great Bahama Bank: Jour. Geology, v. 65, p. 364–375.

LOWENSTAM, H.A., 1963, Biologic problems relating to the composition and diagenesis of sediments: *in* Donnelly, T.W., ed., The Earth Sciences—Problems and Progress in Current Research: Univ. Chicago Press, Chicago, IL; p. 137–195.

MATTHEWS, R.K., 1974, A process approach to diagenesis of reefs and reef-associated limestones: *in* Laporte, L.F., ed., Reefs in Time and Space, Soc. Econ. Paleontologists and Mineralogists, Spec. Publ. 18, p. 234–256.

MAZZULLO, S.J., 1980, Preface—carbonate facies:

in Mazzullo, S.J., ed., Stratigraphic Traps in Carbonate Rocks: Amer. Assoc. Petroleum Geologists Reprint Ser. 23, p. i–ix.

MORROW, D.W., 1982a, Diagenesis 1. Dolomite Part 1, The chemistry of dolomitization and dolomite precipitation: Geoscience Canada, v. 9, p. 5–13.

MORROW, D.W., 1982b, Diagenesis 2. Dolomite Part 2, Dolomitization models and ancient dolostones: Geoscience Canada, v. 9, p. 94–107.

MURRAY, R.C., 1960, Origin of porosity in carbonate rocks: Jour. Sedimentary Petrology, v. 30, p. 59–84.

NEWELL, N.D., J.K. RIGBY, A.J. WHITEMAN, and J.S. BRADLEY, 1951, Shoal-water geology and environments, eastern Andros Island, Bahamas: Amer. Museum Nat. Hist. Bull., v. 97, p. 1–30.

PRAY, L.C. and P.W. CHOQUETTE, 1966, Genesis of carbonate reservoir facies (abst.): Amer. Assoc. Petroleum Geologists Bull., v. 50, p. 632.

PRAY, L.C., and R.C. MURRAY, eds., 1965, Dolomitization and Limestone Diagenesis—a symposium, Soc. Econ. Paleontologists and Mineralogists, Spec. Publ. 13, 180 p.

PURCELL, W.R., 1949, Capillary pressures—their measurement using mercury and the calculation of permeability therefrom: Amer. Inst. Mining and Metallurgical Engineers Trans., v. 186, p. 39—48.

PURDY, E.G., 1968, Carbonate diagenesis—an environmental survey: Geologica Romana, v. 7, p. 183—228.

REECKMAN, A., and G.M. FRIEDMAN, 1982, Exploration for Carbonate Petroleum Reservoirs (transl. of *Elf Aquitaine* report by R. Cussey *et al.*): John Wiley and Sons, New York, 213 p.

ROEHL, P.O., 1967, Stony Mountain (Ordovician) and Interlake (Silurian) facies analogs of Recent low-energy marine and subaerial carbonates, Bahamas: Amer. Assoc. Petroleum Geologists Bull., v. 51, no. 10, p. 1979–2032.

SCHLAGER, W., and N.P. JAMES, 1978, Low-magnesian calcite limestones forming at the deep-sea floor, Tongue of the Ocean, Bahamas: Sedimentology, v. 25, p. 675–702.

SCHLANGER, S.O., and R.G. DOUGLAS, 1974, Pelagic ooze-chalk limestone transition and its implications for marine stratigraphy: *in* Hsü, K.J. and H.C. Jenkyns, Sedimentologists Spec. Publ. 1, Blackwell Scient. Publ., p. 117–148.

SCHOLLE, P.A., 1977, Chalk diagenesis and its relation to petroleum exploration—oil from chalks, a modern miracle? Amer. Assoc. Petroleum Geologists Bull., v. 61, p. 982–1009.

SHINN, E.A., 1969, Submarine lithification of Holocene carbonate sediments in the Persian Gulf: Sedimentology, v. 12, p. 109–144.

SHINN, E.A., R.M. GINSBURG, and R.M.

LLOYD, 1965, Recent supratidal dolomite from Andros Island, Bahamas: *in* Pray, L.C. and R.C. Murray, eds., Dolomitization and Limestone Diagenesis—a symposium, Soc. Econ. Paleontologists and Mineralogists, Spec. Publ. 13, p. 112–123.

SHEARMAN, D.J., 1963, Origin of marine evaporites by diagenesis: Inst. Mining and Metallurgy Trans. (B), v. 75, p. 208–215.

STEINEN, R.P. and R.K. MATTHEWS, 1973, Phreatic vs. vadose diagenesis—stratigraphy and mineralogy of a cored borehole on Barbados, W.I.: Jour. Sedimentary Petrology, v. 43, p. 1012–1020.

STOUT, J.L., 1964, Pore geometry as related to carbonate stratigraphic traps: Amer. Assoc. Petroleum Geologists Bull., v. 48, p. 329–337.

THOMAS, G.E., and R.P. GLAISTER, 1960, Facies and porosity relationships in some Mississippian carbonate cycles of Western Canada basin: Amer. Assoc. Petroleum Geologists Bull., v. 44, p. 569–588.

THOMEER, J.H.M., 1960, Introduction of a pore geometrical factor defined by the capillary pressure curve: Jour. Petroleum Technology, Tech. Note 2057.

WARDLAW, N.C., 1976, Pore geometry of carbonate rocks as revealed by pore casts and capillary pressure: Amer. Assoc. Petroleum Geologists Bull., v. 60, p. 245–257.

WARDLAW, N.C., 1980, The effects of pore structure on displacement efficiency in reservoir rocks and in glass micromodels: First Joint Soc. Petroleum Engineers and U.S. Dept. of Energy Sympos. on Enhanced Oil Recovery, Proc., Tulsa, OK, p. 345–352.

WEYL, P.K., 1960, Porosity through dolomitization—conservation-of-mass requirements: Jour. Sedimentary Petrology, v. 30, p. 85–90.

WILSON, J.L., 1980a, Limestone and dolomite reservoirs: *in* Hobson, G.D., ed., Developments in Petroleum Geology 2, Applied Science Publishers Ltd, London, p. 1–51.

WILSON, J.L., 1980b, A review of carbonate reservoirs: *in* Miall, A.D., ed., Facts and Principles of World Petroleum Occurrence, Canadian Soc. Petroleum Geologists Mem. 6, p. 95–119.

ZENGER, D.H., 1972, Dolomitization and uniformitarianism: Jour. Geol. Education, v. 20, p. 107–124.

ZENGER, D.H., J.B. DUNHAM, and R.L. ETHINGTON, eds., 1980, Concepts and Models of Dolomitization—a symposium, Soc. Econ. Paleontologists and Mineralogists, Spec. Publ. 28, 320 p.

Paleozoic Reservoirs

1
Puckett Field

Robert G. Loucks and James H. Anderson

RESERVOIR SUMMARY

Location & Geologic Setting	Pecos Co., 90 mi (145 km) SW of Midland, west Texas, USA
Tectonics	Faulted anticline structure on shelf transitional between Val Verde and Delaware basins
Regional Paleosetting	Cambro-Ordovician cratonic Tobosa Basin; transgressive sequence deposited on Precambrian basement
Nature of Trap	Structural; faulted anticline; principal seal provided by overlying Simpson Fm shales
Reservoir Rocks	
Age	Early Ordovician
Stratigraphic Units(s)	Ellenburger Formation
Lithology(s)	Dolomite
Dep. Environment(s)	Shallow subtidal marine to supratidal, shoaling upward, cyclically repeated
Productive Facies	Dolomite, especially in solution-collapse zones
Entrapping Facies	Shales of Simpson Formation; also dolomotized mudstone of Ellenburger
Diagenesis Porosity	Repeated dolomitization, emergence, and wholesale solution (diagenetic terranes)
Petrophysics	
Pore Type(s)	Fracture and breccia, also interparticle, intercrystal moldic, vuggy
Porosity	0–12%, 3.5% avg
Permeability	0–169 md, 10–50 md avg range
Fractures	Tectonic and solution-collapse fractures
Source Rocks	
Age	Ordovician Simpson Formation
Lithology(s)	Marine shales overlying post Ellenburger unconformity
Migration Time	NA
Reservoir Dimensions	
Depth	12,000–15,000 ft (3650–4570 m)
Thickness	<1 ft to >150 ft (<0.3->45 m)
Areal Dimensions	9 × 5 mi (14.4 × 8 km)
Productive Area	17,000 acres (68 km^2)
Fluid Data	
Saturations	NA
API Gravity	NA
Gas-Oil Ratio	NA
Other	—
Production Data	
Oil and/or Gas in Place	NA
Ultimate Recovery	3.3 trillion CFG (0.93 trillion m^3); 550 million equivalent BO
Cumulative Production	2.73 trillion CFG (0.77 trillion m^3), 455 million equivalent BO; and 1.64 million BGC (0.26 m^3) through 1982

Remarks: Discovered 1952. IP 12 billion CFGPD plus 38 BOPD (Phillips No. 1 Glenna). 35 producing wells (Jan. 1979) at 640-acre spacing. Production mechanism depletion drive. Gas column 1600 feet (~490 m) thick gross.

1
Depositional Facies, Diagenetic Terranes, and Porosity Development in Lower Ordovician Ellenburger Dolomite, Puckett Field, West Texas

Robert G. Loucks and James H. Anderson

Location and Discovery

The Puckett Field is located in Pecos County, west Texas, approximately 220 miles (360 km) east-southeast of El Paso and 90 miles (145 km) southwest of Midland (Fig. 1-1). The discovery well, the Phillips Petroleum Co. No. 1 Glenna, was completed in 1952 at a total depth of 16,575 feet (5052 m), with an initial production of 12 million cubic feet of gas per day (340,000 m^3) and 38 barrels of oil per day, from perforations in the Ellenburger Dolomite (Lower Ordovician) at 13,700 to 14,285 feet (4165–4343 m) (Blanchard, 1977). As of January 1979, the field had 35 producing wells

Reservoir Trap Characteristics

Puckett Field lies along a faulted, somewhat elongate anticlinal structure (Blanchard, 1977) with more than 2500 feet (>760 m) of closure (Figs. 1-2, 1-3). Normal faults and anticlinal closure have combined to create the trap. Marine shales of the overlying Simpson Formation provide the reservoir seal and, according to Momper (1979), were probably the source of hydrocarbons in the Ellenburger reservoir.

The field has a productive area of about 17,000 acres (69 km^2) and a gas column that was originally 1600 feet (490 m) thick. The main producing zone is between 12,000 and 15,000 feet (3600–4600 m), and comprises a series of relatively porous, permeable intervals ranging in thickness from less than a foot to more than 150 feet (~0.3-45 m), as illustrated by the petrophysical profile shown in Figure 1-4.

Production and Reserves

From 1952 to January 1979, the Ellenburger Dolomite reservoir in Puckett Field produced approximately 2.6 trillion cubic feet of gas (65 billion m^3) and 1.8 million barrels of gas condensate. Ultimate recoverable gas reserves have been estimated by Blanchard (1977) to be 3.3 trillion cubic feet (93 billion m^3).

Approaches in This Study

Cores from two gas-producing wells in the field, the Phillips Petroleum Co. No. 1 Glenna and No. 1 Puckett, provide 1700 feet (520 m) of continuously cored section from the Precambrian basement through the Bliss Sandstone to the top of the Ellenburger Dolomite (Figs. 1-2–1-4). Most of the data for this study came from these cores and from thin sections and core analyses of the cores. To aid in understanding subsurface relationships, and especially breccia zones, cores from other wells in west Texas were also examined, as were exposures of the Ellenburger

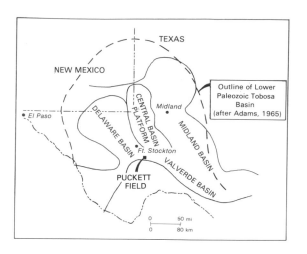

Fig. 1-1. Map showing major geologic features of West Texas Permian Basin and location of Puckett Field. Outline of Tobosa Basin shown by dashed lines.*

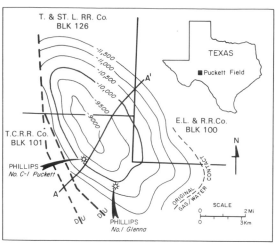

Fig. 1-2. Structure map on top of the Ellenburger Dolomite. Holland (1966) used over 40 wells to define structure, but only wells with cores used in this study are shown. Modified from Holland (1966). Published by permission, West Texas Geological Society.

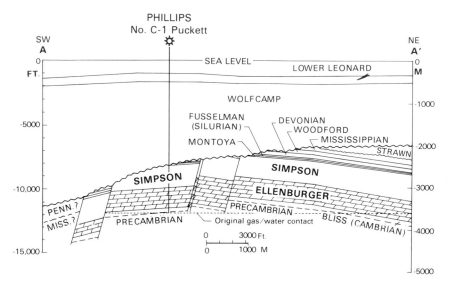

Fig. 1-3. Generalized structural cross section A–A', upon which the Phillips Petroleum Co. No. C-1 Puckett is projected.

*Figures 1-1, 1-4, 1-5, 1-6, 11-7A, 1-9, 1-10, 1-11, 1-13, 1-14 from Loucks and Anderson, 1980. Published by permission, Society of Economic Paleontologists and Mineralogists.

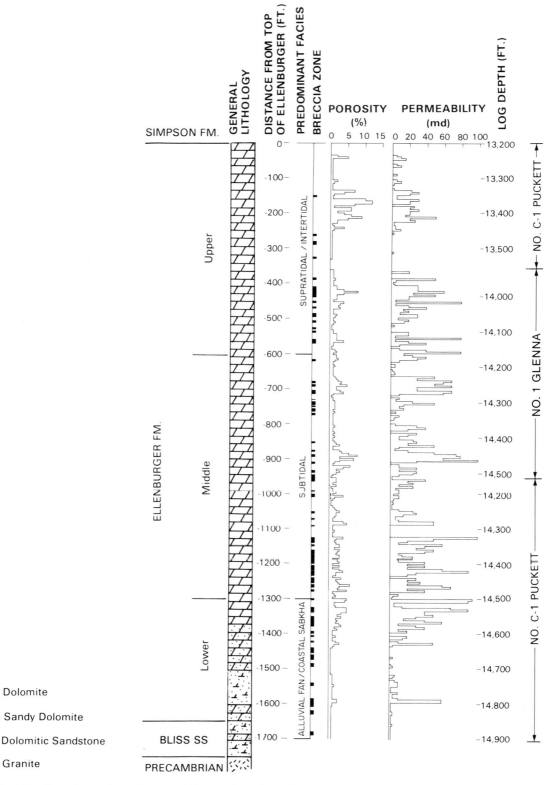

Fig. 1-4. Profiles of porosity and permeability vs. depth in the Phillips No. C-1 Puckett well. Both well-log depths and depth below top of Ellenburger are shown. *Predominant facies* refers to the facies comprising largest proportions of individual cycles.

Dolomite in the Marble Falls area of central Texas.

Regional Setting

In west Texas, carbonates of the Ellenburger Dolomite and underlying Bliss Sandstone (Cambrian) were deposited in the Tobosa Basin (Fig. 1-1) on a broad, shallow-marine shelf up to 500 miles (800 km) wide, during a major Late Cambrian/Early Ordovician marine transgression over Precambrian basement rocks (Barnes *et al.*, 1959; Ross, 1976). During early Pennsylvanian tectonism, the west Texas and adjoining regions were divided by the uplifted Central Basin Platform into the Delaware, Val Verde, and Midland basins. Puckett Field, where the Ellenburger Dolomite is approximately 1600 feet (~490 m) thick, is located at the transition between the Val Verde and Delaware Basins (Fig. 1-1).

Main Facies and Inferred Depositional Models

In the Puckett Field area, the carbonates of the Ellenburger Dolomite and uppermost Bliss Sandstone have been altered to dolomite. In the remainder of this study, the carbonate rocks are differentiated into various textures and facies, for which the word dolomite is used only in places but is implicit everywhere.

In general, the Ellenburger Dolomite and upper Bliss Sandstone can be divided into two major sequences (Fig. 1-4): a lower sequence of arkosic, conglomeratic sandstones interbedded with dolomites (upper Bliss/lower Ellenburger), and an upper sequence entirely composed of dolomites (middle and upper Ellenburger). Contrasting depositional models are inferred for these two sequences, as shown in Figures 1-5A and B, respectively. All of the facies that comprise these models occur in the two cored wells described in this paper. As shown in the petrophysical profiles of Figure 1-4, intervals with relatively high and low porosity and permeability are cyclically repeated.

Bliss and Lower Ellenburger Sequence

During deposition of this lower sequence, the sedimentation regime in the Puckett Field area was dominated by carbonates deposited on sabkhas and tidal flats as well as the adjoining shallow-subtidal shelf. Carbonate deposition was interrupted at times and in places by influxes of clastic sediments of alluvial-fan and fan-delta origin derived from ancestral Precambrian highlands to the north and northwest (Fig. 1-5A).

This general model is inferred from a sequence of arkosic sandstones and conglomerates interbedded with fine-crystalline dolomite mudstones in the upper Bliss/lower Ellenburger sequence that were cored in the No. 1 Puckett well (Fig. 1-6). The main lithologies found in the *lower* part of that sequence are shown diagrammatically in Figure 1-6A and illustrated by photographs in Figure 1-7. The main lithologies found in the *upper* part of that sequence are shown diagrammatically in Figure 1-6B and illustrated by photographs in Figure 1-8.

Conglomerates in the upper Bliss/lower Ellenburger sequence consist of poorly sorted quartz pebbles, unweathered feldspar grains, and rare granitic cobbles in a pyrite-bearing, clay-rich matrix (Fig. 1-7A). Some intervals consist of laminated, algal-mat dolomites in which dark-colored, pyrite-rich laminations are common (Fig. 1-7B). Dolomite mudstones commonly contain pseudomorphs of chalcedony, quartz with anhydrite inclusions, and open-space crystals of baroque dolomite (Figs. 1-7C,D).

Algal stromatolites, similar to stromatolites in modern sediments of Shark Bay, Australia (Logan *et al.*, 1964), are developed in dolomites interpreted as lower intertidal to subtidal, first in this part of the sequence (Figs. 1-6B ,1-8) and then in the middle and upper Ellenburger sequence (Figs. 1-9, 1-10 B,C). Units of oolite grainstone, commonly with peloids and/or intraclasts, also occur in this part of the sequence (Fig. 1-6B) and probably accumulated in high-energy shoals or tidal bars (see Fig. 1-5A,B). Photographs of silicified ooid grainstone (chert) are shown in Figure 1-8C,D.

Middle and Upper Ellenburger Sequence

Sedimentation during the time interval represented by this upper sequence was dominated by a shallow subtidal shelf containing a complex system of tidal flats, channels, levees, and ponds probably similar to the broad western part of Andros

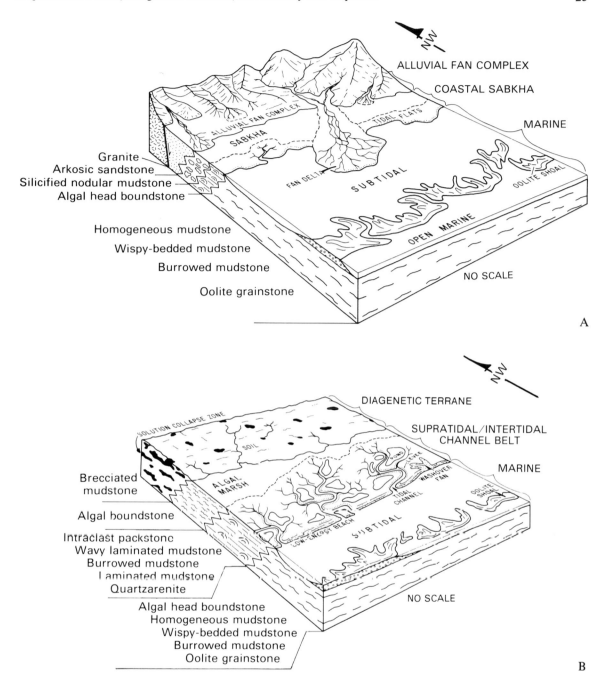

Fig. 1-5. A. Depositional model for the Bliss and lower Ellenburger. *B.* Depositional model for the middle and upper Ellenburger.

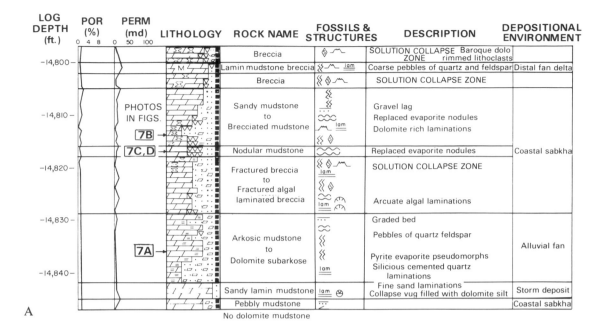

A

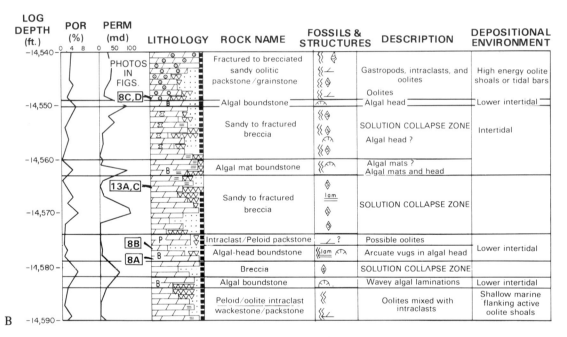

B

Fig. 1-6. A. Core description, environmental interpretation, and porosity-permeability profile of a section of the alluvial fan/sabkha sequence in the lower Ellenburger and upper Bliss. Note stratigraphic positions of photographed rocks. Phillips No. C-1 Puckett well, 14,797 to 14,846 feet (4510–4525 m). *B.* Core description, environmental interpretation, and porosity-permeability profile of a section of the high-energy intertidal sequence. Sequence is in the lower Ellenburger/upper Bliss. Phillips No. C-1 Puckett well, 14,540 to 14,590 feet (4432–4447 m).

Fig. 1-7. A. Core photograph of alluvial-fan facies, an arkosic conglomerate containing grains of pebble-sized quartz (1) and feldspar (2). Matrix is pyritic and clay-rich. Phillips No. C-1 Puckett well, 14,837 feet (4522 m). Bar scale equals 2 cm. *B.* Core photograph of alluvial-fan facies, a laminated arkosic conglomerate. Argillaceous laminations are rich in sulfur (as sulfide in pyrite). Phillips No. C-1 Puckett well, 14,814 feet (4515 m). Bar scale equals 2 cm. *C.* Core photograph of sabkha facies, a microcrystalline dolomite (formerly lime mud or mudstone) with silicified nodules (1) interpreted as replaced evaporites. Nodules now consist of pseudocubic quartz with anhydrite inclusions, micro-flamboyant chalcedony, and white baroque dolomite. Phillips No. C-1 Puckett well, 14,816 feet (4515 m). Bar scale equals 2 cm. *D.* Photomicrograph of silicified evaporite nodule. Large quartz crystal (1) contains "ghosts" of anhydrite crystal laths (2). Phillips No. C-1 Puckett well, 14,816 feet (4515 m). Plane light, bar scale equals 1 mm.

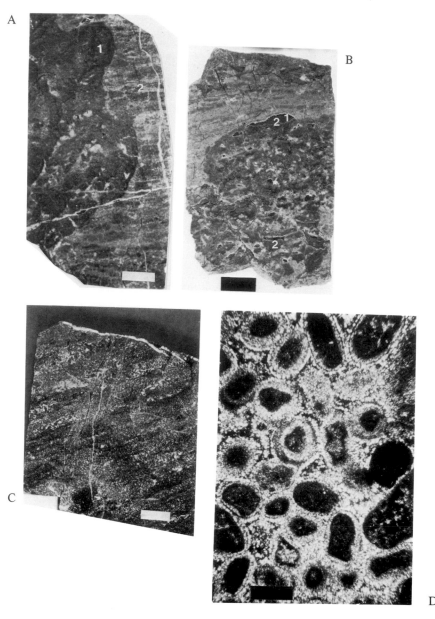

Fig. 1-8. A. Core photograph of lower intertidal to subtidal facies, an algal-head boundstone. Note the irregular shape and distinct boundary of the algal head (1) with the laminated facies (2). Phillips No. C-1 Puckett well, 14,578 feet (4442 m). Bar scale equals 2 cm. *B.* Core photograph of lower intertidal to subtidal facies, an algal-head boundstone. Darker area is the algal head. Note distinct boundary between algal head and light-gray intraclast-pelletal packstone. This sample is 10 cm above sample shown in *A*. Arcuate pore spaces (1) are characteristic of the algal boundstones and are often filled with detrital dolomite showing geopetal structure and with pyrite (2). Phillips No. C-1 Puckett well, 14,576 feet (4443 m). *C.* Core photograph of high-energy oolite shoal facies, a silicified oolite grainstone. Phillips No. C-1 Puckett well; 14,548 feet (4434 m). Bar scale equals 2 cm. *D.* Photomicrograph of silicified oolite grainstone showing texturally well preserved (but replaced) oolites and cements. Silicified grainstone is from an area in the core shown in *C*. Phillips No. C-1 Puckett well, 14,548 feet (4434 m). Plane light, bar scale equals 1 mm.

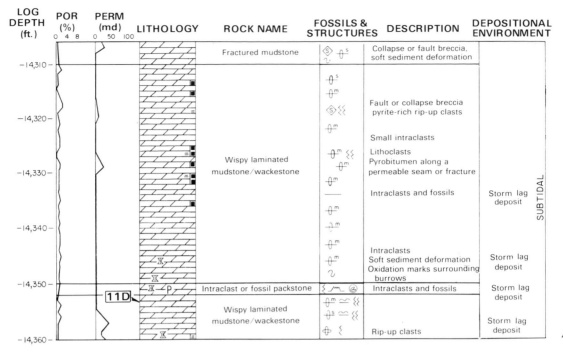

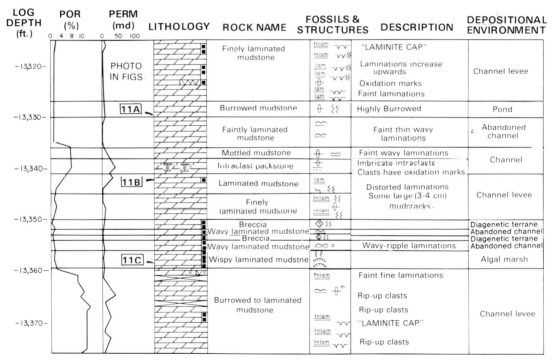

Fig. 1-9. A. Core description, environmental interpretation and porosity-permeability profile of a section of the low-energy subtidal sequence. Sequence is in the middle Ellenburger. Phillips No. 1 Glenna well, 14,306 to 14,359 feet (4361–4377 m). *B.* Core description, environmental interpretation, and porosity-permeability profile of a section of the supratidal/channel belt sequence. Sequence is in the upper Ellenburger. Phillips No. C-1 Puckett well, 13,315 to 13,375 feet (4058–4076 m).

Island in the Bahamas (Hardie and Garrett, 1977). Facies of subtidal, channeled tidal-flat, and supratidal origins occur repeatedly in both study wells (Fig. 1-9), as shallowing-upward cycles up to 30 feet (9 m) thick. These cycles are poorly developed in the middle Ellenburger, where they are dominated by subtidal carbonates (Fig. 1-9A). They are better developed in the upper Ellenburger where they are dominated by intertidal and supratidal carbonates (Figs. 1-9B, 1-10A–C, 1-11). The idealized cycle shown in Figure 1-12 is based on detailed sequences illustrated in the upper sequence in both cored wells. Commonly, this shallowing-upward cycle terminates in clayey, silty dolomites containing paleosoils, associated solution-collapse breccias, and other features strongly suggesting subaerial exposure. Exposure resulted in the repeated development of diagenetic terranes, as defined by Roehl (1967), in which low-relief, solution-dominated topography was extensive (Fig. 1-5B).

Subtidal facies are characterized by thick units of mudstone (~40 ft, ~12 m) and thinner units of wackestone (Fig. 1-9A). Channeled tidal-flat facies are characterized by thinly laminated and thin units (Fig. 1-9B), rapid vertical and lateral changes in lithology, and a wide variety of textures (Figs. 1-10A–C, 1-11). These include mudstones, wackestones, and packstones; poorly sorted intraclast packstones interpreted as channel-lag or channel-edge slump deposits (Fig. 1-11A), overlain by rippled to wavy-laminated mudstones interpreted as low-energy tidal channel deposits (Fig. 1-11B,C); and thinly laminated mudstones interpreted as channel-levee deposits (Fig. 1-11D). Other lithologies suggestive of environments in this complex setting are illustrated in Figure 1-10A–C.

Diagenetic-Terrane Lithologies

As indicated in the idealized shallowing-upward cycle for the middle and upper Ellenburger (Fig. 1-12), soil zones are interpreted to occur at the tops of many of these cycles. These paleosoils are characterized by dark, highly weathered, poorly sorted, dolomitized lithoclast packstones in which the lithoclasts are commonly rounded and range in size from silt to gravel (Fig. 1-10D). More abundant throughout virtually all of the Ellenburger and uppermost Bliss are intervals of dolomite breccia interpreted as solution-collapse breccia (Fig. 1-13). Lucia (1971) described similar collapse breccias in the Lower Ordovician in west Texas and southwestern New Mexico. The brecciated intervals in Puckett Field range up to 50 feet (15 m) in thickness (Figs. 1-4, 1-6, 1-9). The "clasts" may show very little displacement (Fig. 1-13A) to random displacement, are sharply angular, and range in size from several millimeters to several meters. Some breccias are infilled by detrital silt and clay probably washed in from the contemporaneous exposure surface by circulating ground water (Fig. 1-13A–D). Other breccias retain some interclast pore space due to incomplete filling by sediment or no sediment at all. Open-boxwork breccias and breccia-related fractures are in some intervals filled or lined by coarse-crystalline, white baroque dolomite (Figs. 1-13C,D). Numerous paleosoil horizons and solution-collapse breccias throughout the middle and upper Ellenburger Dolomite in Puckett Field indicate many cycles of sedimentation capped by subaerial diagenetic terranes, forming along the lines suggested by the model in Figure 1-5B.

The most compelling evidence supporting a near-surface, solution-collapse origin for the breccia is in the presence of detrital sediment with geopetal structure between fragments (Fig. 1-13A,C). The sediment is composed of mechanically abraded dolomite sand and silt, rounded quartz grains, and/or clay. Some of the detrital clay has a red to brown color similar to modern *terra rossa* deposits. These features support a near-surface origin for the breccias and also indicate that turbid waters flowed rapidly through the breccias.

The process of brecciation in the Ellenburger Dolomite is not clearly defined. There is strong evidence that the brecciated zones formed by collapse near the surface and were later modified by tectonic fractures at depth. It is speculative to state what materials were dissolved to produce the collapse or how these materials were dissolved.

Exposure to weathering between shallowing-upward cycles can lithify and even dolomitize carbonate sediments, as suggested by Mueller (1975) for shallow-water cyclic carbonates in the Cretaceous Edwards shallow-water carbonates on the South Texas Shelf. Continued exposure might well have caused dissolution of evaporites and/or carbonates and could have formed paleosoils and collapse breccias. There is evidence of

Fig. 1-10. Core photographs. All bar scales equal 2 cm. *A.* Channel-lag or storm deposit facies, an intraclast packstone. Intraclasts are poorly sorted but rounded, suggesting transport while soft. Phillips No. C-1 Puckett well, 13,236 feet (4034 m). *B.* Mud-rich tidal-channel fill facies, a ripple-laminated mudstone. Original sediment may have been pelleted, but texture is altered due to compaction and dolomitization. Phillips No. C-1 Puckett well, 13,472 feet (4106 m). *C.* Mud-rich tidal-channel fill facies, a wavy laminated mudstone. Faint wavy laminae may have been caused by rippled pellet layers with mud drapes. Several layers contain intraclasts (1). Phillips No. C-1 Puckett, 13,477 feet (4108 m). *D.* Tidal-channel levee deposit facies, a finely laminated mudstone highly fractured with local solution enlargement of fractures. Black material (1) is pyrite and pyrobitumen. Vertical burrow (2) disrupts laminations. Dark gray laminae are reduced layers. Phillips No. C-1 Puckett well, 13,457 feet (4102 m).

Fig. 1-11. Core photographs. All bar scales equal 2 cm. *A.* Pond facies, a highly burrowed mudstone. Phillips No. C-1 Puckett well, 13,329 feet (4063 m). *B.* Intertidal algal-mat facies, an algal-laminated boundstone. Lack of desiccation features and presence of channel scour (1) suggest deposition in the intertidal zone. Phillips No. C-1 Puckett well, 13,342 feet (4067 m). *C.* Algal-marsh facies, a massive algal boundstone. In the bottom section of the slab is a storm bed composed of a poorly sorted mixture of pellets and intraclast. Note open, cement-lined fractures (1). Phillips No. C-1 Puckett well, 13,359 feet (4072 m). *D.* Diagenetic terrane soil, a highly weathered dolomite with silt-to-gravel-size detrital dolomite litholasts. Phillips No. C-1 Puckett well, 14,353 feet (4375 m).

Fig. 1-12. Idealized Ellenburger depositional cycle. Cycles range up to 30 feet (9 m) thick.

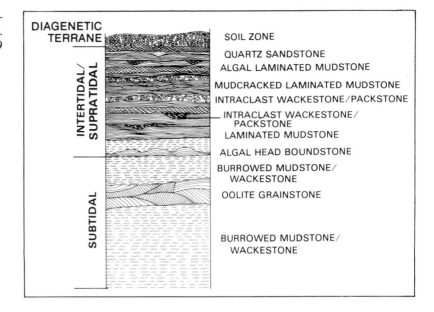

DIAGENETIC TERRANE

SOIL ZONE
QUARTZ SANDSTONE
ALGAL LAMINATED MUDSTONE
MUDCRACKED LAMINATED MUDSTONE
INTRACLAST WACKESTONE/PACKSTONE
INTRACLAST WACKESTONE/PACKSTONE
LAMINATED MUDSTONE
ALGAL HEAD BOUNDSTONE
BURROWED MUDSTONE/WACKESTONE
OOLITE GRAINSTONE

BURROWED MUDSTONE/WACKESTONE

INTERTIDAL/SUPRATIDAL

SUBTIDAL

former evaporites in the lower Ellenburger Dolomite (Figs. 1-7C, D), but not in the middle and upper Ellenburger Dolomite in the Puckett Field. Barnes *et al.* (1959), however, described preserved evaporites in the Ellenburger of West Texas and southeastern New Mexico.

If the climate was arid during early Ellenburger deposition, as indicated by the signs of evaporites and suggested for the Early Ordovician of this vast shelf area (Ross, 1976), there is the question of how dissolution occurred. One possibility is that active flow of dilute ground water took place via alluvial-fan sediments into the underlying or "down-dip" carbonate muds and mudstones of the sabkhas. A modern analog for such a situation has been reported by Amiel and Friedman (1971), in a continental sabkha near the Gulf of Aqaba.

Moreover, a constantly arid climate over the 20 million years of the Early Ordovician seems unlikely; there may have been intermittent wet periods that caused the dissolution of evaporites and/or carbonates. By the close of Early Ordovician time, a well-developed karst surface existed (Barnes *et al.*, 1959), indicating such a change toward a more humid climate.

Limestone dissolution is common in the upper meteoric phreatic zone (Thrailkill, 1968), but it is also possible that dissolution of limestone occurred in a mixing zone between fresh water and saline water, producing solution-collapse breccias. Holland (1964) and Runnells (1969) noted that undersaturation with respect to calcite may result when two fluids, saturated with respect to calcite, are mixed. Vernon (1968) suggested this dissolution mechanism as the cause of the cavernous limestone in the Boulder zone of the Tertiary of Florida. The mixing zone may not only cause solution-collapse brecciation by dissolving limestone but may also promote dolomitization (Vernon, 1968; Hanshaw *et al.*, 1971). Carbonate dissolution in a mixing zone would account for broader regions of cavernous layers at greater depths (tens of feet) below sea level than dissolution along a shallow (few feet) upper phreatic zone of the relatively low-relief Ellenburger diagenetic terrane surface.

All of the intervals in the Ellenburger/upper Bliss sequence that have significant amounts of porosity (more than a few percent) and permeability (more than 1–2 md) are dolomites, but tend strongly to be intervals of solution-collapse breccia that affected all rock types (Fig. 1-4), or tidal-flat and tidal-channel dolomites with combinations of moldic, interparticle, intercrystal, and vuggy porosity. Thinner porous intervals occur in some subtidal dolomite mudstone and wackestone. Intertidal/subtidal algal-boundstone dolomites may have a few percent of arcuate-vug porosity in algal heads (Fig. 1-6A,B). The more common types of reservoir rocks are illustrated in Figures 1-13 and 1-14.

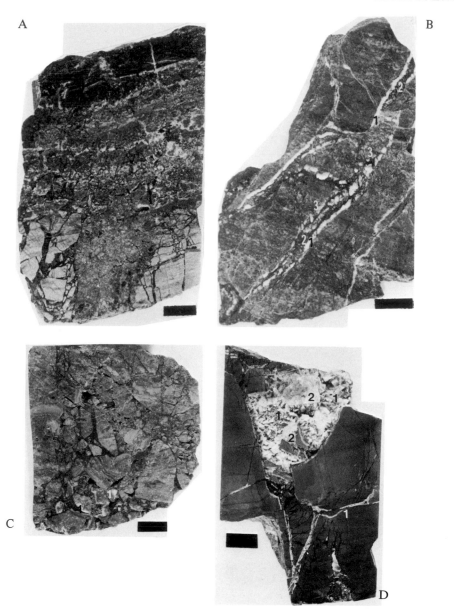

Fig. 1-13. Core photographs of solution-collapse breccias. All bar scales equal 2 cm. *A.* Solution-collapse
dolomite breccia. Fractures extend out from initial
collapse zone (bottom of slab) into above bedded
dolomite. Some fractures are open, and geopetal
dolomite silt occurs between fragments. Some porosity
can be seen in matrix between fragments. Phillips No.
C-1 Puckett well, 14,563 feet (4439 m). *B.* Solution-
collapse dolomite breccia. Breccia and inclined breccia
fractures were cemented first by coarse-crystalline
white baroque dolomite (1), followed by detrital fill of
clay, pyrite, and broken dolomite rhombs (2). Later
tectonic fractures (3) cut across fragments and sediment fill. These tectonic fractures are solution-en-

larged. Phillips No. C-1 Puckett well, 14,408 feet
(4392 m). *C.* Solution-collapse dolomite breccia. This
sample is located immediately beneath the one shown
in A. Some dolomite fragments are partially rimmed by
baroque dolomite (1). Dark fill between clasts is clay,
pyrite, and broken dolomite rhombs. Breccia porosity
occurs where matrix did not totally occlude pores between clasts. Phillips No. C-1 Puckett well, 14,563 feet
(4439 m). *D.* Solution-collapse dolomite breccia.
Large breccia vug contains geopetal silt and clasts and
is lined by coarse-crystalline, white baroque dolomite
(1) and quartz crystals (2). Phillips No. 1 Glenna well,
14,055 feet (4284 m).

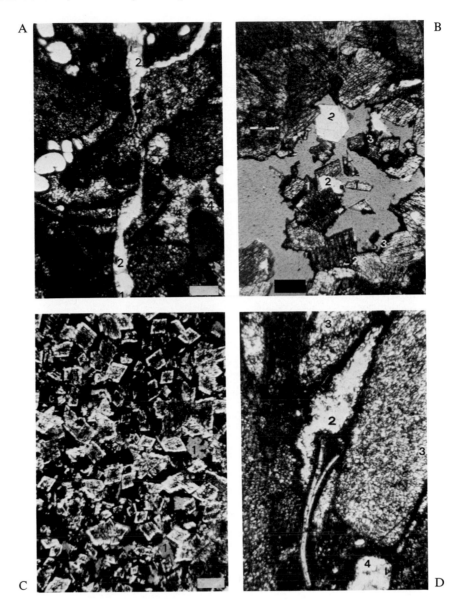

Fig. 1-14. Photomicrographs of pore types. All photos in plane light and bar scales equal 1 mm. *A.* Solution-enlarged fracture porosity (1). Fractures are partly filled with quartz (2). Minor moldic (3) and interparticle (4) porosity is connected by the fractures. Phillips No. C-1 Puckett well, 13,237 feet (4035 m). *B.* Breccia porosity. Note fragments of broken dolomite rhombs (1) and authigenic quartz (2). Black material between rhombs is pyrobitumen (3). Phillips No. 1 Glenna well, 14,124 feet (4305 m). *C.* Intercrystal porosity (1) between zoned dolomite rhombs. Abundant black material between rhombs is pyrobitumen. Phillips No. 1 Glenna well, 14,447 feet (4403 m). *D.* Moldic (1) and interparticle porosity (2). Interparticle porosity is between dolomite intraclasts (3). Moldic pore space is partially cemented with authigenic quartz (4). Phillips No. C-1 Puckett well, 13,237 feet (4035 m).

Fracture porosity (Figs. 1-13B and 1-14A) is most abundant in the upper 200 feet (60 m) of the Ellenburger Dolomite. Porosity in small fractures (Figs. 1-10C, 1-11B,D, and 1-14A), commonly solution-enlarged, probably improves permeability by connecting isolated pores and porous zones. Breccia porosity is the dominant type in solution-collapse breccias, and occurs as voids up to 8 centimeters in apparent size between breccia fragments. These pores are commonly filled partially by detrital sediment and/or baroque dolomite, as noted earlier. In general, fracture and breccia porosity are the most important types visually, and by inference volumetrically, in Ellenburger dolomites. Fabric-selective porosity, including interparticle and moldic porosity in packstone and grainstone facies, and intercrystal porosity in many formerly mud-rich carbonates, may add significantly to storage capacity.

Average porosity from core analyses in the Phillips No. 1 Glenna and No. 1 Puckett wells is 3.5 percent, with a range of nil to 12 percent. Permeability ranges from nil to 170 millidarcys, but the common range is between 10 and 50 millidarcys (Fig. 1-4).

Summary

Puckett Field produces gas and gas condensate from structurally entrapped reservoirs in porous dolomites occurring throughout some 1700 feet (520 m) of the Ellenburger Dolomite and the underlying uppermost Bliss Sandstone. The carbonate sequence in the Ellenburger Dolomite and uppermost Bliss Sandstone is dominantly a product of cyclically repeated, laterally shifting shallow subtidal to supratidal deposition, with adjoining low-relief diagenetic terranes. In general, porosity occurs as solution-collapse breccia and intercrystalline porosity interconnected by fracture porosity. Fracture porosity is most abundant in the upper 200 feet (60 m) of the Ellenburger section. Shales of the overlying Simpson Formation appear to form the major seal for trapped hydrocarbons, but the presence of many isolated brecciated zones suggests the potential for individual diagenetic traps throughout the section and not just beneath the Ellenburger-Simpson contact.

Acknowledgments Publication was authorized by the Director, Bureau of Economic Geology, The University of Texas at Austin, where this work was done. This case study is based on core workshop notes published earlier by Loucks and Anderson (1980). The authors gratefully acknowledge the contributions of P. A. Mench, V. E. Barnes, C. W. Achauer, C. R. Handford, D. A. Budd, and D. G. Bebout. Phillips Petroleum Company supplied cores and core analyses.

References

ADAMS, J.E., 1965, Stratigraphic-tectonic development of Delaware Basin: Amer. Assoc. Petroleum Geologists Bull., v. 49, p. 2140–2148.

AMIEL, A.J., and G.M. FRIEDMAN, 1971, Continental sabkha in Arava Valley between Dead Sea and Red Sea: significance for origin of evaporites: Amer. Assoc. Petroleum Geologists Bull., v. 55, p. 581–592.

BARNES, V.E., P.E. CLOUD, JR., L.P. DIXON, R.L. FOLK, E.C. JONAS, A.R. PALMER, and E.J. TYNAN, 1959, Stratigraphy of the pre-Simpson Paleozoic subsurface rocks of Texas and southeast New Mexico: Univ. Texas, Austin, Bur. Econ. Geology Publ. 5924, 2 vols., 836 p.

BLANCHARD, K., 1977, Puckett (Ellenburger): *in* Gas fields in west Texas—a symposium: West Texas Geol. Soc., v. 3, p. 76–78.

HANSHAW, B.B., W. BACK, and R.G. DEIKE, 1971, A geochemical hypothesis for dolomitization by groundwater: Econ. Geology, v. 66, p. 710–724.

HARDIE, L.A., and P. GARRETT, 1977, General environmental setting: *in* Hardie, L.A., ed., Sedimentation on the Modern Carbonate Tidal Flats of Northwest Andros Island, Bahamas: John Hopkins Univ. Studies in Geology, No. 22, p. 12–49.

HOLLAND, H.D., 1964, Solubility of calcite in NaCl solutions between 50° and 200° C (abst.): Geol. Soc. America Spec. Paper 82, p. 94–95.

HOLLAND, R.R., 1966, Puckett, Ellenburger, Pecos County, Texas: *in* Oil and gas fields in west Texas—a Symposium: West Texas Geol. Soc. Publ. 66–52, p. 291–295.

LOGAN, B.W., R. REZAK, and R.N. GINSBURG, 1964, Classification and environmental significance of algal stromatolites: Jour. Geology, v. 72, p. 68–83.

LOUCKS, R.G., and J.H. ANDERSON, 1980, Depositional facies and porosity development in Lower Ordovician Ellenburger Dolomite, Puckett

Field, Pecos County, Texas: *in* Carbonate Reservoir Rocks: notes for SEPM Core Workshop No. 1, Denver, Colorado, June 1980: Soc. Econ. Paleontologists and Mineralogists, p. 1–31.

LUCIA, F.J., 1971, Lower Paleozoic history of the western Diablo Platform, west Texas and south-central New Mexico: *in* Cys, J.M., ed., Robledo Mountains and Franklin Mountains—1971 Field Conf. Guidebook: Permian Basin Section, Soc. Econ. Paleontologists and Mineralogists, p. 174–214.

MOMPER, J.A., 1979, Characterization of oil types in the Permian Basin: Geochemistry for Geologists Course Notes, Amer. Assoc. Petroleum Geologists, 1979, Keystone, Colorado.

MUELLER, H.W., 1975, Centrifugal progradation of carbonate banks: a model for deposition and early diagenesis, Ft. Terrett Formation, Edwards Group, Lower Cretaceous, Central Texas: Ph.D. Dissert., Univ. Texas, Austin, 300 p.

ROEHL, P.O., 1967, Stony Mountain (Ordovician) and Interlake (Silurian) facies analogs of Recent low-energy marine and subaerial carbonates, Bahamas: Amer. Assoc. Petroleum Geologists Bull., v. 51, no. 10, p. 1979–2032.

ROSS, R.J., JR., 1976, Ordovician sedimentation, western U.S.A.: *in* Bassett, M.G., ed., The Ordovician System: Proceedings of a Paleontological Assoc. Symp., Birmingham, England, Sept. 1974, Univ. of Wales Press and National Museum of Wales, Cardiff, p. 73–105.

RUNNELLS, D.D., 1969, Diagenesis, chemical sediments and the mixing of natural waters: Jour. Sedimentary Petrology, v. 39, p. 1188–1201.

THRAILKILL, J., 1968, Chemical and hydrologic factors in excavation of limestone caves: Geol. Soc. America Bull., v. 79, p. 19–46.

VERNON, R.O., 1968, The geology and hydrology associated with a zone of high permeability (Boulder zone) in Florida: Society Mining Engineers, Preprint 69-AG-12, 24 p; (abs): Mining Engineers, v. 20, no. 12, p. 58, 1968.

2
Cabin Creek Field
Kenneth Ruzyla and Gerald M. Friedman

RESERVOIR SUMMARY

Location & Geologic Setting	Fallon Co., SE Montana, in SW Williston Basin, USA
Tectonics	Crest of Cedar Creek Anticline, a highly elongate faulted fold episodically uplifted
Regional Paleosetting	SW shelf of cratonic basin
Nature of Trap	Structural plus stratigraphic (paleostructural closures, facies variations, hydrodynamics)
Reservoir Rocks	
Age	Late Ordovician[1]
Stratigraphic Units(s)	Red River Formation (also Silurian Interlake Fm[1])
Lithology(s)	Dolomite
Dep. Environment(s)	Peritidal and shallow carbonate bank
Productive Facies	Finely crystalline dolomite, microcrystalline dolomite, and skeletal dolomite
Entrapping Facies	Mottled dolobiomicrite/wackestone-packstone; black shale stringers
Diagenesis Porosity	Dolomitization, vadose-zone(?) dissolution of evaporite minerals, local cementation by anhydrite and silica
Petrophysics	
Pore Type(s)	Vug, moldic, interparticle, intercrystal
Porosity	1–25%, 13% avg (dolomite)
Permeability	0–142 md, 7.9 md avg (dolomite)
Fractures	Vertical & horizontal fractures common, many healed by anhydrite
Source Rocks	
Age	Ordovician
Lithology(s)	Winnipeg shale
Migration Time	Middle Late Cretaceous
Reservoir Dimensions	
Depth	9000 ft (2750 m) avg
Thickness	50 ft (15.2 m) net
Areal Dimensions	7.5 × 2 mi (12 × 3.2 km)
Productive Area	7620 acres (30.5 km^2)
Fluid Data	
Saturations	S_o = 30% avg, S_w = 30% avg
API Gravity	33°
Gas-Oil Ratio	NA
Other	
Production Data	
Oil and/or Gas in Place	224 million BO, 29,420 B/acre-ft
Ultimate Recovery	75 million BO (original estimate); URE = 33.5%
Cumulative Production	75.4 million BO thru 1979 (commingled Ordovician and Silurian production)

Remarks: IP 100–300 BOPD. IP mechanism depletion drive. Capillarity: $P_D \approx 92.3$ psi (65 Kg/cm^2), $G \approx 1.0$ (0.1–5.0 range), BVp∞ ≈ 0.1. Discovered 1953, water-flooding commenced 1964.

[1] See also Roehl, this volume.

2
Factors Controlling Porosity in Dolomite Reservoirs of the Ordovician Red River Formation, Cabin Creek Field, Montana

Kenneth Ruzyla and Gerald M. Friedman

Location and Discovery

Cabin Creek Field is located in eastern Montana approximately midway along the Cedar Creek Anticline in the southwestern sector of the Williston Basin (Fig. 2-1). Discovered in 1953 by seismic methods, the field is currently producing from about 90 wells, from reservoirs in the Silurian Interlake Formation (Roehl, 1967 and this volume) and the Upper Ordovician Red River Formation. Another case study discusses Red River reservoirs in Pennel Field to the southeast along the anticline (Clement, this volume).

Regional Geology and Geologic History

The Williston Basin is a broad, roughly circular, indistinctly bounded, shallow depression on the southeastern side of the Canadian Shield (Fuller, 1961). At the basin center, Lower Paleozoic rocks attain maximum thicknesses in excess of 3000 feet (~900 m) and are overlain by about 12,000 feet (~3700 m) of Upper Paleozoic, Mesozoic, and Tertiary strata (Fig. 2-1).

The Cedar Creek Anticline is an asymmetrical, linear feature some 100 miles long (160 km) that came into existence in the early Paleozoic, apparently as a result of basement faulting (Roehl, 1967). Subsequent periods of uplift have occurred during the Devonian, Mississippian to Triassic, and post-Paleocene. The anticline is accentuated by northwest–southeast linear faults, and the axis of the anticline rises and falls along its length. Figure 2-2, a structural map on the Red River Formation, illustrates the closure of about 250 feet (~75 m) in the area of Red River production along the axis of the anticline. Production along the northeast flank of the Red River structure comes from Silurian reservoirs.

Hydrocarbon Source Rocks

Previous studies indicate that oil migration from Ordovician source rocks began in the middle Late Cretaceous, and that accumulation was controlled by paleostructural closures, faulting, geomorphology, facies variations, and hydrodynamic conditions (Clement, 1976). According to Williams (1974), the Winnipeg Shale (Middle Ordovician) is the most probable source of oil in the Red River Formation because the soluble organic matter extracted from the Winnipeg Shale correlates with Red River oil, as well as most other lower Paleozoic oil in the Williston Basin. The Winnipeg Shale lies stratigraphically below impermeable limestones of the Red River Formation, and any oil originating in Winnipeg source rocks must have been emplaced in the Red River by vertical migration through fracture systems (Dow, 1974).

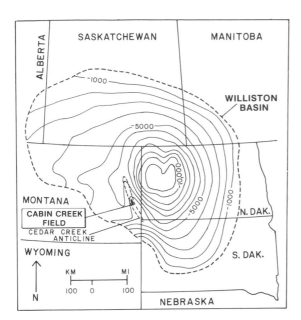

Fig. 2-1. Location map showing outline of Williston Basin, Cedar Creek Anticline, and Cabin Creek Field. Contours show structure on top of Red River Formation. Contour interval 1000 feet (300 m).*

Methods

Approximately 1450 feet (444 m) of Red River carbonate core from 12 different wells was studied to delineate the physical stratigraphy, distribution of facies, and depositional environments, and to compare these attributes with log characteristics and core analysis data (Ruzyla, 1980). Diagenesis, porosity types, and origin of porosity were evaluated by petrographic analysis of thin sections. Values for porosity, permeability, and fluid saturations are from standard core plug analysis. Size and shape of pore throats were determined by estimation from mercury capillary-pressure data (see also papers by Roehl, 1967, and this volume) and from scanning-electron micrographs of resin pore casts, respectively (Ruzyla and Friedman, 1982).

Production Characteristics

Within the field area, the Red River Formation averages about 500 feet (153 m) in thickness and

*All figures in this chapter are from Friedman, Ruzyla, and Reeckmann, 1981, and Ruzyla and Friedman, 1982. Published by permission of the United States Department of Energy and by permission of the Society of Petroleum Engineers of AIME, © 1982 SPE-AIME, respectively.

consists of a sequence of alternating limestones and dolomites. Production is from the U_2-U_3, U_4-U_5, and U_6-U_7 informal stratigraphic units (zones) in the upper 150 feet (46 m) of the Red River Formation. These units, which are dolomites, are stratigraphically equivalent to the A through C zones used widely by industry operators in the Williston Basin (Clement, 1978), and also to the Herald Formation (Kendall, 1976). The interstratified U_1-U_2, U_3-U_4, and U_5-U_6 are limestone units and are non-productive. Lateral and vertical variations in amount of dolomite are mainly responsible for variations in reservoir properties and effective pay thicknesses (Clement, 1978).

Red River production is commingled with production from Silurian carbonates. Combined Ordovician and Silurian primary oil recovery is estimated at about 75 million barrels, with remaining reserves recoverable by secondary water-flooding of about 14 million barrels. The field has been on waterflood since April 1964.

Reservoir Geology

Study of 12 wells indicates that the upper 150 feet (46 m) of the Red River Formation consists of a cyclical sequence of three limestone and three dolomite units defined by log markers and of rather uniform thickness across the field (Fig. 2-3). The log markers at the top of limestone units are designated as the U_1, U_3, and U_5 (youngest to oldest), and those at the top of dolomite units are designated as the U_2, U_4, and U_6 (youngest to oldest). Based on these markers, which follow industry usage along the Cedar Creek Anticline, we define the following stratigraphic intervals: the U_1-U_2 limestone, averages 25 feet (7.6 m) thick, ranges from 10 to 28 feet (3.0–8.5 m) thick; the U_2-U_3 dolomite, averages 6 feet (1.8 m) thick, ranges from 2 to 10 feet (0.6–3.0 m) thick; the U_3-U_4 limestone, averages 20 feet (6.1 m) thick, ranges from 12 to 23 feet (3.7–7.0 m) thick; the U_4-U_5 dolomite, averages 20 feet (6.1 m) thick, ranges from 8 to 36 feet (2.4–11.0 m) thick; the U_5-U_6 limestone, averages 35 feet (10.7 m) thick, ranges from 24 to 46 feet (7.3–14.0 m) thick; and the U_6-U_7 dolomite, averages 55 feet (16.8 m) thick, ranges from 41 to 71 feet (12.5–21.6 m) thick. The upper surfaces of the dolomite intervals are erosional; thus the contacts between the dolomite intervals and the overlying limestone in-

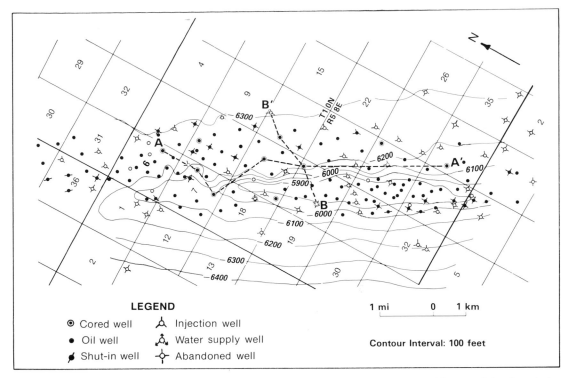

Fig. 2-2. Structure-contour map of Cabin Creek Field. Contour interval 100 feet (~30 m) and depths relative to sea level.

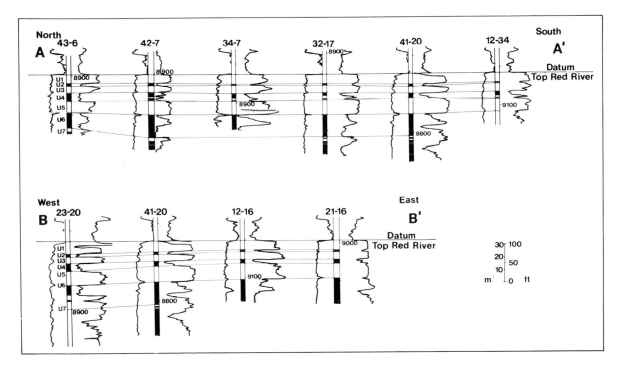

Fig. 2-3. Stratigraphic cross-sections A–A' and B–B' through the Cabin Creek Field (see Fig. 2-2). Dolomite is indicated in black, limestone in white. All logs shown are gamma ray-neutron. Datum is top of Red River.

Table 2.1. Description and Occurrence of Upper Red River Facies for Cabin Creek Field

	Lithology	Texture and Grain Types	Fossils	Sedimentary Structures	Environmental Setting	Stratigraphic Occurrence
FACIES 1	Mottled dolomitic biomicrite-wackestone to packstone	Mainly skeletal grains in pelletal-mud matrix. Peloids, intraclasts, and ooids locally	Echinoderm, brachiopod, bryozoa, trilobite, mollusk, ostracod, and gastropod common. Rare solitary corals and stromatoporoids	Massive to poorly bedded, intensely burrowed, abundant horizontal stylolites. Local cross-bedding and graded bedding	Subtidal to intertidal zones	Comprises U_1-U_2, U_3-U_4, and U_5-U_6 limestones. Occurs locally in lower part of U_4-U_5 and U_6-U_7 dolomites
FACIES 2	Mottled dolomitic biomicrite-mudstone	Skeletal grains in pelletal-mud matrix	Same as Facies 1	Massive to poorly bedded, intensely burrowed, horizontal stylolites	Subtidal zone	Locally present in U_2-U_3 dolomite
FACIES 3	Finely crystalline dolomite (10–60 μ)	Hypidiotopic dolomite	Traces of small shell debris, mostly ostracode or small brachiopod	Wavy laminae predominate. Flat-pebble breccias, desiccation cracks, truncation surfaces, oncolites, nodular anhydrite, and solution breccias common	Supratidal zone	Occurs in upper part of U_2-U_3 and U_4-U_5 dolomites
FACIES 4	Microcrystalline dolomite (less than 10 μ)	Xenotopic dolomite	Same as Facies 3	Same as Facies 3	Supratidal zone	Occurs in upper part of U_6-U_7 dolomite
FACIES 5	Skeletal dolomite	Dolomitized biomicrite, wackestone to packstone. Locally abundant peloids and intraclasts	Same as Facies 1	Same as Facies 1	Subtidal to intertidal zones	Occurs in lower part of U_4-U_5 and U_6-U_7 dolomites
FACIES 6	Laminated black shale stringers	Fissile and carbonaceous	Unfossiliferous	Locally contain flat-pebble breccia clasts of microcrystalline dolomite. Graded bedding	Supratidal zone marsh	Occurs locally atop U_4-U_5 and U_6-U_7 dolomites

tervals are abrupt and disconformable. In contrast, the bases of the dolomite intervals are transitional and conformable with the underlying limestone intervals.

Depositional Facies

Six facies are recognized within the cored portion of the Red River Formation at Cabin Creek Field. The lithology, environmental setting, stratigraphic occurrence, and other features of the facies are summarized in Table 2-1. Figure 2-4 shows the lateral distribution of facies across Cabin Creek Field and relates them to the stratigraphic intervals.

Facies 1 is mottled dolomitic biomicrite lime wackestone to packstone (Fig. 2-5A). The major allochems are skeletal grains consisting of echinoderm, brachiopod, bryozoan, trilobite, mollusk, ostracod, gastropod, coral, and stromatoporoid fragments. Concentrations of peloids, intraclasts, and ooids are present locally. Matrix material consists of pelletal lime mud. Pellets occurring within or adjacent to large shells survived the effects of compaction. Bedding within this facies is poorly preserved because of intense burrowing activity. Burrows have been dolomitized selectively, resulting in mottling. Most burrows exhibit preferred horizontal orientation, although vertical

burrows are common. Burrow shapes have been modified greatly by the combined effects of selective dolomitization and pressure solution. Stylolites are common throughout and are generally irregular and of low amplitude. Cross-bedding and graded bedding occur locally. Facies 1 constitutes essentially all of the U_1-U_2, U_3-U_4, and U_5-U_6 limestones. Its thickness varies from 14 to 40 feet (4.3–12.2 m) across the field. Facies 1 also is present locally within the U_4-U_5 and U_6-U_7 dolomites, attaining maximum thickness of 9 feet (2.7 m) in these intervals (e.g., Fig. 2-4, Shell Cabin Creek 43-6 well).

Facies 2 is mottled dolomitic biomicrite lime mudstone. Allochems consist of skeletal grains very similar to those in facies 1. Matrix material consists of pelletal lime mud. Facies 2 is massive to poorly bedded and intensely burrowed. Burrows have been dolomitized selectively, resulting in mottling similar to that of facies 1. Horizontally oriented stylolites are common. Facies 2 is present only locally, within the U_2-U_3 dolomite, and has a maximum thickness of only 4 feet (1.2 m). Laterally it interfingers with facies 3 (Fig. 2-4).

Facies 3 is finely crystalline dolomite 10 to 60 micron size, having predominantly subhedral-shaped crystals (hypidiotopic fabric). Wavy laminae are the predominant structure (Fig. 2-5B). Other common structures of this facies are flat-pebble breccias, desiccation cracks (Fig. 2-5C),

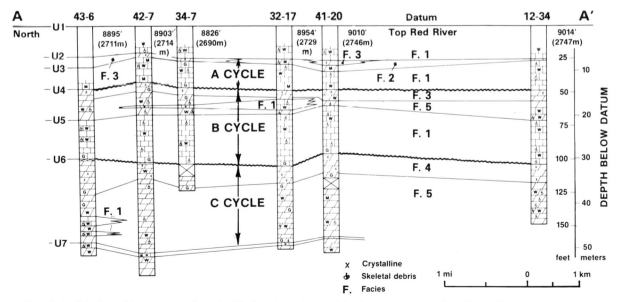

Fig. 2-4. Stratigraphic cross-section A–A′ showing relationships between upper Red River lithofacies, cycles, and informal zones or stratigraphic units (see Fig. 2-2 for section location). Datum is top of Red River.

truncation surfaces, oncolites (Fig. 2-5D), nodular anhydrite, and solution breccias healed by secondary anhydrite. The pelletal fabric is preserved locally. Facies 3 is mostly unfossiliferous, although scattered, small bits of skeletal debris consisting of ostracod and brachiopod shells are present. Facies 3 is present locally in the U_2-U_3 dolomite and also in the upper part of the U_4-U_5 dolomite (Fig. 2-4) and has a maximum thickness of 8 feet (2.4 m).

Facies 4 is microcrystalline dolomite (less than 10 micron size) with predominantly anhedral-shaped crystals (xenotopic fabric). Structures are similar to those of facies 3 (Fig. 2-5E). Fossil remains are rare. Facies 4 is restricted to the upper part of the U_6-U_7 dolomite and varies in thickness from 10 to 25 feet (3.1–7.6 m).

Facies 5 is skeletal dolomite (Fig. 2-5F). Allochems consist of skeletal grains largely derived from the same faunal assemblage as that of facies 1. In addition to skeletal grains, facies 5 contains locally abundant peloids and intraclasts. Bedding is poorly developed because of burrowing activity (Figs. 2-6A, B). Other structures include local cross-bedding and graded bedding (Figs. 2-6C, D), stylolites, chert nodules, and anhydrite nodules (Fig. 2-6E). Before being dolomitized, facies 5 was a biomicrite wackestone. Facies 5 is present in the lower part of the U_4-U_5 and U_6-U_7 dolomites and interfingers locally with facies 1. Its thickness varies from 5 to 54 feet (1.5–16.5 m).

Facies 6 consists of laminated black shale stringers. The shales are fissile, carbonaceous, and unfossiliferous. Locally, they contain flat-pebble breccia clasts of microcrystalline dolomite (Fig. 2-6F) that exhibit graded bedding. Facies 6 is distributed locally atop the U_4-U_5 and U_6-U_7 dolomites. Its thickness is less than 1 foot (0.3 m). It is recognized on gamma-ray logs by a characteristic sharp increase in radioactivity, which is correlated readily. Such shales have been reported to be laterally persistent across the Williston Basin (Carroll, 1978).

Depositional Environments

Depositional environments of the upper Red River Formation are shown in Figure 2-7. Three cycles of subtidal through supratidal deposition

occurred on a broad, shallow-marine, carbonate bank. The environments inferred for the facies are summarized in Table 2-1. A summary of sedimentary features occurring within these cycles appears in Figure 2-8. Each cycle represents a shoaling-upward sedimentary sequence. A fully developed cycle consists of the following units from the base upward: (1) a basal skeletal wackestone deposited under subtidal conditions; (2) cross-bedded and ripple-marked skeletal packstone containing intraclasts, peloids, and ooids deposited under intertidal conditions; (3) a laminated unfossiliferous dolomite containing anhydrite and solution-collapse breccias, deposited under high intertidal to supratidal conditions; and (4) a thin black laminated fissile shale containing flat-pebble breccias of microcrystalline dolomite, deposited under supratidal conditions.

Subtidal

The dominance of mud-supported sediments, poorly developed bedding, and the abundance of bioturbation together indicate that subtidal conditions of relatively low energy predominated during deposition of much of facies 1, 2, and 5. The relative abundance and diversity of fossils, and the abundance of burrow mottling suggest normal salinity and open-marine conditions.

Indications of currents, such as cross-bedding or ripple marks, are generally absent in subtidal deposits of the upper Red River Formation, suggesting that subtidal deposition occurred under relatively quiet conditions below effective wave base. Rate of sedimentation did not exceed the rate of sediment reworking by burrowing organisms. Thin beds of packstone, which generally contain imbricated skeletal debris as well as skeletal debris, peloids, and some concentrations of intraclasts and ooids, occur locally and indicate periodic winnowing and sorting of bottom sediments by storm waves. Most skeletal debris is intact and unabraded, as expected under low-energy conditions. Fragmentation of shells and other skeletal remains can be attributed to effects of boring organisms and sediment churning by burrowing organisms. Occurrences of pyrite and local thin black kerogen within the skeletal wackestones indicate periods of euxinic bottom conditions (Kohm and Louden, 1978). Stratigraphically, subtidal conditions predominated during

Fig. 2-5. Photographs of slabbed cores. *A.* Mottled dolomitic biomicrite-wackestone (facies 1) containing small brachiopods with micrite geopetal fill, and various fragments of skeletal debris. Specimen was stained with alizarin red-S. The dolomite mottling is light-colored. Cabin Creek 42-7 well, 8936 feet (2724 m), U_3-U_4 limestone subtidal zone. *B.* Finely crystalline dolomite (facies 3) with wavy laminae. In many cases, the laminae are separated by stylolitic surfaces containing organic matter. Cabin Creek 42-7, 8949 feet (2728 m), U_4-U_5 dolomite supratidal zone. *C.* Laminated, finely crystalline dolomite (facies 3) locally containing small fragments of skeletal debris. Desiccation crack is filled with clasts of laminated dolomite as well as skeletal debris. Cabin Creek 42-7, 8925 feet (2720 m), U_2-U_3 dolomite supratidal zone. *D.* Truncation surface within a laminated, finely crystalline dolomite (facies 3) containing oncolites (arrow) Cabin Creek 34-7, 8851 feet (2698 m), U_2-U_3 dolomite intertidal to supratidal zone. *E.* Laminated microcrystalline dolomite (facies 4) which has undergone solution-collapse and fracturing. Secondary anhydrite now fills fractures and solution voids. Cabin Creek 42-7, 9030 feet (2752 m), U_6-U_7 dolomite intertidal to supratidal zone. *F.* Skeletal dolomite (facies 5) with crude bedding and horizontal stylolites. Many skeletal molds are present, and the larger ones have been filled by secondary anhydrite (arrow). Molds exhibit crude horizontal stratification. Cabin Creek 41-28, 9048 feet (1759 m), U_4-U_5 dolomite intertidal zone.

Fig. 2-6. Photographs of slabbed cores. *A.* Skeletal dolomite (facies 5) containing tabulate coral debris and burrow mottles. Cabin Creek 41-28, 9054 feet (2760 m). U_4-U_5 dolomite subtidal zone. *B.* Skeletal dolomite (facies 5) which has undergone intense burrowing. Note random orientation of burrow mottles. Cabin Creek 42-7, 9064 feet (2763 m), U_6-U_7 dolomite subtidal zone. *C.* Cross bedded, oolitic, skeletal dolomite (facies 5) containing multiple horizontal stylolites. Cabin Creek 42-7, 9016 feet (2748 m), U_6-U_7 dolomite intertidal zone. *D.* Cross-bedded skeletal dolomite (facies 5) consisting of a peloidal grainstone. Bedding has been enhanced by oil stain. Cabin Creek 42-7, 9020 feet (2749 m), U_6-U_7 dolomite intertidal zone. *E.* Skeletal dolomite (facies 5) at contact with a large anhydrite nodule (arrow). Bedding within the dolomite has been displaced downward by nodule growth. The dolomite contains several chert nodules, one of which is cross-cut by a vein of secondary anhydrite (below arrow). Cabin Creek 34-7, 8884 feet (2708 m), U_4-U_5 dolomite intertidal zone. *F.* Black shale (facies 6) containing flat-pebble breccia clasts of microcrystalline dolomite. The breccia clasts exhibit graded bedding and are topped by a layer of fissile organic shale. Note horizontal orientation of the breccia clasts. Cabin Creek 43-18, 8890 feet (2710 m), U_6-U_7 dolomite supratidal zone.

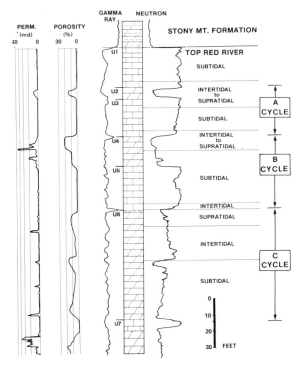

Fig. 2-7. Typical columnar section of the upper Red River showing depositional environments, porosity, and permeability determined by core analysis and gamma ray-neutron curves.

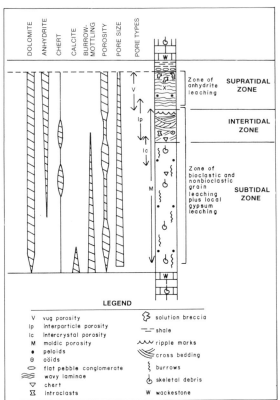

Fig. 2-8. Depositional and diagenetic features of the upper Red River dolomites. Base of dolomite is transitional, with the limestone beneath. Contact between dolomite and uppermost limestone is erosional with black shale containing flat-pebble breccia deposited atop the supratidal zone.

deposition of the lower U_6-U_7 dolomite; again during deposition of the upper U_5-U_6 limestone and lower U_4-U_5 dolomite units; and during deposition of the middle U_3-U_4 and upper U_1-U_2 limestone units (Fig. 2-7).

Intertidal

Parts of facies 1 and 5 were at times deposited under relatively high-energy intertidal conditions, as indicated by the presence of cross-bedding, graded bedding, ripple-bedding, imbricate shells, conglomerates, and oncolites. In addition to skeletal debris, intertidal sediments contain abundant intraclasts, peloids, and ooids. Burrow mottling occurs in the intertidal sediments, but is much less abundant than in the subtidal zone. Intertidal conditions predominated during deposition of the middle to lower-upper U_6-U_7 dolomite, the basal part of the U_5-U_6 limestone, the middle part of the U_4-U_5 dolomite, the basal and upper parts of the U_3-U_4 limestone, and the basal U_1-U_2 limestone.

Supratidal

Supratidal conditions prevailed during deposition of facies 3, 4, and 6, as indicated by the presence of desiccation cracks, truncation surfaces, flat-pebble breccias, wavy laminated dolomite, rare fossil remains, and lack of burrows. Solution-collapse breccias are common in the supratidal sediments and are evident from *in situ* sediment folding, warping, and dislocation. These breccias probably formed by dissolution of evaporitic sulfate minerals followed by collapse of the overlying sediment. Deposition of laminae occurred by periodic flooding of the supratidal surface by marine waters laden with fine-grained calcium-carbonate sediment.

It has been suggested that algae have contributed to the formation of laminated dolomites of the upper Red River Formation (Asquith *et al.*, 1978; Carroll, 1978). Organic-rich laminae are regarded by Carroll (1978) as poorly preserved algal structures. Wavy laminae strongly resembling stromatolites are conspicuous in Red River supratidal sediments at Cabin Creek Field. In many examples, the laminae are separated by stylolitic surfaces.

Organic-rich muds accumulated locally in marshes on the supratidal surface, as indicated by occurrence of laminated black shales (facies 6). The flat-pebble breccias associated with these shales imply desiccation and erosion of adjacent sediment. Carbonate clasts were ripped up and redeposited in low-lying marsh areas by storm waves that periodically inundated the supratidal surface.

Anhydrite occurs as porphyrotopic, anhedral crystals up to 60 microns in size, and generally constitutes less than 1 percent by volume of the supratidal sediments. Its presence indicates the occurrence of evaporitic conditions in the supratidal zone. Displacive nodular anhydrite occurs locally in the lower supratidal sediments. Former abundance of evaporitic sulfate minerals in the supratidal zone is indicated by the occurrence of solution-collapse breccias and also vug porosity, which are interpreted as having formed by leaching of sulfate minerals (see Fig. 2-8). Supratidal conditions prevailed during deposition of the upper parts of the U_6-U_7, U_4-U_5, and U_2-U_3 dolomites.

Diagenesis

Each shoaling-upward cycle of upper Red River subtidal through supratidal sediments displays diagenetic overprinting. Early stages of diagenesis include leaching of evaporitic sulfate minerals and formation of dolomite in the supratidal zone. Later stages of diagenesis include continued dolomitization, followed by leaching of undolomitized calcite, reprecipitation of sulfate minerals as cement, replacement and cementation by silica, and pressure solution.

Leaching of Evaporitic Sulfate Minerals

It is inferred that supratidal sediments of the upper Red River Formation were periodically subjected to leaching by fresh meteoric water percolating down through the vadose zone. Under these conditions, early diagenetic minerals such as calcium sulfate are ephemeral and easily destroyed (Kinsman, 1969). Dissolution is indicated by (1) corroded edges of relict anhydrite crystals, (2) the presence of vugs that closely approximate the size and shape of relict anhydrite crystals (gypsum pseudomorphs), and (3) the occurrence of solution-collapse breccias and fractures.

Porphyrotopic crystals of anhydrite occur in varying stages of dissolution within upper Red River supratidal sediments. Undissolved anhydrite crystals generally average 0.05 mm in size and are anhedral. In many examples, the size and shape of such anhydrite crystals closely approximate the size and shape of vugs that contribute to a large part of the porosity in upper Red River supratidal zones. Partially leached anhydrite is common (Fig. 2-9A).

Gypsum and anhydrite molds occur within supratidal, intertidal, and subtidal sediments. They are recognized by their characteristic straight sides and rectangular re-entrants and projections (Fig. 2-9B) and are on the order of 0.1 to 0.5 mm in size. In many examples, a distinction can be made between gypsum and anhydrite molds on the basis of shape. The occurrence of gypsum and anhydrite in modern sabkha environments has been well documented, as has leaching of such evaporitic sulfate minerals (*e.g.*, Murray 1960).

In addition to leaching of evaporitic sulfate minerals, there is evidence to suggest that skeletal and peloidal grains of calcite or aragonite were also leached from supratidal sediments. Round to ovoid-shaped molds up to 1.0 millimeter in size are locally interlaminated with supratidal sediments and exhibit preferred horizontal orientation (Fig. 2-9C). Breccias and fractures produced by solution-collapse are common in the supratidal sediments and are most likely the result of dissolution of gypsum or anhydrite crystals and nodules. The solution breccias are healed by secondary anhydrite (Fig. 2-9D). Clasts are up to several centimeters in size, angular, ungraded, and very poorly sorted. Criteria for recognition of evaporite solution breccias have been summarized by Beales and Oldershaw (1969). Dissolution of evaporitic minerals from underlying strata resulted in collapse, downwarping, fracturing, and rotation of breccia clasts.

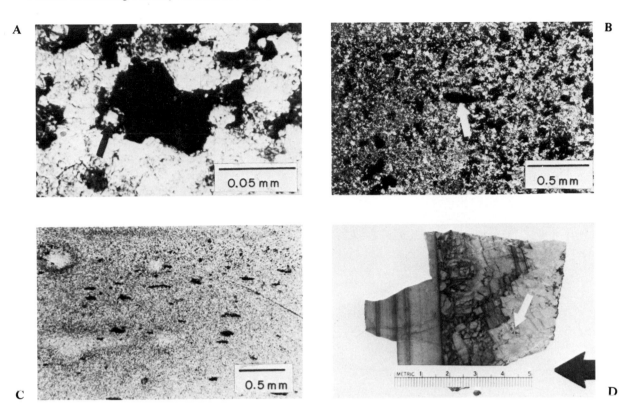

Fig. 2-9. Photomicrographs (*A–C*) showing evaporite molds and cements, and core-slab photo (*D*). *A.* Partly leached anhydrite (arrow) in microcrystalline dolomite matrix. Note the straight-sided, rectangular boundaries of the vug left after the anhydrite was leached. Cross-polarized light. Cabin Creek 34-7, 8944 feet (2726 m), U_6-U_7 dolomite supratidal zone. *B.* Gypsum mold (arrow) in finely crystalline dolomite. It is likely that much of the other pore space is of the same origin, but that dolomitization has altered the shape of the original gypsum molds. This section photomicrograph, cross-polarized light. Cabin Creek 34-7, 8881 feet (2707 m), U_4-U_5 dolomite intertidal zone. *C.* Aligned molds in microcrystalline dolomite. Long dimensions of the molds are aligned parallel to bedding. Cross-polarized light. Cabin Creek 21-8, 9036 feet (2754 m), U_6-U_7 dolomite supratidal zone. *D.* Solution collapse breccia in a laminated microcrystalline dolomite. Secondary anhydrite (arrow) fills the fractures and much void space. Top of breccia zone is bounded by a planar stylolite. Arrow indicates stratigraphic-up direction. Cabin Creek 42-7, 9032 feet (2753 m), U_6-U_7 dolomite intertidal zone.

Dolomitization

Environmental conditions suggest the occurrence of syngenetic dolomite in Cabin Creek Field supratidal sediments, as well as elsewhere in the Williston Basin (Asquith, 1979; Asquith *et al.*, 1978; Carroll, 1978). Original supratidal zone sediment, consisting of pelletal lime-mud, could have been dolomitized under hypersaline conditions resulting from capillary concentration (*e.g.*, Friedman and Sanders, 1967) and an increase in the Mg/Ca ratio of the interstitial brines

(Murray and Pray, 1965). Hypersalinity is indicated by presence of anhydrite within the supratidal sediments. The chemical, climatic, and physiographic setting of upper Red River dolomites may have been analogous to the sabkha environments of the Persian Gulf, as described by Kinsman (1969).

Evidence suggesting that syngenetic dolomitization occurred in the upper Red River Formation at Cabin Creek Field includes (1) preservation of the original pelletal fabric of the sediment, (2) the shape of vugs formed as a result of leaching of anhydrite from the supratidal sediments, and (3) very small crystal sizes. The preservation of pelletal fabric implies that dolomitization occurred soon after deposition, before significant burial and compaction. It is possible that pellets within the upper Red River supratidal sediments survived compaction because they were replaced and cemented by syngenetic dolomite. Shinn *et al.* (1980) suggest that well-preserved pellets in ancient carbonates indicate predepositional hardening or synsedimentary cementation. Vugs formed in upper Red River supratidal sediments as a result of anhydrite leaching typically have rather sharp, distinct boundaries (Fig. 2-9A). In most examples, dolomite rhombs do not invade the vugs, implying that dolomitization of the host sediment occurred before leaching of the anhydrite.

Syngenetic dolomitization occurred in facies 3 and 4. These consist of locally pelletal, finely crystalline hypidiotopic dolomite and microcrystalline xenotopic dolomite, respectively. The textural lamination of the original sediment has been largely preserved in both examples. Size and textural differences between facies 3 and 4 dolomites may have resulted from differences of the original supratidal sediments, or from chemical differences during the dolomitization process. Carbonate crystal morphology may be related to rate of crystallization and also to effects of magnesium and other ions present in the precipitating waters (Folk, 1974).

The occurrence of diagenetic dolomite in the intertidal and subtidal zones is indicated by the selective replacement of former calcium carbonate grains and matrix. In general, the grains that composed the original calcium carbonate sediments are preserved as relics in the dolomite. Such relict grains include pellets, intraclasts, peloids, ooids, and skeletal grains. The abundance of molds of the various grains indicates that the original lime-mud matrix was dolomitized before the grains. Contacts of dolomite units are transitional with underlying limestones. As contacts are approached, fewer and fewer dolomitized skeletal grains occur until only the matrix consists of dolomite.

Dolomitic mottling is common throughout the subtidal limestone facies. The mottling has resulted from selective dolomitization of burrows. Although vertical burrows are common, most have a preferred horizontal orientation. Dolomite within burrows is generally finely crystalline and hypidiotopic. Rhombs are not strictly confined within burrows but tend to be scattered into the surrounding sediment (Fig. 2-10A), except in areas where shells have obstructed the passage of dolomitizing solutions (Fig. 2-10B). Burrow shapes have been modified by dolomitization and also by pressure solution. Where sediments were compacted or partly lithified, burrows acted as a major conduit of dolomitizing solutions.

Pervasive dolomitization occurred locally in the supratidal, intertidal, and subtidal zones, but was most widespread in the uppermost part of the U_6-U_7 subtidal sediments. This bed of pervasive dolomite is about 2 to 4 feet (0.6–1.4 m) thick and occurs in most of the cores studied. It consists of finely crystalline, mostly idiotopic dolomite (Fig. 2-10C). Void-filling dolomite cements occur locally and consist of euhedral rhombs of saddle dolomite up to 1.0 millimeter in size (Fig. 2-10D). Interparticle dolomite cements consist of finely crystalline to microcrystalline xenotopic dolomite between pellets, peloids, or intraclasts of subtidal to intertidal grainstones (Fig. 2-10E). Dolomite cement also heals some fractures.

Anhydrite Cementation

Secondary anhydrite heals solution-breccias and fractures in the supratidal sediments (Fig. 2-10F), and locally fills moldic, interparticle, and intercrystal porosity of the intertidal and subtidal sediments. Several beds of anhydrite-cemented

---

Fig. 2-10. Photomicrographs of features in Red River dolomites. *A.* Edge of a dolomitized burrow protruding into matrix of a biomicrite-wackestone. Additional dolomite rhombs are scattered through the limestone matrix (arrow). Plane light. Cabin Creek 43–6, 8971 feet (2734 m), U_5-U_6 limestone subtidal zone. *B.* Partly dolomitized biomicrite-wackestone. Dolomitization was obstructed by presence of brachiopod shell (arrow). Note how the sharp boundary of dolomitization extends well beyond the end of the shell. Plane light. Cabin Creek 34-7, 8907 feet (2715 m), U_5-U_6 limestone subtidal zone. *C.* Pervasive dolomite occurring in the upper subtidal zone. Crystals are finely crystalline, idiotopic to hypidiotopic. Intercrystal pore space is black. Cross-polarized light. Cabin Creek 41-20, 9150 feet (2789 m), U_6-U_7 dolomite subtidal zone. *D.* Saddle dolomite crystals have filled a large vug in a biomicrite-wackestone. Note undulose extinction and curved crystal faces typical of saddle dolomite. Cross-

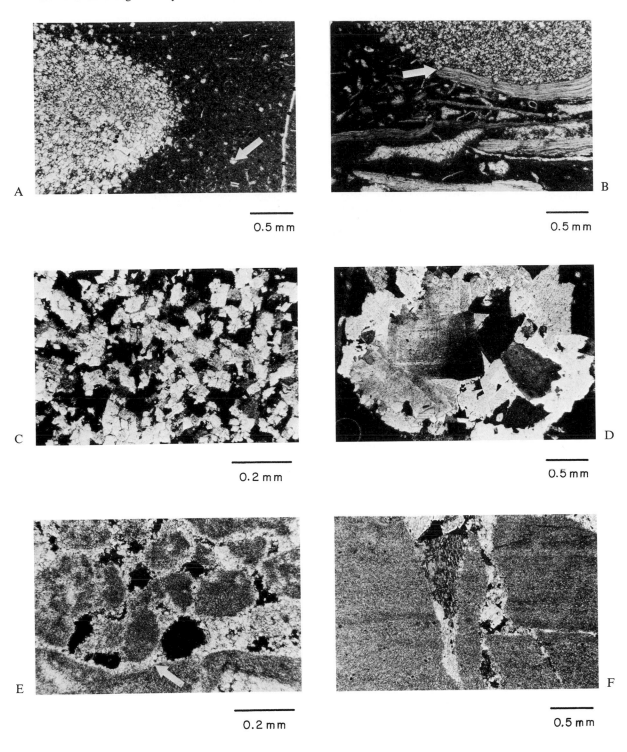

A

0.5 mm

B

0.5 mm

C

0.2 mm

D

0.5 mm

E

0.2 mm

F

0.5 mm

polarized light. Cabin Creek 43-18, 8885 feet (2708 m), U_5-U_6 limestone subtidal zone. *E*. Finely crystalline dolomite (arrow) cementing dolomitized peloids. Interparticle and moldic pores are black. Thin section photomicrograph, cross-polarized light. Cabin Creek 12-34, 9156 feet (2791 m), U_6-U_7 dolomite subtidal zone. *F*. Fractures in a microcrystalline dolomite are healed by secondary anhydrite. The fractures originated from solution-collapse. Cross-polarized light. Cabin Creek 41-20, 9111 feet (2777 m), U_6-U_7 dolomite supratidal zone.

Table 2.2. Summary of Porosity, Permeability, Saturation, and Thin-Section Data for Dolomite Reservoirs in the U_4-U_5 and U_6-U_7 Zones, Upper Red River Formation, Cabin Creek Field. (Relative pore type amounts and pore size are from thin-section point counting of core samples. Remaining data are from core plug analysis.)

Zone & Environment	Porosity %	Perm (md)	N_c	Oil Sat. %	Water Sat. %	Total Sat. %	Pore Size (mm)	Relative Amounts of Each Pore Type[1]				
								V	M	Ip	Ic	N_{ts}
U_4-U_5 supratidal	16.8	11.7	59	29.3	21.4	50.7	0.13	48%	13%	13%	26%	23
U_6-U_7 supratidal	7.4	0.6	110	16.3	47.4	63.7	0.04	95%	5%	—	—	38
U_4-U_5 intertidal	15.3	7.1	49	30.2	23.5	53.7	0.19	19%	81%	—	—	26
U_6-U_7 intertidal	7.1	3.6	156	17.2	39.4	56.6	0.13	31%	31%	26%	11%	57
U_4-U_5 subtidal	10.7	1.4	110	15.3	41.3	56.6	0.19	32%	64%	—	4%	33
U_6-U_7 subtidal	9.4	2.3	156	16.2	44.3	60.5	0.17	25%	44%	10%	21%	64

[1]V = vug, M = moldic, Ip = interparticle, Ic = intercrystal
N_c = number of core analyses
N_{ts} = number of thin-section analyses

dolomite, generally less than a foot in thickness, can be correlated among cored wells. Anhydrite has also replaced fossil fragments, a feature common in dolomites (Murray, 1960). Sources of secondary anhydrite may have been either the primary evaporitic sulfate minerals that underwent earlier leaching, or local brine pools that accumulated in supratidal ponds. Cementation by anhydrite in most cases postdated dolomitization.

Silica Cementation

Silica-cemented dolomite beds occur as chert, megaquartz, and chalcedony. Void-filling silica occurs most commonly in the intertidal and subtidal sediments and generally postdates both dolomitization and anhydrite cementation. Chert nodules, in which original sediment fabric is usually preserved, are up to several centimeters in size and occur as beds less than a foot in thickness. Chert is replacive and infilling rather than displacive, and some beds can be correlated across the field within U_4-U_5 and U_6-U_7 units. Megaquartz and chalcedony occur as void-filling cements, locally occlude porosity, and are most abundant in the intertidal zone.

Pressure Solution

The effects of late diagenetic pressure solution are ubiquitous throughout the upper Red River, as indicated by the abundance of stylolites. The stylolites are mostly irregular, low-amplitude, subhorizontal, and are commonly multiple, although high-amplitude stylolites occur locally. They are most abundant in subtidal biomicrite-wackestones (facies 1) and least abundant in supratidal dolomites (facies 3 and 4). The lower abundance of stylolites in fine-grained dolomites was also noted by Carroll (1978) in the upper Red River Formation of North Dakota. This variation in abundance may reflect the relative resistance of dolomite to pressure solution.

Porosity

Porosity is restricted to the dolomite units and is therefore a function of diagenetic overprint, which in turn is a function of depositional environment (Figs. 2-7, 2-8). Four major types of porosity occur: vug, moldic, intercrystal, and interparticle (Table 2-2). Diagenetic events that controlled porosity development were (1) dolomitization, (2) post-depositional leaching of evaporitic sulfate minerals and calcium carbonate grains, and (3) local precipitation of secondary anhydrite and silica in both primary and secondary pores. Although interparticle porosity still exists, the vast majority of pore space originated as secondary porosity. The processes that were responsible are outlined in Figure 2-11.

Vug porosity is the dominant type in the supratidal sediments, mainly as a result of eogenetic leaching of evaporitic sulfate minerals. This

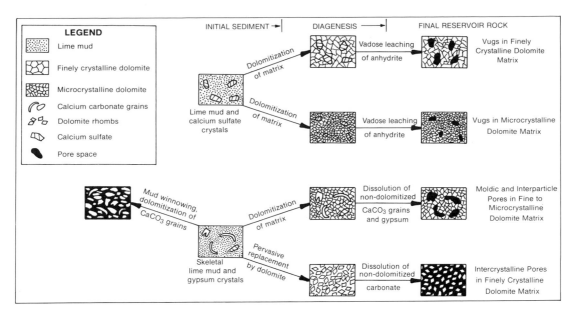

Fig. 2-11. Diagenetic processes involved in the development of upper Red River pore systems. Type I pores are vugs in finely crystalline dolomite matrix. Type II pores are vugs in microcrystalline dolomite matrix.

Type III pores are moldic and interparticle pores in fine-to-microcrystalline dolomite matrix. Type IV pores are intercrystalline pores in finely crystalline dolomite matrix.

mechanism was most effective in the U_4-U_5 interval (Fig. 2-12A), resulting in an average of 16.8 percent porosity. Dolomite composing the supratidal facies in this interval is finely crystalline and hypidiotopic. The initial sediment consisted of laminated mud-to-sand-sized calcium carbonate particles largely devoid of skeletal debris. The size and predominantly subhedral shape of the matrix dolomite crystals have resulted in large, moderately sorted pore throats of rather homogeneous distribution, and consequently, excellent reservoir properties. In contrast, vugs in the supratidal zone of the U_6-U_7 interval occur in microcrystalline, xenotopic dolomite and are much smaller in size and less abundant (Fig. 2-12B). This difference resulted from greater abundance of sulfate minerals removed from the U_4-U_5 interval, and also from textural differences. Reservoir properties of this interval are rather poor, primarily because of the small size of the pore throats.

Moldic porosity occurs most abundantly in the intertidal and subtidal facies and resulted from leaching of calcite and aragonite particles remaining after dolomitization of matrix materials (Fig. 2-12C). Some moldic porosity also developed by leaching of gypsum crystals from sediment prior to dolomitization (Fig. 2-12D). Size and shape of moldic pores depends on those of the original leached particles and crystals. In many examples, moldic pores have been enlarged by solution to form vugs that are much larger than vugs of the supratidal zone. Interparticle porosity occurs most abundantly in the high-energy intertidal zone. Interpelletal porosity is present locally in supratidal sediments. Both moldic and interparticle pores, as well as the associated pore throats, are quite heterogeneous in shape, size, and distribution. This is a result of the variety of particle types dissolved to create the pores and the size and shape of the matrix dolomite crystals. The small size and moderate to poor sorting of the pore throats account for the poor reservoir properties associated with these pore types.

Intercrystal porosity is rare except in the upper part of the U_6-U_7 interval subtidal zone (Fig. 2-12E). Where it is well developed, it constitutes up to 20 percent of the bulk volume, with permeability over 100 millidarcys (Table 2-2). Typically, the dolomite matrix associated with this pore type is finely crystalline and idiotopic to hypidiotopic. The presence of skeletal debris in

rocks immediately above and below the U_6-U_7 interval intercrystal porosity zone suggests that the original sediment was a skeletal lime mud. Pore geometry is rather homogeneous due to the fabric of the matrix dolomite crystals. The relatively large pore-throat size and moderate sorting account for excellent reservoir properties.

Dolomitization accounts for most porosity in the upper Red River. For example, porosity in the finely crystalline dolomite is typically much greater than that in the skeletal or microcrystalline dolomite. Since the thickness of each facies varies across the field, there is a corresponding change in porosity value from well to well. Porosity is locally occluded by secondary anhydrite and silica cements (Fig. 2-12F). Cemented zones are thin and do not account for a major reduction of the total pore space, but may act as permeability barriers and thus influence production.

◄ *Fig. 2-12.* Photomicrographs of Red River porosity types. *A*. Vug porosity is the supratidal zone of the U_4-U_5 reservoir. Pores in black. Matrix is finely crystalline hypidiotopic dolomite. Cross-polarized light. Cabin Creek 34-7, 8876 feet (2705 m). Porosity = 17.6%. Permeability = 2.2 md. *B*. Vug porosity in the supratidal zone of the U_6-U_7 reservoir. Arrow points to a partially dissolved anhydrite lath. Matrix is microcrystalline xenotopic dolomite. Cross-polarized light. Cabin Creek 12-34, 9122 feet (2780 m). Porosity = 3.8%. Permeability = 0 md. *C*. Moldic porosity in the subtidal zone of the U_6-U_7 reservoir. Pores are black. Porosity resulted from dissolution of skeletal material. Note mold of brachiopod shell. Porosity has been partly occluded by anhydrite (see arrow). Matrix is microcrystalline dolomite. Cross-polarized light. Cabin Creek 32-17, 9114 feet (2778 m). Porosity = 13.4%. Permeability = 0.3 md. *D*. Moldic porosity in intertidal zone of the U_4-U_5 reservoir. Pores are black. Arrow points to mold of gypsum rosette. Matrix is microcrystalline dolomite. Cross-polarized light. Cabin Creek 34-7, 8881 feet (2707 m). Porosity = 18.8%. Permeability = 3.7 md. *E*. Intercrystalline porosity in the subtidal zone of the U_6-U_7 reservoir. Pores are black. Note zoning of dolomite crystals. Cross-polarized light. Cabin Creek 41-20, 9176 feet (2797 m). Porosity = 13.1%. Permeability = 92.1 md. *F*. Silica-filled moldic and interparticle porosity in the intertidal zone of the U_6-U_7 reservoir. Matrix consists of peloidal dolomite. Cross-polarized light. Cabin Creek 41-20, 9132 feet (2783 m). Porosity = 2.0%. Permeability = 0 md.

Conclusions

Upper Red River carbonate peritidal sequences at Cabin Creek Field were deposited on a broad, shallow-marine, carbonate bank under arid conditions. Average reservoir zone porosity is 13 percent, and maximum is 33 percent. Porosity developed in several stages and was controlled by depositional environments, dolomitization on evaporitic sabkhas, and post-depositional leaching by meteoric waters of evaporitic sulfate minerals and undolomitized $CaCO_3$ grains to create moldic and vug porosity. Sulfate-rich brines caused local reprecipitation of calcium sulfate as a cement. Parts of the subtidal zone were cemented by secondary silica.

These stages of porosity development and cementation were repeated during each of three main depositional cycles. Transgression of each successive supratidal surface began a renewal of sedimentation, followed in turn by renewed dolomitization, fresh water leaching, and development of additional secondary porosity. Each cycle differs from the others in thickness and relative proportions of subtidal through supratidal sediments, as well as in original particle sizes, fabrics, and sedimentary structures. As a result, each cycle has a different set of porosity and permeability characteristics.

References

ASQUITH, G.B., 1979, Subsurface Carbonate Depositional Models—a Concise Review: The Petroleum Publ. Corp., Tulsa, OK, 121 p.

ASQUITH, G.B., R.L. PARKER, C.R. GIBSON, and J.R. ROOT, 1978, Depositional history of the Ordovician Red River C and D zones, Big Muddy Creek Field, Roosevelt County, Montana: *in* Montana Geol. Soc. Guidebook: Williston Basin Symp., p. 71–76.

BEALES, F.W., and A.E. OLDERSHAW, 1969, Evaporite-solution brecciation and Devonian carbonate reservoir porosity in Western Canada: Amer. Assoc. Petroleum Geologists Bull., v. 53, no. 3, p. 503–512.

CARROLL, W.K., 1978, Depositional and paragenetic controls on porosity development, upper Red River Formation, North Dakota: *in* Montana Geol. Soc. Guidebook: Williston Basin Symp., p. 79–94.

CLEMENT, J.H., 1976, Geologic history—key to accumulation at Cedar Creek (abst): Amer. Assoc. Petroleum Geologists Bull., v. 60, p. 2067–2068.

CLEMENT, J.H., 1978, Depositional sequences and characteristics of Ordovician Red River producing zones, Pennel Field, Cedar Creek Anticline, Fallon County, Montana (abst): in Montana Geol. Soc. Guidebook: Williston Basin Symp., p. 97.

DOW, W.G., 1974, Application of oil-correlation and source-rock data to exploration in Williston Basin: Amer. Assoc. Petroleum Geologists Bull., v. 58, p. 1253–1262.

FOLK, R.L., 1974, The natural history of crystalline calcium carbonate: effect of magnesium content and salinity: Jour. Sedimentary Petrology, v. 44, no. 1, p. 40–53.

FRIEDMAN, G.M., K. RUZYLA and A. REECKMANN, 1981, Effects of porosity type, pore geometry, and diagenetic history on tertiary recovery of petroleum from carbonate reservoirs: U.S. Dept. of Energy DOE/MC/11580-5 Final Report. Morgantown, WV, Sept. 1981.

FRIEDMAN, G.M. and J.E. SANDERS, 1967, Origin and occurrence of dolostones: in Chilingar, G.V., H.J Bissell, and R.W. Fairbridge, eds., Carbonate Rocks—Origin occurrence and classification: Elsevier Publ. Co., Amsterdam, p. 267–348.

FRIEDMAN, G.M., and J.E. SANDERS, 1978, Principles of Sedimentology: New York, John Wiley and Sons, 792 p.

FULLER, J.G.C.M., 1961, Ordovician and contiguous formations in North Dakota, South Dakota, Montana, and adjoining areas of Canada and United States: Amer. Assoc. Petroleum Geologists Bull., v. 45, p. 1334–1363.

KENDALL, A.C., 1976, The Ordovician carbonate succession (Bighorn Group) of southeastern Saskatchewan: Dept. Mineral Resources, Saskatchewan Geol. Surv., Sedimentary Geol. Div., Report 180, p. 1–185.

KINSMAN, D.J.J., 1969, Modes of formation, sedimentary associations, and diagnostic features of shallow-water and supratidal evaporites: Amer.

Assoc. Petroleum Geologists Bull., v. 53, p. 830–840.

KOHM, J.A., and R.O. LOUDEN, 1978, Ordovician Red River of eastern Montana and western North Dakota: relationships between lithofacies and production: Montana Geol. Soc. Guidebook: Williston Basin Symp., p. 99–117.

MURRAY, R.C., 1960, Origin of porosity in carbonate rocks: Jour. Sedimentary Petrology, v. 30, p. 59–84.

MURRAY, R.C., and L.C. PRAY, 1965, Dolomitization and limestone diagenesis—an introduction: in Pray, L.C., and R.C. Murray, eds., Dolomitization and Limestone Diagenesis, a Symposium: Soc. Econ. Paleontologists and Mineralogists, Spec. Publ. 13, p. 1–2.

ROEHL, P.O., 1967, Stony Mountain (Ordovician) and Interlake (Silurian) facies analogs of Recent low-energy marine and subaerial carbonates, Bahamas: Amer. Assoc. Petroleum Geologists Bull., v. 51, no. 10, p. 1979–2032.

RUZYLA, K., 1980, The relationship of diagenesis to porosity development and pore geometry in the Red River formation (Upper Ordovician), Cabin Creek Field, Montana: Ph.D. Dissert., Rensselaer Polytechnic Inst., Troy, NY.

RUZYLA, K., and G.M. FRIEDMAN, 1982, Geological heterogeneities important to future enhanced recovery in carbonate reservoirs of Upper Ordovician Red River Formation at Cabin Creek Field, Montana: Soc. Petroleum Engineers Jour., v. 22, no. 3, p. 429–444.

SHINN, E.A., D.M. ROBBIN, and R.P. STEINEN, 1980, Experimental compaction of lime sediment (abst.): Amer. Assoc. Petroleum Geologists Bull., v. 64, no. 5, p. 783.

WILLIAMS, J.A., 1974, Characterization of oil type in Williston Basin: Amer. Assoc. Petroleum Geologists Bull., v. 58, p. 1243–1252.

3
Killdeer Field

James R. Derby and John T. Kilpatrick

RESERVOIR SUMMARY

Location & Geologic Setting Dunn Co., T145N, R94W, North Dakota, USA: central Williston Basin

Tectonics Intracratonic, basement block-faulted structural basin (late Tertiary)

Regional Paleosetting SE shelf of cratonic Williston Basin (Paleozoic)

Nature of Trap Structural (B zone) and stratigraphic (diagenetic; D zone)

Reservoir Rocks
  Age Late Ordovician (Ashgillian, Cincinnatian)
  Stratigraphic Units(s) Red River Formation, B and D zones
  Lithology(s) Dolomite
  Dep. Environment(s) Peritidal
  Productive Facies Algal boundstone (B) and burrowed skeletal wackestone (D)
  Entrapping Facies Anhydrite (B) and micritic limestones (D)
  Diagenesis Porosity Recrystallization to microsucrosic dolomite (B); dissolution and syngenetic dolomitization (D)

  Petrophysics
  Pore Type(s) Intercrystal only (B); intercrystalline crumbly non-tectonic fractures, vugs (D)

  Porosity 12–15% avg, 7–15% range (B zone); 12–25% avg, 0–25% range (D zone)

  Permeability NA
  Fractures Present (minor)

Source Rocks
  Age Ordovician (Red River D)
  Lithology(s) Kerogenites
  Migration Time Late Mississippian

Reservoir Dimensions
  Depth 13,400 ft (4084 m)
  Thickness 10–12 ft (3–4 m) B zone; variable, 32 ft gross & 22 ft net (9.8 & 6.7 m) in Amoco #1 Carlson (D zone)

  Areal Dimensions NA
  Productive Area 960 acres (3.8 km^2) B zone, 320 acres (1.3 km^2) D zone

Fluid Data
  Saturations S_o = 55–60%, S_w = 40–45%
  API Gravity 43°
  Gas-Oil Ratio 1400
  Other —

Production Data
  Oil and/or Gas in Place 2.1 million BO[1]
  Ultimate Recovery 305,400 BO[1] to abandonment; URE = 14.5%
  Cumulative Production Same as above, 3 wells, B and D zones

Remarks: [1] Data for D zone, NA for B zone. Discovered 1976.

3
Ordovician Red River Dolomite Reservoirs, Killdeer Field, North Dakota

James R. Derby and John T. Kilpatrick

Location and Discovery

Killdeer Field was discovered by Amoco Production Co. in 1976, when the #1 Grant Carlson well drill-stem tested more than 4000 feet (1219 m) of oil from the Ordovician Red River Formation. Located in the Williston Basin of west-central North Dakota (T-145-N, R-94-W, Dunn County), U.S.A., the field has produced oil from two zones in the Red River (Fig. 3-1). As of January 1, 1982, the Red River Formation had produced a total of 335,624 barrels of oil from three wells in Killdeer Field (Fig. 3-2).

General Zonation of the Red River

The Red River Formation has been subdivided into four distinct, easily recognizable zones based on the depositional sequence and log characteristics of the sediments. By common industry practice in this area, these zones are called A, B, C, and D from top to bottom. In Killdeer Field, they are approximately 55, 75, 90, and 80 feet (17, 23, 27, and 24 m) in thickness, respectively. The sequence of lithologies, principal constituents, and interpreted depositional environments in the Red River are shown in Figure 3-3.

In this paper, we wish to emphasize that the D and B porosity units are distinctly different from one another. Both are dolomite reservoirs, but their origin, porosity types, geometry, and lateral continuity all differ significantly. As shown in the cross-section of the field (Fig. 3-2), the major lithologies in the upper three zones are laterally continuous and easily correlatable from well to well. The D zone, however, is laterally discontinuous in lithology type and porosity development. Kendall (1976) in Saskatchewan and Carroll (1978) in North Dakota describe in detail from many wells the lithologies and stratigraphy of the Red River. Although we do not agree in every detail with either author on the precise method of reservoir rock development, we do agree with their observations on the differing character of the D compared with the A, B, and C zones.

Cyclic Deposits of the Upper Red River

During Cincinnatian (Late Ordovician, Ashgillian) time, western North Dakota was approximately astride the equator (Ross, 1976), and the sediments deposited in the Williston Basin during this time reflect that warm, alternately humid and arid climate. The upper three zones of the Red River, A, B, and C are a series of deposits formed during three distinct upward-shoaling carbonate-evaporite hemicycles (see Wilson, 1975, p. 50). Generally, the sediments in each hemicycle are limestone, dolomite, and anhydrite in ascending

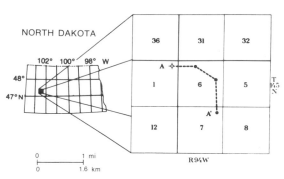

Fig. 3-1. Location map of Killdeer Field, Dunn County, North Dakota.

order. Each cycle started with open-marine conditions, and, with time, became increasingly evaporitic, with increasingly restricted circulation. Rapid incursion of normal marine sea water brought a return to open-marine conditions and the abrupt beginning of the next hemicycle. Kendall (1976), in his study of the Red River in Saskatchewan, Canada, assigned the upper three hemicycles to his Herald Formation.

C Zone Hemicycle

The C zone marine limestone in Killdeer Field is a wackestone-packstone, with abundant fossil fragments and common pellets and burrows. The latter in the C zone limestone consist primarily of Type II burrows, with Type I burrows within both matrix and Type II burrow fillings (see Table 3-1 for descriptions of burrow types). The overlying C zone dolomite is an algal-mat intertidal to supratidal deposit at the base that grades upward into a uniformly fine-grained, tight micritic mudstone with poor intercrystalline porosity. The dolomite was originally a lime mudstone deposited with some mechanical reworking in centimeter-thick graded beds. The dolomitization was penecontemporaneous, as evidenced by reworking and rounding of dolomite rhombs. There are a few algal-mat deposits up to 10 centimeters in thickness in the upper part of the C dolomite. The C zone dolomite is oil-productive elsewhere in the Williston basin, but not in Killdeer Field. Grading upward from the C zone dolomite is the C anhydrite, which consists of three or four anhydrite beds, each 5 to 10 feet (1.5–3 m) thick,

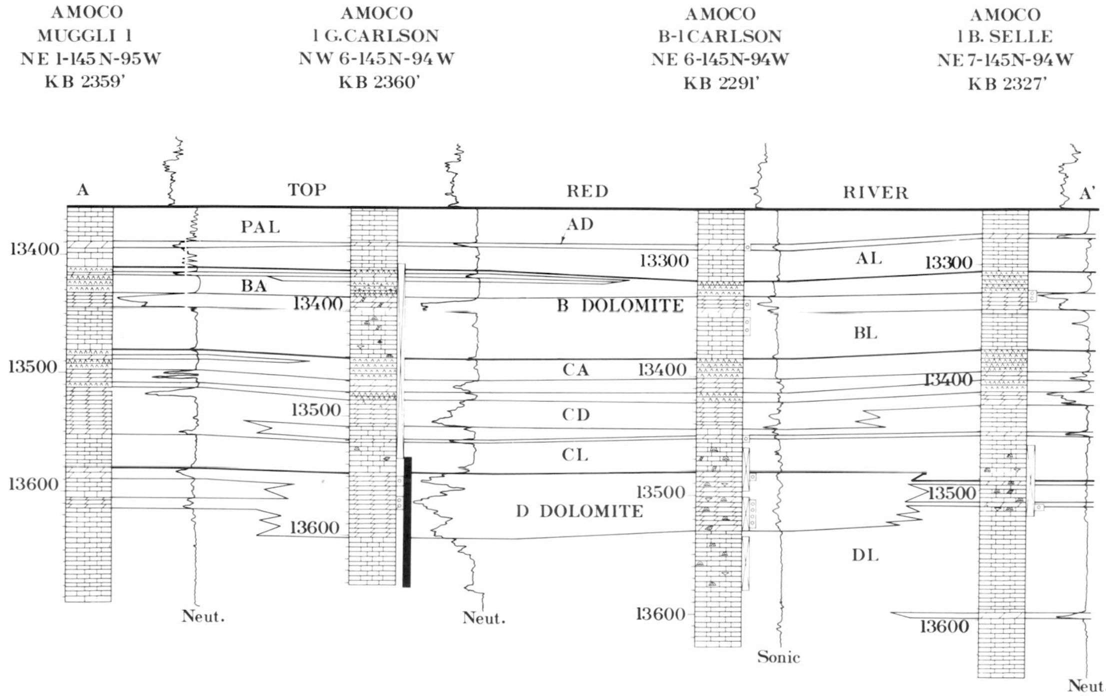

AMOCO MUGGLI 1 NE 1-145 N-95 W KB 2359'	AMOCO 1 G.CARLSON NW 6-145N-94 W KB 2360'	AMOCO B-1 CARLSON NE 6-145N-94W KB 2291'	AMOCO 1 B. SELLE NE 7-145N-94W KB 2327'

A   TOP   RED   RIVER   A'

13400   PAL   AD
                                        13300   AL   13300
         BA
              13400   B DOLOMITE
                                                                        BL
13500   CA   13400
              13500   CD   13400
                             CL
13600   13500   13500
              13600   D DOLOMITE
                                        DL

              Neut.   Neut.   13600

                                        Sonic   Neut.

Fig. 3-2. Red River Stratigraphic cross-section of Killdeer Field. PAL—Post-A Zone Limestone, AD—A Zone Dolomite, AL—A Zone Limestone, BA—B Zone Anhydrite, BL—B Zone Limestone, CA—C Zone Anhydrite, CD—C Zone Dolomite, CL—C Zone Limestone, DL—D Zone Limestone. Depths in feet. ☒ Cored intervals; ☉ Perforated intervals; ■ Drillstem tests.

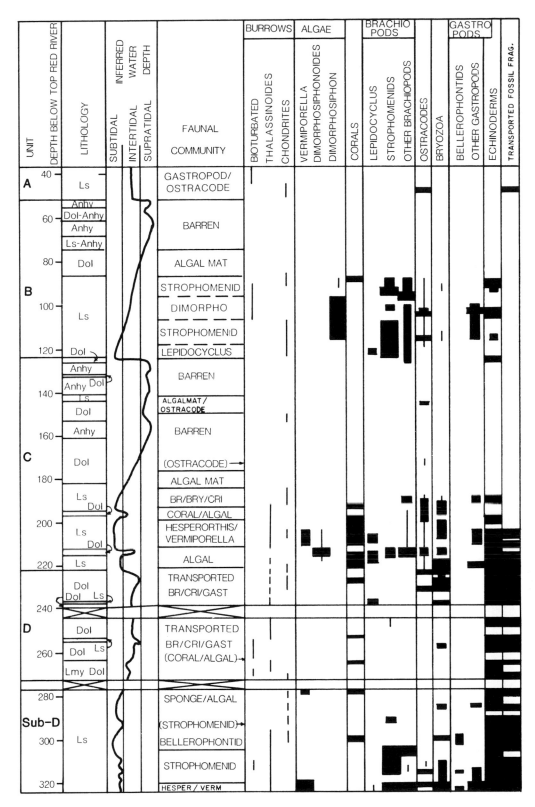

Fig. 3-3. Lithologic sequence, inferred water depth, and principal faunal content of strata represented by cores in Amoco #1 Carlson and #B-1 Carlson showing environmental control on both lithology and fauna.

Table 3.1. Description and Location of Three Important Sedimentary Features of the Red River in Kildeer Field

Feature	Description	Location	Remarks
Type I Burrows	Small, 0.5–1.5 mm diameter Cylindrical-elongate shape Parallel to bedding, or nearly so Commonly straight with few branches Typically back-filled with carbonate mud Rarely contain fossil fragments	Commonly in unbedded mudstones Commonly in interbedded mudstones and wackestones Less commonly within Type II burrows	Similar to *Chondrites* of various authors
Type II Burrows	Large, 4–10mm diameter Cylindrical-elongate shape Horizontal to vertical Abundant branches Distinct outer margins Centers typically vuggy with drusy anhydrite or coarse dolomite fill Commonly contains fossil fragments	Commonly in fossiliferous lime wackestone Rarely in mudstone	Similar to *Thalassinoides* of various authors. Rarely the outline of the burrow may appear enlarged due to later dolomitization of the surrounding matrix
Dimorphosiphon	Calcareous codiacian alga Segmented thallus which breaks up to form cylindrical segments 2–3mm diameter, 10mm length Internally contains 8–14 long circular tubes with short cross tubes	Only in the open-marine wackestones-packstones immediately below the B dolomite porosity zone	Similar to the recent genus *Halimeda* (See Figure 3.4)

separated by dolomite beds, each less than 2 feet (<0.6 m) thick.

B Zone Hemicycle

The B hemicycle conformably overlies the C zone anhydrite and is of particular interest. The dolomite of the B zone is the single most persistent porous dolomite bed in the Red River and is easily correlatable on logs throughout the Williston Basin. The B zone dolomite is also one of the two oil-productive intervals in the Red River at Killdeer Field. In the core of the B zone from the Amoco #1 Carlson well, a wackestone-packstone occurs similar to the C zone limestone. There are, however, significant differences in the types of fossils and burrows found in the two limestones. The B zone limestone has no Type II burrows, but has abundant Type I burrows and distinctive *Dimorphosiphon* algae that are exclusive to the B zone limestone throughout the central part of the Williston Basin (Fig. 3-4).

The B zone dolomite in Killdeer Field is thin, approximately 10 to 12 feet (3—4 m) thick, microsucrosic in textural porosity type (modified Archie Type II-III) and has good intercrystalline porosity. There is some anhydrite infill toward the top. In Killdeer Field, the B dolomite has a distinctive log character of higher porosity at the top and bottom, with a slightly lower porosity in the middle of the dolomite. Despite dolomitization, bi-lobed response on porosity logs is controlled by the original rock-sediment type. The higher-porosity zones are algal-mat deposits with small

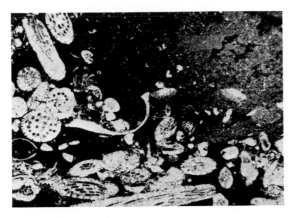

Fig. 3-4. Dimorphosiphonalgae in B Limestone. This alga is an "index fossil" of the B unit and may account for considerable porosity within the B limestone. Thin-section, high-contrast positive print, 2.5X.

Fig. 3-5. Algal-mat dolomite with thrombolitic texture from B Dolomite. Thin-section positive print, 2.5X.

wavy laminations, and the intervening zone is a tufted, clotted thrombolite (Fig. 3-5). This is a typical succession of rock type and porosity in a shoaling-upward, high-intertidal carbonate sequence. This log expression is also common in other parts of the cyclic section of the Red River at the transition between marine and supratidal rocks.

The B dolomite is areally widespread, with good effective porosity of 10 to 20 percent throughout and as such requires structural closure for oil entrapment. Two wells in Killdeer Field, #1 Selle and #B-1 Carlson, have produced oil from the B zone dolomite. The production from both wells is marginal, but indicative of a Red River structural closure, as supported by seismic mapping. Cumulative production from the #1 Selle is 19,026 barrels of oil, and from the #B-1 Carlson, 11,163 barrels of oil. Both wells are now plugged and abandoned in the Red River interval but have been recompleted in the Devonian Duperow Formation.

Within the B zone, a supratidal deposit of anhydrite overlies the dolomite. It is generally massive, but may have small interbeds of dolomite or even limestone.

A Zone Hemicycle

The B hemicycle deposits are overlain by the A limestone, followed by a thin A dolomite. This limestone is dark gray in color compared with the medium-brown to brownish-gray color of the B and C limestones. The limestone has a sparse fauna, predominantly ostracodes tolerant of salinity variations and transported fossil fragments. The lack of mineralogical evidence for hypersalinity suggests alternatively an abnormally low salinity for the wackestone deposits. The authors interpret this as a lagoonal environment. The overlying dolomite is the thinnest dolomite unit in Killdeer Field, at 4 feet (1.2 m), and is overlain by limestones of the post-A hemicycle, rather than by a supratidal anhydrite as in the B and C evaporitic hemicycles. Although the A dolomite was not cored in any Killdeer Field wells, cores from other wells in the Williston Basin indicate an algal-mat environment and a penecontemporaneous (early diagenetic) origin for this dolomite, like the B dolomite. This origin is believed to apply to Killdeer Field as well. The A dolomite is not productive in Killdeer Field because it lacks effective porosity.

Post-A Hemicycle

Above the A dolomite is the previously mentioned post-A marine limestone. It is a wackestone-packstone, as interpreted from cuttings and cores outside Killdeer Field, and extends to the top of the Red River where it is conformably overlain by the Stony Mountain Shale. In cores from the basin center, an abrupt Red River-Stony Mountain contact is not readily apparent, although bored hardgrounds are present, suggesting slow rates of deposition. The remainder of the post-A hemicycle comprises the Stony Mountain Formation.

Strata Below the D Zone

In the core from the Amoco #B-1 Carlson well, the strata below the D dolomite are limestones— burrow-mottled, slightly fossiliferous packstones and wackestones with little permeability. Partially dolomitized burrow mottles comprise 10 to 30 percent of the rock. Similar rocks form the bulk of the lower Red River (or the B unit of the Yeoman Formation of Kendall, 1976). In the #B-1 Carlson well of Killdeer Field and also in the Amoco #1 Lubke well of nearby Rattlesnake Point Field, the gradual decrease in faunal diversity and abundance up to the base of the D dolomite suggests a corresponding upward increase in salinity. The normal-marine fauna of brachiopods, echinoderms, various gastropods, and occasional corals becomes less diverse until, just below the D dolomite, the indigenous fauna is only codiacean algae, sponges, and gastropods, all of which probably could tolerate high or low, possibly fluctuating, salinity conditions.

D Dolomite: Burrow-Mottled Reservoirs

The D dolomite is the lowest of the commonly occurring porosity intervals of stratigraphic origin in the Red River; it is equivalent to the C and D units of the Yeoman Formation of Kendall (1976) and to the lower part of the C burrowed member of Kohm and Louden (1978). The D is approximately 49 feet (15 m) thick in the #B-1 Carlson, based on cores, and 56 feet (17 m) thick in the #1 Carlson, based on logs. In the two other wells, the interval is much less dolomitized, and consequently, the interval of the D dolomite is difficult to identify. In Killdeer Field, only the #1 Carlson has produced from the D dolomite; in January 1982, it was plugged after producing 305,435 barrels of oil and 41,803 barrels of water.

In the two D zone cores from Killdeer Field, the interval consists of burrow-mottled mudstones, wackestones, and rare packstones interrupted by sparse, thin beds of fossil-fragment packstones or grainstones (Fig. 3-6). Near the base are sparse, thin, kerogen-rich layers probably composed of abundant algal material. Burrows in the D zone are predominantly of Type II, with smaller-diameter Type I burrows that are less strongly dolomitized in both the matrix and the Type II burrow fill.

Type II burrow-mottles are typically fine sucrosic dolomite with fair-to-good intercrystalline porosity (Fig. 3-7). The burrows are commonly oil-stained and therefore appear darker than the surrounding mudstone matrix. The mudstone matrix can be either limestone or dolomite and comprises between 70 and 90 percent of the rock volume; burrow-mottles comprise the remaining 10 to 30 percent. In a limestone matrix, the boundaries of the dolomite burrow-fill range from extremely sharp to transitional. Timing of the dolomitization in the burrows is not completely resolved (Kendall, 1976, 1977), but most workers agree that it is an early diagenetic event shortly following deposition. We believe that the model proposed by Morrow (1978) for penecontemporaneous dolomitization of burrow fillings by seasonal salinity changes in a shallow marine-shelf environment adequately explains the origin of this early dolomite.

Rock matrix surrounding the burrows is mostly micritic lime wackestone, as seen in the core of the Amoco #1 Selle well. Three thin intervals of dolomite totaling 6.5 feet (2 m) in the #1 Selle contrast sharply with the nearly 56 feet (17 m) of dolomite in the #B-1 Carlson well less than 1 mile away (Fig. 3-2). Faunal and lithologic characteristics indicate that the thin dolomites represent shoaling events that preceded and followed deposition in a restricted lime-mud pond or lagoon. Above the thin dolomite, no clear boundary with the overlying C limestone is evident. A gradual increase in normal-marine faunas marks the transition.

In contrast to the D interval in the #1 Selle, the originally burrowed limestone of the D interval in the #B-1 Carlson is now 80 percent dolomite. Dolomitization is pervasive, not limited to burrows, and obscures burrow-matrix rock relationships (Fig. 3-8). Many intervals contain numerous crumbly fractures and fabrics, suggesting

Fig. 3-6. Partially dolomitized Type II (Thalassinoides) burrows in fossiliferous lime wackestone with later (smaller) Type I (hondrites) burrows. Samples from low in D Dolomite. Thin-section positive print 2.5X.

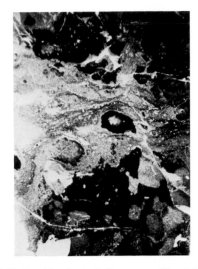

Fig. 3-8. Dolomitized, partially recrystallized, burrowed lime wackestone with crumbly fracturing in the D dolomite. Thin-section positive print, 2X. Relationship between dolomitized burrow-mottles and dolomitized matrix is obscured. Some early burrow-mottles appear dark because of organic staining. Dolomites of this type are effective reservoirs in the D dolomite.

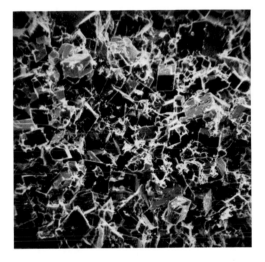

Fig. 3-7. Dolomitized burrow from the D dolomite. Note the euhedral dolomite crystals and intracrystal porosity. Scanning-electron photomicrograph, 500X.

Fig. 3-9. Fractured recrystallized D dolomite. Light areas are secondary voids filled by internal sedimentation. Thin-section positive print, 2X.

dissolution and recementation, with subsequent internal sedimentation in secondary voids (Fig. 3-9). Part of these altered fabrics may be the result of the formation, boring, infill, and subsequent differential compaction of submarine hardgrounds. At least seven clearly defined hardgrounds are recognizable in the D zone of the #B-1 Carlson, with fabrics comparable to the hardground pseudobreccias described by Bromley in

the Chalk of England (Bromley, 1975). The extent of dissolution and recrystallization as well as pervasive dolomitization of the matrix strongly suggests post-depositional emergent conditions with the development of a low-salinity shallow lens groundwater system in the area following deposition of the D. Leaching by undersaturated ground water in a freshwater phreatic zone and dolomitization in a mixing zone (Badiozamani,

1973), as a freshwater-saltwater interface fluctuated across the area, could have produced the complex of observed fabrics in a geologically short period of time.

The marine fauna of the D dolomite interval suggests highly restricted conditions. In the #B-1 Carlson (and the #1 Lubke from Rattlesnake Point Field), very few indigenous fossils are present, and nearly all fossils are transported fragments. Elevated or seasonally fluctuating salinities could account for the lack of indigenous fauna and also would have led to early dolomitization of the burrows by the mechanism proposed by Morrow (1978) for similar facies in the Irene Bay Formation.

D Zone Permeability

Effective permeability evidenced by production tests and core analyses is present in the D

dolomite only where the matrix has been dolomitized. Presence of dolomitized burrow-mottles alone does not form an effective reservoir and is only faintly detectable on mechanical logs, although the dolomite of the burrows may constitute 30 percent of the rock. Studies of cuttings from the Amoco #1 Carlson and of cores from other wells throughout the basin indicate that strongly altered fabrics comparable to those found in the #B-1 Carlson core are required to form a D zone reservoir. Optimum reservoir characters include pervasive dolomitization of the matrix limestone without recementation of incipient voids.

Comparison of Reservoirs

The fact that all four zones in the Red River are productive of hydrocarbons and are dolomites has led many workers to assume a single controlling

Table 3.2. Comparison of B-zone vs D-zone Dolomites

OBSERVATIONS

Zone	Lithology	Stratigraphy Position vs Marker Beds	Thickness	Regional Occurrence	Continuity	Porosity
B	Dolomite	Fixed	Constant 3 to 5 meters	Basin-wide	Continuous	Low to moderate
D	Mottled ls or dolomite	Relatively fixed	Variable 0 to 17	Basin-wide	Abruptly discontinuous	Extremely variable 0 to 25%

INTERPRETATIONS

Zone	Originally Deposited Rock	Depositional Environment	Faunal Community	Penecontemporaneous Diagenesis	Early Secondary Diagenesis	Cause of Reservoir Porosity
B	Silt-size dolomite algal boundstone	High intertidal	Algal mat	Dolomitization; slight anhydrite crystal growth	Not significant	Syngenetic dolomitization in original depositional environment
D	Limestone; burrowed fossil-fragment wackestone	Subtidal or low intertidal and restricted ponds	Burrowers; rare opportunistic brachiopods and gastropods	Dolomitization of burrows	Dolomitization of matrix in mixing zone— local dissolution and recementation	Early secondary dolomitization in mixing zone localized by development of freshwater lens

mechanism and a single stage of dolomitization for the entire Red River Formation. The fact that abrupt lateral variations occur in the D dolomite and in the porosity development of the C dolomite has led some workers to assume a structural or vertical control of dolomitization along major fractures in the Red River (Dow, 1971; Kohm and Louden, 1978). As summarized in Table 3-2, the authors believe that two different styles of diagenesis have affected the different productive zones of the Red River in Killdeer Field. The depositional environments of the A, B, and C dolomites are similar to each other, but are distinctly different from that of the D dolomite. The type of diagenesis that resulted in the essentially primary A, B, and C dolomites is distinctly different from that which resulted in the D dolomite.

In other fields, the A through C hemicycles contain locally thick dolomites of *secondary* origin. These secondary dolomites are extremely important as Red River reservoirs, especially near the margins of the Williston Basin (Kohm and Louden, 1978; Clement, this volume). Subsequent to submission of this manuscript, Longman (1982) and Longman *et al* (1982) proposed a model in which the C and D zone dolomites were dolomitized by gravitational seepage of Mg-enriched (Ca depleted) brines from the overlying anhydrite brine pool. We agree that theirs is an excellent model for the secondary dolomites of the C zone, as intended (Longman, 1982), but we would not apply it to the primary algal mat and supratidal dolomites of the B dolomite reservoir rock nor to the non-reservoir dolomites of the C in Killdeer Field. Further, their proposed application of the model is not consistent with the evidence in Killdeer Field nor with the geographic distribution of D dolomites throughout the basin.

Acknowledgments Most of the work reported here was done while the authors were with Amoco Production Co., Tulsa (JRD) and Denver (JTK). Permission to publish is gratefully acknowledged.

References

BADIOZAMANI, K., 1973, The *dorag* dolomitization model—application to the Middle Ordovician of Wisconsin: Jour. Sedimentary Petrology, v. 43, 965–984.

BROMLEY, R.G., 1975, Trace fossils at omission surfaces: *in* R.W. Frey, ed., The Study of Trace Fossils: Springer-Verlag, Inc., New York, 399–428.

CARROLL, W.K., 1978, Depositional and paragenetic controls on porosity development, upper Red River Formation, North Dakota: Montana Geol. Soc., 24th Annual Conf., Williston Basin Symp., p. 79–98.

DOW, W.G., 1971, Theory of Red River reservoir development in Williston Basin (abst.): Amer. Assoc. Petroleum Geologists Bull., v. 55, p. 536.

KENDALL, A.C., 1976, The Ordovician carbonate succession (Bighorn Group) of southeastern Saskatchewan: Mineral Resources, Saskatchewan Geol. Surv., Sedimentary Geol. Div., Report 180, p. 1–185.

KENDALL, A.C., 1977, Origin of dolomite mottling in Ordovician limestones from Saskatchewan and Manitoba: Canadian Petrol. Geol. Bull., v. 25, p. 480–504.

KOHM, J.A., and R.D. LOUDEN, 1978, Ordovician Red River of eastern Montana and western North Dakota: relationships between lithofacies and production: Montana Geol. Soc., 24th Annual Conf., Williston Basin Symp., p. 99–120.

LONGMAN, M.W., 1982, Carbonate diagenesis as a control on stratigraphic traps: Amer. Assoc. Petroleum Geologists, Education Course Note Series 21, 159 p.

LONGMAN, M.W., T.G. FERTAL, and J.S. GLENNIE, 1982, Dolomitization in the Red River Formation, Richland County, Montana (abst.): Amer. Assoc. Petroleum Geologists Bull., v. 66, no. 5, p. 595–596.

MORROW, D.W., 1978, Dolomitization of lower Paleozoic burrow-fillings: Jour. Sedimentary Petrology, v. 48, p. 295–306.

ROSS, R.J., JR., 1976, Ordovician sedimentation in western United States: *in* Bassett, M.G., ed., The Ordovician System: Proceedings of a Palaeontological Assoc. Symp., Birmingham, England, Sept. 1974, Univ. of Wales Press and National Museum of Wales, Cardiff, p. 73–105.

WILSON, J.L., 1975, Carbonate Facies in Geologic History: Springer-Verlag, Inc., New York, 439 p.

4
Pennel Field
James H. Clement

RESERVOIR SUMMARY

Location & Geologic Setting	Fallon Co., T7-8N, R59-60E, Montana; axis of Cedar Creek Anticline on SW flank of Williston Basin, USA
Tectonics	Basement wrench faulting and differential uplift in linear NW-plunging anticline; episodic uplift
Regional Paleosetting	SW shelf of cratonic basin, shallow marine to sabkha
Nature of Trap	Structural/stratigraphic; along crest of faulted, NW plunging anticline; subtle hydrodynamic controls

Reservoir Rocks

Age	Late Ordovician (other producing levels Silurian and Mississippian)
Stratigraphic Units(s)	Red River Formation
Lithology(s)	Dolomites and minor limestones
Dep. Environment(s)	Shallow-marine to peritidal and/or sabkha, cyclical
Productive Facies	Dolomitized marine mudstone-wackestone, intertidal pellet wackestone and packstone, supratidal laminated mudstone
Entrapping Facies	Supratidal anhydrites and lime mudstone (top seals)
Diagenesis Porosity	Early dolomitization of peritidal and marine facies; dissolution of $CaCO_3$; cementation by anhydrite, dolomite, and silica.

Petrophysics

Pore Type(s)	Intercrystal, moldic-vuggy, fracture
Porosity	2–22%, avg 11%
Permeability	0<0.1–35 md, avg 9 md
Fractures	Common vertical fractures, open and closed; significant contribution to productivity

Source Rocks

Age	Ordovician (Red River Fm)
Lithology(s)	Carbonates and shales
Migration Time	Late Cretaceous through post-Paleocene

Reservoir Dimensions

Depth	8700–9100 ft (2650–2770 m)
Thickness	10 ft (3 m) and 20 ft (6 m) avg net pay (U_4–U_5 and U_6–U_7 zones)
Areal Dimensions	NA
Productive Area	22,300 acres (89.2 km^2)

Fluid Data

Saturations	S_w = 30–45%
API Gravity	33°
Gas-Oil Ratio	282:1 (original)
Other	Black oil with 0.47% sulfur, paraffinic

Production Data

Oil and/or Gas in Place	279 million BO
Ultimate Recovery	58 million BO primary and secondary; URE = 20.8%
Cumulative Production	57 million BO through 1982

Remarks: IP 205 BOPD pumping. Discovered 1955.

Carbonate Petroleum Reservoirs (Roehl & Choquette), Springer-Verlag New York, Inc.

4
Depositional Sequences and Characteristics of Ordovician Red River Reservoirs, Pennel Field, Williston Basin, Montana

James H. Clement

Location

Pennel Field in Fallon County, Montana, is one of 15 oil fields located along the crestal portion of the Cedar Creek anticline in eastern Montana and southwestern North Dakota (Fig. 4-1). Cedar Creek is the major anticlinal structure demarcating the southwestern periphery of the intracratonic Williston Basin; the structure extends for nearly 145 miles (233 km) on a mean strike of S30°E from northwest of Glendive, Montana, to just west of Buffalo, South Dakota. Since establishment of commercial oil production on the anticline in late 1951, more than 300 million barrels of oil have been produced from Mississippian, Silurian, and Ordovician carbonate reservoirs in the Cedar Creek fields.

Discovery and Production History

Discovery of the Pennel Field by Shell Oil Company in September 1955 resulted from the drilling and completion of Shell well State 22X-36 (section 36, T8N, R59E) pumping 205 barrels of oil and 39 barrels of water per day from carbonate reservoirs in the Ordovician Red River at depths from 8625 to 8800 feet (2629–2682 m). The site of this initial exploratory well in the Pennel area was determined after extensive reflection seismic and geologic surface mapping.

Primary development of the field was accomplished during the late 1950s and early 1960s—production was established from carbonate reservoirs (principally dolomites) in Mississippian, Silurian, and Ordovician strata at depths from 7000 to 9100 feet (2133–2774 m). Pennel has produced more than 57 million barrels of oil (avg 33°API gravity); more than 40 million barrels of this has come from dolomite reservoirs in the upper 225 feet (68 m) of the Upper Ordovician Red River Formation (Fig. 4-2). Within a developed area of nearly 22,000 acres (89 km²) more than 150 wells have been drilled. Primary phases of production from the Siluro-Ordovician reservoirs resulted principally from depletion drive; active water drives are effective only in the Mississippian reservoirs and in the lowermost intervals of the Ordovician Red River. Supplemental recovery through waterflooding was initiated in the Red River and Stony Mountain reservoirs in 1969.

Mode of Entrapment

Oil entrapment in the Pennel area, typical of many of the hydrocarbon accumulations along the Cedar Creek trend, is the result of a combination of geologic conditions. Regional structure position along the anticlinal culmination of the northwest-plunging Cedar Creek structure is undoubtedly the dominant factor, and local structural closures are important (Fig. 4-1). However,

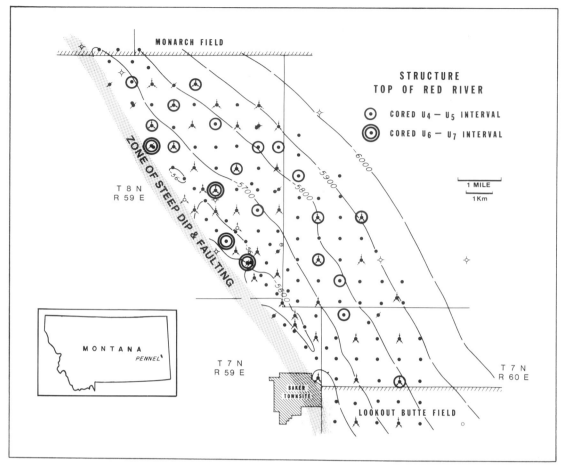

Fig. 4-1. Pennel Field, structure contours on top of Ordovician Red River Formation. Contour interval 100 ft.

significant control of entrapment and productivity results from rapid lateral and vertical changes in porosity and permeability.

The field is limited structurally to the west by the steeply dipping down-faulted west flank of the Cedar Creek anticline. Variable high-water saturation and "free-water" levels occur 250 to 300 feet (76–91 m) structurally downdip on the east flank. The north and south limits of Pennel are only "legally" defined because the field is contiguous with production in Monarch Field to the north and Lookout Butte Field to the south.

General Stratigraphy—Ordovician Red River Formation

Geologists concerned with the Paleozoic of the Williston Basin have long recognized that the region was not necessarily a bathymetric basin during most of the Ordovician (Fuller, 1961; Macauley, 1964; Roehl, 1967). Major sediment deposition occurred in shallow epicontinental seas, bordered to the west by the Cordilleran miogeosyncline. The Williston area was periodically a shallow sedimentary intracratonic basin and, at stages, widespread areas emerged and broad tidal-flat regions existed. The Cedar Creek area, including Pennel Field, occupied a somewhat intermediate position between the shallow basin area and the oceanic platform to the west-southwest. Therefore, the area was an optimum site for rather cyclic marine, intertidal, and supratidal depositional sequences typical of the upper Red River producing intervals at Pennel Field.

In the Pennel Field, the Red River Formation is 500 to 525 feet (152–160 m) thick and is predominantly a limestone-dolomite sequence with minor, but significant, anhydrite beds in the upper

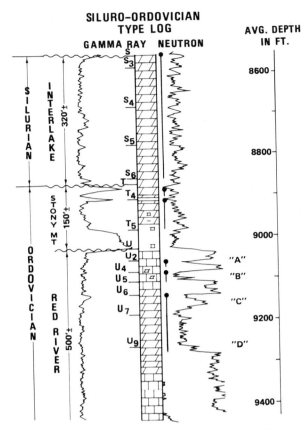

Fig. 4-2. Typical log of Silurian and Ordovician carbonate reservoirs, Cedar Creek Anticline, Montana.

and extractable heavy hydrocarbons, were found in cores of these sediments. Studies of levels of organic metamorphism (LOM) in the basin indicate that the source beds probably attained thermal maturation and generation levels by mid-Late Cretaceous time and that the major phases of migration into the Cedar Creek region, including Pennel Field, occurred during the late Cretaceous through post-Paleocene (Clement, 1976).

Depositional Sequences of Reservoirs

Extensive coring of 21 wells included for study not only the productive dolomite zones (U_4-U_5, U_6-U_7), but also significant portions of overlying and underlying strata (Fig. 4-1). These studies show that the deposition of these strata resulted from several marine transgressions, stillstands, and minor regressions that deposited cyclic complexes of normal and restricted marine, intertidal, and supratidal carbonate sequences. Although producible reservoirs exist in all facies, they are best developed in the dolomitized restricted marine, intertidal, and supratidal sequences.

The principal producing zones are the U_4-U_5 and U_6-U_7 intervals (Fig. 4-3). Zone U_4-U_5 is generally about 20 feet (6 m) thick, and zone U_6-U_7 averages 50 feet (15 m) (Fig. 4-4). Although the two zones differ significantly in some facies and deserve separate detailed interpretations, they are sufficiently similar in depositional cyclicity to permit summarizing the sequences in this paper. Each sequence begins with rocks best interpreted as normal-marine sediments passing stratigraphically upward through restricted marine, intertidal, and supratidal strata, each culminated by an evaporite facies. Generally, very thin, shaly lithoclastic beds overlie the evaporite facies, which is overlain in turn by transgressive normal-marine sediments. As in most complex carbonate depositional settings, it is not unusual for some facies to be interpreted as totally absent because of (1) our inability to clearly identify the facies, (2) extreme diagenesis (dolomitization) of the primary sediment, or (3) the non-deposition or penecontemporaneous removal of the facies in the tidal-intertidal environment.

250 feet (76 m). The major productive porosity zones are dolomites and dolomitic limestones in several readily identifiable zones designated by log interval markers: U_2-U_3, U_4-U_5 and U_6-U_7 to U_9 (Figs. 4-2, 4-3). Today, industry operators most commonly designate these zones in the Williston Basin by alphabetical sequence—A, B, C, D. Generally, the lower non-productive 250 feet (76 m) below the U_9 marker are lime mudstones and wackestones with discontinuous dolomite strata.

Source rocks for the oils produced in the Ordovician and Silurian reservoirs in Pennel Field, and elsewhere in the Williston Basin, occur principally in the carbonates of the Red River Formation; minor contribution is ascribed to the shales of the underlying Winnipeg Formation (Clement, 1976; Kendall, 1976; Williams, 1974; Dow, 1974). Concentrations of minute kerogen masses, probably of algal origin, rich in organic carbon

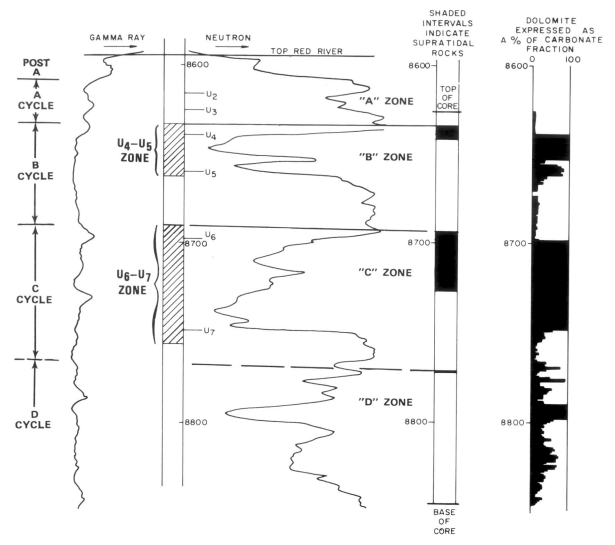

Fig. 4-3. Gamma-ray/neutron log showing depositional cycles and the principal upper Red River producing zones related to percent dolomite and supratidal rocks. Shell Sodergren-Oler No. 33–26 well.

Marine Facies

Mottled, fossiliferous, pelletal lime wackestones (occasionally packstones) and dolomite wackestones characterize this rock type (Fig. 4-5). The sediment was extensively burrowed and churned. Dissolution-compaction structures are commonly dolomitized. A diverse faunal assemblage is common and is comprised of brachiopods, corals, bryozoa, gastropods, echinoderms, and algae. This facies in the U_4-U_5 zone tends to be more commonly dolomitized than in the U_6-U_7 zone. Because of the diverse fauna and evidence of significant bioturbation of the original sediment,

these rocks are interpreted as a shallow-marine facies.

Burrowed Mud Facies

These rocks are vaguely laminated to bedded, fossiliferous and pelletoidal lime or dolomite mudstones grading to wackestones, or rarely packstones (Fig. 4-6). Extensive elliptical, near-horizontal burrows are present, and organisms intensely reworked the primary sediment. The faunal suite, though fewer in occurrences, is generally similar to that in the underlying marine facies. Although shown as laterally associated

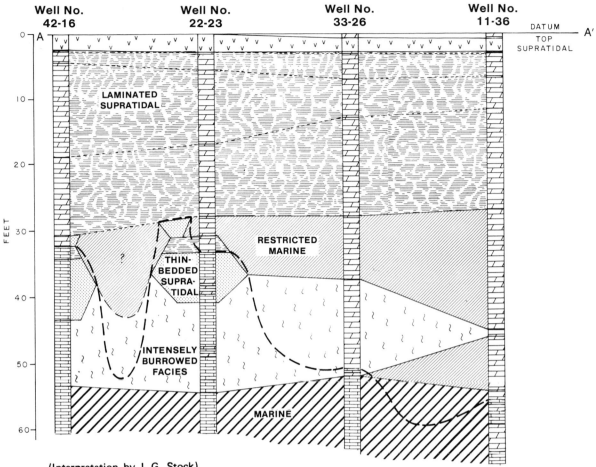

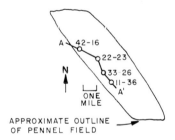

(Interpretation by L.G. Stock)

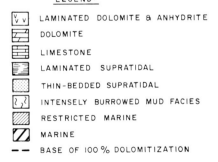

Fig. 4-4. Relationship of lithology to depositional facies, U_6-U_7 zone, Red River Formation. Cross-section incorporates data from logs of wells located between cored sections.

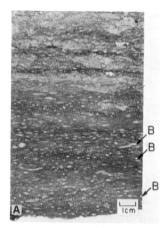

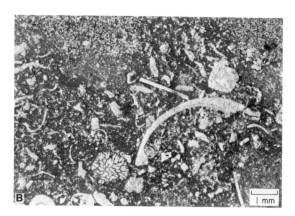

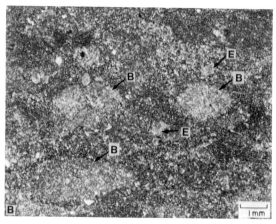

Fig. 4-5. Marine facies, U_6-U_7 zone: *A*. Core-slab photographs of fossiliferous, pelletal, slightly dolomitized lime wackestone, burrowed and churned. *B*. Photomicrograph of fossiliferous wackestone; skeletal grains are principally bryozoa, corals and brachiopods; matrix is microcrystalline calcite. Dolomite crystals average 34 microns. Plane light.

Fig. 4-6. Burrowed mud facies, U_6-U_7 zone: *A*. Core-slab photographs of vaguely bedded pelletal, fossiliferous dolomite lime mudstone; note extensive burrowing. *B*. Photomicrograph of burrowed mudstone, carbonate fraction 80 percent dolomite. Note burrows (B) and echinoderm fragments (E) in matrix. Plane light.

with "restricted marine" strata in Figure 4-4, these burrowed sediments may represent subtle mud buildups in the shallow marine environments. This facies is most common in the U_6-U_7 interval and is generally absent in the U_4-U_5 sequence.

Restricted Marine Facies

Fragmental-appearing, fossiliferous, pelletoidal dolomite wackestones with common interlayers of mudstones and packstones characterize these rocks (Fig. 4-7). They exhibit vague bedding and some burrowing. Encrusting organisms are numerous in the lower portions of the facies, generally

decreasing in abundance upward. Stromatoporoids are common, echinoderms and brachiopods are rare to common, and bryozoa and corals are rare. Limited fauna and minimal bioturbation suggest that this is a "restricted marine" facies. It may, however, include both subtidal and intertidal deposits.

Thin-Bedded Supratidal Facies

These sediments appear to be laterally restricted and occur only in the U_6-U_7 zone; they were observed in only two of four cored sequences. They are thin-bedded, pelletal, microfossiliferous lime packstones and wackestones with very common

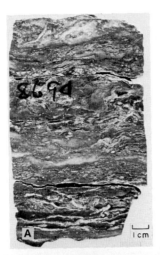

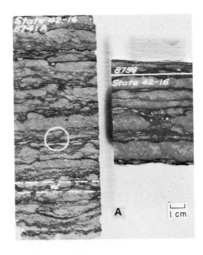

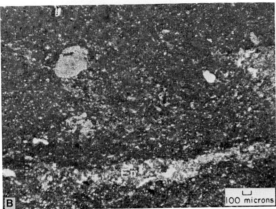

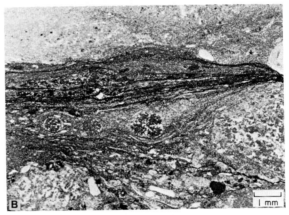

Fig. 4-7. Restricted marine facies, U_4-U_5 zone: *A.* Core-slab photograph of dolomite wackestone with encrusting organisms and echinoderms. *B.* Photomicrograph of dolomite wackestone, echinoderms, and encrusting organisms (En). Plane light.

Fig. 4-8. Thin-bedded supratidal facies, U_6-U_7 zone: *A.* Core-slab photographs of pelletal, microfossiliferous lime packstone to wackestone; shaly laminae contain fossil debris and organic flecks. *B.* Photomicrograph of fossiliferous wackestone with shaly laminae. Plane light.

shale laminae that often contain fragments of marine fauna and flakes of organic debris (Fig. 4-8). Ostracods are numerous in the lower part of the facies. These local facies seem to have acted as significant permeability barriers and inhibited the downward movement of dolomitizing solutions (Fig. 4-4; Stock, 1965). The strata bear a strong similarity to recent supratidal rocks present on Sugar Loaf Key in Florida.

Intertidal Facies

Rocks that seem most diagnostic of intertidal environments are generally lacking in the U_6-U_7 zone, but are common in the U_4-U_5 cycle. These strata are pelletal, lithoclastic, dolomitized wackestones and packstones (Fig. 4-9). Fossil fragments are present; pellet molds and burrows are common. Figure 4-9B displays leached-pellet moldic porosity typical of the facies. Pore filling and replacement by silica and anhydrite, though randomly distributed, occur principally in burrow fillings and interparticle pores and replacements.

Supratidal Facies

Well-laminated to thin-bedded dolomite mudstones, grading rarely to pelletal, microfossiliferous dolomite packstones, compose a significant portion of the Red River reservoir zones

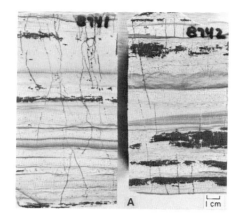

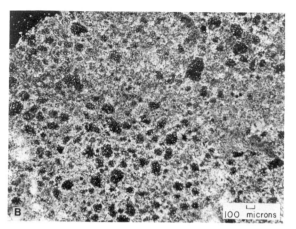

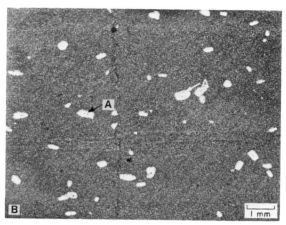

Fig. 4-9. Intertidal facies, U_4-U_5 zone: *A*. Core-slab photographs of lithoclastic, pelletal dolomite packstone; silicified section in small core section. *B*. Photomicrograph of pelletal dolomite packstone with leached-pellet moldic porosity (dark). Cross-polarized light.

Fig. 4-10. Supratidal facies, U_6-U_7 zone: *A*. Core-slab photographs of mudstone with generally planar laminae and bedding planes. *B*. Photomicrograph of dolomite mudstone with random replacement anhydrite crystals (A). Dolomite crystals average 10 microns. Plane light.

(Fig. 4-10). In the U_6-U_7 zone, the lowermost of these strata commonly contain lithoclasts and wavy or irregular laminae; no marine fossils are present. The middle and upper laminated rocks are generally thin-bedded dolomite mudstones with planar laminae and occasional cross-laminations. Lithoclasts and ostracods are rare, and typical marine fossils are absent.

Deformed laminae in the upper portion of the facies are probably due to nodular gypsum growth in the primary sediment. Anhydrite-filled pores and small borings occur in places, and intervals of secondary silification are present.

Rocks assigned to this facies in the U_4-U_5 interval are similar well-laminated mudstones or pelletal lime and dolomite wackestones (Fig. 4-11). They are devoid of fossils and contain only rare burrow structures. All of the features noted in

these strata in both U_4-U_5 and U_6-U_7 zones are characteristic of supratidal and high intertidal facies associated with prograding tidal flats.

Evaporite Facies

Thinly interlaminated and/or interbedded anhydrites and dolomite mudstones culminate the depositional sequences in the zones and are representative of the regressive geometry of prograding tidal flats (Fig. 4-12). These strata in the U_6-U_7 zone average 3 feet (1 m) in thickness and appear to occur throughout the field. Most of the interval appears to have originated as a bedded evaporite, although nodular characteristics do occur in the lower portion. Cross-laminations of dolomite and anhydrite are also present.

Similar strata, ranging in thickness from zero to

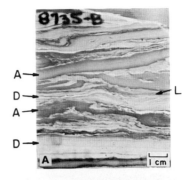

Fig. 4-11. Supratidal dolomite facies, U_4-U_5 zone: A. Core-slab photograph of laminated dolomite mudstone with desiccation features. B. Photomicrograph showing small crystal size variations in laminae. Average crystal size 30 microns. Plane light.

Fig. 4-12. Evaporite facies, U_6-U_7 zone: A. Core-slab photograph of interlaminated anhydrite (A) and dolomite (D) showing cross-lamination (L) and soft sediment deformation features. B. Photomicrograph of interlaminated dolomite (D) and anhydrite (A), possible desiccation feature. Plane light.

approximately 4 feet (1.2 m), culminate the U_4-U_5 sequence in about half of the wells in the field. The U_4-U_5 facies consists of bedded, interlaminated dolomite mudstones and anhydrites that often contain desiccation features (Fig. 4-13). The lower part appears nodular in origin; the nodules are enclosed in dolomite mudstones or are separated by thin sheaths of carbonate sediment.

The U_4-U_5 evaporite unit coincides with variations in thickness of the underlying supratidal rocks. The unit appears to have been deposited in restricted depressions, possibly shallow channels or drainage systems that coalesced towards the very shallow regional depression to the east, where bedded evaporites occur more commonly in the Williston Basin (Fuller, 1961).

Very thin lithoclastic shaly units generally overlie the evaporite or supratidal facies. Generally, in an interval of 2 to 3 feet ($\sim$1m), the rocks grade upward from peritidal dolomitic limestones into lime mudstones and wackestones of normal marine facies, indicating marine transgression.

Concepts of Dolomitization

The dolomite sequences in the Red River zones, considered in light of their depositional settings, are believed to be consistent with the concepts of tidal-flat reflux dolomitization, a process wherein relatively dense waters with Mg^{2+}/Ca^{2+} originate in the supratidal sediments and reflux downward, dolomitizing both the supratidal and underlying sediments. The depth of dolomitization in the U_6-U_7 zone is variable, seemingly inhibited downward by the local occurrence of shaly, thin-bedded "supratidal" rocks (Fig. 4-4). Where these rocks are absent, dolomitization depth is

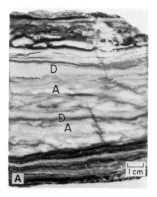

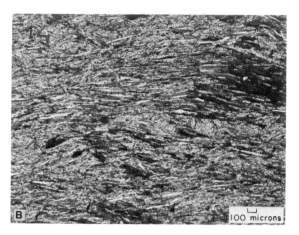

Fig. 4-13. Evaporite facies, U_4-U_5 zone: *A*. Core-slab photograph of interlaminated dolomite (D) and anhydrite (A) with cross-lamination and desiccation features. *B*. Photomicrograph of needle-lathlike anhydrite crystals paralleling laminae planes. Plane light.

probably a function of the permeability of original rock fabric and/or the duration of diagenetic processes.

In the U_4-U_5 zone, dolomitization extends principally down through the supratidal, intertidal, and restricted marine rocks. Depth of significant dolomitization appears to have been restricted downward by a very thin lithoclastic, shaly facies occurring at the base of the restricted marine rocks.

Reservoir Characteristics

Important relationships exist among porosity and permeability, lithology, and fabrics and dolomite crystal size of the depositional sequences. However, over-generalization of average porosity and permeability values for the different facies can be meaningless in view of the common heterogeneity of such rocks. Fractures contribute significantly to reservoir productivity. Vertical fractures are common, particularly in the dolomites, and are open and healed with anhydrite, calcite, and silica.

U_6-U_7 Zone

Porosity in the U_6-U_7 zone consists primarily of intercrystalline pore space (*i.e.*, vugs usually contribute less than 5 percent of total porosity). An increase in crystal size is usually accompanied by a corresponding increase in porosity and permeability (Fig. 4-14). There are, however, many exceptions generally attributable to pore plugging by anhydrite and dolomite. Some cores lack permeability as a result of very small intercrystalline pores rather than low total porosity. In general, dolomite crystal size in the U_6-U_7 zone gradually increases with depth through the restricted marine facies. However, an inverse relationship with depth also occurs, thus causing difficulty in predicting vertical and lateral optimum reservoir conditions.

U_4-U_5 Zone

Oil production from the U_4-U_5 zone is principally from intertidal and supratidal rocks. Marine-facies mudstones and wackestones of the zone generally have crystal sizes of less than 2 microns, and although porosities average up to 16 percent, the extremely small intercrystal pore size results in low permeabilities averaging less than 1 millidarcy.

Restricted marine rocks of the U_4-U_5 zone also are typified by dolomite crystal sizes usually less than 2 microns (Fig. 4-15). Once again, extremely small intercrystalline pore size results in generally low average permeabilities of 2.5 millidarcys, with an average total porosity of 20 percent.

Intertidal reservoirs of pelletoidal and lithoclastic packstones generally exhibit intraparticle dolomite crystal sizes less than 2 microns, whereas the interparticle dolomite crystal size is 15 to 20 microns. Intertidal wackestones typically have dolomite crystal sizes of 20 microns. Permeability ranges from 1 to 35 millidarcys and porosities, from 13 to 22 percent. Pellet-mold

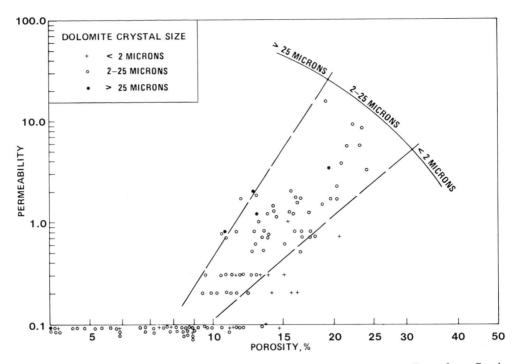

Fig. 4-14. Permeability, porosity and dolomite crystal size relationships in U_6-U_7 zone. Data from Stock (1965).

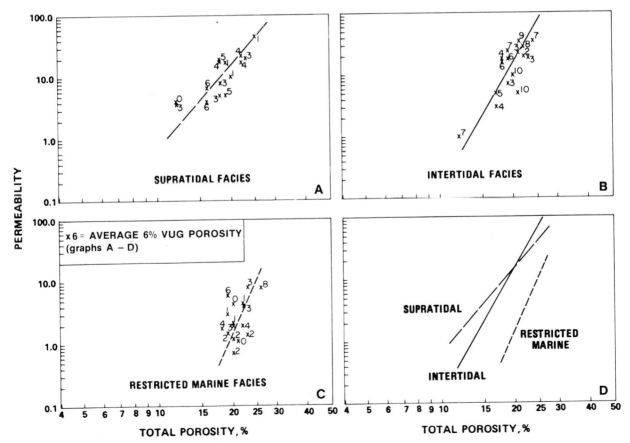

Fig. 4-15. Permeability, total porosity and vug porosity relationships in U_4-U_5 zone: *A*. Supratidal facies. *B*. Intertidal facies. *C*. Restricted marine facies. *D*. Comparison. Data points are average porosities and permeabilities for facies for each well. Data from Stock (1965).

porosity contributes about 5 percent of the total.

Pore space in the supratidal sequences is primarily intercrystalline, with an average matrix dolomite crystal size of 15 microns. Total porosity ranges from 12 to 22 percent and permeabilities, from less than 1 to 35 millidarcys. Vugs contribute an average of only 3 percent porosity.

Summary

Cores of Ordovician Red River producing zones in Pennel Field exhibit cyclic sequences of marine, intertidal, and supratidal rocks. Each cycle begins with normal-marine wackestones passing stratigraphically upward through: burrowed mudstone and wackestone facies; a restricted marine facies of mudstones, wackestones, and packstones; an intertidal facies of pelletal wackestones and packstones; and a supratidal facies of laminated mudstones. The sequence is capped by an evaporite facies which effectively seals the reservoir intervals. Dolomites provide the best reservoir rocks and are developed in peritidal intervals. Significant relationships exist among facies based on porosity, permeability, and dolomite crystal size.

Acknowledgments   The author wishes to thank Shell Oil Company for permission to present these data. I am particularly grateful to Lewis G. Stock, Shell Development Company, who graciously provided many of the materials and data.

References

ABERG, R., and R. SMITH, 1956, A petrographic study of the Red River (U_4-U_5) Reservoir, Cabin Creek Field, Fallon County, Montana. TS Report 92, Shell Development Technical Services Division, Houston, TX.

BETHEL, F., and A. HOLS, 1957, Discussion of reservoir characteristics, Cedar Creek Anticline fields, Montana. Jour. Petroleum Tech., v. IX, no. 12, p. 23–30.

CLEMENT, J.H., 1976, Geologic history—key to accumulation at Cedar Creek (abst.), Amer. Assoc. Petroleum Geologists Bull. v. 60, p. 2067–2068.

DOW, W.G., 1974, Application of oil-correlation and source-rock data to exploration in Williston Basin. Amer. Assoc. Petroleum Geologists Bull., v. 58, p. 1253–1262.

FULLER, J.G.H., 1961, Ordovician and contiguous formations in North Dakota, South Dakota, Montana, and adjoining areas of Canada and United States. Amer. Assoc. Petroleum Geologists Bull., v. 45, p. 1334– 1363.

KENDALL, A.C., 1976, The Ordovician carbonate succession (Bighorn Group) of southeastern Saskatchewan: Mineral Resources, Saskatchewan Geol. Surv., Sedimentary Geol. Div. Report 180, p. 1– 185.

MACAULEY, G., 1964, Regional framework of Paleozoic sedimentation in western Canada and northwestern United States. Billings Geol. Soc., Third International Williston Basin Symp., p. 7– 36.

ROEHL, P.O., 1967, Stony Mountain (Ordovician) and Interlake (Silurian) facies analogs of Recent low-energy marine and subaerial carbonates, Bahamas. Amer. Assoc. Petroleum Geologists Bull., v. 51, no. 10, p. 1979– 2032.

STOCK, L.G., 1965, Tidal flat sedimentation and associated reflux dolomitization in the Red River Formation in the Pennel Field, Fallon county, Montana. EPR Report 903, Shell Development Co., Houston, TX.

WILLIAMS, J.A., 1974, Characterization of oil types in the Williston Basin. Amer. Assoc. Petroleum Geologists Bull., v. 58, p. 1243– 1252.

5
Cabin Creek Field
Perry O. Roehl

RESERVOIR SUMMARY

Location & Geologic Setting	Fallon Co., SE Montana, in SW Williston Basin, USA
Tectonics	Crest of Cedar Creek Anticline, a highly elongate faulted fold episodically uplifted
Regional Paleosetting	SW shelf of cratonic basin
Nature of Trap	Structural plus stratigraphic (unconformity, capillarity)
Reservoir Rocks	
Age	Silurian (Cincinnatian)[1]
Stratigraphic Units(s)	Interlake Formation (also Ordovician Red River Fm[1])
Lithology(s)	Dolomite
Dep. Environment(s)	Peritidal; supratidal and intertidal dominant
Productive Facies	Intertidal and supratidal dolomites (18 oil-bearing zones)
Entrapping Facies	See *Nature of Trap*
Diagenesis Porosity	Dolomitization and selective dissolution; some pore filling by anhydrite
Petrophysics	
Pore Type(s)	Intercrystal, moldic, some vugs; fracture, breccia
Porosity	6–23%, 15% avg (best reservoir rocks, Archie II-III)
Permeability	00.1–? md, 5 md avg (same as above)
Fractures	Extension fractures add to communication between zones; also solution breccia associated with unconformity
Source Rocks	
Age	Ordovician
Lithology(s)	Shales of Winnipeg Formation
Migration Time	Mio-Pliocene (late Laramide)
Reservoir Dimensions	
Depth	8250–8525 ft (2515–2600 m)
Thickness	Contingent (number of zones)
Areal Dimensions	9 × 2 mi (14 × 3 km)
Productive Area	8100 acres (32.4 km^2)
Fluid Data	
Saturations	Variable by zone and by porosity increments
API Gravity	31°
Gas-Oil Ratio	300:1
Other	—
Production Data	
Oil and/or Gas in Place	NA
Ultimate Recovery	NA
Cumulative Production	75.4 million BO through 1979 (commingled Silurian-Ordovician production)

Remarks: IP 206 BOPD through 10/64″ choke. IP mechanisms gas expansion and natural water drive. 125 wells at 80-acre spacing. Discovered 1953.
[1] See also Ruzyla and Friedman, this volume.

5
Depositional and Diagenetic Controls on Reservoir Rock Development and Petrophysics in Silurian Tidalites, Interlake Formation, Cabin Creek Field Area, Montana

Perry O. Roehl

General Setting and History

The Cedar Creek Anticline is an elongated structure occupying the southwest margin of the present Williston Basin in eastern Montana, southwestern North Dakota, and northwestern South Dakota (Fig. 5-1). Pine, Cabin Creek, and Wills Creek fields are contiguous and occupy a midcrestal position on the structure. The latter two fields are independently named because of the manner in which production has been developed and because of unitization commitments.

The anticline is an ancient structure that came into being in the early Paleozoic, apparently as a result of basement faulting. Three known periods of uplift have been largely responsible for the continued existence of the anticline: pre-Mississippian—post-middle Silurian, Mississippian-Triassic, and post-Paleocene. The present closure is due to moderate folding of a few hundred feet, accentuated by high-angle reverse faults along the southwest flank.

In May 1953, the Cabin Creek discovery well, Shell-Unit 22-23, was drilled to a depth of 9062 feet (2755 m) and was completed in the Ordovician Red River Formation (Fig. 5-2). Subsequent tests of the Silurian Interlake Formation yielded an average of 290 barrels of oil and 1 barrel of water per day through a 10/64-inch (0.396-cm) choke. Since discovery, Ordovician and Silurian productive intervals have commonly been commingled (Fig. 5-3). In addition, there is some Mississippian production at Cabin Creek Field. Total undifferentiated field production as of January 1980 was 75.5 million barrels. Development is on 80-acre spacing, with 8100 productive acres (32.4 km²) delineated by structure, oil-water contact, and porosity pinchouts (Matthews, 1971). The field initially produced by gas expansion and water drive. Since 1959 it has been under waterflood.

Geology

The subsurface of Montana, the Dakotas, and the Canadian portion of the Williston Basin contains numerous oil reservoirs developed in dolomitized tidalites of Ordovician and Silurian age. The extent of porosity and permeability development in these rocks depends on two factors: (1) initial environment of deposition (including climatic effects), and (2) diagenesis, especially that related to sabkha interludes and associated dolomitization. The control of reservoir development within the constraints imposed by these depositional and diagenetic features is governed by petrophysical properties, especially capillarity, and by formation fluid properties.

The writer has developed and documented a conceptual depositional model inspired by modern properties and lateral relationships of carbonate sediment fabrics and microfacies (Roehl, 1961, 1967). The model, Figure 5-4, portrays

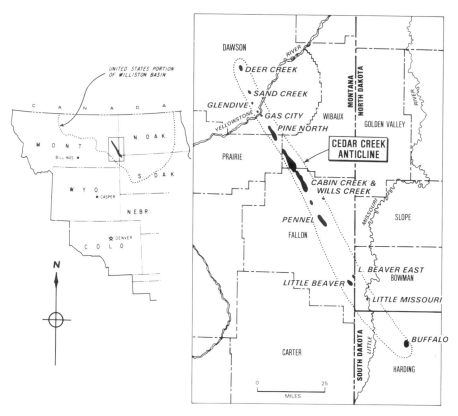

Fig. 5-1. Index map of eastern Montana, and Western North and South Dakota showing location of Cabin Creek Field on Cedar Creek Anticline.*

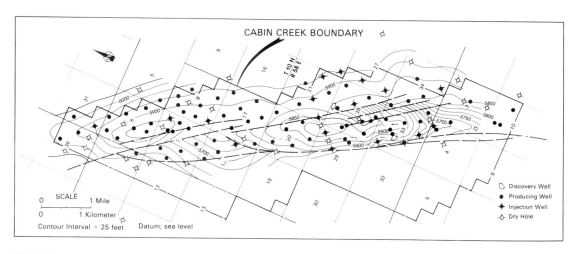

Fig. 5-2. Cabin Creek Field Unit. Structure contours are drawn on the Silurian Interlake C-Marker, which is equivalent to the top of Zone 8. See Figure 5-3 for zonation.

*Figures 5-1, 5-3, 5-4, 5-5B, 5-6, 5-9, 5-10, and 5-14 from Roehl, 1967. Published by permission, American Association of Petroleum Geologists.

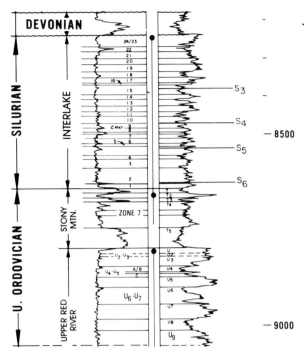

Fig. 5-3. Type radioactivity log of lower Paleozoic formations. Letters to left of lithologic log are engineer log-markers. Solid circles and thickness lines to the right indicate productive intervals.

first, an underlying subtidal and intertidal facies typified by the Stony Mountain Formation (Ordovician). This lower facies is composed of mud-dominant, bioturbated carbonates immediately overlying a dark, illitic shale-rich, fossiliferous member, the so-called Stony Mountain Shale. Both of these units are now dolomite. *Second*, these tidalites pass upward into supratidal deposits of diverse sedimentary fabrics and early post-depositional modifications, including subaerial diagenesis and dolomitization (Fig. 5-5). These modified intertidal and supratidal deposits characterize the Interlake Formation of Silurian age.

There are representatives of infratidal, intertidal, and supratidal sediments in both the Stony Mountain and Interlake formations, products of several transgressions and regressions. However, there is also a progressive change in the paleoenvironmental predominance whereby supratidal deposits most typify the upper Interlake Formation despite several minor episodes of renewed marine transgression, which served mainly to reinitiate the offlap sequences. Offlap deposition is thus predominant, and includes many intervals of hardgrounds, isolated evaporites, and non-

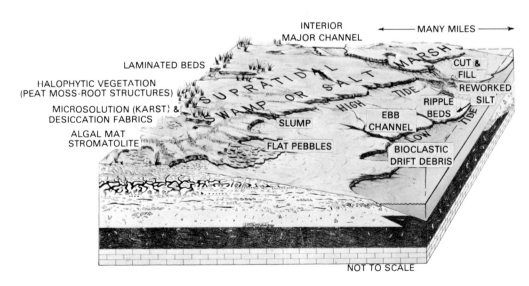

Fig. 5-4. Conceptual schematic of epeiric peritidal deposition and post depositional modifications, Silurian Interlake Formation. Vertical and horizontal scales unequal.

Perry O. Roehl

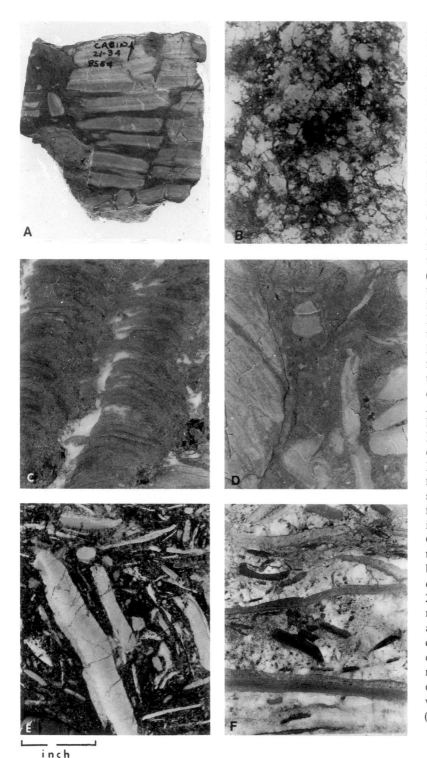

A

B

C

D

E

F

inch

Fig. 5-5. Photographs of slab-bed cores showing diagnostic rock fabrics in the Interlake Formation. *A.* Culminating diagenetic (karst) fabric at the top of the eroded Interlake Formation showing detached dolomite slabs in a red-brown terra rossa matrix. Cabin Creek well 21-34, depth 8554 feet (2600 m). *B.* Epigenetic solution breccia. Original character of sediment is obscured by early diagenesis, thus porosity and volumetric evaluations are difficult. Dolomite, Cabin Creek well 32-18, depth 8518 feet (2589 m). *C.* Linear stromatolites that extend more than 4 feet in a dolomite rock core. Differential growth habit is portrayed by porous, oil-stained hemispheroidal crests that were formerly occupied by filamentous algae. Pine well 34-25A, depth 8708 feet (2647 m). *D.* Flat pebbles derived from large algal stromatolite (left). Pebbles are suspended in porous matrix (dark). Dolomite, Pine well 34-25A, depth 8712 feet (2648 m). *E.* Dense, unsorted flat-pebble breccia in finer grainstone with oil-stained interparticle porosity (dark). Crude size grading and orientation of chips suggest single, brief, high-energy supratidal event. Dolomite, Pine well 34-25A, depth 8711 feet (2648 m). *F.* Porous, oil-bearing algal mat chips suspended in dense matrix. Orientation of chips and fine-grained dense matrix suggest low-energy event. Dolomite, Cabin Creek well 32-7, depth 8447 feet (2568 m).

tectonic breccias. The sequence of Silurian strata culminates with the development of a diagenetic terrane surface, attesting to a long subsequent exposure history that carries into the Devonian period.

Representative Depositional Rock Types

The Interlake Formation is essentially all dolomite, with varying but minor amounts of illitic shale, quartz silt and sand stringers, and anhydrite. Chert is important at the base and top of the section. In spite of pervasive dolomitization, these strata contain a remarkable record of their original sedimentary fabrics. Collectively, they suggest the persistence of strand line deposition: intertidal and supratidal.

The sediment source was substantially the same for all the local environments. Mud, pellets, and bioclastic particles were moved shoreward by high tidal events and/or storms and hurricanes. Abatement of these forces resulted in the "offlap" accretion of marine sediment in each tidal subenvironment.

Laminated Mudstones and Siltstones

Thick successions of very thin-bedded dolomitized mudstones and siltstones are predominant at Cabin Creek field. They are counterparts of recent laminated sediments prevalent throughout the Bahamas, most commonly at or near the strandline just above or below mean tide (Figs. 5-6A, B). However, they also occur far within the supratidal region, where they compete with many other complex sedimentary microfacies. These reservoir rocks are difficult to characterize volumetrically since their porosity is of secondary origin (intercrystal, in dolomite), which is most commonly delineated by primary sedimentary structures and stratigraphic boundaries. Through the dolomitic crystal overprint, one can occasionally discern the characteristic sedimentary structures, such as graded laminae, pellets, and faint ripple layers.

Algal Mats and Stromatolites

The depositional and diagenetic relics of filamentous blue-green algae are also widespread. They commonly occur as columnar, linear stromatolites (Fig. 5-5C), isolated "cabbage" heads (Fig. 5-5D, left side), undulatory encrustations (Fig. 5-6E) or widespread, flat mats, several centimeters thick, commonly represented by broken slabs (chips), as in Figures 5-5D–F and 5-6E. The filamentous blue-green algae and their subsequent constructional (stromatolitic) forms are important for at least three reasons:

1. They tie down vast areas of fine sediment, both along the shore zone and in interior areas of restricted circulation.
2. They protect primary porosity from subsequent destruction by firmly binding the incoherent material.
3. They may themselves contribute to porosity by the subsequent decay of their organic constituents (Fig. 5-6C).

Most of the preserved "depositional" porosity is retained during dolomitization and even enhanced where selected bioparticles are leached out to form secondary microvugs (Fig. 5-5C). The algae preferentially entrap the larger sediment particles, which are carried in suspension during periods of tidal flooding. Finer carbonate mud fills in the intervening troughs on ebb flow. This mud becomes the dense, non-porous dolomite "matrix" between porous oil-stained stromatolitic columns.

Flat Pebble Breccias

Large areas of modern algal mats and stromatolites generally desiccate, exfoliate, and crack open during protracted periods of subaerial exposure. They are then ripped up and transported by tidal flow to form new accumulations of flat pebbles (Figs. 5-5D–F). From a reservoir viewpoint, the most important characteristics of a flat-pebble breccia are the porosity and continuity of the host matrix rock. For example, the breccia shown in Figure 5-5E is interpreted as a high-energy event because the matrix is a grainstone supporting a crudely graded suspension of flat pebbles. Obviously the thick, large pebbles required the greatest transport energy to impart a high angle of repose. The dark matrix is porous and oil-bearing.

In contrast, the float breccia in Figure 5-5F probably represents a low-energy event. The pebbles are unsorted and there is no grading. The largest pebbles are horizontally arrayed. Finally,

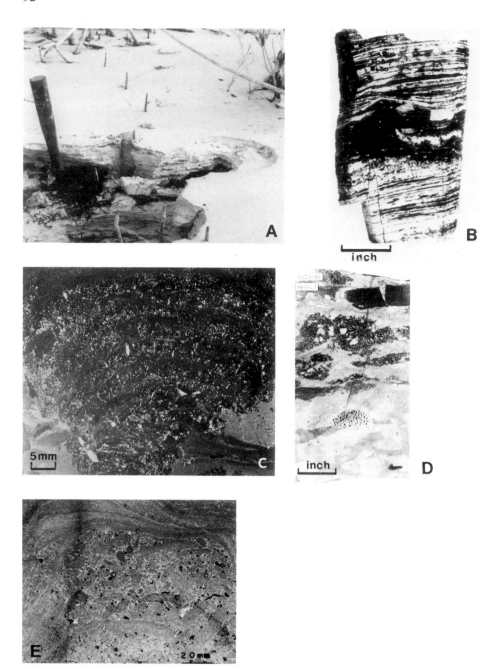

Fig. 5-6. A,B. Photographs comparing modern and ancient oil-bearing laminated tidalites. *A.* Laminated carbonate mudstones and siltstones at the juncture of the intertidal and supratidal zone, Williams Island, Andros, Bahamas. Geological hammer for scale, left center. *B.* Inferred Interlake Formation counterpart to *A.* The multiple fine laminae are now dolomite. Porosity 15%. Cabin Creek well 21-34, depth 8630 feet (2624 m). *C.* Photomicrograph of algal stromatolite showing multiple growth layers of trapped pellet-size grains and mud. Light areas are pores between sediment and elongated traces of decayed algal filaments. Plane light. Shell Northern Pacific Well 12-3, depth 8940 feet (2718 m). *D.* Core-slab photograph of dolomite illustrating commonly observed transition from bioturbated, favositid coral-bearing zone to overlying burrowed and lenticular packstones and mudstones, including fractured algal mat chip (dark, oil-stained). Cabin Creek well 21-34, depth 8644.5 feet (2628 m). *E.* Photomicrograph of dolomitized algal laminae forming stromatolitic encrustations that bind a small mound of flat pebbles and chips. Detritus is druse-coated and porous (dark). Petrophysical values: $0 = 8.9\%$, $G = 0.7$, P_d 20 psia. Cabin Creek well 21-34, depth 8689 feet (2641 m).

the matrix is a dense mudstone. The only porosity preserved occurs within the individual pebbles and chips, thus exhibiting poor pore continuity and low volumetric storage capacity.

Combinations

Admixtures of the aforementioned intertidal and supratidal sedimentary rock fabrics are of course common, and they in turn are transitional with bioturbated infratidal and intertidal fabrics. Figure 5-6D illustrates a prevalent couplet, where either an *in situ* stromatolite, or flat fragments thereof, overlie bioturbated muds that commonly contain benthic faunal forms such as small isolated colonial corals.

Paragenesis

The Interlake carbonates are almost universally dolomite at Cabin Creek Field. Anhydrite, though regionally important, is not stratigraphically significant at Cabin Creek. However, it is locally important as massive anhydrite, and as a secondary pore filling and replacement mineral, being most prevalent beneath diastems, beneath the post-Interlake unconformity, and as a component of solution breccias. Such occurrences are probably related to sabkha exposure surfaces (Figs. 5-7A, B) and the final post-Interlake hiatus represented by a regional subaerial diagenetic terrane (Figs. 5-8A, B) (Roehl, 1967).

Repeated exposure of Interlake sediment to shallow hypersaline waters, general aridity, and frequent subaerial exposure probably led to penecontemporaneous lithification, prior to and during dolomitization. The mechanism of dolomite formation is considered to be related to the displacement and reactivity of subsurface hypersaline brines caused by a combination of lateral hydraulic gradients, sea water flood recharge (Fig. 5-7A), and surface evaporative pumping (Fig. 5-8A) (Butler, 1970; Hsü and Schneider, 1973; McKenzie *et al.* 1980).

The process requires an increase in salinity via capillary evaporation, leading to a drop in the water table. Ionic concentrations then shift in favor of Mg^{++} versus Ca^{++} through the first diagenetic stage of $CaSO_4$ deposition from solution at or beneath the water table. According to McKenzie *et al.* (1980) the piezometric surface evokes the active participation of the regional hydraulic head when the ground water table has been depleted to the level of this surface or below it. The dynamics of this subsurface recharge will depend on the presence or absence of one or more intervening aquicludes and the exchange capacity of the brines through time.

Porosity

There are several types of porosity in the Interlake dolomites at Cabin Creek. Intercrystal porosity is the result of the dolomitization of several sediment types and is predominant (Figs. 5-9–5-12). There are, therefore, few Interlake strata devoid of porosity. A second, quite subordinate porosity type is interparticle. This pore type is preserved from the pre-existing sediment. It is usually free of secondary pore-filling dolomite rhombs, except for rinds on the perimeter of grains (Fig. 5-6E). The third type is fracture porosity, largely but not exclusively confined to dense, fine-crystalline dolomites (Type I) and porous, fine-crystalline dolomites (Type II), as shown in Figures 5-9 and 5-10 (Archie, 1952). Vugs constitute the last pore type, most often moldic, after anhydrite dissolution (Fig. 5-11B), but also common as primary cavities in fossils (Fig. 5-6C,D).

It is the fine intercrystal porosity, especially Type II-III, that constitutes the most volumetrically significant reservoir for hydrocarbons (Figs. 5-10, 5-13). There are other intercrystal pore sizes that also contain important amounts of oil. These have been characterized and classified by Archie (1952) and modified by the writer (Roehl, 1959) (Fig. 5-13).

Matrix Classification and Zonation

It has been demonstrated by Archie (1952) that compact crystalline, chalky, and sucrose-appearing carbonates reflect component crystal or particle size and the interrelation between particles sufficiently to indicate also the association and magnitude of pore size, porosity, and capillary properties. The Interlake dolomites can be so described and form the basis for subdivision throughout the Cabin Creek area (Figs. 5-3, 5-14). With a refinement of size grades and

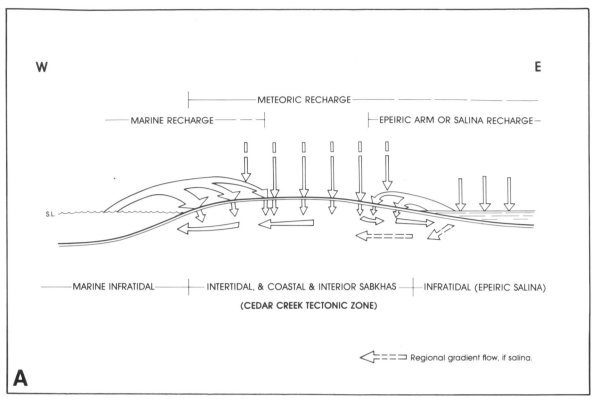

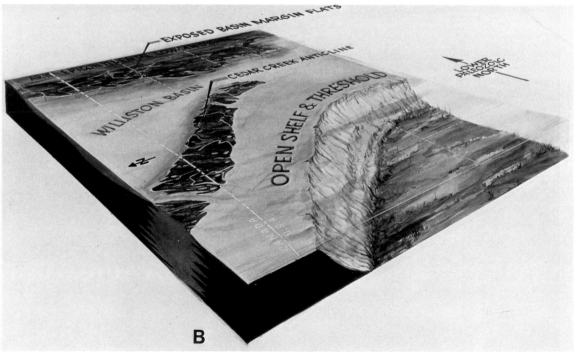

Fig. 5-7. A. Diagram illustrating the proposed origin and movement of Interlake recharge waters during a depositional interlude when peritidal and sabkha conditions reflected the tectonic physiography of the regional Cedar Creek structure. Meteoric, high marine tidal, and storm interludes need not coincide. (Modified from McKenzie et al., 1980. Published by permission of Society of Economic Paleontologists and Mineralogists.) *B.* Block diagram of Williston Basin region illustrating proposed regional depositional setting during period of emergence of western metastable lineament and simultaneous off lap deposition of intratidal, intertidal, and supratidal (sabkha) sediments. Present-day north is toward lower left.

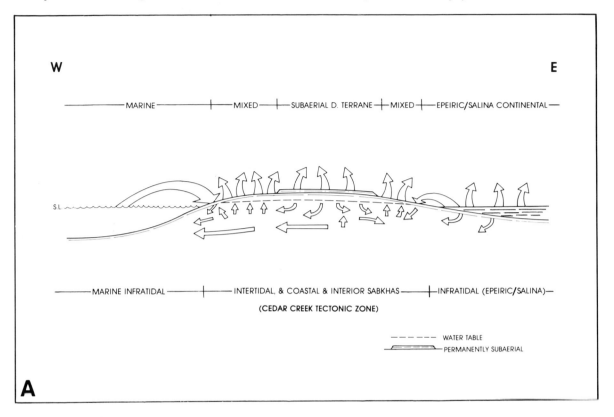

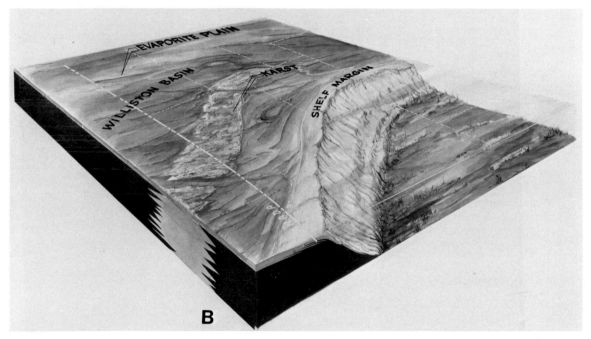

Fig. 5-8. A. Diagram illustrating the proposed origins of evaporative waters and their circulation to and from the sabkha proposed for the Cedar Creek Anticline area at Cabin Creek field. (Modified from McKenzie et al., 1980. Published by permission of Society of Economic Paleontologists and Mineralogists.) *B*. Block diagram suggesting fully emergent stage of Williston Basin region in Late Silurian time. Early differential uplift along a metastable lineament led to the development of microkarst topography, which probably separated a large interior evaporite plain with interludes of salina infilling or epeiric sea encroachments from the north.

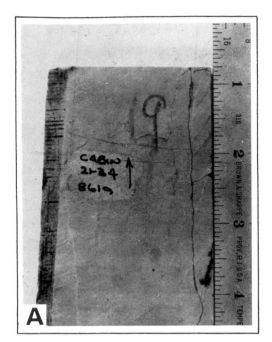

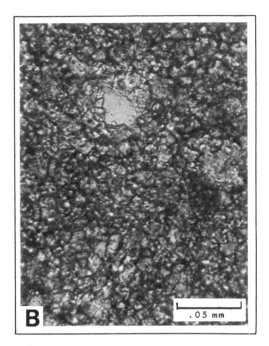

Plane Light

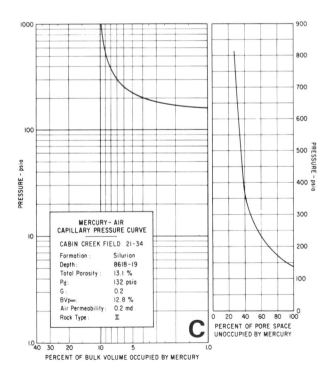

Fig. 5-9. A. Core-slab photographs of rock Type II. Laminated, cream-colored, extra fine-grained, chalky dolosilitite. Cabin Creek well 21-34, depth 8619 feet (2620 m). *B.* Photomicrograph of rock shown in *A,* displaying anhedral to subhedral, 5–20-micron dolomite crystals. *C.* Capillary pressure curve using Thomeer (1960) notation.

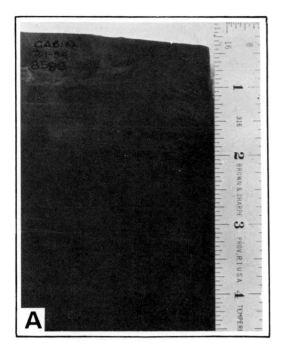

Plane Light

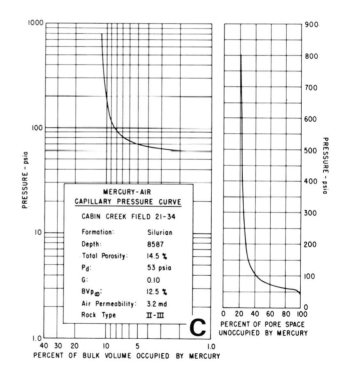

Fig. 5-10. A. Core-slab photograph of rock Type II-III. Dolosiltite similar to Type II (Fig. 5-9) except it is uniformly microsucrosic and oil-stained. Cabin Creek well 21-34, depth 8588 feet (2611 m). *B.* Photomicrograph of rock shown in 10A, displaying subhedral to euhedral, 15–35-micron dolomite crystals. *C.* Capillary pressure curve using Thomeer (1960) notation. Pores are larger and better connected than Type II.

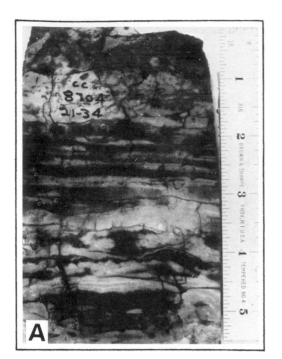

Crossed Nicols

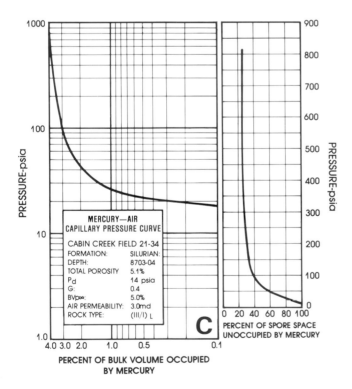

Fig. 5-11. A. Core-slab photograph of rock Type II/I, Low G Factor. An algal, pelletal dolomite with short, vertical irregular fractures. Dark laminae are oil-stained. Cabin Creek 21-34, depth 8704 feet (2646 m). *B.* Photomicrograph of rock shown in *A*, displaying 20–90-micron crystals enclosing a large anhydrite mold. *C.* Capillary pressure curve using Thomeer (1960) notation. Low G Factor documents good pore size distribution.

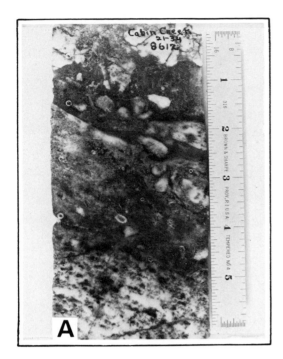

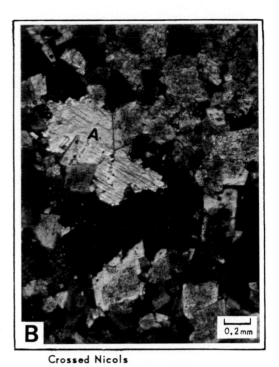

Crossed Nicols
A = Anhydrite

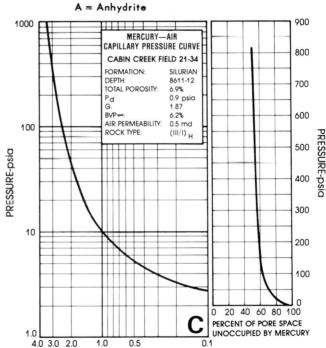

MERCURY—AIR
CAPILLARY PRESSURE CURVE
CABIN CREEK FIELD 21-34

FORMATION:	SILURIAN
DEPTH:	8611-12
TOTAL POROSITY:	6.9%
P_d	0.9 psia
G.	1.87
BVP∞:	6.2%
AIR PERMEABILITY:	0.5 md
ROCK TYPE:	(III/I)$_H$

Fig. 5-12. *A.* Core-slab photograph of rock Type III/I High G Factor. A skeletal wackestone with nodular and pore-filling anhydrite. Cabin Creek well 21-34, depth 8611 feet (2618 m). *B.* Photomicrograph of rock shown in *A*, displaying larger crystal sizes (100-250 microns) and a pore-filling anhydrite crystal (*A*). *C.* Capillary pressure curve using Thomeer (1960) notation. The G factor (curve shape) is quite high—a reflection of divergent pore sizes and shapes, and anhydrite plugging.

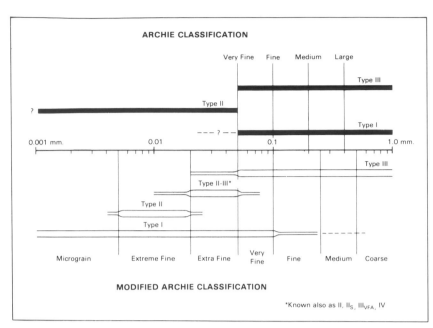

capillary parameters, it is possible to restrict such typing so as to establish definitive petrophysical limits. These permit an estimate of productive capacity, which can be inferred by employing a slightly modified and refined crystalline rock classification, as shown in Figures 5-9–5-13 (Roehl, 1959).

Reference is made to capillary pressure curves (Figs. 5-15, 5-16) employing notation developed by Thomeer (1960). The capillary data permit porosity and connate water saturation to be related, since petrophysical limits have been placed on each rock type. The shapes of curves (G) are related to two asymptotic end members: (1) the extrapolated bulk volume occupied at infinite pressure ($BV_{p\infty}$) and (2) the extrapolated entry pressure (P_d), (Fig. 5-15).

The shape and entry pressure values have been used to define four porous, modified Archie rock types:

Type	P_d (psia)	G
II	>100	<0.4
II-III	20–100	<0.4
(III/I)L	<20	≤0.7
(III/I)H	<30	>0.7

It is obvious that these are end-member rock types and that many combinations are possible and common in the Interlake.

Type II rocks (Figs. 5-9, 5-13) often have moderately high porosity, up to 16 percent. The effective porosity is also quite good, with uniform pore sizes and interconnections giving a well-defined capillary pressure curve plateau of low G value. However, these rock types normally have low oil saturation in the Interlake because of their high entry pressure, P_d. They also have a lower permeability for equivalent porosity than other rock types (Fig. 5-17). This reflects the very fine matrix crystal size and thus the small pore sizes.

Microsucrosic Type II-III rocks (Figs. 5-10, 5-13) are quite similar to Type II, with one very important exception: crystal sizes and intercrystal pores are larger. Type II-III crystals tend toward euhedralism (Fig. 5-10B) and thus impart better pore continuity and higher porosity. Many fine-grained sediments are diagenetically converted into this crystal size range (20–50 μ) and thus become important reservoir rocks. Examples are laminated dolostones, algal mats, and stromatolites (Figs. 5-5C, F; 5-6B; 5-10A).

Type III/I rocks are discrete mixtures of Type I (dense, fine-grained) and Type III, large crystal masses of sucrosic texture. These usually result from the dolomitization of complex sedimentary structures. They lack uniformity in both crystal and particle size, and in pore structure. They are thus characterized by divergent G values. Type

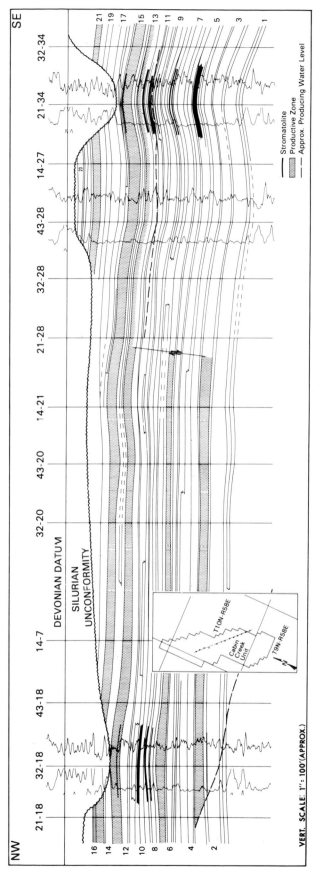

Fig. 5-14. Northwest-southeast cross-section through Cabin and Wills Creek fields showing rock type zonation (1–23) based on crystalline texture classification. Pre-Middle Devonian positive axes underlie channels that have cut out some upper Interlake porous zones.

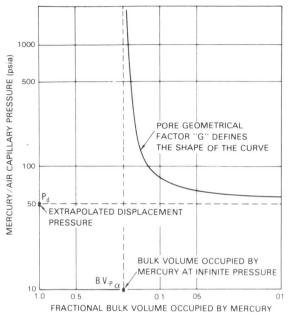

Fig. 5-15. Log-log characterization of a capillary pressure curve employing hyperbolic notation. (After Thomeer, 1960) © 1960 SPE-AIME

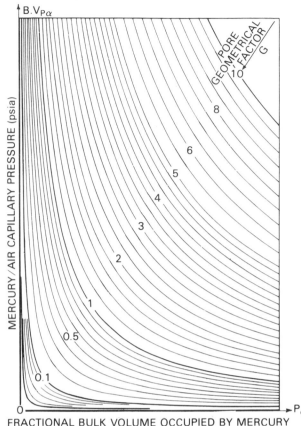

Fig. 5-16. Family of computed capillary pressure curves. (After Thomeer, 1960) © 1960 SPE-AIME

III/I rocks are less persistent laterally and vertically because of the types, admixtures, and general distribution of original sedimentary materials. These rocks may have a wide porosity range and their uniformly low entry pressure pores assure a contribution of hydrocarbons in productive intervals. However, stratigraphic variability and wide range of pore sizes of III/I types severely limit their volumetric contribution. In association with Type II-III, they provide important permeable conduits for production (Fig. 5-17).

Reservoir Quality of Silurian Carbonate Types

From a considerable amount of raw data, four families of coordinate capillary pressure curves have been constructed to correspond to each rock type (Figs. 5-18A–D). The data used are from rocks, the log-log capillary pressures of which closely approximated Thomeer's hyperbola and which had porosities greater than 4 percent (III/I_H types) and 6 percent (all other types). The typical curves are essentially hyperbolas with composite averages of dependent variables P_d, G, and $BV_{p\infty}$ expressed in terms of 2 percent porosity increments. The succession of curves for each type indicates the following:

a. The shape factor G decreases as porosity increases for all rock types. This is due to the direct relation of crystal size to total porosity, and results from a more uniform pore interconnection and distribution in higher porosity rocks. (Note plateau development in Figs. 5-18A, B.)

b. Entry pressure has no consistent relationship to porosity for III/I rock types, but is nevertheless low for all cases.

c. For low G rocks II, II-III, and III/I_L, the 500-psi displacement volume exceeds 75 percent of total pore volume. This probably relates to the good pore interconnection when effective

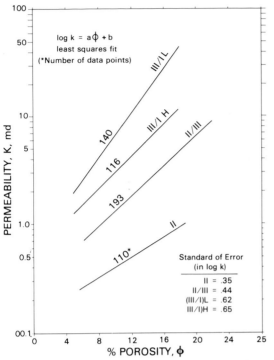

Fig. 5-17. Porosity versus permeability by rock types for the Interlake Formation.

$\phi \equiv$ total ϕ. This is not true with III/I$_H$ rocks, for although at lower values of total porosity these rocks gave fairly good agreement between BV$_{p\infty}$ and total porosity, it was necessary to utilize hyperbolas with BV$_{p\infty}$ greater than total porosity to accurately characterize capillary pressure data for higher porosity rocks.

Capillary characteristics become critical in proximity to field water levels. Such a case occurs on the eastern margin of the Interlake reservoir. In an attempt to learn what quantities of oil and water might be encountered at intervals above the free water level, the curves of Figures 5-19A and B are of some value. The average curves for the scatter plots are first-order equations, as determined by regression analysis checked to fourth order. The points from which the equations were derived were read from actual capillary pressure curves of appropriate rock types. For the reading, a predetermined standard conversion factor of 1.2 psi per foot was used to obtain height in feet from pressure (psia) as measured on the ordinate axis. This same factor on the graphs in Figures 5-18A–D, revealed the following comparison of mini-

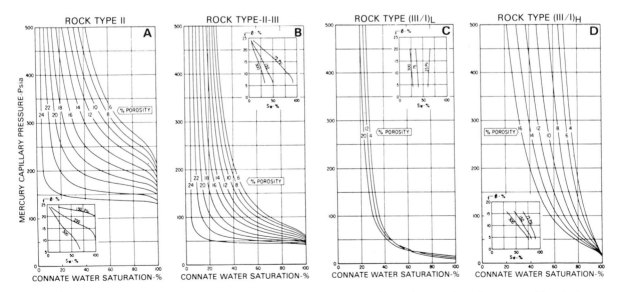

Fig. 5-18. A-D. Family of capillary pressure curves for the four major rock types based on crystalline texture classification of Figure 5-13.

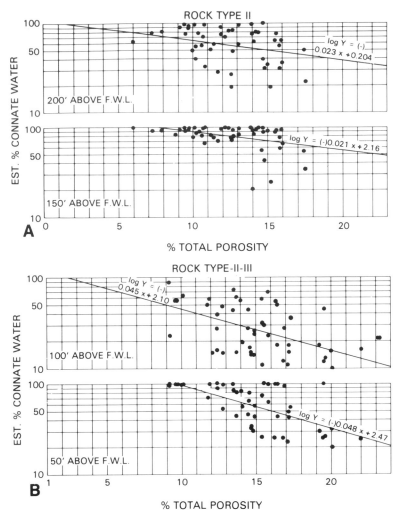

Fig. 5-19. A-B. Percent total porosity of rock types II and II-III, respectively, versus percent total estimated connate water.

mum porosity values for 50 percent water saturation for Types II and II-III rocks:

Type	Height (ft.)	Value Family of Curves	Value Scatter Diagrams
II	200	15	15
II	150	20+	22
II-III	100	10	9
II-III	50	20	16

The values are in good agreement, except for II-III rock types around 50 feet above free water level. The reason for the discrepancy is not known. Scatter diagram point readings taken from the slope plateaus of II-III rocks may be in error.

However, this is a very critical area on the family of capillary pressure curves since large changes in saturation occur over a very limited vertical distance.

Subaerial Diagenetic Terrane

Emergence occurred throughout the Williston Basin at the end of Silurian time, leading to the development of a *subaerial diagenetic terrane* with associated deep laterite (*terra rossa*) throughout the region of the Cedar Creek Anticline (Fig. 5-5A). Because of this emergence, a long period of non-deposition ensued for the area of the Cedar Creek Anticline. In addition, several important reservoir beds were removed by erosion at Cabin Creek Field (Fig. 5-14) (Roehl, 1967). Erosion centered on local paleostructural highs and resulted in the local removal of many of the upper Interlake zones. Erosion reached down to the top of zone 13 at Cabin Creek well 32-18. Thus, well productivity is markedly influenced by the number and type of productive zones as well as their connate water characteristics and their proximity to the oil-water contact. It is to be expected, therefore, under long-term conditions of arid tidal-flat regression that the stratigraphically youngest reservoir zones will most probably be (1) supratidal (sabkha) in origin, (2) dolomitized, and (3) subject to eogenetic removal by subaerial erosion.

Acknowledgments The early, fundamental research for this paper was conducted at Shell Oil Company, Denver Area, and Shell Development Company, Houston, Texas. Several production department staff members made valuable contributions in compiling and interpreting the field data, petrophysics, and statistics. I am particularly grateful for the generous help of J. E. Tuttle, J. A. Pederson, D. W. Solmonson, J. J. Pickell, D. C. Jones, C. N. Tinker, and A. D. Mitchell. Special thanks are due M. M. Ball and R. N. Ginsburg for their critical help during the early days of the Recent sediments research for Shell out of Coral Gables, Florida. J. H. Clement provided field history data.

References

ARCHIE, G.E., 1952, Classification of carbonate reservoir rocks and petrophysical considerations: Amer. Assoc. Petroleum Geologists Bull., v. 36, no. 2, p. 278–298.

BUTLER, G.P., 1970, Holocene gypsum and anhydrite of the Abu Dhabi Sabkha, Trucial Coast: an alternative explanation of origin: *in* Third Symposium on Salt, v. 1, Northern Ohio Geol. Soc., p. 120–152.

HSÜ, K.J., and J. SCHNEIDER, 1973, Progress report on dolomitization—hydrology of Abu Dhabi sabkhas, Arabian Gulf: *in* Purser, B. H., ed, The Persian Gulf, Springer-Verlag, Inc., Heidelberg, p. 409–422.

MCKENZIE, J.A., K. J. HSÜ, and J. F. SCHNEIDER, 1980, Movement of subsurface waters under the sabkha, Abu Dhabi, UAE, and its relationship to evaporative dolomite genesis: *in* Zenger, D.F., J.B. Dunham, and R.L. Ethington, eds., Concepts and Models of Dolomitization, Soc. Econ. Paleontologists and Mineralogists, Special Publ. 28, p. 11–30.

MATTHEWS, N.J., 1971, Progress report on the Cabin Creek waterflood, Fallon County, Montana, Shell Oil Company Production Department.

ROEHL, P.O., 1959, A final report on the production geology and associated petrophysics of the Silurian (Interlake) reservoir, Cabin and Wills Creek Fields, Fallon County, Montana, Shell Oil Company, Denver Area.

ROEHL, P.O., 1961, The Stony Mountain Formation reservoir, Pine Field, Montana—geology and petrophysics of certain epeiric-tidal flat sediments, Shell Oil Company, Production Dept., Denver Area.

ROEHL, P.O., 1967, Stony Mountain (Ordovician) and Interlake (Silurian) facies analogs of Recent low-energy marine and subaerial carbonates, Bahamas: Amer. Assoc. Petroleum Geologists Bull., v. 51, no. 10, p. 1979–2032.

THOMEER, J.H.M., 1960, Introduction of a pore geometrical factor defined by the capillary pressure curve: Jour. Petroleum Technology, Tech. Note 2057.

Mt. Everette and Southwest Reeding Fields
William A. Morgan

RESERVOIR SUMMARY

Location & Geologic Setting	Kingfisher Co., T15-16N, R6W, Oklahoma, NE Anadarko Basin, USA
Tectonics	Remnant of southern Oklahoma aulacogen
Regional Paleosetting	Shallow to deep shelf (skeletal buildups) and ramp (oolite shoals)
Nature of Trap	Stratigraphic; reservoir facies subcropping at Devonian unconformity
Reservoir Rocks	
Age	Silurian
Stratigraphic Units(s)	Hunton Group (Clarita and Henryhouse Fms)
Lithology(s)	Limestone and dolomite
Dep. Environment(s)	Crinoid-rich banks and oolite shoals
Productive Facies	Crinoid packstone and grainstone, ooid grainstone, and skeletal wackestone and packstone
Entrapping Facies	Arthropod wackestone and draping effect of Devonian shale
Diagenesis Porosity	Dolomitization and dissolution of skeletal debris
Petrophysics	
Pore Type(s)	Solution-enlarged biomoldic*
	Solution-enlarged interparticle**
Porosity	0–15% and 8% avg*; 0–20% and 7% avg**
Permeability	0<1.0–560 md and 93% avg*; 0.1–30+ md**
Fractures	Locally important; high angle, open or calcite filled
Source Rocks	
Age	Ordovician (Sylvan Shale) and Mississippian and Devonian (Woodford Shale)
Lithology(s)	Gray to green calcareous shale; black non-calcareous shale
Migration Time	Permian(?)
Reservoir Dimensions	
Depth	7676–8075 ft (2340–2461 m)
Reservoir Dimensions	
Depth	0–50 ft (0–15 m)*; 0.35 ft (0–11 m)**
Areal Dimensions	3.9 × 1.7 mi (6.2 × 2.7 km)
Productive Area	3179 acres (5.1 km²)
Fluid Data	
Saturations	S_o = 76%, S_w = 24% (avg)
API Gravity	39–45°
Gas-Oil Ratio	0–16,667:1
Other	
Production Data	
Oil and/or Gas in Place	12 million BO
Ultimate Recovery	3 million BO; difficult to estimate because production from Hunton and younger reservoirs commingled; URE ≈ 25%
Cumulative Production	NA

Remarks: +Fields are contiguous. *Clarita reservoir, **Henryhouse oolite reservoir IP's 20–200 BOPD and 0–750 MCFGPD (0–2.4 million m³). Discovered 1965.

6

Silurian Reservoirs in Upward-Shoaling Cycles of the Hunton Group, Mt. Everette and Southwest Reeding Fields, Kingfisher County, Oklahoma

William A. Morgan

Location and Discovery

Mt. Everette and Southwest Reeding are two small contiguous oil fields situated on the north-eastern shelf of the Anadarko Basin in Kingfisher County, central Oklahoma, approximately 28 miles (45 km) northwest of Oklahoma City (Fig. 6-1). In 1965, the Jones and Pellow No. 1 Britton well (Sec. 11, T15N, R6W) discovered hydrocarbons in the Hunton Group at Mt. Everette-Southwest Reeding. Hydrocarbon production from the Hunton in these fields is from skeletal-buildup, oolite-shoal, and sub-unconformity reservoirs. Ultimate recoveries per well from these three Hunton reservoirs are roughly 42,500, 7000, and 7000 barrels of oil, respectively.

Regional Setting

The Anadarko Basin is a WNW–ESE-trending remnant of the Southern Oklahoma aulacogen (Hoffman *et al.*, 1974; Pruatt, 1975). The southern margin of the basin is sharply defined by the Amarillo Mountains, which formed during the Pennsylvanian (Morrowan-Atokan) Wichita orogeny (Ham *et al.*, 1964; Ham, 1969). Within the Anadarko Basin, the Hunton Group is a sequence of limestones and dolomites of Late Ordovician through Early Devonian age (Amsden, 1975) up to 1300 feet (396 m) in thickness. The Hunton Group depositional basin extended south beyond the present southern margin of the Anadarko Basin. However, most of the Hunton Group was completely removed south of the Amarillo-Wichita Mountains by the combined erosional effects of Devonian and Pennsylvanian uplifts (Maxwell, 1959). Similarly, north, east, and west of the basin, the depositional extent of Hunton strata has been severely modified by erosion.

The Hunton Group conformably overlies the Sylvan Shale (Late Ordovician), and throughout most of the basin is overlain unconformably by either the Woodford Shale (Late Devonian-Early Mississippian) or, locally, the Misener Sandstone (Middle-Late Devonian). The depositional extent of Hunton units was severely modified by erosion at interludes during Hunton Group sedimentation, especially during the Early Devonian. Extensive erosion also occurred during the Middle Devonian, preceding Misener Sandstone and Woodford Shale deposition. The Woodford Shale may lie unconformably on any of the Hunton units, and in some areas the entire Hunton Group was removed prior to Woodford deposition.

Petroleum Geology of Hunton Group Reservoirs

Hydrocarbon production from the Hunton Group in Mt. Everette and Southwest Reeding fields is mainly from skeletal buildups within the Clarita

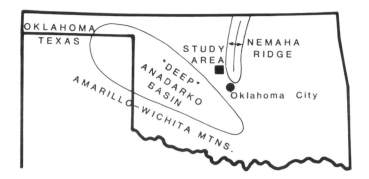

Fig. 6-1. Location map of the Mt. Everette and Southwest Reeding study area.

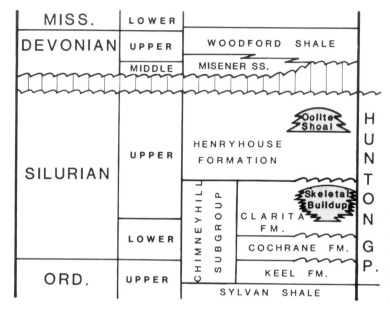

Fig. 6-2. Ordovician-Mississippian stratigraphy within the study area. The Hunton Group in the study area consists of the Chimneyhill Subgroup and the Henryhouse Formation. No Devonian Hunton formations are present.

Formation (upper Chimneyhill Subgroup) and from an oolite shoal within the Henryhouse Formation (Fig. 6-2). Hydrocarbons also are locally trapped in other facies at the top of the Henryhouse, where these facies subcrop at the unconformity below the Woodford Shale. All wells penetrating the Hunton Group are shown in Figure 6-3.

In the area of the two fields (Fig. 6-3), the Hunton Group comprises approximately 275 to 300 feet (84-107 m) of section representing both the Chimneyhill Subgroup and the Henryhouse Formation; apparently, no Lower Devonian formations have been preserved, and the Misener Sandstone or Woodford Shale directly overlies the Henryhouse. Biostratigraphic control for this interpretation is provided by Amsden and Rowland (1967, 1971) and Amsden (1975): cores from the Jones and Pellow No. 1 Farrell (Fig. 6-4) contain the Late Silurian brachiopod *Kirkidium*, which is

diagnostic of the Henryhouse Formation, above the oolite shoal described in this paper and 18 feet (5.5 m) below the base of the Woodford Shale. Amsden (1975) also recorded *Kirkidium* directly below the base of the Misener Sandstone in the Cleary No. 1-21 Gilbert well (Sec. 21, T17N, R6W). Correlation of these cores to the gamma-ray logs and extrapolation via log correlations suggest that no Lower Devonian formations are present within the study area. However, the occurrence of small, thin Lower Devonian outliers cannot be ruled out without additional biostratigraphic control.

The Clarita Formation (upper Chimneyhill Subgroup) consists of light-colored, skeletal carbonates. It can generally be distinguished from the basal Henryhouse Formation because the latter is darker and less fossiliferous and characteristically contains thin, green, argillaceous, silty laminae and wisps. The increase of siliciclastics in the basal Henryhouse Formation corresponds

Fig. 6-3. Map of well control in the Mt. Everette and Southwest Reeding study area. Only wells that penetrated the Hunton Group are shown.

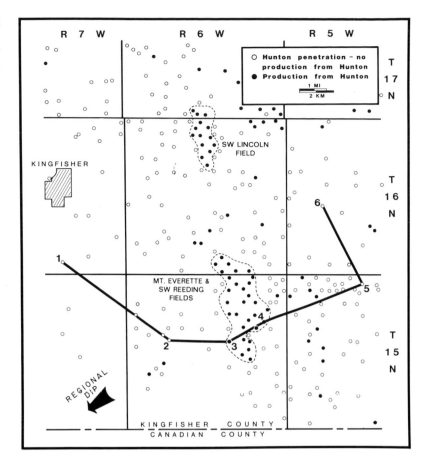

to an increased gamma-ray response (Fig. 6-4). Because differentiation of the Clarita Formation from the Cochrane Formation and the Cochrane Formation from the Keel Formation is generally difficult in the subsurface, especially from well cuttings, no attempt is made in this study to define their contacts. However, their color, bedding style, and weathering characteristics usually make them readily discernible in outcrop (Amsden, 1957, 1960). The skeletal buildup facies, productive at Mt. Everette and Southwest Reeding fields, is assigned to the Clarita Formation based on a "Clarita-age" fauna described by Amsden (1975) directly below this facies in a core from the Jones and Pellow No. 1 Reherman well (Sec. 25, T15N, R6W).

Regional dip within the study area is to the southwest and was imparted during the Wichita orogeny. Drilling depth to the Hunton Group averages 7800 feet (2378 m) or −6600 feet (−2012 m) subsea.

Clarita Formation: Skeletal-Buildup Reservoirs

Porosity within the Clarita buildup facies ranges from 0 to 15 percent and is mainly biomoldic and solution-enhanced. Areas of thickest porosity correspond to thickest skeletal accumulations and most pervasive dolomitization (Fig. 6-5). The trap is formed by a combination of updip loss of porosity (associated with a facies change from porous, dolomitized crinoid-dominated packstones and wackestones to non-porous arthropod packstones and wackestones of the "shallow" shelf facies) and an overlying seal provided by the non-porous "deep" ramp facies of the overlying Henryhouse Formation (Fig. 6-4).

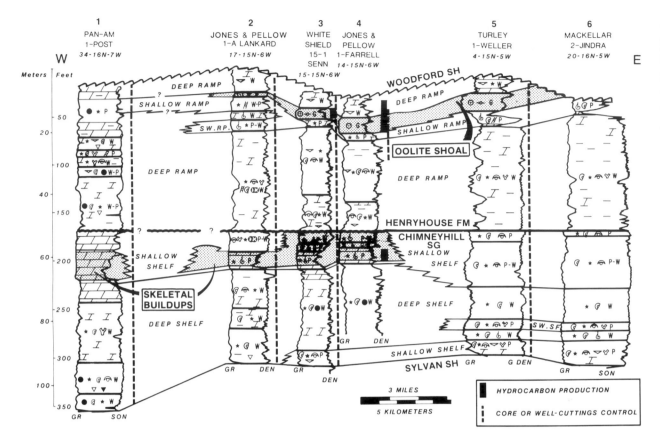

Fig. 6-4. East-west stratigraphic cross-section of the Mt. Everette and Southwest Reeding fields. Datum is a gamma-ray peak near the base of the Henryhouse Formation. Refer to Figure 6-3 for cross-section location.

Geophysical well-log abbreviations: DEN = bulk density, GR = gamma ray, G = DEN gamma density, SON = sonic. Porosity increases to the left. Gamma radiation increases to the right.

Henryhouse Formation: Oolite Shoal Reservoir

The oolite shoal facies is readily identified on gamma logs by its relatively low gamma response, presumably reflecting the high-energy nature of this facies and the lack of terrigenous mud (Fig. 6-4). In this facies, solution-enlarged porosity, mostly inter-ooidal, is most common and ranges from 2 to 20 percent. Local stratigraphic thinning of the shoal or local truncation of the shoal, combined with the draping effect of the overlying Woodford Shale, appears to form the trap and seal at Mt. Everette and Southwest Reeding fields. Local faults are present, but structural closure cannot be demonstrated. East of the fields, erosion of the shoal has resulted in its gradual thinning to a zero edge. However, this truncated margin is not productive, probably because of the gradual erosional thinning of the oolite facies coupled with the presence of porosity in the underlying skeletal packstones, as exemplified by the Turley No. 1 Weller well (Fig. 6-4). Where this combination exists, there is no basal seal for the oolite reservoir, and hydrocarbons have probably migrated updip via the porous packstones.

Henryhouse Formation: Unconformity Reservoir Beneath the Woodford Shale

Hydrocarbon production not associated with the oolite shoal facies is also found at the top of the Henryhouse Formation, directly below the Woodford Shale. Porosity at the unconformity is highly variable, ranging from 0 to 15 percent, and is dependent on the lithology of the subcropping

Fig. 6-5. Thickness map of porosity ≥4% in the Clarita Formation. Skeletal buildups and porosity are generally coincident; therefore, this map reflects the relative thickness of skeletal buildups. Refer to Figure 6-3 for well control.

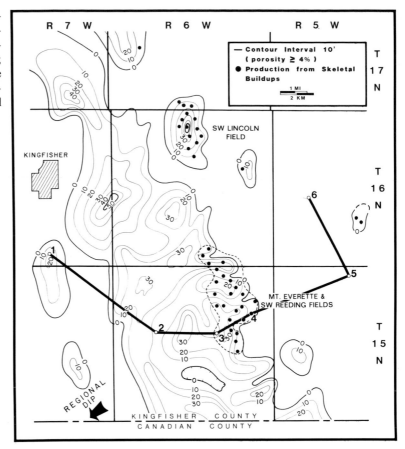

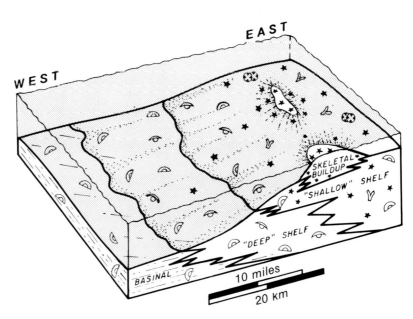

Fig. 6-6. Depositional model of the Clarita Formation. During the Late Silurian, crinoid-rich skeletal buildups accumulated near a shelf edge. Subaerial exposure resulted in local solution brecciation of some buildups, and probably was influential in their dolomitization. Vertical scale is exaggerated.

facies. Local relief on the unconformity surface and drape of the overlying Woodford Shale provide the trap.

Production from this type of reservoir tends to be marginal unless a skeletal-rich facies subcrops below the Woodford Shale, is abruptly truncated, and does not thin to a "feather edge." Dolomitized skeletal wackestones and packstones, exposed at the unconformity, contain better solution porosity and permeability through leaching of skeletal grains than do mudstones, especially where only the lime mud has been dolomitized. Unless there is sufficient relief on the unconformity surface, a good trap is unlikely to be present, and hydrocarbons will have migrated updip.

Hydrocarbon Reserves and Production

At Mt. Everette and Southwest Reeding fields, production from the Hunton is often commingled with that from other formations. However, some abandoned wells that produced solely from the Hunton Group may be used to estimate the *relative* production of the various Hunton Group reservoirs as follows:

Ultimate production from the Clarita skeletal-buildup reservoir ranges from 20,700 to more than 71,000 barrels of oil (42,500 average) per well. Production from the oolite shoal reservoir is more difficult to estimate, but one well, now abandoned, produced 7,000 barrels of oil; another well, which is near its economic limit, has produced 8,000 barrels of oil. Based on initial production data, reserves within the Henryhouse unconformity reservoir lying beneath the Woodford Shale are probably comparable to those of the oolite shoal reservoir.

Source Rocks

Both the overlying Woodford Shale and the underlying Sylvan Shale are potential source rocks in the Anadarko Basin, although the Woodford Shale is thought to be considerably superior (Oehler and Ho, 1982). The Woodford Shale is a black, carbonaceous, locally silty (quartz) shale that produces a very high gamma-log response (300++ API units)—a characteristic commonly associated with a high organic content (e.g., see Schmoker, 1978). Woodford Shale samples from four wells within 34 miles (54 km) of the study area yielded a range of total organic carbon content of 1.3 to 4.5 percent, whereas extractable C_{15+} hydrocarbons, a measure of source richness for oil, ranged between 764 and 2203 parts per million (Oehler and Ho, 1982). The Woodford Shale is considered thermally mature in that area.

The Sylvan Shale is a gray to green, slightly calcareous, locally fossiliferous shale, which produces a much lower gamma-log response (120 API units). Sylvan Shale samples from two wells within 47 miles (76 km) of the study area yielded relatively low total organic carbon contents of 0.26 and 1.02 percent, and extractable C_{15+} hydrocarbons of 212 ppm and 752 ppm, respectively (Oehler and Ho, 1982). The Sylvan Shale is thermally mature in this area.

Chimneyhill Subgroup: Depositional Facies and Shelf-to-Basin Profile

Within the study area, the Chimneyhill Subgroup comprises three major depositional facies— "deep" shelf, "shallow" shelf, and skeletal-buildup—in a continuous upward-shoaling sequence. "Deep" and "shallow" are employed here, as well as in the descriptions of Henryhouse Formation facies, in a relative sense only. A basinal facies, not represented within the study area, is present to the west at the top of the Chimneyhill. Thin, green calcareous shales, interbedded with argillaceous, silty (quartz) lime mudstones and skeletal wackestones, characterize that facies. Ostracodes and trilobite fragments are the main skeletal components.

Deep Shelf Facies

Skeletal wackestones compose most of this facies, and ostracodes and trilobite fragments are the dominant faunal constituents. Siliciclastic silt and clay are disseminated within the wackestones, but their quantity is much reduced compared with that in the basinal facies. Peloids are locally abundant. The deep shelf facies constitutes most of the lower Chimneyhill Subgroup in the western study area and thins to the east, shelfward (Fig. 6-4).

Shallow Shelf Facies

The shallow shelf facies is composed of higher-energy, more skeletal-rich rocks than those of the deep shelf. Ostracode-crinoid packstones are predominant. Trilobites and bryozoans are less abundant constituents, and tabulate corals are sparse.

Skeletal-Buildup Facies

Skeletal fossil-fragment buildups consisting mostly of crinoid packstones and less abundant grainstones are present in the Clarita Formation. Areas of thickest skeletal buildups are coincident with thickest porosity in the Clarita. Therefore, buildups can be delineated with a porosity-isopach map of the interval in which they are known to occur (Fig. 6-5). The buildups are arranged in a northwest-southeast band.

Shelf-to-Basin Profile

During deposition of the Clarita Formation, skeletal buildups accumulated near the basinward margin of a shallow shelf (Fig. 6-6). There is no evidence for a "drop off" shelf-to-basin profile. However, it is presumed that a low gradient for the shelf-slope transition nevertheless provided optimum conditions for crinoid growth and accumulation.

Henryhouse Formation: Depositional Facies and Shelf-to-Basin Profile

Deposition of the Henryhouse Formation commenced with a transgression over the study area, resulting in the onlap of deeper water facies over the skeletal buildups of the Clarita Formation. Evidence is presented below that, following the transgression, progradation of shallower water facies produced an upward-shoaling sequence consisting first of "deep" then "shallow" ramp, and oolite shoal facies. Basinward, to the west,

deeper water conditions persisted throughout Henryhouse deposition, and calcareous shales, silty (quartz) lime mudstones, and fine-grained skeletal wackestones accumulated.

Deep Ramp Facies

Rocks of this facies are mostly skeletal wackestones, but also include packstones. The main constituents are ostracodes, trilobites, brachiopods, sparse crinoid debris and, locally, peloids. Greater faunal diversity and abundance than that of the basinal facies attest to a more hospitable environment. Pentamerid brachiopods of the genus *Kirkidium* are locally abundant and impart a "plums-in-the-pudding" texture to the rock. Siliciclastics are less abundant than in the basinal facies.

Shallow Ramp Facies

The shallow and deep ramp facies are gradational. In general, the shallow ramp facies contains a more diverse fauna, and mixed skeletal packstones are prevalent, suggesting a higher-energy environment. The dominant fauna are crinoids and bryozoans. Tabulate corals and the blue-green alga *Girvanella* are locally found.

Oolite Shoal Facies

Capping the shallow ramp facies of the Henryhouse Formation is an oolite shoal complex, whose maximum thickness was at least 40 feet (12 m). The shoal trends NNW–SSE (Fig. 6-7), and consists mainly of ooid packstones and grainstones. Less abundant components include oncolites, peloids, and crinoid debris. The configuration of the seaward margin of the shoal suggests that tidal currents formed a NE–SW-trending channel in the western part of T16N, R6W, and that longshore drift reworked the ooids into a strike-oriented spit in the northwest corner of T15N, R6W (Fig. 6-7).

Shelf-to-Basin Profile

Facies contacts, both lateral and vertical, within the Henryhouse Formation are commonly gradational and hence difficult to define. Regional mapping outside of the study area indicates that Hen-

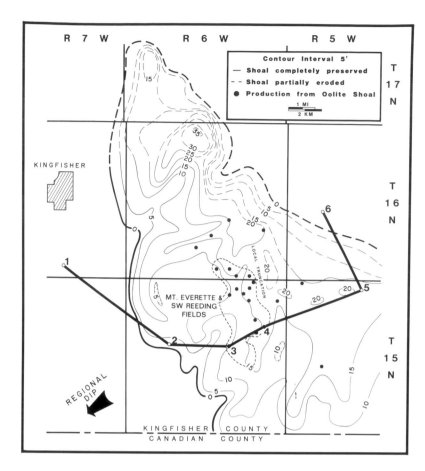

Fig. 6-7. Thickness map of the oolite-shoal complex in the Henryhouse Formation. Note the probable tidal channel in the west half of T16N, R6W. A strike-oriented spit, presumably formed by longshore drift, is present in the northwest corner of T15N, R6W. Refer to Figure 6-3 for well control.

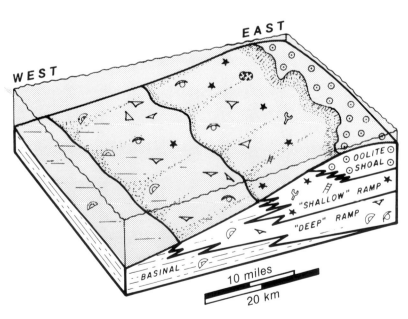

Fig. 6-8. Depositional model of the Henryhouse Formation during sedimentation of the oolite-shoal complex. Facies belts were broad, and the depositional slope probably resembled a ramp with gentle westward dip. Vertical scale is exaggerated.

ryhouse facies belts were broad and extended over several miles as a gently sloping ramp (Fig. 6-8). Locally, however, topographic irregularities may have influenced sedimentation. For example, the basinward margin of the Henryhouse oolite shoal is nearly coincident with the basinward extent of Clarita skeletal buildups, suggesting that any topographic relief may have persisted upward through Henryhouse deposition and affected the basinward extent of the oolite shoal. The shoal was eroded east of the study area following Henryhouse deposition; hence, its depositional extent cannot be completely determined.

Depositional and Diagenetic History of Hunton Group

Chimneyhill Subgroup

Facies throughout the Chimneyhill Subgroup contain a relatively diverse fauna suggestive of well-oxygenated, normal-marine waters. As Chimneyhill sedimentation progressed, the study area gradually shallowed, and during Clarita deposition, crinoid-dominated skeletal buildups accumulated, possibly in response to a stillstand of sea level.

Adjacent to the buildups, sedimentation appears to have been relatively continuous. Skeletal packstones of the Clarita Formation grade into arthropod-rich silty (quartz) wackestones of the overlying Henryhouse Formation. Basinward (west) of the study area, silty (quartz) mudstones, ostracode wackestones, and thin shales in both the upper Chimneyhill Subgroup and lower Henryhouse Formation suggest uninterrupted sedimentation. Growth of the buildups, however, was punctuated by at least one period of subaerial exposure late in their depositional history. This resulted in local solution-induced fracturing and incipient brecciation (Fig. 6-9A).

Most skeletal buildups are partially dolomitized. Dolomite is prevalent in the matrix of skeletal wackestones and packstones, suggesting that it is a replacement of lime mud. Away from the buildups and across facies boundaries, the intensity of dolomitization progressively decreases, so that beyond a mile (1.6 km) of the buildups, only scattered rhombs of dolomite occur. The faunas

in both the dolostone and the limestone are the same and indicate a subtidal, normal-marine origin. This suggests that dolomitization was a post-depositional, diagenetic event.

A model that seems consistent with both the extent of dolomitization and its relationship to depositional facies within the Chimneyhill Subgroup involves the mixing of meteoric water and sea water in the subsurface, as proposed in general by several workers (e.g., Hanshaw et al., 1971; Badiozamani, 1973; Folk and Land, 1975; Land et al., 1975). Dolomitization of existing sediments can occur in a mixing zone formed at the "contact" of a freshwater lens with normal sea water. In order for this model to operate, an area of freshwater recharge must exist so that meteoric waters can percolate into the subsurface and form a freshwater lens. Subaerial exposure of the Clarita skeletal buildups would have provided such recharge areas, creating subjacent mixing zones where sediments originally deposited in a normal-marine, subtidal environment could have been dolomitized. Reduced residence time and vigor of the seaward-migrating mixing zones would result in gradually reduced dolomitization outward from the skeletal buildups, consistent with the observed decrease in dolomitization away from the skeletal buildups of the Clarita Formation.

Porosity within the buildup facies is associated with dolomitization and is mostly biomoldic and solution-enhanced (Fig. 6-9B,C).

Henryhouse Formation

Following the post-Clarita transgression mentioned previously, a gradual shallowing occurred with the deposition of an upward-shoaling sequence. An oolite shoal complex closed out this sequence and was at least 40 feet (12 m) thick over portions of the study area. Onlapping by a deep ramp facies followed.

Petrographic evidence for early compaction of the oolite shoal in the study area is locally present in the form of interpenetrating ooids and ooids with spalled cortices. Primary porosity within the shoal was occluded by granular and coarse, blocky calcite. Existing porosity is secondary, mostly inter-ooid, and related to solution (Fig. 6-9D,E). Biomoldic and solution-enhanced poro-

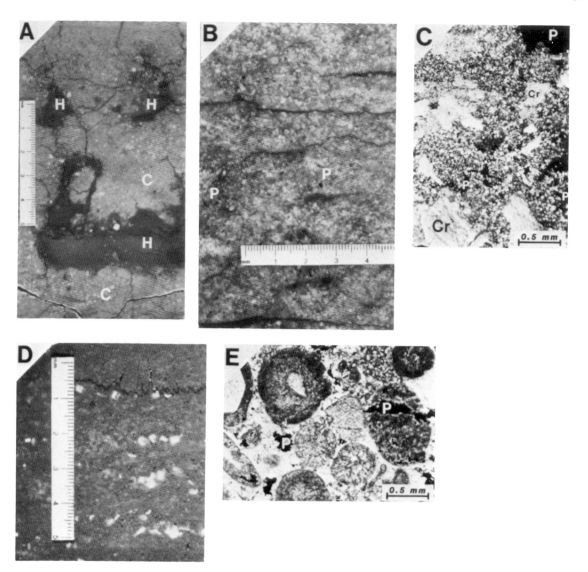

Fig. 6-9. Photographs of depositional and diagenetic features of the Clarita Formation and Henryhouse Formation. All photos oriented with stratigraphic "up" toward top of page. *A*. Incipient solution breccia in the Clarita Formation. C = Clarita crinoid packstone. H = Henryhouse silty lime mudstone filling solution cavities. Polished slab from the Jones and Pellow No. 1 Reherman well (Sec. 25, T15N, R6W), 7881.5 feet (2403 m). Scale divided into mm. *B*. Clarita Formation skeletal-buildup facies, consisting mostly of crinoid packstones. P = solution porosity. Polished slab from the Jones and Pellow No. 1 Reherman well, 7898 feet (2408 m). Scale divided into mm. *C*. Photomicrograph of Clarita Formation skeletal-buildup facies. Crinoid fragments (Cr) in dolomitized lime mudstone matrix. P = solution porosity. Location and depth same as in *B*. Cross-polarized light. *D*. Henryhouse Formation oolite-shoal facies. An oolite grainstone with scattered peloids and crinoid fragments. Polished slab from the Jones and Pellow No. 1 Farrell well (Sec. 14, T15N, R6W), 7778 feet (2371 m). Scale divided into mm. *E*. Photomicrograph of ooid grainstone exhibiting solution porosity (P), oolite-shoal facies, Henryhouse Formation. Location and depth same as in *D*. Cross-polarized light.

sity occurs locally at the top of the formation and is related to solution at the unconformity and to dolomitization.

Dolomite in the Henryhouse Formation is most pervasive at the top of the formation beneath the unconformity and is not facies-specific. Dolomite occurs below this zone, but only as scattered rhombs.

Summary

Two stratigraphic units in the Hunton Group contain reservoirs in Mt. Everette and Southwest Reeding fields: the Chimneyhill Subgroup (Early-Late Silurian), and the Henryhouse Formation (Late Silurian). The Chimneyhill Subgroup is an upward-shoaling sequence, at the top of which are hydrocarbon-bearing, crinoid-rich, skeletal-build-up reservoirs of the Clarita Formation. These buildups contain biomoldic and other solution porosity created during subaerial exposure and dolomitization prior to Henryhouse sedimentation. Away from the buildups, time-equivalent, non-porous arthropod wackestones provide updip barriers to hydrocarbon migration. Transgressive argillaceous lime mudstones and wackestones of the Henryhouse Formation provide the overlying seal for the Clarita skeletal-buildup reservoir.

During Henryhouse deposition, progradation of shallow-shelf carbonate sediments produced a second upward-shoaling sequence that ended with the deposition of an oolite shoal complex. Inter-ooid solution porosity in the shoal grainstones, created during subaerial exposure prior to Woodford Shale sedimentation (Late Devonian-Early Mississippian), provides a second reservoir. Minor depositional thickness variations and local truncation, combined with drape of the Woodford Shale, provide the trap. A third reservoir locally occurs at the top of the Henryhouse and is related to dolomitization and leaching (especially of skeletal-rich facies) at the unconformity. Local relief on the unconformity and drape of the Woodford Shale form the trap.

Acknowledgments The author expresses his gratitude to Conoco Inc. for granting permission to publish, and for providing drafting and secretarial services. B. A. MacPherson, Peggy J. Rice, and K. Swirydczuk reviewed the manuscript and offered many helpful suggestions. Stimulating discussions with T. W. Amsden have helped to clarify my thoughts on various aspects of Hunton Group geology. D. Z. Oehler and T.T.Y.Ho provided source-rock information. John Warren provided core analyses for the Jones and Pellow wells.

References

AMSDEN, T.W., 1957, Introduction to stratigraphy, pt. 1 of Stratigraphy and paleontology of the Hunton Group in the Arbuckle Mountain region: Oklahoma Geological Surv., Circ. 44.

AMSDEN, T.W., 1960, Hunton stratigraphy, pt. 6 of Stratigraphy and paleontology of the Hunton Group in the Arbuckle Mountain region: Oklahoma Geol. Surv., Bull. 84.

AMSDEN, T.W., 1975, Hunton Group (Late Ordovician, Silurian, and Early Devonian) in the Anadarko Basin of Oklahoma: Oklahoma Geol. Surv., Bull. 121.

AMSDEN, T.W., and T.L. ROWLAND, 1967, Geologic maps and stratigraphic cross sections of Silurian and Lower Devonian Formations in Oklahoma: Oklahoma Geol. Surv., Map GM-14.

AMSDEN, T.W. and T.L. Rowland, 1971, Silurian and Lower Devonian (Hunton) oil- and gas-producing formations: Amer. Assoc. Petroleum Geologists Bull., v. 55, p. 104–109.

BADIOZAMANI, K., 1973, The *dorag* dolomitization model—application to the Middle Ordovician of Wisconsin: Jour. Sedimentary Petrology, v. 43, p. 965–984.

FOLK, R.W., and L.S. LAND, 1975, Mg/Ca ratio and salinity: two controls over crystallization of dolomite: Amer. Assoc. Petroleum Geologists Bull., v. 59, p. 60–68.

HAM, W.E., 1969, Regional geology of the Arbuckle Mountains, Oklahoma: *in* Ham, W.E., ed., Regional geology of the Arbuckle Mountains, Oklahoma: Oklahoma Geol. Surv., Guidebook 17.

HAM, W.E., R.H. DENNISON, and C.H. MERRITT, 1964, Basement rocks and structural evolution of southern Oklahoma: Oklahoma Geol. Surv., Bull. 95.

HANSHAW, B.B., W. BACK, and R.G. DEIKE, 1971, A geochemical hypothesis for dolomitization by groundwater: Econ. Geology, v. 66, p. 710–724.

HOFFMAN, P., J.F. DEWEY, and K. BURKE, 1974, Aulacogens and their genetic relation to geosynclines, with a Proterozoic example from Great Slave Lake, Canada: *in* Dott, R.H., Jr. and R.H. Shaver, eds., Modern and Ancient Geosynclinal

Sedimentation: Soc. Econ. Paleontologists and Mineralogists, Spec. Pub. 19, p. 38–55.

LAND, L.S., M.R.I. SALEM, and D.W. MORROW, 1975, Paleohydrology of ancient dolomites: geochemical evidence: Amer. Assoc. Petroleum Geologists Bull., v. 59, p. 1602–1625.

MAXWELL, R.W., 1959, Post-Hunton pre-Woodford unconformity in southern Oklahoma: *in* Petroleum geology of southern Oklahoma, v. 2—a symposium by the Ardmore Geol. Soc.: Amer. Assoc. Petroleum Geologists, p. 101–125.

OEHLER, D.Z., and T.T.Y. HO, 1982, Source potential of the Woodford and Sylvan Shales in the Anadarko Basin: Conoco Exploration Research Report No. 660-10-1-1-82.

PRUATT, M.A., 1975, The southern Oklahoma aulacogen: A geophysical and geological investigation: M.S. thesis, Univ. Oklahoma, Norman, 59 p.

SCHMOKER, J.W., 1978, The relationship between density and gamma ray intensity in the Devonian shale sequence, Lincoln County, West Virginia: *in* Proceedings first eastern gas shales symposium, Morgantown, WV 1977: U. S. Dept. Energy, Morgantown Energy Technology Center, MERC/SP-77/5, p. 355–360.

7
Belle River Mills Gas Field
Dan Gill

RESERVOIR SUMMARY

Location & Geologic Setting	St. Clair Co., T4n, R16E, SE Michigan, Michigan Basin, USA
Tectonics	Stable cratonic basin
Regional Paleosetting	SE shelf of cratonic basin
Nature of Trap	Stratigraphic
Reservoir Rocks	
Age	Silurian (Niagaran)
Stratigraphic Units(s)	Guelph Formation
Lithology(s)	Dolomite
Dep. Environment(s)	"Pinnacle" reefs, persistent within shelf facies
Productive Facies	Algal stromatolite boundstone, stromatoporoid-coralgal boundstone, crinoid-bryozoan wackestone
Entrapping Facies	Anhydrite, salt beds, and dense, micrite carbonate, capping and laterally adjacent
Diagenesis Porosity	Primary porosity, occluded by marine cement; secondary porosity by karst-related dissolution, and *dorag* and reflux dolomitization, occluded by evaporite cements; reopened by fresh ground water.
Petrophysics	
Pore Type(s)	Dissolution vugs and larger cavities
Porosity	3–30%, 10% avg
Permeability	05–1000 md, 7.9 md avg
Fractures	Nontectonic, solution-collapse
Source Rocks	
Age	Late Silurian Salina Group (Cain Fm and possibly Ruff Fm)
Lithology(s)	NA
Migration Time	NA
Reservoir Dimensions	
Depth	2500 ft (762 m)
Thickness	431 ft (131 m)
Areal Dimensions	At base 10,500 × 3937 ft (3200 × 120 m). At top 8530 × 1312 ft (2600 × 400 m)
Productive Area	250 acres (1 km^2)
Fluid Data	
Saturations	NA
API Gravity	Gas
Gas-Oil Ratio	NA
Other	87% methane, 5% ethane, 2% propane, 2% butane, 3% nitrogen
Production Data	
Oil and/or Gas in Place	52.5 billion CFG (1.48 billion m^3); 8.75 million equivalent BO
Ultimate Recovery	NA (= cum. prod.)
Cumulative Production	21.4 billion CFG (0.61 billion m^3); 4.08 million equivalent BO; URE = 40.8%

Remarks: Converted in 1965 to an underground gas storage facility. Discovered 1961.

7
Depositional Facies of Middle Silurian (Niagaran) Pinnacle Reefs, Belle River Mills Gas Field, Michigan Basin, Southeastern Michigan

Dan Gill

Introduction

Niagaran reefs have been the main source of oil and gas production in the state of Michigan for the past three decades. In 1980, the reefs produced about 28 million barrels of oil (82 percent of the state's production) and almost 150 billion cubic feet of gas (94 percent of the state's production). At the end of 1980, cumulative production from the reefs reached 190 million barrels of oil and 1.12 trillion cubic feet (31.8 billion m^3) of gas (Bricker *et al.*, 1982). Estimates of the primary recoverable reserves contained in the reefs range between 300 to 400 million barrels of oil and 3 to 5 trillion cubic feet (85–140 billion m^3) of gas (Mantek, 1976; Caughlin *et al.*, 1976; Yelling and Tck, 1976). The main purpose of this study is to present a detailed analysis of the petrography and sedimentary facies of one of the largest and best core-sampled Niagaran pinnacle reefs in the Michigan Basin, and evaluate their spatial disposition.

Pinnacle reefs of the Michigan Basin are isolated carbonate buildups completely encased by salt, anhydrite, and fine-grained carbonate deposits. This juxtaposition between rocks of different densities and seismic velocities gives rise to lateral and vertical contrasts readily detectable by geophysical methods. Until the late 1960s, the gravimetric method was the principal successful tool used for pinnacle reef exploration (Dyer, 1956; Geyer, 1963; Pohly, 1968). In more recent years, the search for reefs has been based almost entirely on seismic methods, with exceptionally good results (Caughlin *et al.*, 1976; McClintock, 1977).

Exploration History

The first discovery of commercial production from Niagaran reef reservoirs in the Michigan Basin was made in southwestern Ontario, Canada, in 1889. Exploration for pinnacle reefs moved from southwestern Ontario into the adjacent areas of southeastern Michigan in the late 1920s, concentrating first in St. Clair County, where a gas-producing pinnacle reef was discovered in 1927 (Ells, 1963). A more intensive phase of exploration for pinnacle reefs along the southern margins of the basin began in 1950. Beginning in 1966, exploration spread also to the northern part of the lower peninsula. The first discoveries in the northern pinnacle reef belt were made in 1969 (Burgess and Benson, 1969). Since then, this area has been the main exploration ground in the state of Michigan in which new discoveries have been made at an impressive pace. At the end of 1982, approximately 1,100 individual reefs had been discovered in the Michigan Basin. Of these, about 900 reefs are productive, including 300 (190 oil, 110 gas) along the southern margin of the basin in Michigan and Ontario, and 600 (365 oil, 235 gas) along the northern margin.

This study is dedicated to the memory of Professor Louis I. Briggs, my teacher and friend, whose untimely death in May 1979 deprived us all of a prominent scholar and sedimentologist.

Paleogeography and Stratigraphy

The general paleogeographic setting of the Michigan Basin area during the growth of the reefs in Middle-Silurian (Niagaran) time is portrayed in Figure 7-1 (Briggs *et al.*, 1980). Three principal contemporaneous carbonate depositional belts can be differentiated: (1) the (outer) carbonate platform, (2) the platform-to-basin (transitional) slope, and (3) the basin. Parts of the platform sediments are presently exposed in various places all around the Michigan Basin. The other two zones are confined to the subsurface. Along the basinward edge of the platform, organic reefs coalesced to form an almost continuous bank, or barrier reef. During the Niagaran, the interior of the basin was a relatively deep-water environment that accumulated only a thin column ($<$ 100 feet, $<$ 30 m) of fine carbonate debris derived from the surrounding shallower, higher-energy belts.

The pinnacle reefs occupy an open-shelf facies zone (Wilson, 1975) intermediate in position between the fore-slope of the platform and the basin. This zone extends as a circular belt all around the basin (Fig. 7-1). It is 15 to 20 miles (24–32 km) wide in the north and up to 50 miles (80 km) wide in the south. This narrow ring is densely populated with mostly very small reefs. The area of the base of reefs ranges from 30 to 850 acres or 0.12–3.4 square kilometers (avg 80 acres, 0.32 km^2). The height of the reefs increases in the basinward direction from 300 feet to about 430 feet (90–130 m) in the southeastern part of the basin (Brigham, 1971), and from 300 to 600 feet (90–180 m) along the northern reef trend (Mantek, 1976). Reef flank slopes are 30 to 45 degrees, and the reefs are commonly referred to as *pinnacle* reefs. Strictly speaking, this term is appropriate only for the smallest reefs, which are also very tall. Across the northern part of the belt, reservoir fluids are systematically partitioned in an updip direction into bands of gas, oil, and water, in full agreement with Gussow's principle of differential entrapment (Gussow, 1954; Gill, 1979).

The stratigraphic subdivision and relationships of the reef rocks and their enclosing sediments are depicted in Figure 7-2. This interpretation departs in certain respects from previous interpretations by the author. Two rock units of group rank are involved: the Middle Silurian reef-bearing Niagara Group and the Upper Silurian reef-enclosing Salina Group. An erosional unconformity separates the two groups over reef crests, whereas the off-reef sequence is continuous and conformable. The crinoidal wackestone of the Lockport Formation provides a uniform substrate, sloping basinward about 0.5 to 1.5 degrees, underlying both reef and off-reef areas. Above the Lockport, two different depositional settings can be distinguished.

The reef buildups are designated as the Guelph Formation. Basinal and inter-reef Guelph equivalents are very thin and hardly discernible from the Lockport. The reefs evolved in two growth stages, an initial bioherm followed by an organic reef phase. Reefs of the Niagara Group are separated from the Salina Group by a subaerial erosional unconformity marked by a caliche hozizon. The caliche is overlain by an algal stromatolite unit that is up to 75 feet (23 m) thick. On stratigraphic and petrographic grounds, it is almost certainly correlative with the Maumee Algal Stromatolite, which caps the Maumee reef in northern Ohio (Kahle, 1974, 1978). This unit is usually correlated with the Ruff Formation (A-1 carbonate; see Budros and Briggs, 1977; Nurmi, 1974, 1975; Huh *et al.*, 1977). The contact between the stromatolite and overlying A-2 evaporite is considered disconformable. From a reservoir engineering standpoint, the Guelph reefs and their overlying Salina algal caps constitute one contiguous mass of carbonate reservoir rocks.

The lower Salina Group below the Ruff Formation contains two additional units in interpinnacle and basinal areas that onlap and wedge out against the lower slopes of the reef masses (Fig. 7-2). In stratigraphic sequence, they include the Cain Formation (Briggs *et al.*, 1980), which here also includes the A-0 carbonate, and the A-1 evaporite. In the vicinity of reefs and along the basin margins, the A-1 evaporite consists of anhydrite. Between reefs and basinward of the basin-margin anhydrite, the A-1 evaporite consists of halite, with a sylvinite-potash facies in the center of the basin (Matthews, 1970; Mesolella *et al.*, 1974; Briggs *et al.*, 1980). In the vicinity of the reefs, the Cain is underlain by a conglomerate composed of caliche pebbles that can be traced to the caliche horizon atop the reefs. This is conclusive evidence for the post-organic reef and post-exposure age of the Cain Formation and the other off-reef Salina units.

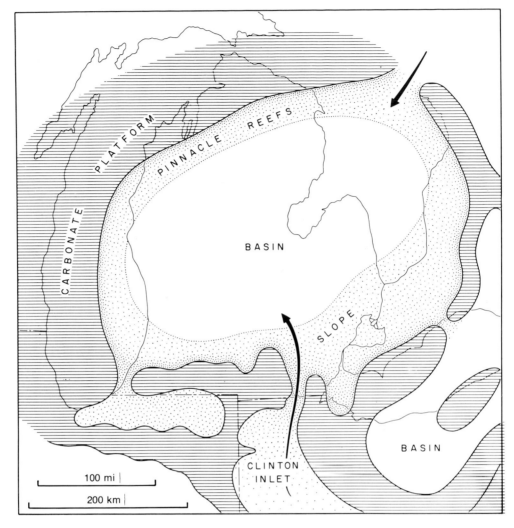

Fig. 7-1. Michigan Basin Middle Silurian paleographic map illustrating main carbonate depositional environments. From Briggs *et al.*, 1980. Published by permission, Elsevier Scientific Publishing Company.

The Belle River Mills Reef Gas Field

A petrographic analysis has been conducted on cores from the Belle River Mills (hereafter called BRM) reef, located in St. Clair County, southeastern Michigan (Fig. 7-3). The BRM gas field was discovered in 1961. In 1965, it was converted to an underground gas storage facility. Until its conversion, the field produced 21.4 of its estimated original 52.5 billion cubic feet (1.48 billion m³) of gas (Lundy, 1968).

The BRM is one of the largest Michigan reefs. It is ellipitical in shape and elongated in a northeasterly direction. Its base is 10,600 feet (3200 m) long and 2935 feet (885 m) wide,

covering an area of 700 acres (2.83 km²) (Fig. 7-4). The corresponding dimensions of the crestal area are 8550 feet (2580 m) by 1325 feet (400 m), covering an area of 250 acres (1 km²). At its tallest point, the top of the Maumee Algal Stromatolite is 431 feet (130 m) above the base of the reef. The reef's flanks are very steep, forming a 330-foot (100-m) escarpment with slopes of 30 to 40 degrees. Fifty-four wells were drilled in the field, 31 of which were cored. The data presented here are based on the examination of about 4000 feet of core from 14 reef wells.

Facies Analysis

Guelph Formation reefs evolved in two growth stages: an initial biohermal stage followed by an

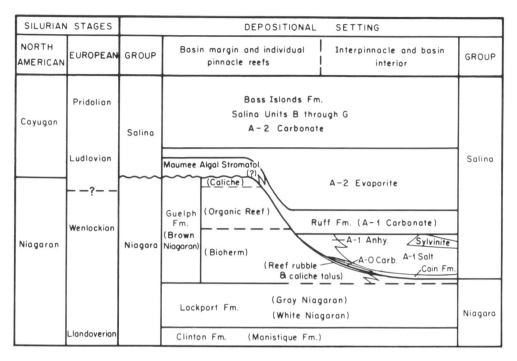

Fig. 7-2. Stratigraphic relations and nomenclature of reefs and enclosing strata, Middle and Upper Silurian Formations, in subsurface of Michigan Basin.

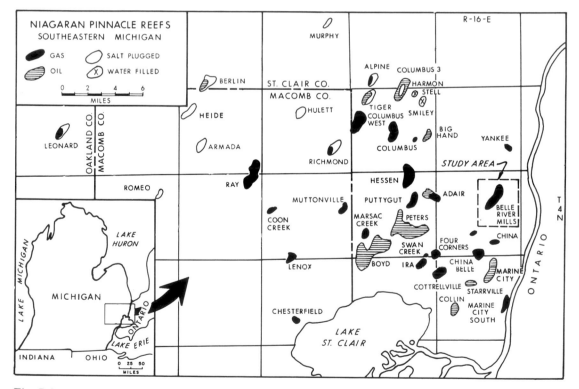

Fig. 7-3. Distribution of Niagaran pinnacle reefs in southeastern Michigan; location of study area.

Fig. 7-4. Belle River Mills reef gas field, isopach map of total carbonate buildup (Guelph Formation and Maumee Algal Stromatolite). Contour interval 50 feet. Line A-A′ marks location of cross section shown in Figure 7-5.

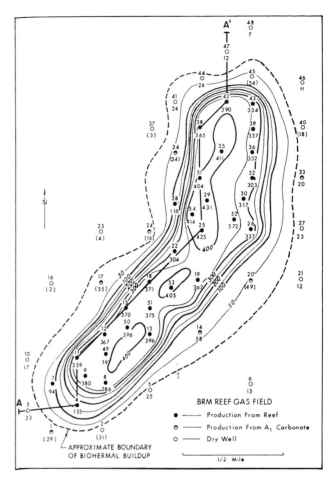

organic reef stage. After a period of subaerial exposure, the Maumee Algal Stromatolite was deposited over the reef crests. Each of these events is recorded in a characteristic suite of carbonate biofacies, here discussed in ascending stratigraphic order. The distribution of the various lithofacies and biofacies is portrayed in a composite cross-section that runs parallel to the long axis of the reef (Figs. 7-4, 7-5). Both the Guelph reef and the Maumee Algal Stromatolite presently consist entirely of dolomite.

Biohermal Stage Facies

Three principal facies have been recognized in the biohermal stage. Stratigraphically, these are skeletal biomicrite, biohermal core, and skeletal-lithoclastic packstone facies.

1. Skeletal Biomicrite

This facies consists of crinoids and bryozoans, which are embedded in non-bedded or massive lime mud matrix (Fig. 7-6A). Variations in rock type depend on the relative abundance and size of skeletal grains in relation to carbonate mud. Size-sorting is variable and abrupt, with major fauna ranging in size up to 4 centimeters. *Favosites,* *Halysites* and tabular stromatoporoids occur sporadically. Other species identified include *Heliolites, Arachnophyllum, Alveolites, Syringopora, Tryplasma,* and *Coenites.*

2. Bioherm Core ("Knoll")

This facies consists of tabular stromatoporoid and branching bryozoa subfacies. In the first, individual platy colonies, each about 0.5 cen-

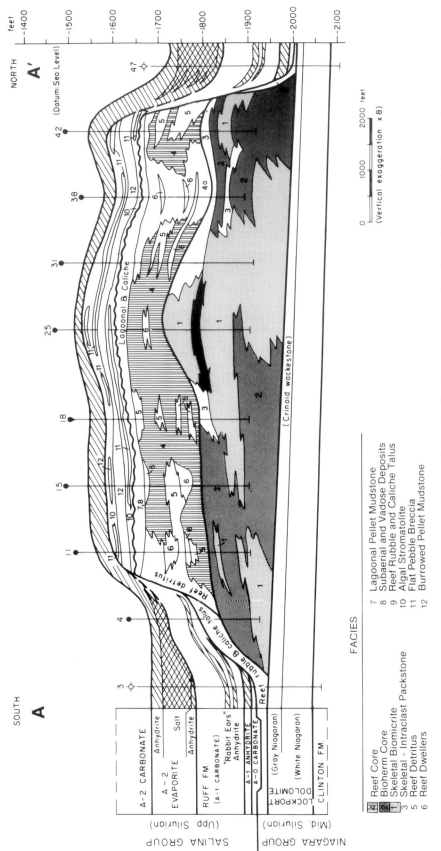

Fig. 7-5. Belle River Mills reef gas field, geologic cross-section and facies distribution pattern along line A-A' of Figure 7-4. Numbers 1 to 12 correspond to facies types discussed in text. Top of Lockport Dolomite dips with a northward component. Datum is sea level.

timeters thick, appear as planar or undulating white streaks, showing occasionally an internal stromatoporoidal organization of fine laminae with intervening vertical pillars (Fig. 7-6B). Large tabular and small rugose corals occur locally in subordinate amounts. This biohermal core may attain vertical growth up to 60 feet (18 m), with some interbedded intervals of facies 1. The second subfacies consists of tan skeletal packstones and wackestones almost exclusively composed of up to 40 percent branching bryozoans (Fig. 7-6C).

Bryozoa species are typically more robust than the delicate plumose bryozoans of facies 1, being characterized by a thicker wall structure and large size (up to 4 cm). Rugose corals and crinoid ossicles occur in subordinate amounts. The facies forms interfingering biostromal units 10 to 20 feet (3-6 m) thick with other rock types of the bioherm.

3. Skeletal-Intraclast Packstone

Skeletal fragments of all the fossil fauna mentioned above, together with lithoclasts, constitute 30 to 60 percent of the rock. In addition, whole and fragmental brachiopods are common (Fig. 7-6D). Crinoid and bryozoan fragments are the most abundant. The lithoclasts are irregular, subrounded to angular, and consist of very fine-grained, dense, brown-gray micritic dolomite, with a size range of 1 to 20 millimeters. Sediment sorting is poor, with many grains coated by thin, dark-brown micritic envelopes. A very fine mosaic of anhedral dolomite crystals fills intergranular pore space, together with some irregular minute patches of dolomitized mud. This facies forms massive beds 10 to 25 feet thick (60 feet in some places) (3–8 m, up to 18 m) and a few bedding planes inclined up to 20 degrees.

Facies Distribution and Environmental Interpretation

The biohermal facies occupies the lower half of the overall reef structure and reaches a maximum thickness of 245 feet (75 m). Its net areal growth configuration is shown in Figure 7-7. Facies 1 and 2 frequently interfinger (Fig. 7-5). The crinoid-bryozoan biomicrite is the most abundant facies in areal distribution and volume. It is estimated to constitute about 70 percent of the total biohermal mound.

Several textural and fabric properties, as well as paleoecologic considerations, attest to a low-energy origin for this facies. Location and extent of carbonate and noncarbonate mud indicates sluggish water movement. Both crinoids and bryozoans are delicate mud-dwelling organisms (Kissling, 1969; Duncan, 1957), and the facies is better-developed in the more protected parts of the bioherm and less widespread in its exposed outer flanks. Biohermal construction was facilitated by the sediment-trapping capacity of the intricate crinoid root system, dense population of bryozoans, and sporadic presence of frame-building corals and tabular stromatoporoids.

Additional constructional support was provided by facies 2, formed either by concentrations of tabular stromatoporoids with or without corals, or by the great abundance of more robust forms of branching bryozoans. The tabular stromatoporoids form local small knolls, which are vertically continuous for tens of feet and in other places spread laterally in a biostromal fashion. They occupy both flanks and central parts of the bioherm and played a significant role in shaping and maintaining the rigidity of the structure. The branching bryozoan facies suggests a slight change in environmental conditions. It is more prevalent in the upper parts of the mound and, where better-developed, is devoid of crinoids and delicate forms of bryozoans. These features and the presence of robust bryozoans probably indicate a gradual increase in water turbulence as the bioherm grew up toward wave base due to its growth above the surrounding sea floor. Greater water turbulence is strongly indicated also by the nature and distribution of the skeletal intraclast facies 3, the constituents of which are most probably rock fragments derived from the surrounding and underlying consolidated biohermal rocks that also sourced the greater part of its skeletal constituents. The scarcity of mud and argillaceous matter indicated by the light-tan color, plus the presence of high-angle bedding planes (20°), further attest to a relatively high-energy origin of this facies.

The most significant expressions of environmental change that resulted from the biohermal growth are those of the faunal assemblage and facies that mark the termination of the biohermal stage and the transition to a new stage in the developmental history of the structure. The main biohermal dwellers and builders, the crinoids and

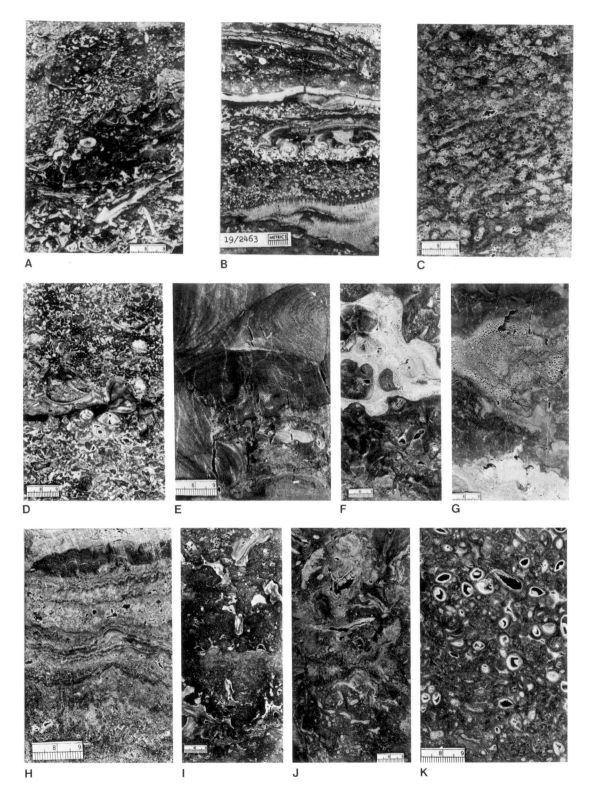

A   B   C

D   E   F   G

H   I   J   K

Fig. 7-7. Belle River Mills reef gas field, configuration of mound at termination of biohermal stage. Contour interval 50 feet.

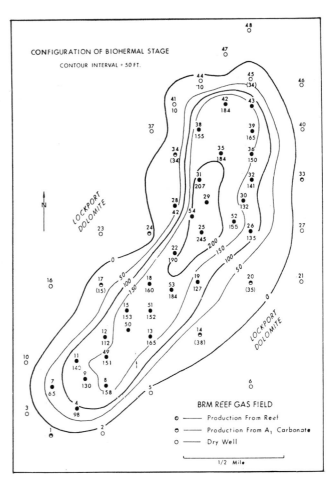

Fig. 7-6. Representative core-slab photographs of facies 1 to 6 (scale bar is 2 cm long). *A.* Facies 1: Crinoid-bryozoa wackestone. Elongated, delicate bryozoa that disintegrate into small rounded grains and coarse fragments in a mud matrix. *B.* Facies 2: Tabular stromatoporoids (upper part) and tabular colonial corals (lower part) in crinoid-bryozoa wackestone. *C.* Facies 2: Large, robust branching bryozoa, locally grain-supported. *D.* Facies 3: Skeletal-lithoclast packstone rudite. Grain-supported fragments of crinoids, bryozoa, brachiopods, and angular clasts of a micritic rock. *E.* Facies 4: Large massive stromatoporoid displaying remnants of original organic reticulated structure. *F.* Facies 4: Encrusting colonial coral (white) in association with encrusting stromatoporoids (dark, irregular patches) binding coarse stromatoporoid debris. *G.* Facies 4: Favositid coral (center), digitate stromatoporoid (top), and encrusting tabulate coral (bottom) in reef core facies. *H.* Facies 4: Algal boundstone (?) in reef core facies. Laterally linked hemispheroid stromatolite at top; clusters of closely spaced thin, undulating algal laminations binding micritic sediment at center. *I.* Facies 5: Coarse stromatoporoid fragments in carbonate mud matrix. *J.* Facies 5: Reef detritus consisting of fragments of tabulate and rugose corals, stromatoporoids, bryozoa, and crinoids. *K.* Facies 6: Brachiopod wackestone with subordinate amounts of fine bryozoa fragments.

bryozoans, were not adaptable to the new turbulent conditions. They were replaced by a new community of organisms, which built the succeeding organic reef stage.

Organic Reef Stage Facies

The main facies comprising the organic reef stage are the reef core, reef detritus, reef dwellers, and lagoonal facies. Each of these is represented by a number of rock types that can be grouped into characteristic facies.

4. Reef Core

The reef core facies consists of boundstones and wackestones in which three groups of frame-building organisms are present: stromatoporoids, corals, and suspected algae (Fig. 7-6E–H). Frame-builders in growth position comprise only 30 percent of the reef core. Stromatoporoids are the most abundant and important frame-builders, and in some places they are vertically continuous for several feet. Their growth, however, was more frequently interrupted by deposits of coarse skeletal wackestone. Recognition of stromatoporoids is difficult because of intensive recrystallization and dolomitization. Commonly, stromatoporoids appear as brown homogeneous masses in which the original reticulated organic structure is only seldom preserved (Fig. 7-6E). On the basis of gross morphological features, massive, digitate, and encrusting stromatoporoids can be recognized. Tabulate corals, predominantly *Favosites* sp., occur sporadically (Fig. 7-6F, G).

Certain portions of the reef core consist of a rock-type facies best described as a brown to tan, massive, mottled ("clotted") mudstone. It is occasionally associated with features resembling algal encrustations and laminations (Fig. 7-6H). Minute thread-like cellular aggregates and microfilaments from this rock type have been interpreted as an algal structure and tentatively identified as cf. *Coactilum* sp. Maslov (R. Rezak and B. Buchbinder, written communication, 1970). Without direct and conclusive proof, however, this facies is interpreted as an algal boundstone (Gill, 1977a).

5. Reef Detritus

The reef detritus facies consists of coarse skeletal wackestone in which broken stromatoporoids are the main fragments (Fig. 7-6I). Fragments are angular and poorly sorted, ranging from a few millimeters up to about 10 centimeters. Large fragments of tabular corals are fairly common, with minor quantities of smaller rugose coral fragments, bryozoans, and lithoclasts (Fig. 7-6J). Coarse skeletal grains commonly constitute about 30 percent of the wackestone, with the remainder consisting of sand to silt-size skeletal debris embedded in a mud matrix. The reef detritus comprises a substantial part of what has been mapped as the reef core facies (Fig. 7-5). In this association, fragments of bryozoa and rugose corals are fairly abundant. In addition, the reef detritus forms mappable rock units that occur as small patches of limited vertical and lateral extent occurring between reef-core zones or as fore-reef deposits, which form an apron around the reef-core facies.

6. Reef Dwellers

The reef dwellers facies is a fine-grained skeletal wackestone that consists of fragments of organisms that apparently formed an intra-reef community. Particle composition is principally crinoids, bryozoans, solitary rugose corals, small colonial tabulate corals, and brachiopods. Fragments of the reef-building organisms are conspicuously absent. Very fine-grained lithoclasts are locally abundant. Included in this facies are several more specific rock types that differ in the dominant type and proportion of their skeletal constituents. These include bryozoa-coral wackestone, fine rudite, skeletal-lithoclast arenite, and brachiopod packstone and wackestone in which whole and broken brachiopods are sometimes the only grains present (Fig. 7-6K). These rock types are commonly interbedded, and are confined to intra-reef areas, except for the bryozoa-coral wackestone, which also occurs in a fore-reef position as interbedded thin tongues.

7. Lagoonal Pelletal Mudstone

The lagoonal facies consists of pelletal mudstones either massively bedded or finely laminated (Fig. 7-8A). Well-rounded and elliptical pellets constitute up to about 20 percent of the rock. Most pellets are less than one millimeter in diameter. The only fossils present are scarce, small shells (ostracods). Some lagoonal deposits are riddled with small pores that are concentrated along laminae. This fabric closely resembles the "birdseye" or fenestral fabric of many authors

(Choquette and Pray, 1970). It is often associated with thin, dark-brown seams and suites of closely spaced, stromatolite-like laminations.

The massive zones contain irregularly shaped, dark-brown patches of micrite. This mottled fabric is probably the result of bioturbation by burrowing (Winder, 1968). Well-defined burrows are also present in the laminated mudstone.

Facies Distribution and Environmental Interpretation

The cumulative thickness of the lithofacies of the organic reef stage ranges between 75 and 200 feet (23–61 m), adding an average height of about 150 feet (46 m) to the reef. At the termination of this growth stage, the buildup formed a reef with a rather flat crestal plateau and steep (30–45-degree) slopes, which rose to between 300 to 350 feet (90–105 m) above the surrounding sea floor (Fig. 7-9). The distribution and interrelationships of the various lithofacies is shown in Figure 7-5. The reef-core facies, occupying the bulk of the central part, is the most widespread lithofacies. Massive and encrusting stromatoporoids are the most important frame-building organisms that provided the wave resistance potential of the mound. Algal boundstone is well developed in the northern part of the reef (Fig. 7-5, facies 4a).

The rigid reef-core zones are interbedded and surrounded by the coarse skeletal, reef detritus facies (facies 5), which occupies a fore-reef position. Tongues of this facies emanate from the reef core area outward at various levels and coalesce to form the fore-reef fronts. The sites of active organic proliferation shifted with time, and the reef elevated itself upward over its own debris. The fine skeletal wackestone (facies 6), interpreted to be a "reef dwellers" facies, occurs as lenses of limited vertical and lateral extent. Most probably, it represents *in situ* accumulation of fossil debris from an assemblage of organisms that inhabited protected pools formed within and between more rigid reef-core prominences. In comparison with the robust reef-building fauna, the fine skeletal facies consists of rather delicate forms of crinoids, bryozoans, brachiopods, and small rugose corals that sought physical shelter from the turbulent waters characterizing the reef.

The laminated and massive pelletal mudstones of facies 7 are interpreted to represent a lagoonal environment. It is confined to the uppermost 20 or 30 feet of the organic reef section and probably represents the development of protected shallow pools throughout the crestal area of the reef. At this stage in its evolution, the reef formed an elevated flat plateau very close to sea level. The fine laminations and the predominance of mud indicate low-energy, quiet-water conditions. The fenestral fabric and algal laminations may indicate the intermittent existence of supratidal conditions. As in the case of the preceding biohermal stage, vertical reefal accretion exceeded the rate of regional subsidence. The reef gradually elevated itself into shallower water until it eventually reached the water surface and reef growth terminated. The cessation of reef growth is linked to a major change in the sedimentological regime brought about by the commencement of strong evaporitic conditions at the end of Niagaran time.

Subaerial Exposure Stage

8. Subaerial and Vadose Deposits

This facies is a complex of both sedimentary and diagenetic features that include vadose pisoliths, laminated and pisolitic caliche, breccias, and infilled veins.

Most pisoliths are between 1 and 5 millimeters in diameter and subspherical, elliptical, or irregularly shaped (Fig. 7-8B,E). Nuclei consist of micrite, silt, or fragments of older pisoliths. Well-developed pisoliths with several concentric layers are associated with monolayered ones and uncoated micritic grains. The pisoliths are often in contact and cemented with coarse sparry dolomite.

The laminated caliche consists of extremely fine, conformable laminations, commonly only 50 microns thick (Fig. 7-8C, left slab). The mode of lamination varies within short distances. Macroscopically, caliche is usually horizontal, forming dark-brown, continuous layers several centimeters thick. The association of laminated caliche and pisoliths results in quite spectacular and exceedingly complicated fabrics (*e.g.*, Fig. 7-8B,C,E). Interbedded are patches of particulate material, consisting of poorly sorted fragments of caliche pisoliths and micritic lithoclasts. The altered zone contains abundant sinuous and vertical cracks (Fig. 7-8C, right slab) infilled with

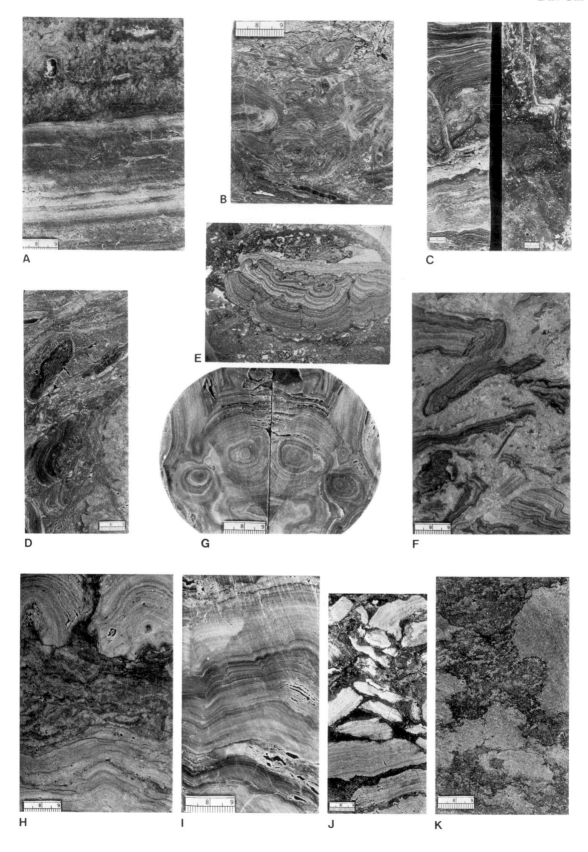

Fig. 7-9. Belle River Mills reef gas field, Guelph Formation isopach map displaying reef configuration at termination of organic reef stage.

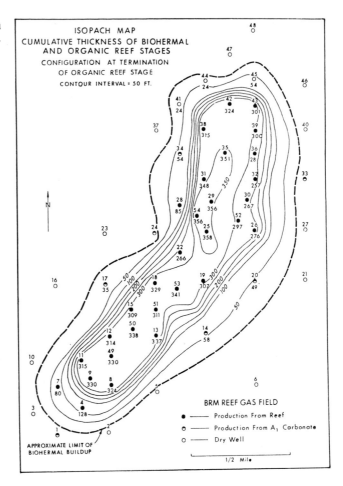

Fig. 7-8. Representative core-slab photographs of facies 7 to 12 (scale bar is 2 cm long). *A.* Facies 7: Laminated mudstone, occasionally pelletal, with birdseye fabric, overlain by massive, mottled and burrowed pelletal mudstone. *B.* Facies 8: Pisolitic caliche and fragments of pisolitic crust. *C.* Facies 8: Left slab—finely laminated caliche with enterolithic fabric (lower part) and cross-cutting infilled fracture (upper part). Right slab—infilled crack across pelletal mudstone lagoonal facies at top of reef. *D.* Facies 8: Pisolitic caliche coating wall of open crack (lower part); fragments of pisolitic and laminated caliche form part of debris infilling crack. *E.* Facies 8: Pisolitic caliche, negative print of thin-section (thin-section is 3 cm long). *F.* Facies 9: Reef talus conglomerate, upper part, consisting of angular pebbles of laminated caliche. *G.* Facies 10: Algal stromatolite, laterally linked hemispheroidal LLH) structure as viewed in section parallel to the bedding. *H.* Facies 10: Undulating small LLH stromatolite at bottom; flat pebble conglomerate at center; two stromatolite hemispheres at top. *I.* Facies 10: Closely packed small-amplitude LLH stromatolite. *J.* Facies 11: Flat pebble breccia composed of algal stromatolite fragments. *K.* Facies 12: Burrowed mudstone. Light-colored patches are presumed to be infilled burrows.

brecciated caliche crusts and often veneered with laminated and pisolithic caliche (Fig. 7-8D).

9. Reef Rubble and Caliche Talus Conglomerate

Based on the composition of its pebbles, the conglomerate can be divided into a lower and upper part. In the lower part, the pebbles consist of dark-brown rock fragments, composed of a homogeneous mass of recrystallized micrite. The fragments are angular and poorly sorted, varying in size from sand to cobbles 8 centimeters long. It is interbedded with horizons of mottled mudstone, in places with scattered crinoid fragments. The upper part of the conglomerate is composed of debris of laminated caliche crusts of facies 8 (Fig. 7-8F). Particles range in size from sand to very large cobbles. Most are elongated, with angular edges. Some of the larger cobbles lie with their long axis at high angles to the horizontal.

Facies Distribution and Environmental Interpretation

The caliche and vadose deposits are best-developed in the uppermost 20 to 30 feet (6–9 m) of the organic reef section, where they replace rocks of the lagoonal facies (Figs. 7-2, 7-5). They result from dissolution and concomitant reprecipitation processes under subaerial and vadose weathering conditions. Analogues of Pleistocene to Holocene age can be cited from the Florida Keys (Multer and Hoffmeister, 1968; Multer, 1972) and indurated calcareous "nari" crusts in Israel (personal observations). However, the closest and most convincing time-equivalent analogues are found in the Maumee reef in northern Ohio, which is, in fact, an exposed Niagaran reef (Kahle, 1974, 1978; and personal observations).

The caliche deposits are a record of a subaerial exposure period that terminated the growth of reefs throughout the surrounding carbonate platforms of the Michigan Basin. It is a manifestation of the evaporative withdrawal of the sea to the central, deeper part of the basin and the initiation of evaporite deposition at the end of Niagaran (Wenlockian) time. This evaporative drawdown is expressed in a regional unconformity recognizable throughout the Michigan Basin (Gill, 1977b). The exposure of the reefs to the dissolution action of meteoric water resulted in the formation of an extensive karstic system within

which is developed the principal porosity of the pinnacle reefs.

The conglomerate of facies 9 accumulated as a talus apron around the lower part and base of the reef slope and wedges out within a mile of the reef. The lower part of the conglomerate represents reef-derived rubble that accumulated as a manifestation of the turbulent water conditions that prevailed during the organic reef growth stage. Thin interbeds of mottled mudstone represent extensively burrowed and reworked sediments deposited contemporaneously around the reef in relatively deep water.

Stratigraphically, the conglomerate is equivalent to the regional interreef "Brown Niagaran" unit of some authors (Ells, 1967). The caliche pebbles in the upper part of the conglomerate can be traced to the caliche deposits capping the reef (facies 8). This provides an important stratigraphic clue for the post-exposure age of the off-reef Salina section.

Maumee Algal Stromatolite Cap

10. Algal Stromatolite

Vertical sections of the stromatolites display a compact succession of extremely fine, undulating, even laminae, with a high degree of congruence (Fig. 7-8I). The most prevalent growth form is the "laterally linked hemispheroid" type LLH-C of Logan et al. (1964) (Fig. 7-8G,H). The basal diameter of individual hemispheroids is between 2 and 3 centimeters. Zones of almost flat laminations are quite common. Petrographically, stromatolites appear as a succession of couplets of micritic laminae, about 0.05 millimeters thick, separated by 0.1- to 1-millimeter-thick dolospar laminae, which commonly display a birdseye pore system occluded by coarse anhydrite crystals. Randomly scattered individual crystals or clusters of needle-like cleavage flakes of anhydrite are fairly common. Coarse rectangular anhydrite crystals, which in places coalesce to form almost continuous horizons, are particularly abundant in the upper part of the unit toward its contact with the overlying A-2 evaporite.

11. Flat Pebble Breccia

Brecciated fragments of the subjacent flat-laminated algal stromatolite comprise this facies. Most of the larger ones are pseudorectangular, with straight edges and sharp corners (Fig. 7-8J). The groundmass consists of sand and silt-sized lithoclasts of micrite and finer stromatolite debris. In places, pebbles are in mutual contact displaying a "pile of bricks" fabric.

12. Burrowed, Pelletal Mudstone

Massive mudstone is the predominant rock-type facies. It displays color mottling of tan patches that have irregular, diffuse boundaries in a darker brown groundmass. Larger mottles are elongated and have well-defined, smooth boundaries. The former could be due to bioturbation, whereas the latter probably represent infilled burrows (Fig. 7-8K). The burrowed mudstone is devoid of skeletal material. Several zones within this facies contain abundant small (2 mm or less) spherical micritic pellets, probably of fecal origin.

Facies Distribution and Environmental Interpretation

The algal stromatolite complex that caps the reef is the first sediment of the Salina Group to be deposited over the reef crest following its exposure in late Niagaran time. It overlies the caliche horizon and underlies the A-2 evaporate (Fig. 7-2). It is probably the equivalent of the Maumee Algal Stromatolite, northern Ohio, and of part of the off-reef Ruff Formation (A-1 carbonate). The complex adds about 60 feet (18 m) of carbonate reservoir rocks to the Niagaran reef proper. In contrast to the massive construction and frequent lateral lithofacies changes characterizing the underlying reef mass, the stromatolite exhibits a well-defined, laterally persistent stratification. From bottom up, the complex is clearly divisible into a lower stromatolite unit, a middle burrowed mudstone unit, and an upper stromatolite unit (Fig. 7-5). The vertical continuity of the algal stromatolite beds (lithofacies 10) is often interrupted by lenses of flat-pebble breccia of lithofacies 11. In places, sharp cut-and-fill unconformable relations between these two lithofacies can be observed. The entire unit indicates the prevalence of peritidal conditions over the reef crest during Ruff Formation time. This interpretation is based on the close similarity of the described facies to other peritidal or tidal-flat deposits, both Recent and ancient (see extended discussion in Roehl, 1967; Kahle and Floyd, 1971; Bathurst, 1971; Walter, 1976; and Gill, 1977b). The algal

stromatolites are indicative of an intertidal to supratidal setting. The middle burrowed mudstone unit represents a short phase of submergence to the shallow subtidal zone. Excessive salinity is indicated by the absence of other marine organisms. The birdseye fabric and anhydrite denote periodic exposure and subjection to supratidal, sabkha-type diagenetic processes. The flat-pebble breccia is a weathering product of the algal stromatolite formed by brecciation and short-distance transportation. Thus, each breccia horizon marks a minor disconformity, indicating that the Ruff Formation reef tops experienced several short episodes of exposure during deposition.

Summary

The Niagaran pinnacle reefs of the Michigan Basin grew inside a narrow, circular, open-shelf facies belt, transitional between an outer stable shallow-water carbonate platform and the relatively deeper water of the basin interior. The rate of vertical reef growth exceeded the rate of subsidence, resulting in progressive shallowing with time. The continuous upward shallowing is reflected in the lithologic, biologic, and overall sedimentologic aspects of the rocks. Twelve different depositional facies are recognized within the reef buildup. These facies furnish a record of a four-stage evolution, entailing: (1) initial biohermal mound, constructed by carbonate mud-trapping crinoids, bryozoans, and corals. Upon reaching wave base, conditions became more favorable for (2) the formation of a genuine wave-resistant organic reef. The principal reef-building organisms were stromatoporoids, corals, and possibly algae. The reef grew upward into surface waters, at which point reef growth was terminated. The termination of reef growth was also linked to the commencement of intensive evaporitic conditions in early Salina Group time (early Upper Silurian). (3) The evaporative lowering of sea level within the basin exposed the reefs to subaerial diagenesis, weathering, and erosional processes. Much of the dolomitization of the reefs can be attributed to this exposure. The reefs remained exposed during the deposition of the lower Salina Cain Formation and A-1 evaporite unit. The secondary porosity generated during this exposure significantly enhanced the reservoir properties of the reef. (4) The reefs were resubmerged under very shallow and highly saline water during Ruff Formation (A-1 carbonate) time. An algal stromatolite cap, containing several minor disconformities indicating frequent return to subaerial conditions, was deposited over the reefs. With the deposition of the A-2 evaporite, the reefs became completely encased in evaporite and dense, fine-grained carbonates, which provided a very effective seal for the entrapment and containment of hydrocarbons.

References

BATHURST, R.G.C., 1971, Carbonate Sediments and their Diagenesis: Developments in Sedimentology 12, Elsevier Scientific Publ. Co., Amsterdam, 620 p.

BRICKER, D.M. *et al.*, compilers, 1982, Michigan Oil and Gas Fields, 1980. Annual Statistical Summary 34: Mich. Geol. Survey, Lansing, Mich., 46 p.

BRIGGS, L.I., D. GILL, D.Z. BRIGGS, and R.D. ELMORE, 1980, Transition from open marine to evaporite deposition in the Silurian Michigan Basin: *in* Nissenbaum, A., ed., Hypersaline and evaporite environments, Elsevier Scientific Publ. Co., Amsterdam, p. 253–270.

BRIGHAM, R.J., 1971, Structural geology of southwestern Ontario and southeastern Michigan: Ontario Dept. Mines and Ministry of Northern Affairs, Petroleum Resources Sect., Paper 71-2, 110 p.

BUDROS, R., and L.I. BRIGGS, 1977, Depositional environment of Ruff Formation (upper Silurian) in southeastern Michigan, *in* reefs and evaporites—concepts and depositional models: Amer. Assoc. Petroleum Geologists, Studies in Geology No. 5, p. 53–71.

BURGESS, R.J., and A.L. BENSON, 1969, Exploration for Niagaran reefs in northern Michigan: Ontario Petroleum Inst. 8th Ann. Conf. Tech. Paper 1; also 1969-70: Oil and Gas Jour., v. 67, no. 51, p. 80–83; v. 67, no. 52, p. 180–188; v. 68, no. 1, p. 122–127.

CAUGHLIN, W.C., F.J. LUCIA, and N.L. MCIVER, 1976, The detection and development of Silurian reefs in northern Michigan: Geophysics, v. 41, p. 646–658.

CHOQUETTE, P.W., and L.C. PRAY, 1970, Geological nomenclature and classification of porosity in sedimentary carbonates: Amer. Assoc. Petroleum Geologists Bull., v. 54, p. 207–250.

DUNCAN, H., 1957, Bryozoa: *in* Treatise on Marine Ecology and Paleontology, v. 2, Paleoecology: Geol. Soc. America Mem. 67, p. 783–800.

DYER, W.B., 1956, Gravity prospecting in southwestern Ontario: Canadian Oil and Gas Industries, v. 9, no. 3, p. 43–44.

ELLS, G.D., 1963, Information on Niagaran Silurian oil and gas pools: Mich. Geol. Surv., 34 p.

ELLS, G.D., 1967, Michigan's oil and gas pools: Mich. Geol. Surv., Rept. Invest., No. 2, 49 p.

GEYER, R.A., 1963, How to locate reefs in Michigan via gravity work: World Oil, April 1963, p. 101–104.

GILL, D., 1977a, The Belle River Mills gas field; productive Niagaran reefs encased by sabkha deposits, Michigan basin: Michigan Basin Geol. Soc., Spec. Paper 2, 187 p.

GILL, D., 1977b, Salina A-1 sabkha cycles and the late Silurian paleogeography of the Michigan basin: Jour. Sedimentary Petrology, v. 47, p. 979–1017.

GILL, D., 1979, Differential entrapment of oil and gas in Niagaran pinnacle reef belt of northern Michigan: Amer. Assoc. Petroleum Geologists Bull., v. 63, no. 4, p. 608–620.

GUSSOW, W.C., 1954, Differential entrapment of oil and gas: a fundamental principle: Amer. Assoc. Petroleum Geologists Bull., v. 38, p. 816–853.

HUH, J.M., L.I. BRIGGS, and D. GILL, 1977, Depositional environments of pinnacle reefs, Niagara and Salina Groups, northern shelf, Michigan Basin: *in* Reefs and evaporites—concepts and depositional models: Amer. Assoc. Petroleum Geologists Studies in Geology 5, p. 1–21.

KAHLE, C.F., 1974, Nature and significance of Silurian rocks at Maumee quarry, Ohio, *in* Silurian reef-evaporite relationships: Michigan Basin Geol. Soc. Field Conf., p. 31–54.

KAHLE, C.F., 1978, Patch reef development and effects of repeated subaerial exposure in Silurian shelf carbonates, Maumee, Ohio: *in* Kesling, R.V., ed., Guidebook to Field Excursions: The North-central Section, Geol. Soc. America, p. 63–115.

KAHLE, C.F., and J.C. FLOYD, 1971, Stratigraphic and environmental significance of sedimentary structures in Cayugan (Silurian) tidal flat carbonates, Northwestern Ohio: Geol. Soc. America Bull., v. 82, p. 2071–2098.

KISSLING, D.L., 1969, Ecology of Silurian crinoid-root bioherms (abst): Geol. Soc. America Abstr. with Prog. 1969, pt. 7, p. 126.

LOGAN, B.W., R. REZAK, and R.N. GINSBURG, 1964, Classification and environmental significance of algal stromatolites, Jour. Geology, v. 72, p. 68–83.

LUNDY, C.L., 1968, Belle River Mills Field: Oil and

Gas Field Symp., Mich. Basin Geol. Soc., p. 49–55.

MANTEK, W., 1976, Recent exploration activity in Michigan: Ontario Petroleum Inst., 15th Ann. Conf. Proc., 29 p.

MATTHEWS, D., 1970, The distribution of Silurian potash in the Michigan Basin: in Sixth Forum on Geology of Industrial Minerals: Mich. Geol. Survey, Lansing, MI, p. 20–33.

MCCLINTOCK, P.L., 1977, Seismic data processing techniques in exploration for reefs, northern Michigan: in Reefs and evaporites—concepts and depositional models: Amer. Assoc. Petroleum Geologists Studies in Geology 5, p. 111–124.

MESOLELLA, K.J., J.D. ROBINSON, and A.R. ORMISTON, 1974, Cyclic deposition of Silurian carbonates and evaporites in Michigan basin: Amer. Assoc. Petroleum Geologists Bull., v. 58, p. 34–62.

MULTER, H.G., 1972, Field guide to some carbonate rock environments, Florida Keys and western Bahamas: Miami Geol. Soc., Rickenbacker Causeway, Miami, FL, 159 p.

MULTER, H.G., and J.E. HOFFMEISTER, 1968, Subaerial laminated crusts of the Florida Keys: Geol. Soc. America Bull., v. 79, p. 183–192.

NURMI, R.D., 1974, The Lower Salina (Upper Silurian) stratigraphy in a desiccated deep Michigan basin: Ontario Petroleum Inst., 13th Ann. Conf. Proc., Paper 13, 25 p.

NURMI, R.D., 1975, Stratigraphy and Sedimentology of the Lower Salina Group (Upper Silurian) in the Michigan basin: Unpubl. Ph.D. Dissert.,Rensselaer Polytech. Inst., Troy, NY, 260 p.

POHLY, R.A., 1968, Seek reefs with gravity work: World Oil, April 1968, p. 81–93.

ROEHL, P.O., 1967, Stony Mountain (Ordovician) and Interlake (Silurian) facies analogs of Recent low energy marine and subaerial carbonates, Bahamas: Amer. Assoc. Petroleum Geologists Bull., v. 51, no. 10, p. 1979–2032.

WALTER, M.R., ed., 1976, Stromatolites: Elsevier Scientific Publ. Co., Amsterdam, 790 p.

WILSON, J.L., 1975, Carbonate Facies in Geologic History: Springer Verlag, Berlin, 471 p.

WINDER, L.G., 1968, Carbonate diagenesis by burrowing organisms: 22nd Internat. Geol. Congr., v. 8, p. 173–183.

YELLING, W.F., JR., and M.R. TEK, 1976, Prospects for oil and gas from Silurian-Niagaran trend in Michigan: Univ. Michigan Inst. Sci. and Technology, 35 p.

8
Rainbow "A" Pool
Volkmar Schmidt, Ian A. McIlreath, and Alfred E. Budwill

RESERVOIR SUMMARY

Location & Geologic Setting	Alberta, Canada, in Western Canada Basin
Tectonics	Cratonic basin, minor salt-solution collapse and differential compaction folding
Regional Paleosetting	Transition from marine shelf to evaporitic shelf at craton margin
Nature of Trap	Stratigraphic; pinnacle reef enclosed by evaporites and bituminous limestones
Reservoir Rocks	
Age	Middle Devonian (Givetian)
Stratigraphic Units(s)	Keg River Formation, Rainbow Member
Lithology(s)	Limestone and dolomite
Dep. Environment(s)	Pinnacle reef
Productive Facies	Reef bioclastic arenites and rudites (grainstone and packstone)
Entrapping Facies	Evaporites and bituminous carbonates
Diagenesis/Porosity	Marine and evaporitic cementation; minor mesogenetic dissolution and cementation
Petrophysics	
Pore Type(s)	Inter- and intraparticle, intercrystal moldic, vuggy, fractures
Porosity	3–15% & 10.1% avg in reef facies
Permeability	K_{max} = 570 md avg, K_{90} = 184 md avg, K_v = md avg
Fractures	Strong, results in excellent vertical interconnection
Source Rocks	
Age	Mainly Middle Devonian
Lithology(s)	Bituminous carbonates of interreef facies
Migration Time	Cretaceous(?)
Reservoir Dimensions	
Depth	6380 ft (1945 m) avg
Thickness	660 ft (200 m) gross, 295 ft (90 m) net oil pay; 200 ft gross, ~100 ft (60 m) net gas pay
Areal Dimensions	Circular, 1.2 mi (2.0 km) in diameter
Productive Area	625 acres (2.5 km^2)
Fluid Data	
Saturations	S_w = 10% avg
API Gravity	40°
Gas-Oil Ratio	88:1
Other	Low sulfur, mixed-base type
Production Data	
Oil and/or Gas in Place	90.3 million BO & 1.4 million CFG (0.039 million m^3); 90.5 million equivalent BO total
Ultimate Recovery	79.3 million BO
Cumulative Production	NA

Remarks: IP 300 BOPD to 1750 BOPD. Presently under solvent flood. Recovery factor 50% primary + 38% enhanced recovery. Discovered 1965.

Carbonate Petroleum Reservoirs (Roehl & Choquette), Springer-Verlag New York, Inc.

8
Origin and Diagenesis of Middle Devonian Pinnacle Reefs Encased in Evaporites, "A" and "E" Pools, Rainbow Field, Alberta

Volkmar Schmidt, Ian A. McIlreath, and Alfred E. Budwill

Introduction

The Rainbow area of petroleum production in Western Canada is located some 400 miles (640 km) northwest of Edmonton, Alberta (Fig. 8-1). The main production is found in Middle Devonian limestone and dolostone reefs, which are encased in evaporites (Figs. 8-2–8-4). The reefs form the Rainbow Member of the Upper Keg River Formation in the Upper Elk Point Subgroup, which is of Givetian Age (Langton and Chin, 1968).

The first productive reef, the "A" Pool, was discovered in 1965 (Figs. 8-5, 8-6). Subsequent exploration encountered oil production in more than 80 individual reefs, at depths ranging from 5000 to 6900 feet (1500–2100 m). These reef pools vary in area from 0.25 to 8.0 square miles (0.6–21.0 km^2). The thickness of individual reefs ranges to about 750 feet (230 m). Maximum thickness of net oil and/or gas pay is 610 feet (185 m). The total initial volume of in-place oil in the reefs and related strata of the combined Rainbow area reservoirs is 1.32 billion barrels. Primary oil recovery from the reefs varies from 1 to 58 percent. Oil recovery has been enhanced, by either water or solvent-flood methods, to an additional 6 to 50 percent. In several cases, total recovery of as much as 80 percent of in-place reserves is anticipated. Recoverable oil reserves in the reef pools total 760.0 million barrels, of which 422.3 million have been produced to

January 1, 1982. Initial volumes of in-place oil for individual reef pools vary from 0.18 to 260.0 million barrels, with the "B" Pool containing the largest reserves. In addition to oil, the combined fields contain more than 1.14 trillion cubic feet (32.2 billion m^3) of total in-place gas reserves.

The reefs of the Rainbow Member were previously interpreted (Langton and Chin, 1968; Barss *et al.*, 1970) to be ecologic reefs that owe their wave-resistant paleotopographic relief primarily to the attached growth of frame-building stromatoporoids and corals. These authors attributed major porosity modifications to both eodiagenesis and mesodiagenesis. Different interpretations were proposed by the authors (Schmidt, 1970, 1971; Schmidt *et al.*, 1977), who concluded that pervasive eogenetic cementation of carbonate debris was the main process of reef growth and preservation, and that eodiagenesis had a paramount influence on reservoir characteristics.

Method of Petrologic Study

The excellent preservation of depositional textures in undolomitized Rainbow reef limestones permits accurate reconstruction of original sediment composition. In addition, reasonable preservation of original textures can be found in most dolostones, allowing extrapolation of the information obtained in limestones. Original sediment

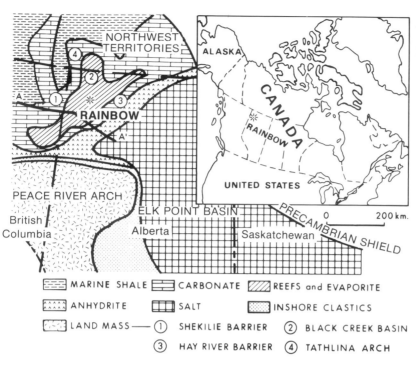

Fig. 8-1. Regional location map with Upper Keg River/ Muskeg facies.

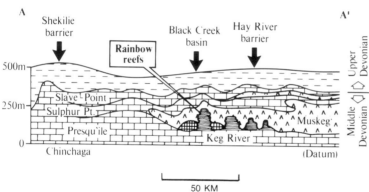

Fig. 8-2. Idealized regional stratigraphic cross-section of Middle Devonian formations. See Figure 8-1 for cross-section location.

composition and subsequent diagenetic alteration were examined in approximately 5000 feet (1500 m) of representative cores from seven reefs (Fig. 8-6). Special emphasis is placed here on the study of two of these, the "*A*" and "*E*" Pools, which consist predominantly of limestone. Thin sections from 240 samples were studied by petrography and cathodoluminescence. The observations provided a basis for interpretive analysis.

Regional Geological Setting

The Rainbow reefs are located in the Middle Devonian Black Creek Basin (Fig. 8-1). This intra-shelf basin developed as a shallow depression behind a barrier reef, shelfward from an area of normal-marine shale sedimentation to the north and west. Maximum water depths most likely were on thc order of 200 to 300 feet (60–90 m). Eventually, restriction at the barrier reef caused intermittent evaporitic conditions, which repeatedly interrupted and finally terminated reef growth. A regional cross-section schematically illustrates the position of the Rainbow reefs within an intrashelf bathymetric low, proximal to the barrier reef, and their encasement within evaporites (Fig. 8-2).

The Middle Devonian strata of the Rainbow area experienced only minor local faulting of

Devonian age along basement lineaments. Regional dip is about 30 feet per mile (6 m/km) to the southwest.

General Stratigraphy

McCamis and Griffith (1967) established the time stratigraphic correlation of Upper Keg River reefs with the off-reef sediments of the Upper Keg River and evaporites of the lower part of the Muskeg in the Zama area of the Black Creek Basin. Similar temporal reef to off-reef relationships are postulated here for the Rainbow area (Figs. 8-3–8-5). Detailed correlations of mechanical logs (Fig. 8-5) and careful observation of lithostratigraphic and ecologic criteria reveal that the upper part of the Upper Keg River reefs at Rainbow was deposited after sedimentation of the evaporitic Muskeg Formation had begun in the inter-reef areas. We propose an informal subdivision of the Upper Keg River and Muskeg formations in the Rainbow area that reflects an inferred diachronous origin of the contact between the two formations (Fig. 8-2).

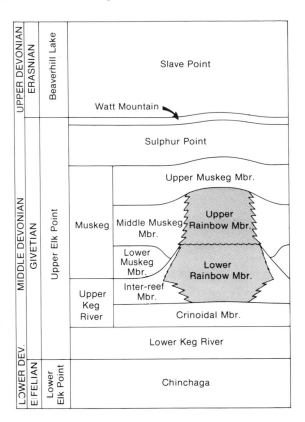

Fig. 8-3. Stratigraphic chart of Devonian formations in Rainbow region. From Schmidt et al., 1980. Published by permission, Society of Economic Paleontologists and Mineralogists.

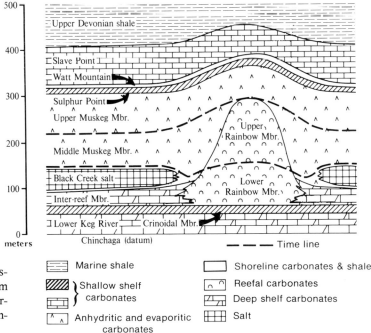

Fig. 8-4. Schematic stratigraphic cross-section of a Rainbow reef. Modified from Schmidt et al., 1980. Published by permission, Society of Economic Paleontologists and Mineralogists.

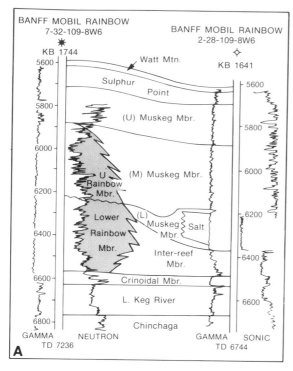

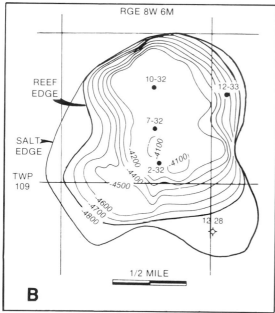

Fig. 8-5. A. Log correlation from "A" Pool reef to off-reef. Location of "A" Pool and the 2-28 off-reef well are indicated in Figure 8-6. *B.* Structure map on top of the Upper Rainbow Member, "A" Pool, showing variably steep flanks and edge of Black Creek Salt. Contours in feet below sea level.

Upper Keg River Formation:

1. *Crinoidal Member*—the basal unit of the Upper Keg River Formation is a blanket deposit, and is characterized by dolomitized crinoidal packstones. It varies in thickness from 40 to 60 feet (11–17 m).
2. *Inter-Reef Member*—consists of off-reef limestones, and is coeval with the Lower Rainbow Reef Member. It varies in thickness from 100 to 150 feet (30–45 m). The distribution of a bituminous argillaceous subfacies of the Inter-reef Member coincides with the area of greatest reef thickness, as shown in Figure 8-6.
3. *Lower Rainbow Reef Member*—see description below.
4. *Upper Rainbow Reef Member*—see description below.

Muskeg Formation:

1. *Lower Muskeg Member*—typically consists of 50 feet (11–15 m) of laminated and nodular anhydrite, and may, in addition, include up to about 270 feet (80 m) of halite (Black Creek Salt).
2. *Middle Muskeg Member*—is characterized by thick evaporitic zones of nodular anhydrite and barren anhydritic dolostone. Intercalated are thin beds of dolomitized fossiliferous mudstones that contain reef-derived debris. Maximum thickness is 300 feet (90 m).
3. *Upper Muskeg Member*—consists of interbedded nodular anhydrite and barren or fossiliferous dolostone. The thickness varies from 150 feet (45 m) above reefs to 330 feet (100 m) in inter-reef areas.

Reef Stratigraphy, Morphology and Distribution

The *Lower Rainbow Reef Member* represents the basal portion of reef growth (Fig. 8-5A). It is characterized by abundant stromatoporoid and coral rudstones, and reaches 350 feet (105 m) in thickness. Stratigraphically, it is syndepositional with the previously described Inter-reef Member. A major diastem surface occurs at the top of this reef unit, which was later onlapped by the Lower Muskeg Member.

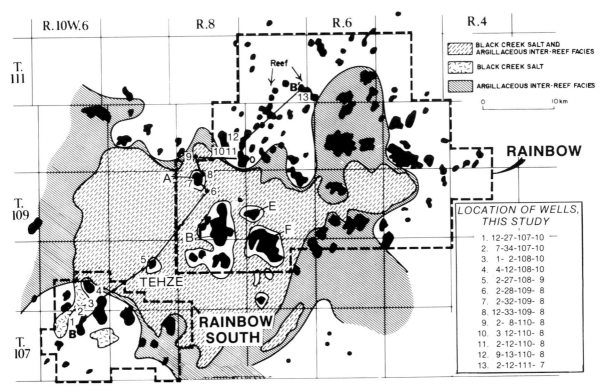

Fig. 8-6. Map of Keg River reefs in Rainbow area, showing distribution of the Black Creek Salt, argillaceous sub-facies of the Interreef Member, and location of the "A" and "E" pools, Rainbow and Rainbow South Fields. Modified from Schmidt *et al.*, 1980. Published by permission, Society of Economic Paleontologists and Mineralogists.

The *Upper Rainbow Reef Member* typically includes abundant stromatoporoid rudstones. The unit interfingers with the Middle Muskeg Member, and is capped by the Upper Muskeg Member. At least four major, evaporitic diastem surfaces are found within the Upper Rainbow Reef Member. The unit attains a maximum thickness of about 450 feet (135 m).

The morphology of Rainbow reefs is well known from seismic data and drilling results, and has been summarized by Barss *et al.* (1970). After initiation of reef growth, individual reefs expanded in area by overstepping inter-reef sediments. This progradation is found only in the lowermost 100 feet (30 m) of the reefs. In places, progradation led to coalescence of closely spaced incipient reefs, forming reefs of larger areal extent. Above the 100-foot (30-m) level, the reefs taper, with an average angle of slope ranging from about 10 to 30 degrees. Maximum angle of slope is about 45 degrees. The steepest slopes occur generally at north and northeast reef perimeters that presumably faced the prevailing winds.

The reefs range from circular to oval or irregular in outline (Fig. 8-6). Areal sizes of reefs, at the level of maximum girth, range from about 0.3 to 9.0 square miles (0.8–23.0 km²). Terminal reef areas vary from 5 acres (0.02 km²) to about 5 square miles (3250 acres, 13 km²). Reefs with maximum or terminal diameters smaller than about 1.2 mile (2 km) did not develop a central lagoon, and are termed *pinnacle reefs*. Larger reefs developed a peripheral reef rim around the central lagoon, and are classified as *atoll reefs*. The peripheral reef is usually best-developed at the north and northeast reef margin, and its width may reach 0.6 mile (1 km).

The basinal facies of the Inter-reef Member indicates that initiation of reef growth coincided with a sudden increase of water depth in the Black Creek Basin. Water depth is also a key factor controlling reef thickness, because the Rainbow reefs appear to have reached sea level repeatedly during their deposition and also at the termination of reef growth. The factors that controlled reef density and location, areal size, and orientation of

Table 8.1. Quantitative Estimates of Present Composition of Rainbow Reefs

Constituent	Average Volume Percent of Total Reef Volume	Average Volume Percent of Individual Reefs
1-1 Primary porosity	2.5	Trace to 10
1-2 Eogenetic porosity	4.5	3 to 10
1-3 Mesogenetic porosity	1 (?)	Trace to 3 (?)
Total porosity	8	3 to 15
2a-1 Depositional carbonates	40	40 to 55
2a-2 Eogenetic carbonate cement	45	40 to 55
2a-3 Mesogenetic carbonate cement	2	1 to 4
2b-1 Calcite	20	1 to 83
2b-2 Dolomite	67	5 to 91
Total carbonate	87	77 to 92
3-1 Eogenetic anhydrite	2	1.5 to 4
3-2 Mesogenetic anhydrite	1	Trace to 2
Total anhydrite	3	1.5 to 6
4 Mesogenetic carbon	1	Trace to 3
5 Other constituents	1	0.5 to 1.5
TOTAL	100	

reefs are poorly understood. Slightly positive areas at the top of the Crinoidal Member are found beneath a number of reefs (Barss *et al.*, 1970), but these may be equally common in inter-reef areas. There is no evidence that structure, especially normal faulting, was responsible for any initial relief.

Present Composition

The estimated average composition of all known Rainbow reefs is listed in Table 8-1, together with the range for each component in individual reefs. A high variability in composition is evident among individual reefs. This is especially noticeable in the amount of porosity, which varies by a factor of five, and the extent of dolomitization, which ranges from 5 to 100 percent.

The present composition of Rainbow reefs is the result of a spectrum of diagenetic processes, superimposed on a great diversity of original sediments. Individual reefs, therefore, consist of an enormous multiplicity of rock types that are

distributed in intricate vertical and lateral patterns. No two Rainbow reefs are alike. By necessity, this paper principally addresses the important aspects of the "*A*" and "*B*" Pools of the Rainbow reefs, and cannot address the very interesting subject of reef-by-reef variability.

Original Sedimentology

Marine Reef Sedimentary Rocks

The depositional composition of the pinnacle reefs is similar to that of the reef rims defined above for atoll reefs. The relative abundance of sediment types in pinnacle reefs, as shown in Figure 8-7A, is therefore representative of all the Rainbow reefs, with the exception of lagoonal areas.

Initially, a skeletal rudite with arenite matrix accumulated as the dominant sedimentary facies. Therefore, it is most typical of the Lower Rainbow Member (Fig. 8-3). Rudites without matrix were deposited in areas of highest turbulence.

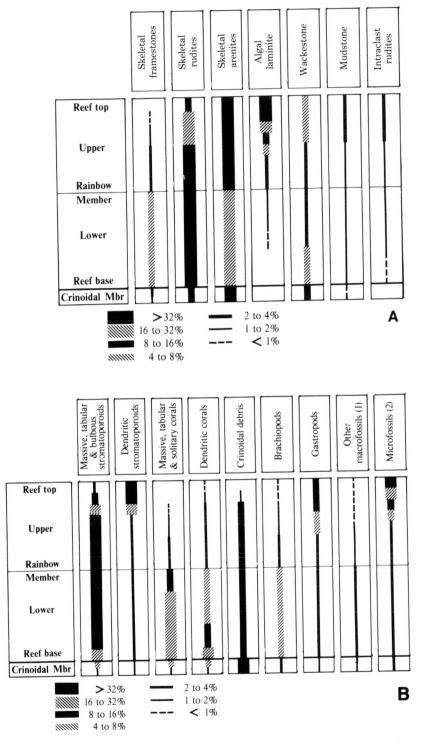

Fig. 8-7. A. Primary sedimentary compositions of Rainbow Reef members. B. Compositions of skeletal fraction in Rainbow Reef members.

Skeletal arenites of mainly grainstone and subordinately packstone texture are second in abundance. The amount of *in situ* skeletal framework is small (about 5%), and consists mainly of framestone constructed of corals and stromatoporoids. These are concentrated in areas of relatively low turbulence in the deeper forereef and in the backreef. A subordinate amount of wackestone occurs, mainly in the basal section of the lower Rainbow Member. Other subordinate sediment types include algal laminites, mudstones, and intraclast rudites and arenites.

The Upper Rainbow Member is principally composed of skeletal rudites and arenites. However, in the uppermost reef interval, algal laminites become abruptly predominant. The textures of skeletal rudites and arenites are similar to those described for the Lower Rainbow Member. Sediments trapped in the algal laminite beds mainly consist of skeletal-pelletal grainstones and packstones. Wackestones, mudstones, and intraclast rudite/arenites are subordinate types.

The abundance, collective growth habit, and morphological traits of *in situ* skeletal framework in the "*A*" and "*E*" pool reefs suggest that they are not true ecologic organic reefs. However, they may be termed "bioherms" in accord with the distribution and predominance of skeletal fractions. The light color of the marine reef sediments suggests oxidizing conditions on the sea floor.

The composition of the skeletal framework and skeletal particles of the pinnacle reefs also shows vertical variation (Fig. 8-7B) that reflects changes in the depositional environment. Stromatoporoids are, by far, the most important skeletal group of both the Lower and Upper Rainbow Members. Corals and brachiopods are mainly concentrated in the Lower Rainbow Reef Member, suggesting normal-marine conditions during their growth and accumulation. Crinoidal debris is significant throughout the pinnacle reefs, except in the uppermost portions. Gastropods are more common in the upper half of the Upper Rainbow Member. Other groups of macrofossils in the "*A*" and "*E*" pool reefs include bryozoans, calcareous algae and very rare trilobites, calcisponges, and cephalopods. Microfossils are principally represented by calcispheres, in association with ostracodes, minor calcareous algae, and foraminifers.

The paucity of corals and brachiopods in the Upper Rainbow Reef Member suggests some restriction of the marine environment. In the upper-

most zone of the reef, *Amphipora*, calcispheres, and gastropods dominate the skeletal fraction, and indicate a significantly restricted marine environment. The abundance of algal laminite beds in this zone is in accord with this interpretation.

Lagoonal Marine Sedimentary Rocks

The estimated average primary compositions of reefal and atoll-lagoon marine sedimentary rocks are listed in Table 8-2. It indicates that the original mix of primary marine sediments attributable to atoll lagoons differs generally from that of pinnacle reefs and reefal margins of atolls. Reduced fossil content of the atoll lagoonal rocks contrasts significantly with reefal facies. Fossil groups, in reduced abundance, include tabular and massive stromatoporoids, corals, crinoid debris, and brachiopods. In contrast, there is an increased abundance of bulbous stromatoporoids, *Amphipora*, gastropods, and calcispheres.

The textural composition of the rocks, as well as the fossil content, indicates that they were deposited in a generally more restricted and less turbulent environment than is indicated for sediments of the reef itself.

Evaporitic Sedimentary Rocks

Unfossiliferous mudstone and intraclast wackestone rocks form the evaporitic lithofacies, which is intercalated with marine reef facies. These barren carbonates are light to dark gray in color, and commonly contain nodular anhydrite. Intraclasts are generally composed of evaporitic mudstone, but may include reworked anhydrite nodules (Fig. 8-8D). In addition, angular clasts of well-cemented marine carbonates occur suspended in an evaporitic sedimentary matrix near diastem surfaces.

Evaporitic sedimentary rocks comprising the reefs are restricted to the Upper Rainbow Member and the very top of the Lower Rainbow Member. They are always associated with diastems within reefal or lagoonal facies, and occur in three relationships: (1) below diastem surfaces as sediment infiltrating vugs and fractures (Figs. 8-8A, 8-9C) to at least a 65-foot (20 m) depth; (2) at diastem surfaces, as the matrix of breccias derived from marine reef sediments (Fig. 8-8B); and (3) as intercalations from 2 inches (5 cm) to

Table 8.2. Estimated Textural Composition and Porosity Percent of Original Marine Carbonates in Rainbow Pinnacle, Atoll Reef Margins, and Atoll Lagoons Reefs, Based Principally on *A* and *E* Pools

Constituent	Pinnacle Reefs and Atoll Reef Margins Volume %		Atoll Lagoons Volume %	
1a Interstitial porosity	40		50	
1b Intraskeletal porosity	15		5	
Total porosity		55		55
2a *In-situ* skeletal framework	2		1	
2b Skeletal rudite debris	17		7	
2c Skeletal arenite grains	13		12	
2d Skeletal coarse silt grains	3		5	
Total skeletal allochems		35		25
3a Rudite intraclasts and oncoids	1		2	
3b Arenite intraclasts, pellets, and oncoids	3		4	
3c Coarse silt pellets	2		5	
Total nonskeletal allochems		6		10
4 Fines		4		10
TOTAL		100		100

6 feet (2 m) thick, which commonly overlie diastem surfaces (Fig. 8-8C, D).

Increased thicknesses of evaporitic sediments are found in the reefs at the northern, eastern, and southern margins of the Rainbow area. The diastem surfaces and associated evaporitic sediments within the reefs correlate with thick intervals of nodular anhydrite and evaporitic carbonate sediments in inter-reef areas. Repetition of sedimentation and eogenetic processes during alternating periods of marine and evaporitic conditions resulted in a chronostratigraphic zonation and correlation of reef to off-reef sections, as schematically shown in Figures 8-10A and B.

Diagenesis

Marine Eodiagenesis

Fabric relationships between diagenetic textures and marine depositional constituents permit the identification of eogenetic changes that occurred in the marine environment (Schmidt, 1971; Schmidt *et al.*, 1977). Early diagenetic marine processes that were followed or accompanied by the internal sedimentation of marine sediments or the growth of marine organisms include (1) cementation by calcium carbonate (Figs. 8-9A, B), (2) leaching of calcium carbonate, and (3) fracturing (Fig. 8-9D).

Textures of the marine calcium carbonate cement include: (1) micrite crusts and pellets, (2) syntaxial spar overgrowths, (3) randomly oriented equant crystals, (4) equant to bladed drusy rims, (5) simple or zoned crusts of bladed to fibrous crystals, which may or may not be isopachous, (6) radiating aggregates of bladed to fibrous crystals, and (7) spherulitic aggregates. The marine cements appear to have been originally composed of magnesian calcite, possibly with minor aragonite.

Marine cements completely lithify the reef and have reduced the average depositional porosity from about 55 percent to less than 10 percent by volume. In lagoonal sediments, marine cementation was considerably less extensive.

The hypothetical paleogeology portrayed by the section in Figure 8-10A illustrates how marine cementation probably transformed mech-

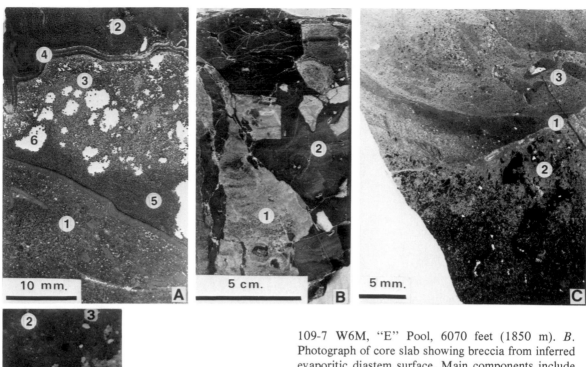

Fig. 8-8. A. Photomicrograph of Lower Rainbow Reef carbonate composed of skeletal calcarenite grainstone (1) at bottom, stromatoporoid limestone (2) and shelter cavity filled by anhydritic dolomite (3). Cavity filling sequence consists of lining of cavity wall (4) with zoned crust of eogenetic marine calcium carbonate cement, and younger, dolomitized sediment composed of evaporitic carbonate mud (5). Internal sediment hosted displacive growth of anhydrite nodules and was partially replaced by patchy-appearing calcite. Selective dolomitization of internal sediment occurred during evaporitic eodiagenesis. Thin section treated with alizarin-red S. Upper Rainbow Reef Member, well 7-8- 109-7 W6M, "E" Pool, 6070 feet (1850 m). *B.* Photograph of core slab showing breccia from inferred evaporitic diastem surface. Main components include light-colored skeletal arenite grainstone, fractured, brecciated and partially dolomitized (1); a dark-colored, fracture-filling evaporitic mudstone (2); a mudstone now completely dolomitized (3). Core slab treated with alizarin-red S. Top of Lower Rainbow Reef Member, well 7-8-109-7 W6M, "E" Pool, 6210 feet (1893 m). From Schmidt *et al.,* 1980. Published by permission, Society of Economic Paleontologists and Mineralogists. *C.* Photomicrograph of the Upper Rainbow Reef Member showing an evaporitic diastem surface (1) developed above a marine skeletal arenite grainstone (2) strongly cemented prior to both evaporitic eogenetic erosion and partial dolomitization. An evaporitic intraclastic wackestone (3), completely dolomitized during evaporitic eodiagenesis, overlies the surface. Thin section treated with alizarin-red S. Well 7-18-109-7 W6M, "E" Pool, 6175 feet (1882 m). From Schmidt *et al.,* 1980. Published by permission, Society of Economic Paleontologists and Mineralogists. *D.* Photograph of core slab of an evaporitic anhydritic dolostone from the Upper Rainbow Reef Member. It consists of intraclastic wackestone (medium gray, 1), overlain by mudstone (dark gray, 2). Anhydrite, white, occurs both as in situ nodules (3) and as reworked abraded pebbles (4). The wackestone is considered a supratidal evaporitic mud flat deposit, and the mudstone a peritidal evaporitic sediment. Note soft sediment fractures (5) and eroded surface (6) of the wackestone. Well 2-27-108-9 W6M, Tehze Pool, 6114 feet (1864 m).

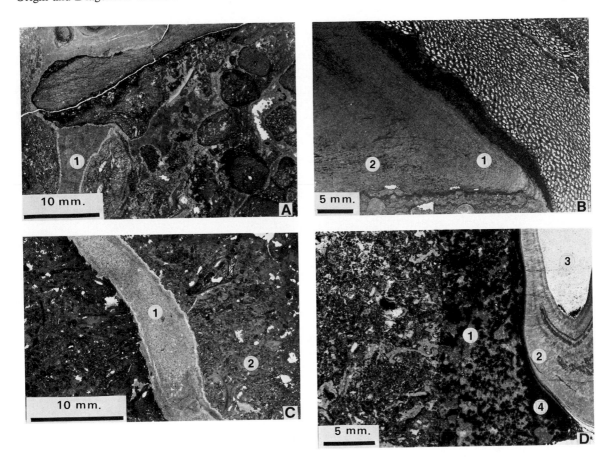

Fig. 8-9. Photomicrographs of reef carbonates stained with alizarin-red S. *A*. Skeletal rudite limestone. Interskeletal space is largely filled with eogenetic, marine, calcium carbonate cements, which include thick zoned isopachous crusts (1). Upper Rainbow Reef Member, well 7-18-109-7 W6M, "E" Pool. Depth 5864 feet (1787 m). *B*. Skeletal rudite limestone. Colonial coral fragment (right) has primary intrafossil porosity. Shelter cavity at left is filled with fibrous, eogenetic, marine, calcium carbonate cement (1) which encloses a small skeletal debris, pellets, and micrite laminae along former cavity floors (2). Lower Rainbow Reef Member, well 7-18-109-7 W6M, "E" Pool, 6106 feet (1892 m). From Schmidt *et al.*, 1980. Published by permission, Society of Economic Paleontologists and Mineralogists. *C*. Eogenetic evaporitic fracture (1) occurring in skeletal calcarenite grainstone (2). Fracture is filled with selec- tively dolomitized, evaporitic mudstone. Pervasive calcium carbonate cementation of eogenetic, marine origin reduced original porosity of the grainstone prior to fracturing. Upper Rainbow Reef Member. Well 7-18-109-7 W6M, "E" Pool, 6196 feet (1889 m). From Schmidt *et al.*, 1980. Published by permission, Society of Economic Paleontologists and Mineralogists. *D*. Eogenetic marine fracture, right, and skeletal cal- carenite packstone, left. Fracture was sequentially in- filled by: (1) spongy-appearing encrustation resembling stromatolite crust; (2) zoned cement of eogenetic marine calcium carbonate; (3) internal sediment com- posed of evaporitic carbonate mud that was selectively dolomitized during evaporitic eodiagenesis, upper right; and (4) pore-filling mesogenetic black carbon. Upper Rainbow Reef Member, well 2-27-108-9 W6M, Tehze Pool, 6250 feet (1905 m).

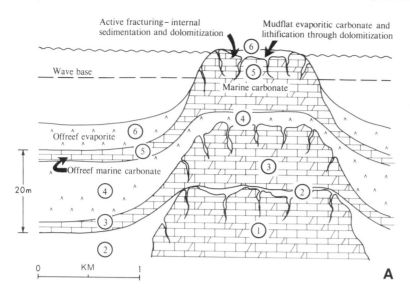

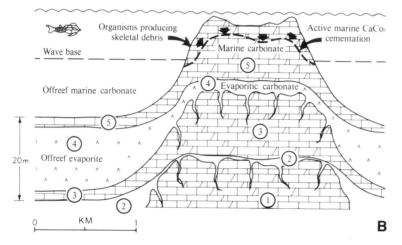

Fig. 8-10. A. Paleogeological section proposed for evaporitic episode. *B.* Paleogeological section proposed for marine episode. From Schmidt *et al.*, 1980. Published by permission, Society of Economic Paleontologists and Mineralogists.

anically deposited skeletal carbonate sediments into wave-resistant rocks, before they were onlapped by evaporitic carbonate. Authigenesis of carbonate was probably enhanced by agitation in shallow waters and high physicochemical gradients on reef slopes. Marine leaching of calcium carbonate is occasionally observed, and resulted in molds and vugs. Marine eogenetic fractures are moderately common. They cut through marine cements and in turn were commonly infilled by marine cement and/or internal sediment (Fig. 8-9D).

Evaporitic Eodiagenesis

Eodiagenesis that occurred during evaporitic periods affected both evaporitic and marine sediments. Diagenetic and evaporitic processes commonly operated in a sequential order (Fig. 8-11): (1) erosion, fracturing, and brecciation of marine carbonates; (2) surficial and internal deposition of evaporitic carbonate; (3) growth of nodular anhydrite; (4) compaction; (5) dolomitization; (6) leaching of molds and vugs; and (7) anhydritization.

During the early stage of evaporitic episodes, sea level dropped and the reef became partially emergent. Emergence brought about nondeposition, minor erosion, surficial brecciation, and intensive fracturing (Figs. 8-8B, C).

With rising sea level, evaporitic carbonate muds were deposited in subaquatic to sabkha environments, onlapping the diastem surfaces and the fractured marine carbonates (Fig. 8-10B). Nodular anhydrite grew in the unconsolidated muds.

Fig. 8-11. Diagram portraying inferred sequence of diagenetic alteration in Rainbow reefs. Modified from Schmidt *et al.*, 1980. Published by permission, Society of Economic Paleontologists and Mineralogists.

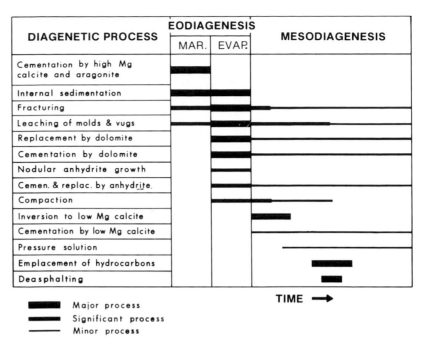

DIAGENETIC PROCESS	EODIAGENESIS		MESODIAGENESIS
	MAR.	EVAP.	
Cementation by high Mg calcite and aragonite	▬		
Internal sedimentation	▬	▬	
Fracturing	▬	▬	
Leaching of molds & vugs	▬	▬	
Replacement by dolomite		▬	
Cementation by dolomite		▬	
Nodular anhydrite growth		▬	
Cemen. & replac. by anhydrite		▬	
Compaction		▬	
Inversion to low Mg calcite			▬
Cementation by low Mg calcite			
Pressure solution			—
Emplacement of hydrocarbons			▬
Deasphalting			▬

TIME ➡

▬▬▬ Major process
▬▬▬ Significant process
——— Minor process

Virtually all the evaporitic carbonate muds are dolomitized and consist of tightly welded, equant dolomite crystal silt (Fig. 8-9C). Early dolomitization is indicated by dolomitized clasts of evaporitic mudstone that are embedded in marine limestone (Fig. 8-12A). Very fine to medium sand-size dolomite crystals, anhedral to euhedral in shape, pervade the marine carbonates and appear to spread downward from evaporitic diastem surfaces (Fig. 8-8C) and laterally from evaporitic fractures. The latter suggests dolomitization by percolating evaporitic brines.

Dolomitization of marine carbonates was accompanied by both dolomite cementation and leaching of calcium carbonate. Cementation by dolomite generally outranked leaching, so that a considerable net loss of porosity resulted.

Medium to coarsely crystalline anhydrite commonly replaced carbonates and has cemented voids after dolomitization and leaching. This replacement is most prevalent within evaporitic intervals and marine carbonates that underlie evaporitic diastems, thus suggesting their eogenetic origin.

Meteoric Eodiagenesis (?)

There is no clear evidence that meteoric water caused significant eogenetic alterations. Prolonged subaerial exposure of the upper parts of reefs appears to have occurred during evaporitic periods. Most likely, some areas of the reef crest also were intermittently exposed during marine periods. However, eogenetic anhydrite is preserved in several zones within the reef bodies, which suggests that meteoric water did not have a sustained influence on eodiagenesis.

Mesodiagenesis

Mesodiagenesis affected the reef carbonates to a much lesser extent than eodiagenesis, possibly because of low hydrodynamic activity, since the reefs have been encased in evaporites since burial (Fig. 8-11).

We infer that an inversion of postulated magnesian calcite and aragonite to calcite took place during the early stages of mesodiagenesis. Pressure solution at grain contacts, and stylolitization is common in the lagoonal sedimentary rocks, but rare in reefal deposits. Another major mesogenetic event was the emplacement of powdery black carbon in the porosity of certain zones of oil-bearing reefs (Fig. 8-9C). The carbon appears to be the product of de-asphalting of the present oil, because the regional thermal maturation history does not suggest that an earlier oil accumulation was destroyed.

Mesogenetic leaching of carbonate and sulfate minerals is deduced from corroded late-stage cements. Corrosion is fairly common and is typified by molds, vugs, enlarged fractures, and

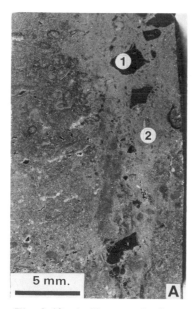
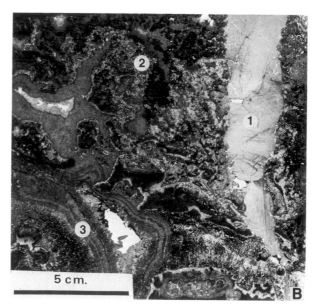

Fig. 8-12. A. Photograph of core slab lightly treated with alizarin-red S. Very dark gray lithoclasts of dolomitized evaporitic mudstone (1) are enclosed in a matrix of medium to light gray, dolomitic, marine skeletal calcarenite grainstone (2). Upper Rainbow Reef Member, well 2-27-108-9 W6M, "E" Pool, 5978 feet (1822 m). *B.* Photomicrograph of reef carbonate consisting of very dark stromatolitic structures (2) and eogenetic marine cement crusts (3) that were partially dolomitized during evaporitic eodiagenesis. Mesogenetic sparry calcite cement (light gray, 1) fills a mesogenetic fracture. Thin section lightly treated with alizarin-red S. Upper Rainbow Reef Member, well 3-5-108-9 W6M, Rainbow South Field, 6275 feet (1913 m).

"chalky" microporosity. Much dissolution may have taken place shortly before and during hydrocarbon migration, if acid conditions prevailed in the formation water. Other mesogenetic processes include: (1) fracturing, (2) mechanical compaction, (3) cementation by calcite spar (Fig. 8-12B), (4) dolomitization, and (5) anhydritization.

Trap Configuration

The Crinoidal Member provides hydrodynamic communication between Lower Rainbow reefs that are separated by the Inter-reef Member. At the margin of the Rainbow Sub-basin, the Lower Rainbow Reef Member coalesces and forms a carbonate platform that permits the transmission of fluids between reefs (Fig. 8-13). Therefore, reef spillpoints are located either at the contact between the Crinoidal Member and Inter-reef Member or at the contact between the Lower Rainbow Reef Member and the Muskeg Formation.

The Lower Rainbow Reef Member is an important oil producer in a few reef pools. However, most reef traps are incompletely filled, and the oil/water contact is found in the Upper Rainbow Reef Member that forms the main reservoir for oil and gas.

The basinal carbonates of the Inter-reef Member and the evaporites of the Muskeg Formation act as effective lateral seals for the reef reservoir. The top seals are formed by the anhydrite beds of the Upper Muskeg Formation that are draped over the reefs. These sealing anhydrite beds, however, are intercalated with porous dolostone intervals that contain commercial accumulations of oil and gas in some locations. In such cases, hydrocarbons may have leaked from the underlying reef or may have been sourced within the Upper Muskeg Member.

Porosity

The average porosity of selected Rainbow reefs is about 8 percent, and is interpreted as follows:

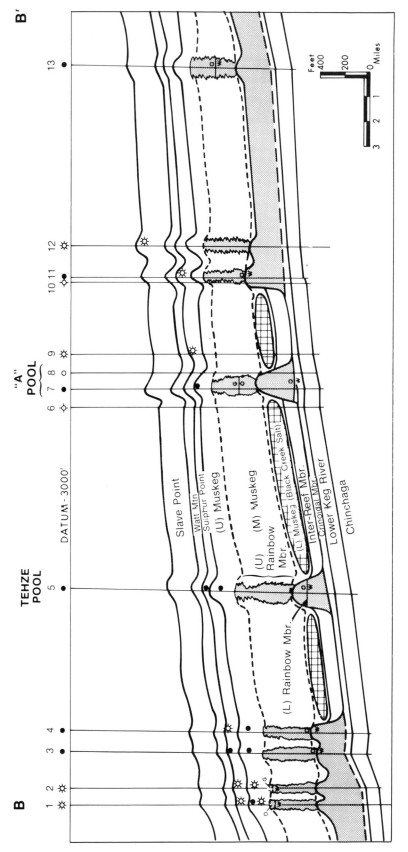

Fig. 8-13. Structural cross-section through Rainbow area. See Figure 8-6 for cross-section location.

Table 8.3. Estimated Composition of Porosity in Rainbow Reefs

Porosity Type	Genetic Mode*			Fraction of Limestone ϕ	Fraction of Dolostone ϕ
Interframe megapores	*Pd*	Se	Sm	Minor	Minor
Interparticle megapores	*Pd*			Significant	Significant
Interparticle mesopores	*Pd*			Very major	Major
Interparticle micropores	*Pd*			Significant	Very minor
Intraframe pores	*Pd*			Minor	Very minor
Intraparticle pores	*Pd*			Very major	Significant
Fenestral pores	*Pd*	Se	Sm	Significant	Major
Shelter pores	*Pd*			Minor	Very minor
Burrows and borings	*Pd*	Se		Very minor	Minor
Intercrystalline pores	*Se*	Sm		Significant	Very major
Micro-lamellar pores	Pp	*Se*	Sm	Very major	Very major
Moldic pores	*Se*	Sm		Very major	Very major
Fracture porosity	*Se*	Sm		Significant	Major
Breccia	*Se*	Sm		Significant	Major
Channel porosity	Pd	*Se*	Sm	Very minor	Significant
Vug mesopores	*Se*	Sm		Significant	Major
Vug megapores	*Se*	Sm		Significant	Major

Relative Abundance	Genetic Mode
Very major = > 8% of ϕ	Pp = pre-depositional
Major = 4—8% of ϕ	Pd = depositional
Significant = 2—4% of ϕ	Se = secondary eogenetic
Minor = 1—2% of ϕ	Sm = secondary mesogenetic
Very minor = < 1% of ϕ	

*Predominant mode is italicized.

primary porosity (2.5%), eogenetic porosity (4.5%), and mesogenetic porosity (1%?). Mesogenetic porosity may possibly be quite significant in some reefs. Average porosity for individual reefs ranges from about 3 to 15 percent, and the average porosity of individual wells varies from 1 percent (some lagoonal wells) to about 20 percent. Individual lithologic zones can average as much as 35 percent porosity. Average matrix permeability of individual reservoirs ranges from about 40 to 200 millidarcys.

The textural composition of porous reefal rocks shows a consistent pattern of more abundant primary porosity types in limestones and more abundant secondary porosity types in dolostones (Table 8-3). The various porosity types occur in almost any possible combination and thus provide a nearly infinite number of reservoir rock types with distinctive pore geometry. The permeability of those pore systems is correspondingly diverse.

The relationship between diagenesis and present porosity is illustrated in Figures 8-14, 8-15, and 8-16. Figure 8-14 is a diagram showing the average compositional history of the marine carbonates in a typical limestone reef. Following marine eogenetic cementation, only subordinate net reductions and additions to porosity occurred, resulting in a present average porosity of about 10 percent.

The average compositional history of the marine carbonates in a dolomitized reef followed at first the same path as that of limestone reefs (Fig. 8-15). However, complete dolomitization during evaporitic eodiagenesis resulted in a significant loss of porosity, so that present porosity averages only about 5 percent. This confirms primary porosity reduction by infilling to exceed

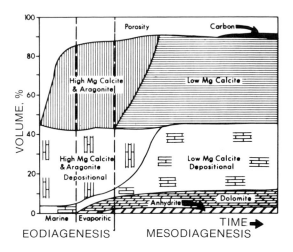

Fig. 8-14. Composition-diagenesis diagram for marine limestones in "A" and "E" pool pinnacle reefs.

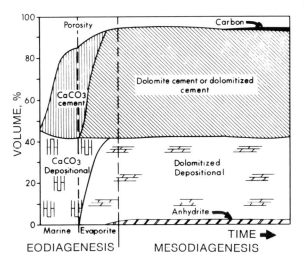

Fig. 8-15. Composition-diagenesis diagram for marine dolostones in pinnacle reefs.

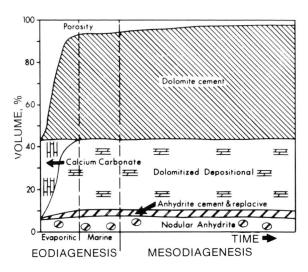

Fig. 8-16. Composition-diagenesis diagram for evaporitic carbonates in "A" and "E" pool pinnacle reefs.

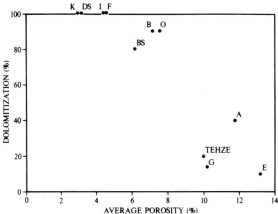

Fig. 8-17. Porosity-dolomitization cross-plot for pinnacle and atoll reefs in eleven pools as determined from core analysis.

concurrent secondary porosity enhancement. As a result, the porosity of limestone reefs generally is about twice that of dolostone reefs, as demonstrated by the general decrease in average porosity with increased dolomitization for eleven reefs (Fig. 8-17).

The compositional history of evaporitic carbonates in Figure 8-16 shows the alteration of a moderately compacted lime mud into relatively tight microcrystalline dolostone. Thus, the dolomitization of this unlithified sediment resulted in the lowest porosities in Rainbow carbonates.

Hydrocarbon Source and Migration

Hydrocarbons could have originated from either local or distant sources. The carbonate rocks of the Lower and Upper Keg River formations are effectively sealed at the bottom by the anhydrites of the Chinchaga Formation and at the top by the anhydrites of the Upper Muskeg Member. The Crinoidal Member is the only stratigraphic unit that provides hydrodynamic communication between reefs and distant areas.

Oil-prone local source beds, with predominantly amorphous kerogen, could have drained during compaction into the Crinoidal Member or directly into the "*A*" and "*E*" pool reefs and presumably other Rainbow reefs as well. Possible source rocks include: (1) the bituminous carbonates of the Lower Keg River Formation; (2) the bituminous carbonates of the Inter-reef Member; and (3) the bituminous carbonate intercalations and algal laminite beds of the Muskeg Formation.

Distribution of the hydrocarbons in the Rainbow area suggests that each of the three local sources contributed substantially to existing oil and gas reservoirs. Significantly, the highest oil columns are found where the Inter-reef Member is developed as a bituminous, argillaceous facies (Figs. 8-6, 8-13), which has the best source characteristics. However, distant sources such as Devonian shales and Middle Devonian bituminous carbonates, west of the Rainbow area, may also have contributed some of the hydrocarbons. It is conjectured that the incomplete filling of most reef traps resulted predominantly from a shortage of migrated hydrocarbons and only subordinately from trap leakage.

Acknowledgments   Research for this paper was conducted first at the geological laboratory of Mobil Oil Canada Ltd., and since 1976 at the geological research laboratory of Petro-Canada. The authors thank both companies for their generous support in the preparation of this paper, and for permission to publish. The authors appreciate the expert assistance of Brian Miller and Susan Peters with assembly of the manuscript.

References

BARSS, D.L., A.B. COPELAND, and W.D. RITCHIE, 1970, Geology of Middle Devonian reefs, Rainbow area, Alberta, Canada: Amer. Assoc. Petroleum Geologists Mem. 14, p. 14–49.

LANGTON, J.R., and G.E. CHIN, 1968, Rainbow Member facies and related reservoir properties, Rainbow Lake, Alberta: Canadian Soc. Petroleum Geologists Bull., v. 16, p. 104–143.

MCCAMIS, J.D., and L.S. GRIFFITH, 1967, Middle Devonian facies relationships, Zama area, Alberta: Canadian Soc. Petroleum Geologists Bull., v. 15, p. 434–467.

SCHMIDT, V., 1970, Diagenesis of Keg River bioherms, Rainbow Lake, Alberta (abst.): Amer. Assoc. Petroleum Geologists Bull., v. 54, p. 868.

SCHMIDT, V., 1971, Early carbonate cementation in Middle Devonian bioherms, Rainbow Lake, Alberta: *in* Bricker, O.P., ed., Carbonate Cements: Johns Hopkins Univ. Studies in Geology, No. 19, p. 209–215.

SCHMIDT, V., D.A. MCDONALD, and I.A. MCILREATH, 1977, Growth and diagenesis of Middle Devonian Keg River cementation reefs, Rainbow Field, Alberta: *in* I.A. McIlreath and R.D. Harrison (eds.), supplemental to The Geology of Selected Carbonate Oil, Gas, and Lead-Zinc Reservoirs in Western Canada: Canadian Soc. Petroleum Geologists, Calgary, Alberta, p. 1–21.

SCHMIDT, V., D.A. MCDONALD, and I.A. MCILREATH, 1980, Growth and diagenesis of Middle Devonian Keg River cementation reefs, Rainbow Field, Alberta: *in* Halley, R.B. and R.G. Loucks, eds., Carbonate Reservoir Rocks: notes for SEPM Core Workshop No. 1, Denver, Colorado, June 1980: Soc. Econ. Paleontologists and Mineralogists, p. 43–63.

9
Northwest Lisbon Field
James A. Miller

RESERVOIR SUMMARY

Location & Geologic Setting San Juan Co., T30S, R24E, SE Utah, central part of Paradox Basin fold and fault province, USA

Tectonics Contemporaneous folding; late Mississippian to early Pennsylvanian block faulting

Regional Paleosetting NW shelf of Paradox Basin (intermontane)

Nature of Trap Structural; faulted anticline

Reservoir Dimensions
  Depth Middle Mississippian
  Stratigraphic Units(s) Leadville Formation
  Lithology(s) Dolomite
  Dep. Environment(s) Shallow-marine to peritidal; local skeletal and oolite shoals
  Productive Facies Sucrosic dolomite (wackestone-packstone); local solution-collapse breccia
  Entrapping Facies Evaporites, tight limestones and dolomites
  Diagenesis/Porosity Multistage dolomitization; dissolution of $CaCO_3$ and evaporites, associated with karsting

  Petrophysics
    Pore Type(s) Moldic and intercrystal local breccia and fracture
    Porosity 1–12%, 5.5% avg
    Permeability 00.01–100 md, 22 md avg
    Fractures Vertical and also fractures associated with collapse brecciation

Source Rocks
  Age Mississippian(?)
  Lithology(s) Shales and dark fine-grained carbonates(?)
  Migration Time Early to middle Pennsylvanian(?)

Reservoir Dimensions
  Depth 7600–9000 ft (2317 2743 m)
  Thickness 225 ft (69 m) avg net pay
  Areal Dimensions NA
  Productive Area 5120 acres (20.5 km^2)

Fluid Data
  Saturations S_w = 39% avg
  API Gravity 54°
  Gas-Oil Ratio 1200:1 in oil zone
  Other Oil sour, yellow to red; gas sour, with 21% CO_2 and 1.2% H_2S, 0.97 gm/cm^3 specific gravity

Production Data
  Oil and/or Gas in Place 91.2 million BO and 357 billion CFG (56.7 billion m^3); 150.7 million equivalent BO
  Ultimate Recovery 42.9 million BO and 250 billion CFG (40 billion m^3); 84.6 million equivalent BO; URE = 56.1%
  Cumulative Production NA

Remarks: IP mechanism expanding gas cap and gravity drainage. Discovered 1960. Production data from Clark (1978).

9
Depositional and Reservoir Facies of the Mississippian Leadville Formation, Northwest Lisbon Field, Utah

James A. Miller

Location and Setting

Northwest Lisbon Field is located in the salt anticline belt on the southwest edge of the Paradox Basin in San Juan County, Utah (Fig. 9-1). The field was discovered by the Pure Oil Company in January 1960 from seismic data and stratigraphic analysis of the Pure NW Lisbon No. 1 (B-610) well. Initial production was from Pennsylvanian, Mississippian, and Devonian strata, but only the Mississippian (Leadville) carbonate sequence, by far the major producing unit, will be discussed in this paper.

Northwest Lisbon is situated on the southern flank of a faulted anticline and produces from a sequence of dolomites with intercrystalline, moldic, and fracture porosity (Figs. 9-2, 9-3). The Mississippian interval is 326 to 548 feet (~100–167 m) thick in the field area, ranges in composition from 50 to 90 percent dolomite, and is capped by a major unconformity. This study is based on cores, core chips, and cuttings from 11 wells within or adjacent to the field area (Fig. 9-2).

Reservoir Development and Characteristics

The Mississippian carbonates lie at depths of greater than 7500 feet (2286 m), or 800 feet (243.8 m) below sea level, in the field area (Fig. 9-2). The field covers an area of about 8.0 square miles (20.7 km^2) and the Mississippian interval has a proven productive area of 5120 acres (20.7 km^2).

The reservoir facies is extremely variable (Fig. 9-4), but has an average net pay of 225 feet (68.6 m), with average porosity and permeability values of 5.5 percent and 22 md, respectively. Production is complicated by several major and minor faults that displace portions of the reservoir interval.

Dolomitization, leaching, and fracturing are responsible for the development of porosity and permeability in the Lisbon Field. Seven types of porosity have been observed in the Mississippian section: intercrystalline, interparticle, moldic, fenestral, vuggy, fracture, and breccia. However, the major reservoir intervals are dominated by intercrystalline and moldic porosity in dolomite (Figs. 9-5A, B and 9-6A, B), with local enhancement by solution brecciation or fracturing (Fig. 9-5). As with most carbonate reservoirs, porosity is often discontinuous both laterally and vertically, and porosity and permeability are not necessarily developed together.

Depositional and diagenetic facies control the distribution of tight and porous intervals in the Mississippian carbonate section at Lisbon. Porosity is generally developed in discrete zones 5 to 25 feet (1.5–7.6 m) thick, which are separated by relatively impermeable streaks (Fig. 9-4). The tight intervals are usually related either to areas of

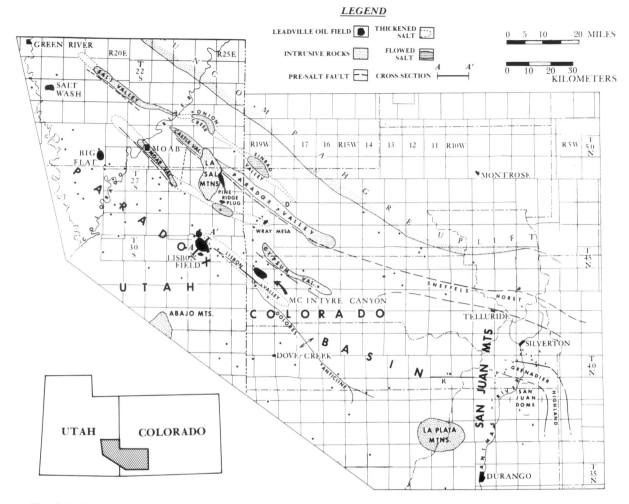

Fig. 9-1. Index map of the eastern Paradox Basin and structural cross section of the Lisbon Field. Modified from Baars (1966) and Parker (1968). Published by permission of American Association of Petroleum Geologists.

early cementation or to stylolitization-cementation. However, near the upper and lower boundaries of the Mississippian, porosity is also reduced by detrital clay that occurs as disseminated particles and discontinuous shaly streaks.

It is not possible to recognize specific porosity types on the acoustic and gamma-ray logs, but in most cases, specific lithologic units or porous streaks can be projected from adjacent wells. Overall, the acoustic logs provide a fair indication of non-fracture porosity distribution in the section, but are not reliable for obtaining specific porosity values.

Reserves and Production Mechanisms

Ultimate recoverable field reserves are currently estimated to be 42.85 million barrels of oil and 250 billion cubic feet (40 billion m^3) of gas. The oil and gas are sour, with the oil having an API gravity of 54°, and the gas, a CO_2 content of approximately 21 percent. As of 1978, the average daily production for the carbonate section was 2667 barrels of oil per day and 55.6 million cubic feet (1.58 million m^3) of gas per day from 11 wells, and 49.2 million cubic feet (1.395 million

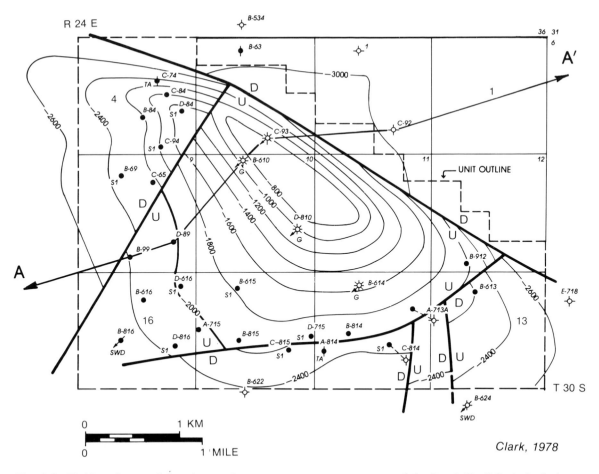

Fig. 9-2. Field outline, well location, and structure-contour map on top of the Leadville (Mississippian) carbonates, NW Lisbon Field, San Juan Co., Utah. Modified from Clark (1978).

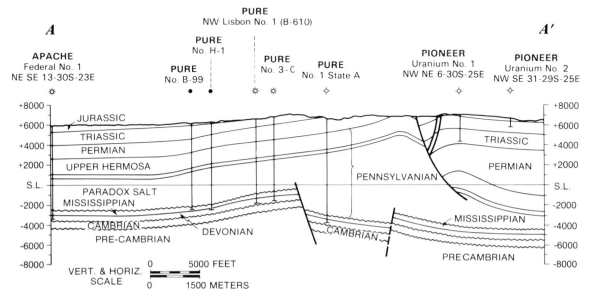

Fig. 9-3. Structural cross-section of NW Lisbon Field. Vertical exaggeration 2X. The westernmost well and the two eastern wells are situated off the map in Figure 9-2.

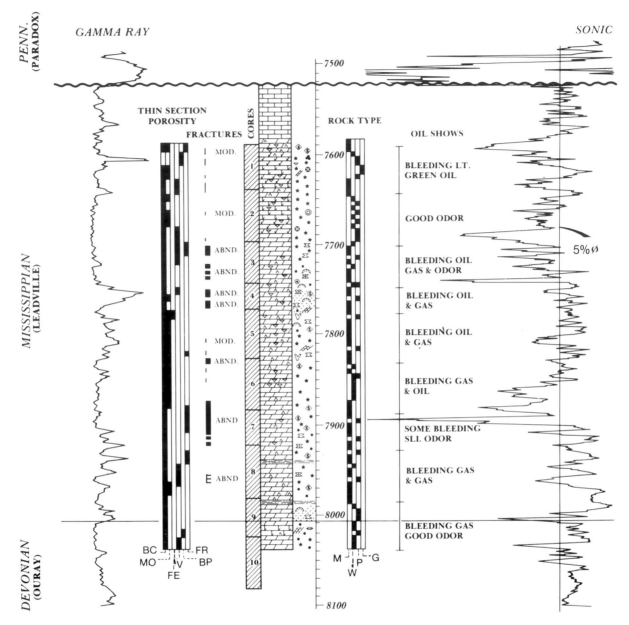

Fig. 9-4. Lithology and porosity log for the Leadville (Mississippian) carbonates in the NW Lisbon #1 well (B-610), NW Lisbon Field, Section 10-T30S-R24E, San Juan Co., Utah.

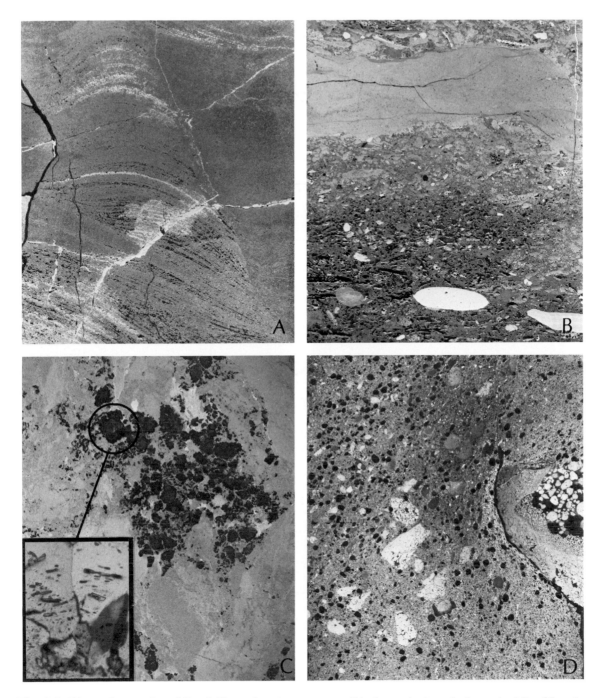

Fig. 9-5. Photomicrographs of Leadville carbonates. *A.* Dolomitized algal stromatolite sequence with fenestral porosity. Well B-614, 8551.5 feet (2606.5 m). Negative print, 2X. *B.* Dolomitized tidal sediments with micritic drapes and intraclasts and bioclast moldic porosity. Well B-614, 8554 feet (2607.3 m). Negative print, 2X. *C.* Mottled dolomite mudstone with bladed quartz (black patches) replacing anhydrite. Negative print, 2X. *INSET*—bladed authigenic quartz containing remnant anhydrite laths. Positive print, crosspolarized light, 160X. *D.* Dolomite packstone containing well-rounded quartz sand grains (black) and carbonate lithoclasts. Well B-610, 8002.2 feet (2439.1 m). Negative print, 2X.

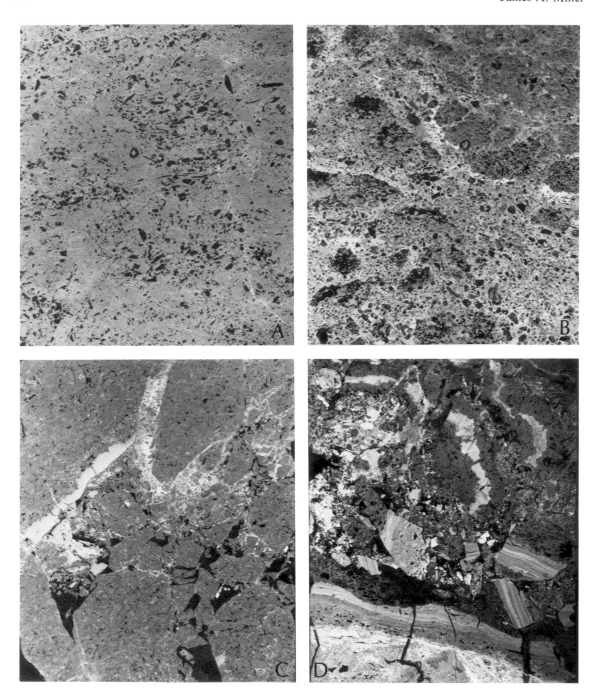

Fig. 9-6. Photomicrographs of Leadville carbonates. Negative prints, 2X. *A.* Bioclastic dolomite wackestone/packstone with abundant moldic porosity (black). *B.* Porous, nodular dolomite packstone with extensive moldic and intercrystalline porosity (black). *C.* Brecciated (solution breccia) dolomite mudstone/ wackestone with intercrystalline, fracture, and minor moldic porosity. Partial filling of fractures. Pores are black. *D.* Brecciated dolomite mudstone (solution breccia) with a variety of clasts, including laminated crusts and possible tidal-flat sediments. Partial filling of fractures with cement and finely crystalline dolomite.

m^3) of gas per day injected back into the reservoir (Clark, 1978).

The Mississippian carbonate reservoir produces by a combination of expanding gas cap and gravity drainage, with limited natural water drive. The interval was unitized on May 1, 1962, after which, on the basis of calculated reservoir performance, a pressure-maintenance program was begun (Parker, 1968).

Depositional History

The Mississippian (Leadville) and Upper Devonian (Ouray) carbonates in the Lisbon Field area represent a very shallow and partially restricted marine shelf that periodically built above sea level and produced broad supratidal flats. The presence of sandy and shaly intervals suggests that there may have been minor tectonic uplift and erosion in the area, which provided a source of terrigenous clastic sediment. Dolomitization, neomorphism, and leaching have altered or destroyed many of the allochems, cements, and textures of the original sediments. However, sedimentary structures such as current-bedding, burrowing, and breccia are usually visible.

The change from Ouray to Leadville facies is transitional, and is marked in most wells by a 10 to 20 foot (3–6 m) interval of shaly and sandy carbonates that produce a characteristic response on gamma-ray and acoustic logs (Fig. 9-4). The lower part of the Mississippian section is dominated by barren to sparsely fossiliferous dolomite mudstone deposited in a shallow subtidal to supratidal setting. The subtidal and intertidal facies are represented by bioclastic and pelletal mudstone to current-bedded grainstone, rip-up sedimentary clasts (Fig. 9-5), and submarine cements. The supratidal sediments are massive to well-laminated, contain relatively few fossil fragments, and have some irregularly laminated intervals that may represent algal stromatolites (Fig. 9-5A). Primary evaporites are rare or absent, but there are nodular masses of calcite cement, anhydrite, and chalcedony that appear to replace primary evaporites or fill cavities resulting from the leaching of nodular anhydrite or gypsum (Fig. 9-5C).

Higher in the sequence, at about 7880 feet (2401.8 m), a marine transgression initiated a new cycle of deposition (Fig. 9-1), as indicated by the presence of laterally extensive crinoidal mudstones, wackestones, and local accumulations of ooid and bioclastic grainstone (beach and tidal carbonate sand bodies). Tidal-flat and evaporitic facies developed locally and are represented by laminated mudstones, stromatolitic structures (Fig. 9-5A), evaporite relicts and replacement chalcedony, and solution-collapse breccias (Fig. 9-6C,D).

As deposition continued, upward shoaling of the sedimentary sequence resulted in intertidal and supratidal sedimentation over a large part of the area (Fig. 9-7). Lenses of crinoidal wackestone and packstone developed in intershoal areas, while algal laminae and evaporite minerals accumulated within the supratidal environment. Dolomitization obscured many of the primary structures, and leaching resulted in the formation of evaporite-solution breccias.

Overall, Mississippian deposition occurred on a shallow, partially restricted, medium-low energy shelf, in a moderately arid climate, as suggested by the presence and remnants of evaporite minerals (Fig. 9-7). Throughout most of the area, the sediment consists of shallow-marine bioclastic wackestone containing areally restricted developments of bioclastic sand and evaporitic supratidal dolomite mudstone. However, the discovery well (B-610) is dominated by intertidal and supratidal facies and solution-collapse breccias (Figs. 9-5, 9-6), suggesting that there may have been some early growth on the Lisbon structure that affected local facies distribution, and possibly also accelerated some of the diagenetic events.

Carbonate deposition ended during late Mississippian or early Pennsylvanian time; tectonic activity increased, producing the present Lisbon structure, and a major marine regression resulted in the development of a prominent unconformity on top of the Leadville carbonates. The uppermost part of the Mississippian section was severely eroded and an undetermined amount of section lost, possibly more than 100 feet (>30 m). All these events were followed by the development of the Paradox evaporite basin, which was subsequently filled by interbedded salt and shale units that butt against and completely overlie the Lisbon structure, producing an impervious hydrocarbon seal.

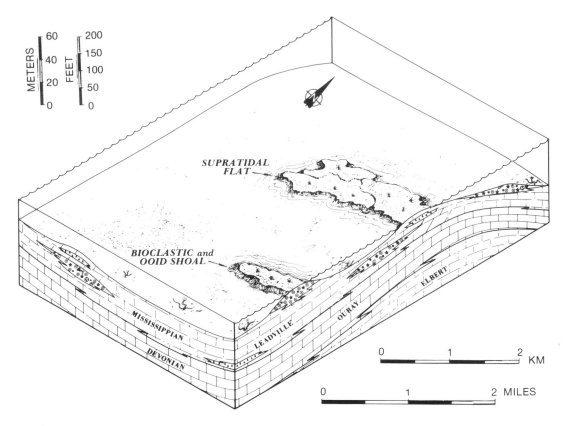

Fig. 9-7. Depositional model for the shallow-marine to supratidal Leadville carbonates, NW Lisbon Field area, San Juan Co., Utah.

Diagenesis

Intense leaching, dolomitization, and neomorphism (recrystallization and inversion) have severely altered many of the original sediments and cements. However, relict features ("ghosts") and areas of minimal alteration provide some insight into the early history of the sediments, and can be extrapolated to other parts of the section.

Cementation

Since the Mississippian section of this area is predominantly a low-energy sequence composed of mudstone to packstone, well-sorted carbonate sand bodies with good interparticle porosity capable of accommodating distinct cement fillings are rare. However, some of the better-sorted bioclastic and ooid grainstones do contain recognizable bladed and micritic cements. They are very

similar to cements described by James *et al.* (1976), and occur as micrite alone, bladed spar alone, and two-part cements composed of micrite (inner) and bladed spar (outer). These cements are probably submarine, and were precipitated shortly after deposition of the sediment. The final pore-filling cement in most of the grainstone is coarse equant calcite that probably was precipitated following deeper burial.

Lithification of the undolomitized micritic sediments probably occurred both by growth of neomorphic spar, and by the precipitation of interstitial calcite cement. These sediments probably were poorly lithified initially, and retained their fine-grain size and much of their original intercrystalline porosity.

Some of the fractured and vuggy or moldic porosity in the Lisbon section was partially or completely filled by anhydrite or sparry calcite and dolomite cements that were precipitated during deposition of the overlying Pennsylvanian evaporites.

Dolomitization

Geologic evidence suggests multiple episodes of dolomitization in the Lisbon Field carbonates. In the evaporitic tidal-flat parts of the section, dolomitization probably occurred almost simultaneously with deposition or reaction with solutions present in the depositional environments, whereas most dolomitization in the open-marine bioclastic intervals probably took place postdepositionally as a result of interaction with ground water.

The presence of evaporite facies (nodular anhydrite, etc.), solution breccias, and silicified zones indicates that concentrated brines were sometimes present during or shortly after deposition of the Mississippian (Figs. 9-5C and 9-6C,D). These brines were probably responsible for much of the very finely crystalline (15–30 μ) dolomite mudstone associated with the evaporitic intertidal and supratidal sediments. This type of dolomitization may have taken place intermittently as the sediments built above sea level and developed supratidal flats.

At least one and probably two or more later periods of dolomitization occurred. Near the end of Mississippian deposition, or during the subsequent uplift and erosion, percolating brines (ground water) selectively dolomitized the micritic fraction of many mudstones and wackestones, producing a finely crystalline dolomite matrix and leaving the bioclasts basically unaltered. The migration paths for these dolomitizing brines were probably determined by residual primary porosity and permeability in the micritic sediments.

Coarsely crystalline dolomite, zoned dolomites, and white sparry dolomite cement were observed in a few thin sections. Although no geochemical data are available, these types of dolomite resemble those described by Folk and Land (1975) and may be related to fresh water, or to the mixing of fresh water and marine or hypersaline brines.

The dolomitizing fluids may also have been responsible for the leaching of bioclasts and evaporite minerals, but it seems more likely that this leaching occurred during the period of uplift and erosion that followed deposition. Regardless of the exact mechanism or timing, selective leaching of crinoidal fragments produced widespread moldic and vuggy porosity, and the solution of evaporite minerals created solution-collapse breccias (Fig. 9-6).

An even later phase of dolomitization postdates most of the leaching and brecciation. This event, which probably occurred during the late stages of uplift and erosion, produced dolomitization along fractures and caused some infilling of moldic and fracture porosity by coarse dolomite.

Silicification

Minor intervals (usually less than 5%) of the carbonate section were completely replaced by silica, and some of the vug and fracture porosity was filled by quartzine (length-slow) chalcedony. Much of the silicification seems to have occurred shortly after deposition, and before dolomitization, because the structures of the original allochems and matrix are almost perfectly preserved in silica.

There is an excellent example of quartzine chalcedony replacing nodular anhydrite in core from well B-610. In this particular case, the nodular form and texture is well preserved, and some of the original anhydrite laths are included within the bladed silica (Fig. 9-5C).

Nearly all of the silicification and vug or fracture-filling chalcedony occurs within, or adjacent to, the solution breccia zones. This association has previously been noted by Folk and Pittman (1971) and Folk (1973), and suggests that there is a relationship between original evaporite facies and silicification in these sediments.

Fractures and Breccias

Compaction and mild tectonic activity seem to have been responsible for some of the closely spaced vertical fractures common throughout the Mississippian section at Lisbon. More than half of these fractures are filled with either shale, calcite mosaic, or anhydrite. The unfilled fractures are extremely fine and usually no more than six inches long. Since these fractures are discontinuous and are most abundant in the tighter, more dense portions of the section, they probably contribute little to reservoir porosity but may provide communication between porous units.

Evaporite solution-collapse breccias are present throughout the section at Lisbon. The distribution of the evaporites and resulting breccia is related to sabkha and restricted-lagoon facies that existed in the area throughout much of the Mississippian (Fig. 9-6C,D). Solution breccias are recognized by the following criteria:

1. Breccia zones may vary greatly in thickness and extent.
2. Clasts are very poorly sorted and vary greatly in size (<1 mm–1 m). They are usually angular or subangular, with very sharp surfaces devoid of coatings or overgrowths. Many clasts are fractured or veined.
3. Within a particular zone, the clasts are of similar lithology.
4. Clasts may be floating or grain-supported, and have a matrix or cement of dolomite, calcite, or anhydrite.
5. The matrix material usually has a high insoluble residue content.
6. Fracturing is rare beneath the breccia zones, but extends some distance upward, gradually decreasing in intensity and scale.

These breccias are important for various reasons: (1) their recognition and proper interpretation can be very helpful in identifying highly altered evaporitic sequences, (2) they may be confused with tectonic breccias or erosion surfaces, and (3) the breccia zones commonly have excellent porosity and permeability.

Erosion Surfaces

A karst surface and the overlying Molas Formation regolith clearly mark the unconformity at the top of the Mississippian. However, within the Mississippian rocks that were studied, there is no substantial evidence of a disconformity or even a recognizable hardground or erosion surface.

Baars (1966) postulated a major disconformity in the Lisbon area, occurring at 7760 feet (2365 m) in well B-610, on the basis of a thin zone of intraformational conglomerate. Megascopic and microscopic examination of the rock in this interval confirms that there is a "conglomeratic" zone, but the evidence indicates that it is a solution-collapse breccia rather than a depositional breccia. The matrix in this brecciated zone contains a very minor amount (<5%) of the fine (50–100 μ) quartz sand and silt, which seems to be an insoluble residue from the solution process, since

similar material occurs in other brecciated zones.

With the exception of the Mississippian-Pennsylvanian unconformity, no evidence was found in the Leadville Formation of erosion surfaces, karst features, or caliche.

Source and Distribution of Petroleum

The existence of the Lisbon Field in the midst of a largely unproductive region has caused considerable difficulty in interpreting the factors responsible for the accumulation of oil and gas. It has generally been assumed that any Mississippian petroleum was lost during uplift and erosion that occurred at the end of the Mississippian Period, and that the petroleum being produced from the reservoir migrated in from adjacent Pennsylvanian or Devonian rocks.

Although petroleum is present in the associated Pennsylvanian and Devonian rocks, various considerations suggest that oil and gas is unlikely to have migrated into the Lisbon reservoir across either depositional or fault contacts:

1. The top and bottom of the Mississippian carbonates are quite tight and show little or no oil staining, thus discouraging the idea of migration from above or below.
2. The fracturing observed in these rocks is of limited lateral and vertical extent and, as such, does not provide communication throughout the field.
3. The laterally discontinuous nature of porous intervals and presence of prominent faults within the field argue against extensive lateral migration.
4. Fault contacts between the Lisbon carbonates and Pennsylvanian shale are just as likely to have been seals, which would have prevented large-scale lateral migration of fluids.
5. A geochemical study of Pennsylvanian, Mississippian, and Devonian crude oil samples shows that the petroleum being produced from the Mississippian section at Lisbon is distinctly different from that present in the adjacent Pennsylvanian and Devonian rocks.

On the basis of these observations, I believe that the Lisbon oil and gas are unique to the Mississippian section, and that there has been no significant contribution from Pennsylvanian or Devonian sediments in the field area.

Crawford, John Dunham, and Tom Elliott, who provided suggestions, technical assistance, and editorial review.

Summary

Northwest Lisbon Field lies on the flank of a faulted anticline on the southern edge of the Paradox Basin in southeastern Utah. The field produces primarily from intercrystalline, moldic, and fracture porosity in Mississippian (Leadville) carbonates, and has estimated recoverable reserves of 42.85 million barrels of oil and 250 billion cubic feet of gas (7 billion m^3).

The carbonate sediments were deposited in shallow-marine to evaporitic supratidal environments. During and following deposition, the sediments were subjected to a variety of diagenetic events, including cementation, dolomitization, solution, uplift, and a major period of subaerial erosion that followed deposition. The carbonates were subsequently buried and sealed by a thick Pennsylvanian evaporite sequence.

Geological and geochemical evidence suggests that the petroleum was derived from the Mississippian sediments and migrated into the reservoir facies during mid- to late-Pennsylvanian time.

Acknowledgments   I thank the Union Oil Company of California for allowing this study to be published, and I acknowledge my colleagues, Al

References

BAARS, D.L., 1966, Pre-Pennsylvanian paleotectonics—key to basin evolution and petroleum occurrences in Paradox Basin, Utah and Colorado: Amer. Assoc. Petroleum Geologists Bull., v. 50, no. 10, p. 2082–2111.

CLARK, C.R., 1978, N.W. Lisbon Field production data: in Oil and Gas Fields of the Four Corners Area, v. II, Four Corners Geol. Soc., p. 662–665.

FOLK, R.L., 1973, Evidence for peritidal deposition of Devonian Caballos Novaculite, Marathon Basin, Texas: Amer. Assoc. Petroleum Geologists Bull., v. 57. no. 4, p. 702–725.

FOLK, R.L., and L.S. LAND, 1975, Mg/Ca ratio and salinity: two controls over crystallization of dolomite: Amer. Assoc. Petroleum Geologists Bull., v. 59, no. 1, p. 60–68.

FOLK, R.L., and J.S. PITTMAN, 1971, Length-slow chalcedony–a new testament for vanished evaporites: Jour. Sedimentary Petrology, v. 41, no. 4, p. 1045–1958.

JAMES, N.P., R.N. GINSBURG, D.S. MARSZALEK, and P.W. CHOQUETTE, 1976, Facies and fabric specificity of early subsea cements in shallow Belize reefs: Jour. Sedimentary Petrology, v. 46, no. 3, p. 523–544.

PARKER, J.M., 1968, Lisbon Field area, San Juan County, Utah: in Natural Gases of North America: Amer. Assoc. Petroleum Geologists Mem. 9, v. 2, p. 1371–1388.

10
Little Knife Field
Robert F. Lindsay and Christopher G. St. C. Kendall

RESERVOIR SUMMARY

Location & Geologic Setting Billings, Dunn and McKenzie Co.'s (T144 & 145N, R97 & 98W, Williston Basin, North Dakota, USA

Tectonics Linear block and wrench-fault tectonics, recurrent basement movement

Regional Paleosetting SE shelf of cratonic Williston Basin

Nature of Trap Structural/stratigraphic

Reservoir Rocks
  Age Mississippian
  Stratigraphic Units(s) Mission Canyon Formation
  Lithology(s) Dolomite and calcareous dolomite
  Diagenesis/Porosity Transitional open-restricted marine and restricted marine, shoaling upward, cyclically repeated
  Productive Facies Dolomitized skeletal wackestone and pellet wackestone-packstone
  Entrapping Facies Anhydrite, intertidal packstone-grainstone, and subtidal dense packstone
  Diagenesis/Porosity Partial replacement by anhydrite, followed by dolomitization, then dissolution of anhydrite.

Petrophysics
  Pore Type(s) Moldic and intercrystal
  Porosity 8.5–27% avg 14% dolomite
  Permeability 01.0–167 md range, 30 md avg
  Fractures Common but widely spaced, closed to hairline vertical

Source Rocks
  Age Late Devonian to Early Mississippian
  Lithology(s) Bakken Shale, organically rich (Lodgepole limestone secondary source)
  Migration Time Cretaceous

Reservoir Dimensions
  Depth 9800 ft avg ($\sim$3000 m)
  Thickness Max 109 ft (33 m) net, avg 24 ft (7.3 m) net, B–D zones
  Areal Dimensions 12 $\times$ 2.5–6.0 mi (19.2 $\times$ 4–10 km)
  Productive Area $\sim$24,000 acres (96 km^2)

Fluid Data
  Saturations S_o = 60%, S_w = 40% (averages)
  API Gravity 41°
  Gas-Oil Ratio 1250:1
  Other C_1-C_3 = 51.7%, C_4–C_6 = 10%, C_6 = 27%, H_2S = 8%, other 3.3%

Production Data
  Oil and/or Gas in Place 195 million BO
  Ultimate Recovery NA
  Cumulative Production 28 million BO through November 1982

Remarks: IP 480 BOPD flowing from 8 ft (2.4 m) of perforated casing. IP mechanism solution-gas (depletion) drive with limited water drive. Discovered 1977.

10
Depositional Facies, Diagenesis, and Reservoir Character of Mississippian Cyclic Carbonates in the Mission Canyon Formation, Little Knife Field, Williston Basin, North Dakota

Robert F. Lindsay and Christopher G. St. C. Kendall

Location and Discovery

Little Knife Field is located near the structural center of the Williston Basin in west central North Dakota, south of the Nesson Anticline, in parts of Billings, Dunn, and McKenzie Counties, T144-145N, R97-98W (Fig. 10-1). Production is from the Mission Canyon Formation (lower Mississippian). The Mission Canyon lies between the Lodgepole Limestone below and the Charles Formation above, and all three combine to form the Madison Group. At this location, the Mission Canyon is 465 feet thick (142 m) and contains several porous hydrocarbon-bearing zones. The field lies within the Little Knife Anticline, a broad, low, northward-plunging anticlinal nose with less than 1 degree of structural relief (Fig. 10-2). The top of the Mission Canyon at the crest of the structure lies at a depth of 9600 to 9700 feet (≈ 3 km). The field is approximately 12 miles (19.2 km) long and 2.5 to 6 miles wide (4–9.3 km).

Little Knife was discovered in January 1977 by the Gulf 1-18 State wildcat well, which flowed 480 barrels of oil per day of undersaturated sour crude (API 43° gravity) from 8 feet (2.4 m) of perforations. As of June 1984, the field included 179 wells drilled on 160-acre spacing and had produced 31 million barrels of oil. The primary reservoir drive mechanism is solution gas drive (depletion drive).

Forty-one Mission Canyon wells were cored inside the field boundaries, and seven outside the field boundaries (Fig. 10-2), yielding 5490 feet (1674 m) of core.

Major Zone Lithologies

Key beds were identified on the basis of open-hole well logs, which permitted subdivision of the section into six informal zones designated as A, B, C, D, E, and F (Fig. 10-3). Facies of the complete Mission Canyon section were studied using several overlapping cores to construct a composite section. Lithologies of these zones, within the Mission Canyon Formation at Little Knife Field, are summarized in this section. Net pay thickness distributions are shown in Figure 10-4.

Zone F, the lowest log/lithology zone in the Mission Canyon, is 170 feet (52 m) thick and composed of alternating medium to thick beds of undolomitized (to slightly dolomitized) lime mudstone interbedded with skeletal packstone/grainstone. The basal 10 to 15 feet (3-5 m) are slightly argillaceous, interlaminated skeletal lime packstones and pelletal lime packstones, forming a transition into the underlying Lodgepole Limestone. Zone F is non-hydrocarbon bearing.

Zone E, 50 feet (15 m) thick, is composed of thick-bedded, slightly to completely dolomitized

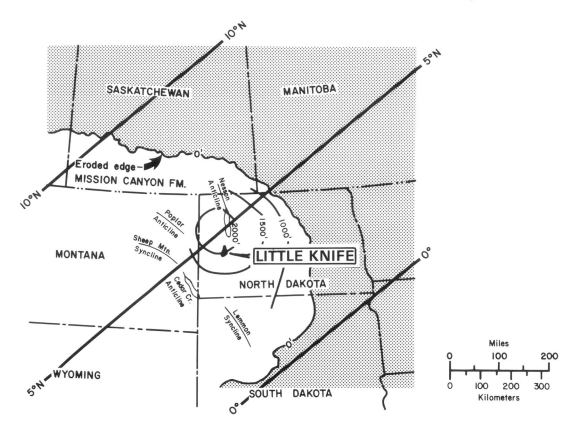

Fig. 10-1. Index map of Williston Basin, showing: eroded edge of Mission Canyon Formation (Proctor and Macauley, 1968); major surface and subsurface structural features (Willstrom and Hagemeier, 1978, 1979); isopach thickness of Madison Group (Carlson and Anderson, 1965); and generalized Carboniferous paleolatitude lines (Habicht, 1979).*

skeletal mudstone/wackestone with irregularly shaped incipient siliceous nodules. Porous beds in this zone also are non-hydrocarbon bearing.

Zone D, 50 feet (16 m) thick, constitutes the lowermost interval of the reservoir itself at Little Knife. It is medium to thick-bedded, partially to completely dolomitized, burrowed, mudstone and skeletal wackestone, which exhibits local replacement by anhydrite that was later leached.

Zone C, 65 feet (20 m) thick, also forms a portion of the porous reservoir interval. It is medium to thick-bedded, becoming mostly nonporous in its mid- to upper portions. Dolomitized skeletal

wackestone beds, containing some replacement anhydrite, grade upward into variable dolomitized pelletal wackestone/packstones. Nonporous beds have small amounts of chert occluding intercrystal pores. Upper portions of the zone have rare quartz-silt laminations in dense microcrystal dolostone beneath porous reservoir beds.

Zone B, 70 feet (21 m) thick, forms the uppermost interval of the reservoir. It is subdivided into upper and lower portions:

1. The *mid-lower* portion consists of beds of dolomitized, burrowed, sparsely skeletal, pelletal wackestone/packstones. These were partially replaced by anhydrite and later leached, producing moldic pores in intercrystalline dolostone porosity. This facies thins southward.

*Figures 10-1, 2, 3 modified from and 10-5, 6, 7, 10 from Lindsay and Kendall, 1980. Published by permission, Society of Economic Paleontologists and Mineralogists.

2. The *upper* portion, 5 to 40 feet (1-1/2–12 m) thick, consists of sparse, thin, lenticular, discontinuous beds of porous, dolomitized skeletal wackestone, also partly replaced by anhydrite which is locally leached, and interbedded with dense, cemented wackestone/grainstones. Constituents include peloids, clotted and lumped micrite, ooids, pisolites, calcispheres, and skeletal detritus. This facies thickens southward and becomes partially anhydritic.

Zone A is a 60-foot (18-m) cap of thin-to thick-bedded anhydrite at the top of the Mission Canyon. It is characterized by a variety of structures that include: (1) chicken-wire mosaic, (2) thin-bedded mosaic, (3) laminated to medium-bedded, (4) ropy displacive, and (5) burrowed replacive (Maiklem *et al.*, 1969). Both a dolomite matrix and interbeds of laminated dolostone are associated with the anhydrite beds. In upper portions of zone B and at the base of zone A, there are pseudomorphs of anhydrite after selenite crystals, anhydrite porphyroblasts, and local laminated crusts.

Depositional Setting

The Mission Canyon as a whole is a shallowing-upward, regressive sequence characterized by a general upward change in lithology from carbonates to anhydrite (Lindsay, 1982; Lindsay and Roth, 1982) (Figs. 10-3, 10-5). It is analogous to the lime mud-to-sabkha cycle of Wilson (1975, p. 297–298). Most of the carbonates are interpreted as subtidal, deposited in five recognizable subenvironments from the base upward: (1) basinal "deeper water" carbonates in basal zone F (zone X of Irwin, 1965); (2) open shallow-marine in zone F (zone X-Y of Irwin, 1965); (3) transitional marine between open marine and a shallow protected shelf in zones E, D, and C (zone Y-Z of Irwin, 1965); (4) a protected restricted marine shelf in zones C and B (zone Z of Irwin, 1965); and (5) a thin, narrow marginal marine belt in mid-upper zone B (zone Z of Irwin, 1965).

This subtidal setting was interrupted by several subsidiary shallowing-upward carbonate cycles beautifully displayed in cores (Lindsay and Kendall, 1980; Lindsay and Roth, 1982). In the

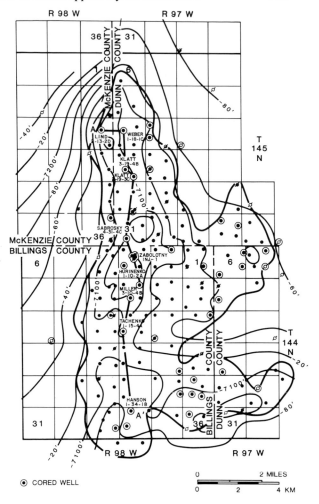

Fig. 10-2. Structure on top of the Mission Canyon Formation in Little Knife Field. Cross-section line A–A' refers to Figure 10-7.

lower two-thirds of the Mission Canyon (zones F, E, D, and C), carbonate cycles are repeated several times, probably in response to subtle eustatic fluctuations in sea level, which formed major cycles in the open marine (zone F) and minor cycles in the transitional open-restricted marine (zones E, D, and C). Major carbonate cycles consist of medium-to thick-bedded, burrowed mudstone grading up into skeletal packstone/grainstones deposited in the open-marine environment (Fig. 10-3).

Minor carbonate cycles consist of medium-to thick-bedded, burrowed, pelletal mudstones grading up into skeletal wackestones. Skeletal fragments were washed in from the open marine and

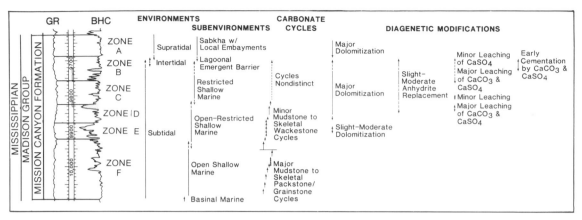

Fig. 10-3. Type gamma-ray/sonic log of Mission Canyon Formation in central portion of Little Knife Field. Depositional environments, carbonate cycles, and diagenetic modifications are from a composite section of several cores. Stippled areas indicate hydrocarbon accumulations.

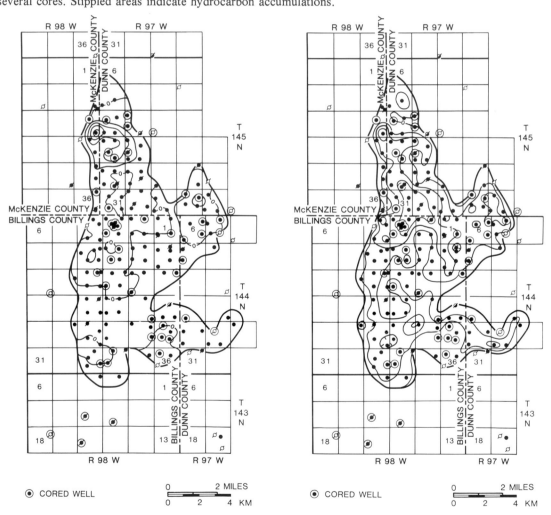

Fig. 10-4. Net pay isopach maps of the Mission Canyon Formation in Little Knife Field. *A.* Uppermost restricted marine interval in middle part of zone B, in northern part of the field. *B.* Remainder of restricted marine interval, in lower zone B and upper zone B, throughout the field.

deposited with accumulating lime mud in the transitional open-to-protected shallow-shelf subenvironment, where further shallowing, wave energy reduction, and loss of circulation occurred. These minor cycles eventually gave way to burrowed, sparsely skeletal, pelletal muds in the restricted inner parts of the protected shelf subenvironment (mid-lower zone B and upper zone C).

The barrier island buildup facies (mid-upper zone B) forms a thin veneer of variable thickness, which terminated subaqueous deposition. A low-energy environment with little tidal exchange is suggested, with barrier beaches reworked by

storms that periodically washed over them. Dense, thin lagoonal limestone beds are immediately behind and interfinger with the buildup facies. These beds then interfinger with, and are overlain by, supratidal anhydrite beds.

Anhydrite cements, pseudomorphs of anhydrite after selenite crystals, anhydrite prophyroblasts, and localized laminated crusts (upper zone B and base of zone A) are the first indications of higher salinities and subaerial exposure as supratidal sequences prograded toward the depocenter. Anhydrite beds overlie these first indicators, displaying subfelted to felted textures. Thin interbeds of laminated dolomite are the only other lithology present, and record storm-washover ponds or playas and local embayments onto the sabkha.

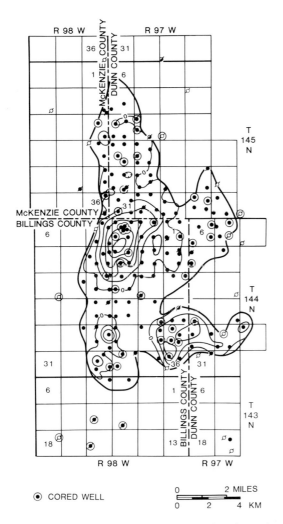

Fig. 10-4C. Transitional, open-to-restricted marine facies in productive central and southern portions of the field (upper zone D and basal zone C). See Figure 10-7 for stratigraphic relationships.

Reservoir—General

A typical log from the central part of the field displays hydrocarbon accumulations in the upper half of the Mission Canyon (stipple pattern, zones B, C, and D; Fig. 10-3). The Little Knife anticline involves average dips of 1/4 degree east, and 1/2 degree west, converging in a structural nose to the north. Hydrocarbon accumulations disappear progressively up- and off-structure. To the south, closure is by lateral facies changes (described later), which create stratigraphic entrapment.

Diagenetic modifications to the protected shelf and transitional marine sediments rich in lime mud created what are now the principal reservoirs, as illustrated by core photos in Figure 10-6. Porosity of these facies was enhanced by partial to complete dolomitization. Skeletal constituents were partially replaced by anhydrite, which was later dissolved. In restricted marine beds (zone B) the result is an abundance of ghosts and molds of pellets preserved as dolomite-rhomb rimmed "necklaces" grading into dolomite intercrystal pores. In transitional open-restricted marine beds (zone D), porosity occurs in dolomitized skeletal wackestone, which consists of both skeletal fragments and skeletal molds partially replaced by anhydrite and leached and intercrystal pores within the dolomite matrix.

Cores reveal discontinuous vertical, planar, hairline fractures that may have been initiated at the present depth of the field, as tensional release

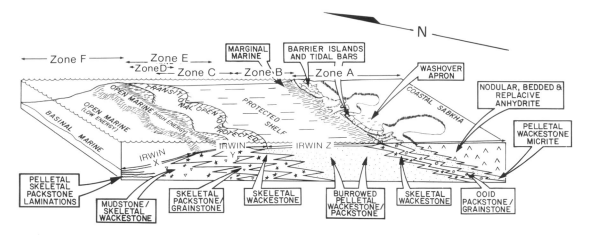

Fig. 10-5. Idealized depositional setting of Mission Canyon Formation at Little Knife Field. Informal log zonations A–F (top) and Irwin's (1965) epeiric sea energy zones X, Y, and Z (base) illustrate respective positions occupied in the depositional system.

features (Narr and Burruss, 1982; W. Narr, personal communication, 1981). Fractures are found in all facies except in anhydrite beds. No displacement along fractures can be demonstrated, and fractures continue, for example, through skeletal and oolitic-pisolitic constituents with no lateral offset. Fractures at Little Knife Field increase overall permeability two-fold (Nettle *et al.*, 1981; Desch *et al.*, 1982, 1983, 1984).

Reservoir Development

The various hydrocarbon-bearing porous beds associated with Little Knife Field have locally well-developed porosity that deteriorates laterally. Major intervals of porous carbonate occur in restricted marine beds in middle to lower portions of zone B, and transitional open-restricted marine beds in portions of zone D (Fig. 10-3). For simplicity, lower zone C porosity is included in the discussion of zone D porosity. Other porous intervals are much thinner and more lenticular, and either cannot be correlated from log to log, or else form only thin persistent sections (upper zone B and mid-portions of zone C). Other beds of porous rock occur lower in the Mission Canyon (zones E and F) but do not contain hydrocarbons.

Transitional Open-Restricted Marine Reservoir Interval (Zone D)

The lowest hydrocarbon-bearing beds in the reservoir interval are within the transitional open to restricted marine facies in dolomitized skeletal wackestones, at the top of the last well-developed minor carbonate cycle (zone D; Fig. 10-7). This portion of the reservoir is confined to central portions of the field, where there is slight additional structural flexure and better porosity development. The reservoir is structurally confined on the east, west, and north, and is bounded by edge water on the northeast, north, and west. A facies change from beds of porous dolomitized skeletal wackestone to dense beds of skeletal packstone forms a southward stratigraphic trap (Fig. 10-8C). Reduction in porosity is due to limited amounts of carbonate mud available to be dolomitized. Skeletal detritus is composed of crinoid columnals, less common brachiopod shells, and sparse bryozoan fronds.

Porosity Generation

Porosity was generated where more carbonate mud associated with skeletal wackestone was

Fig. 10-6. Photographs of core slabs of porous Mission Canyon beds in central part of Little Knife Field. LEFT is restricted marine facies from zone B in Gulf Sabrosky well 4-31-4C, 9733 feet (2967 m); 17% porosity and 10 md permeability. Rock is a dolomitized, thoroughly burrowed, sparsely skeletal, pelletal wackestone/packstone, with coral fragment in upper right. RIGHT is transitional open-restricted marine facies from zone D in Gulf Sabrosky well 4-31-4C, 9816 feet (2992 m); 20% porosity and 98 md permeability. Rock is a partially dolomitized, burrowed, skeletal wackestone. Bar scale is 2 cm long.

available to be converted into dolomite with intercrystal porosity. Once most mud had been dolomitized, local sources of carbonate ions for dolomitization were the crinoid and brachiopod skeletal constituents (see Murray, 1960). At approximately 60 to 70 percent conversion to dolostone, some skeletal fragments were dissolved, leaving larger crinoid columnals and brachiopod fragments to "float" in a matrix of porous, sucrosic dolostone, ringed by dolomite-rhomb "necklaces." Laterally, where skeletal detritus is more abundant, only incipient dolomitization of finer muds (less than 50%) has occurred, with little or no porosity enhancement.

Apparently, where larger volumes of lime mud were available, fluids moving through the subtidal

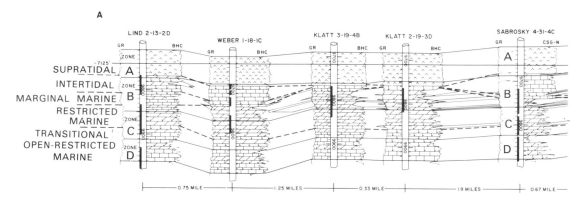

Fig. 10-7. North-south structural cross-section of upper half of the Mission Canyon Formation through cored wells in Little Knife Field. Black vertical bars represent cored intervals; shaded areas are porous hydrocarbon-bearing intervals; and dashed lines are boundaries between depositional settings. Intertidal includes low-lying

beds became less concentrated with respect to magnesium, and nucleation sites were spread farther apart, promoting the creation of porous, fine-grained dolomite (Asquith, 1979, p. 22–23). We speculate that the lack of dolomitization in places is the result of: (1) smaller volumes of carbonate muds; (2) a lack of more homogeneous permeable pathways, due to incomplete homogenization by burrowing organisms; and (3) increased amounts of skeletal detritus which, when their mud matrix was compacted, formed a mud-supported (grains nearly compacted together) texture that inhibited fluid flow. It is probable that all these effects are interrelated so that, at certain locations within the field, lateral sweep efficiency of dolomitizing solutions through the same horizon produced different styles or types of dolomitization and degrees of porosity development. Later, leaching of anhydrite-replaced skeletal detritus further increased porosity (Lindsay and Roth, 1982).

Restricted Marine Reservoir Interval (Zone B)

Restricted marine portions of the reservoir interval (zone C and B) are composed of beds of burrowed, muddy, sparsely skeletal, pelletal wackestone/packstone that have been partially to completely dolomitized. Porosity commonly decreases laterally as in the transitional open-res-

tricted marine beds (zone D), but with a gentler gradient and over greater areas (Fig. 10-7). These sediments were apparently deposited in a restricted, protected subtidal setting with low amounts of circulation. This more restricted environment promoted more accumulation of carbonate mud and pellet-rich sediment. The original sediments were homogenized by burrowing. Only sparse skeletal detritus occurs, as fine, broken crinoid columnals. Laterally, away from the well-developed porous intervals, a facies change into less porous beds of sparsely dolomitized, burrowed, slightly less muddy, skeletal to pelletal wackestone/packstones of marginal marine subtidal origin occurs. Further lateral shifts in facies commonly involve the lowest beds of "dense," cemented, ooid-pisolitic packstone/grainstones, which represent the first sediments to emerge above the subtidal realm at the strand line, where localized waves and storms produced barrier island buildups (mid-upper zone B).

Porosity Generation

Best porosity developments are associated with partial to complete dolomitization of both mud and particles (Figs. 10-8A, B, D, E), in a sequence like that described for the transitional open-restricted marine beds (zone D). Pellets have dolomite-rhomb "necklaces" that seem to have formed prior to grain leaching, producing

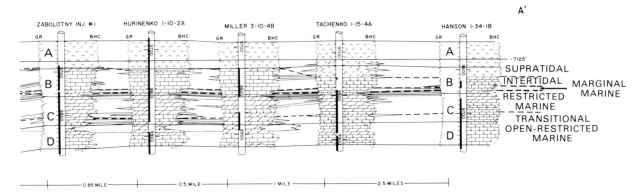

barrier island buildups, intertidal beaches, and lagoonal limestones. Note thickening of anhydrite beds to the south (A') (zone A–uppermost zone B). Also note southward changes in restricted marine (zone B) and transitional open-restricted marine (zone D) beds from porous dolostone to dense limestone.

dolomite-rhomb rimmed, small moldic pores of oval shape (Fig. 10-8E). Schmidt (1965, p. 143–144) noted similar dolomite rim cements, as did Kaldi and Gidman (1982), and concluded that the rim cement was early diagenetic.

Changes in porosity in restricted marine beds (zone B) are far more complex than in transitional open-restricted marine beds (zone D). Vertical as well as lateral changes in porosity development subdivide restricted marine beds (zone B) into more than one porosity unit. These expand laterally and join thicker porous beds or contract to bifurcate or pinch out. This increase in complexity is thought to be due to the relationship between the original subtidal muddy wackestone/packstones and the intertidal rocks that both directly overlie them and pass laterally into them (Figs. 10-2, 10-7). These intertidal rock suites were cemented early by calcite and anhydrite, thus occluding interparticle porosity and forming the lowest beds of discontinuous caprock over the field. The true seal is the overlying anhydrite beds.

Besides being dolomitized, the porous beds of zone B have also undergone slight to moderate levels of anhydrite replacement and later leaching (Figs. 10-8A, 10-B). Skeletal constituents, namely very fine crinoid fragments, though diminished in volume and size, are preferentially replaced. This enhancement of porosity development by anhydrite dissolution is common at Little Knife.

Presumably, brine-enriched solutions concentrated Ca^{+2} which combined with SO_4^{-2} as beds of anhydrite-isolated nodules and replaced skeletal fragments. This effectively reduced the SO_4^{-2} and Ca^{+2} concentration and in increased the Mg^{+2}/Ca^{+2} ratio to promote dolomitization, though at much slower rates (Bathurst, 1975, p. 531–532). Certain beds are composed of dense microcrystalline dolostone. These may reflect areas where multiple, closely spaced dolomite-crystal nucleation occurred.

Pore Types and Geometries

Four pore types and sizes are recognized in the Mission Canyon (Lindsay, in press) (Fig. 10-9). *Moldic* pores (Choquette and Pray, 1970), produced by leaching of anhydrite-replaced skeletal constituents, are largest and measure approximately 30 to 300 microns wide. These large moldic pores are surrounded by dolomitized muddy matrix housing *intercrystal* pores and associated pore throats, between sets of dolomite crystals (Fig. 10-8). *Polyhedral* pores are the largest intercrystal pores, surrounded by several dolomite crystal faces, and form complex polyhedron shapes approximately 10 to 50 microns wide. *Tetrahedral* pores are intermediate-sized intercrystal pores, which form triangular shapes reduced in size to 3 to 10 microns. *Interboundary-sheet* pores are the smallest pores and

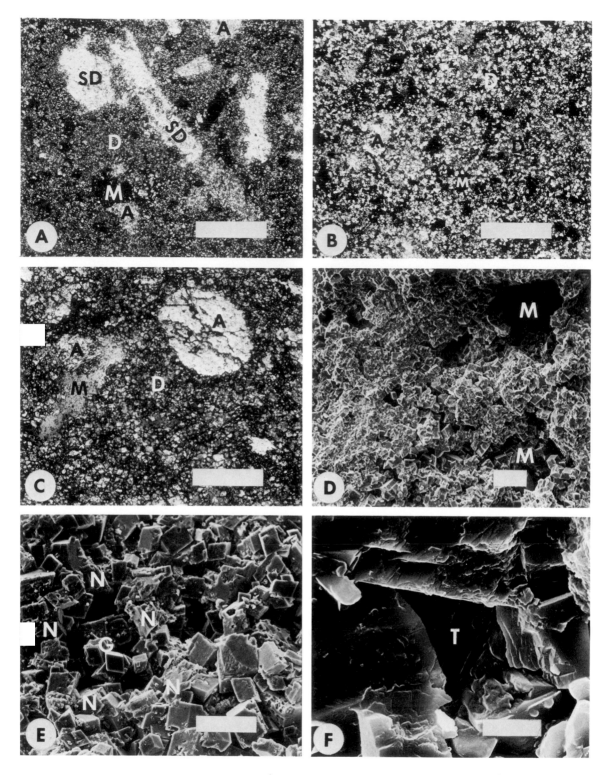

Fig. 10-8. Photomicrographs (*A,B,C*) and scanning electron micrographs (*D,E,F*) of the Mission Canyon Formation at Little Knife Field. *A*. Porous marginal marine dolomite facies (uppermost zone B). Originally skeletal wackestone with completely dolomitized matrix (D), with dolomite impinging into skeletal detritus (SD). Moldic pore space (M) formed where anhydrite replaced skeletal detritus and was later leached, leaving remnants of anhydrite (A). Gulf Lind well 2-13-2D, 9783 feet (2982 m), 20% porosity and 6 md permeability. Cross-polarized light. Bar is 500 μ long. *B*. Porous restricted marine facies (middle zone B). This rock is completely dolomitized (D). Replacement anhydrite (A) is mostly leached, producing moldic (M)

Fig. 10-9. Generalized diagram and scanning electron ➤ micrographs (SEM) of pore casts from porous dolostones of Mission Canyon Formation at Little Knife Field *A*. Various pore types and throats in porous portions of the Mission Canyon. Stippled area is pore space surrounding dolomite crystals. (P-M) is a small moldic pore transitional to a large intercrystal polyhedral pore. Polyhedral pores (P) have complex polyhedron shapes. Tetrahedral pores (T) are intermediate-sized intercrystal pores, formed where three dolomite crystals impinge together forming a triangular shape in two-dimensional view. Interboundary-sheet pores (S) are the smallest and form between closely spaced dolomite crystals. Bar is 30 μ long. *B*. Porous transitional open-restricted marine facies (zone D). Large polyhedral pores (P) give way here to smaller polyhedral, tetrahedral (T) and interboundary-sheet (S) pores. Both large and narrow throats connect the individual pores. Gulf Sabrosky well 4-31-4C, 9820 feet (2992 m), 23.6% porosity and 167 md permeability. Bar is 20 μ long. *C*. Porous, transitional open-restricted marine facies (zone D). Narrow interboundary-sheet pores (S) still provide some effect on permeability, although pore space is highly reduced. Gulf Sabrosky well 4-31-4C, 9820 feet (2993 m), 23.6% porosity and 167 md permeability. Bar is 10 μ long.

pores associated with smaller intercrystal pores. Gulf Klatt well 3-19-4B, 9781 feet (2981 m), 15% porosity and 36 md permeability. Cross-polarized light. Bar is 500 μ long. *C*. Porous transitional open-restricted marine facies (upper zone D). Originally skeletal wackestone with a dolomitized (D) matrix and replacement anhydrite (A). Crinoid columnal left of center was replaced by anhydrite and later partially leached, leaving a moldic pore (M). Gulf Sabrosky well 4-31-4C, 9821 feet (2993 m), 24% porosity and 133 md permeability. Plane light. Bar is 500 μ long. *D*. Porous restricted-marine facies (lower zone B). Two moldic pores (M) are surrounded by a completely dolomitized matrix containing intercrystal polyhedral pores. Gulf Miller well 3-10-4B, 9812 feet (2991 m), 26% porosity and 36 md permeability. Cross-polarized light. Bar is 100 μ long. *E*. Porous restricted marine facies (mid-lower zone B). A pellet "ghost" (G) has been leached to form a moldic pore surrounded by a "necklace" of dolomite rhombs (N). Gulf Sabrosky well 4-31-4C, 9733 feet (2967 m), 17% porosity and 10 md permeability. Bar is 100 μ long. *F*. Porous restricted marine facies (mid-lower zone B). Dolomite crystals have partially grown together reducing pore space to a tetrahedron-shaped tetrahedral pore (T). Gulf Sabrosky well 4-31-4C, 9733 feet (2967 m), 17% porosity and 10 md permeability. Bar is 10 μ long.

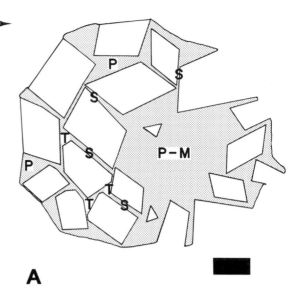

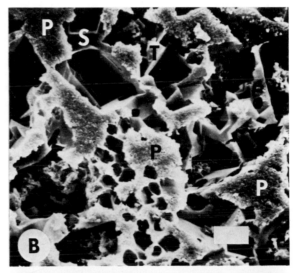

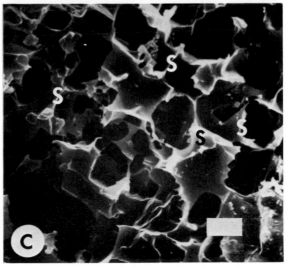

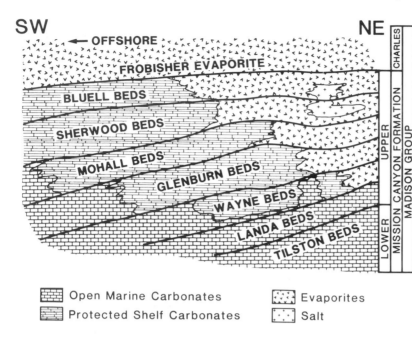

Open Marine Carbonates   Evaporites
Protected Shelf Carbonates   Salt

Fig. 10-10. Regional cross-section of Mission Canyon Formation across North Dakota portion of the Williston Basin. Modified after Harris et al. (1966) and Malek-Aslani (1977).

pore throats, found between individual dolomite-rhombs, where pores appear to be reduced to long, very thin spaces approximately 1.6 microns wide. These intercrystal pore types are similar to those described by Wardlaw (1976) and Wardlaw and Taylor (1976).

Pore-to-throat size ratios are highly variable, ranging from 40:1 where larger moldic pores are well-developed, down to 4:1. The average pore-to-throat coordination number, or average number of throats connected to each pore in a two-dimensional view, is 3 to 5 (Lindsay, in press). Permeability development is dominated by two pore throat sizes in the Mission Canyon: (1) interboundary-sheet pores/pore throats, with 0.8 micron radii, and (2) 1.6 to 4 micron pore-throat radii. Both sizes of pore throats interconnect all four pore types in three dimensions.

coastal sabkha, aragonite and gypsum or anhydrite precipitated out, increasing the Mg^{2+}/Ca^{2+} ratio to favor dolomitization (Butler, 1969). A sabkha, prograding toward the Williston depocenter, may have provided a regional, long-continued, predominantly lateral and downward movement of enriched brines (Jacka and Franco, 1974). In a regional view of the Mission Canyon provided by Harris *et al.* (1966) and Malek-Aslani (1977), an eastward thickening of evaporitic beds suggests that brines in adequate volumes were derived from subaerially exposed areas at the time the carbonate sediments were accumulating (Figs. 10-5 and 10-10). Sediment compaction on the coastal sabkha, as well as the density of the brines, may have provided needed energy to force the brines seaward into the subtidal muds (Lindsay and Roth, 1982).

Brine Source

A source is needed for the large volumes of water containing calcium, magnesium, and sulfate ions to form the thick sequence of anhydrite beds, and at the same time to dolomitize the subtidal muddy sediments below. A coastal sabkha with a seaward flow of brines enriched by evaporation could have been such a source. As these brine solutions moved through the inner recharge zones of the

Source Beds

The source for hydrocarbons now entrapped in the Mission Canyon Formation is widely thought to be the Bakken Shale (Williams, 1974). Stratigraphically, the Bakken Shale straddles the Devonian-Mississippian boundary and underlies the Lodgepole Limestone. It is a highly organic-rich shale with high organic carbon content, especially toward the basin center. Migration from maturing organic matter is calculated to have begun at approximately 7000 feet (2150 m) of burial, with maximum expulsion of hydrocarbons during the Cretaceous (Dow, 1974). A second possible source for the generation of hydrocarbons is the Lodgepole Limestone, which in parts of the Williston Basin is reported to have sufficient quantities of organic matter (Williams, 1974). Migration paths were probably along vertical fractures and then laterally beneath the Mission Canyon evaporites until entrapment occurred.

Conclusions

Reservoir development and hydrocarbon accumulation in Little Knife Field are primarily the result of structural flexure and stratigraphic entrapment, coupled with diagenetic porosity development. The Mission Canyon Formation represents a prograding carbonate sequence punctuated by upward-diminishing cyclic carbonate sedimentation as a result of an overall regression. Depositional environments changed with time from dominantly open shallow-marine settings into transitional and restricted, low-energy, shallow-marine environments on a protected shelf, to emergent barrier island and back-lagoon sections overridden by prograding supratidal flats. Within the mud-enriched protected shelf setting, the transitional open shallow-marine to restricted-shallow marine facies (zones B, C, and D) were dolomitized and contain all hydrocarbon-bearing beds. Subtle facies variations in original lime-mud content, burrowing, and compaction combined to cause parts of the formation to be more susceptible than others to the development of porosity through anhydrite replacement, dolomitization, and later leaching of anhydrite.

References

ASQUITH, G.B., 1979, Subsurface Carbonate Depositional Models—a Concise Review: The Petroleum Publ. Corp., Tulsa, OK, 121 p.

BATHURST, R.G.C., 1975, Carbonate Sediments and Their Diagenesis: Developments in Sedimentology 12: Elsevier Scientific Publ. Co., Amsterdam, p. 531–532.

BUTLER, G.P., 1969, Modern evaporite deposition and geochemistry of coexisting brines, the sabkha, Trucial Coast, Arabian Gulf: Jour. Sedimentary Petrology, v. 39, p. 70–89.

CARLSON, C.G., and S.B. ANDERSON, 1965, Sedimentary and tectonic history of North Dakota part of Williston Basin: Amer. Assoc. Petroleum Geologists Bull., v. 49, p. 1833–1846.

CHOQUETTE, P.W., and L.C. PRAY, 1970, Geologic nomenclature and classification of porosity in sedimentary carbonates: Amer. Assoc. Petroleum Geologists Bull., v. 54, p. 207–250.

DESCH, J.B., W.K. LARSEN, R.F. LINDSAY, and R.L. NETTLE, 1982, Enhanced oil recovery by CO_2 miscible displacement in the Little Knife Field, Billings County, North Dakota: SPE/DOE paper 10696, Third Joint Symp. Enhanced Oil Recovery, p. 329–339.

DESCH, J.B., W.K. LARSEN, R.F. LINDSAY, and R.L. NETTLE, 1983, Little Knife CO_2 minitest, Billings County, North Dakota—Final Report: DOE/MC/08383-45, v. 1, 260 p.

DESCH, J.B., W.K. LARSEN, R.F. LINDSAY, and R.L. NETTLE, 1984, Enhanced oil recovery by CO_2 miscible displacement in the Little Knife Field, Billings County, North Dakota: Jour Petroleum Technology, v. 36, p. 1592–1602.

DOW, W.G., 1974, Application of oil-correlation and source-rock data to exploration in Williston Basin: Amer. Assoc. Petroleum Geologists Bull., v. 58, p. 1253–1262.

HABICHT, J.K.A., 1979, Paleoclimate, paleomagnetism and continental drift: Amer. Assoc. Petroleum Geologists Studies in Geol. 9, 31 p.

HARRIS, S.H., C.B. LAND, JR., and J.H. MCKEEVER, 1966, Relation of Mission Canyon Stratigraphy to oil production in north-central North Dakota: Amer. Assoc. Petroleum Geologists Bull., v. 50, p. 2269–2276.

IRWIN, M.L., 1965, General theory of epeiric clear water sedimentation: Amer. Assoc. Petroleum Geologists Bull., v. 49, p. 445–459.

JACKA, A.D., and L.A. FRANCO, 1974, Deposition and diagenesis of Permian evaporites and associated carbonates and clastics on shelf areas of the Permian Basin: in Fourth Symposium on Salt, v. 1: Northern Ohio Geol. Soc., p. 67–89.

KALDI, J., and J. GIDMAN, 1982, Early diagenetic dolomite cements: Examples from the Permian Lower Magnesian Limestone of England and the Pleistocene carbonates of the Bahamas: Jour. Sedimentary Petrology, v. 52, p. 1073–1085.

LINDSAY, R.F., 1982, Anatomy of a dolomitized carbonate reservoir, the Mission Canyon Formation at Little Knife Field, North Dakota: (abst.): Amer. Assoc. Petroleum Geologists Bull., v. 66, p. 594.

LINDSAY, R.F., and C.G.ST.C. KENDALL, 1980, Depositional facies, diagenesis and reservoir character of the Mission Canyon Formation (Mississippian) of the Williston Basin at Little Knife Field, North Dakota: Soc. Econ. Paleontologists and Mineralogists, Core Workshop No. 1, p. 79–104.

LINDSAY, R.F., and M.S. ROTH, 1982, Carbonate and evaporite facies, dolomitization and reservoir distribution of the Mission Canyon Formation, Little Knife Field, North Dakota: Fourth Internat. Williston Basin Symp., p. 153–179.

LINDSAY, R.F., in press, Pore and throat systems of the Mission Canyon Formation at Little Knife Field, USA—a preliminary SEM view: in Krinsley, D. and B.W. Whalley, eds., Scanning Electron Microscopy in Geology—a symposium: Geo. Abstracts Ltd., Norwich, England.

MAIKLEM, W.R., D.G. BEBOUT, and R.P. GLAISTER, 1969, Classification of anhydrite—a practical approach: Canadian Soc. Petroleum Geologists Bull., v. 17, p. 194–233.

MALEK-ASLANI, M., 1977, Plate tectonics and sedimentary cycles in carbonates: Gulf Coast Assoc. Geol. Soc. Trans., v. 27, p. 125–133.

MURRAY, R.C., 1960, Origin of porosity in carbonate rocks: Jour. Sedimentary Petrology, v. 30, p. 59–84.

NARR, W., and R.C. BURRUSS, 1982, Origin of reservoir fractures in Little Knife Field, North Dakota (abst.): Amer. Assoc. Petroleum Geologists Bull., v. 66, p. 611–612.

NETTLE, R.L., R.F. LINDSAY, and J.B. DESCH, 1981, Well test report and CO_2 injection plan for the Little Knife Field CO_2 minitest, Billings County, North Dakota: First Ann. Rept., DOE/MC/08383-26, 87 p.

PROCTOR, R.M., and G. MACAULEY, 1968, Mississippian of western Canada and Williston Basin: Amer. Assoc. Petroleum Geologists Bull., v. 52, p. 1956–1968.

SCHMIDT, V., 1965, Facies, diagenesis, and related reservoir properties in the Gigas beds (Upper Jurassic), northwestern Germany: in Pray, L.C. and R.C. Murray, eds., Dolomitization and Limestone Diagenesis—a symposium: Soc. Econ. Paleontologists and Mineralogists, Spec. Publ. 13, p. 124–168.

WARDLAW, N.D., 1976, Pore geometry of carbonate rocks as revealed by pore casts and capillary pressure: Amer. Assoc. Petroleum Geologists Bull., v. 60, p. 245–257.

WARDLAW, N.C., and R.P. TAYLOR, 1976, Mercury capillary pressure curves and the interpretation of pore structure and capillary behavior in reservoir rocks. Canadian Soc. Petroleum Geologists Bull., v. 24, p. 225–262.

WHITE, T.M., and R.F. LINDSAY, 1979, Enhanced oil recovery by CO_2 miscible displacement in the Little Knife Field, Billings County, North Dakota: DOE 5th Symp. Enhanced Oil and Gas Recovery and Improved Drilling Tech., v. 2, p. N-5/1–19.

WILLIAMS, J.A., 1974, Characterization of oil types in Williston Basin: Amer. Assoc. Petroleum Geologists Bull., v. 58, p. 1243–1252.

WILSON, J.L., 1975, Carbonate Facies in Geologic History: Springer-Verlag, Inc., New York, 439 p.

WITTSTROM, M.D., JR., and M.E. HAGEMEIER, 1978, A review of Little Knife Field development, North Dakota: Williston Basin Symp., Montana Geol. Soc., p. 361–368.

WITTSTROM, M.D., JR., and M.E. HAGEMEIER, 1979, A review of Little Knife Field development: Oil and Gas Jour., v. 77, no. 6, p. 86–92.

11
Glenburn Field
Lee C. Gerhard

RESERVOIR SUMMARY

Location & Geologic Setting	Bottineau & Renville Co.'s, R81 & 82W, North Dakota, northern interior high plains, USA
Tectonics	Gentle structural and paleobathymmetric arch across paleoshelf with underlying salt solution
Regional Paleosetting	SE shelf of cratonic Williston Basin
Nature of Trap	Structural/stratigraphic
Reservoir Rocks	
Age	Mississippian
Stratigraphic Units(s)	Mission Canyon Formation (Frobisher-Alida interval)
Lithology(s)	Limestone
Dep. Environment(s)	Shallow shelf to sabkha, cyclically repeated vertically
Productive Facies	Pisolitic wackestone and grainstone-packstone
Entrapping Facies	Anhydrite caps and lateral seals, structurally assisted
Diagenesis/Porosity	Subaerial and vadose dissolution and pisolite and caliche formation; local dolomitization and cementation by anhydrite
Petrophysics	
Pore Type(s)	Fenestral-vug, interparticle (between pisolites)
Porosity	15–20% (visual estimate in pisolite cores), 17.3% avg
Permeability	23 md avg
Fractures	Uncommon
Source Rocks	
Age	Devonian
Lithology(s)	Bakken Shale
Migration Time	NA
Reservoir Dimensions	
Depth	4464 ft (1360 m)
Thickness	100 ft (30 m) oil column
Areal Dimensions	7 × 4 mi (11.2 × 6.4 km)
Productive Area	11,840 acres (47.6 km^2)
Fluid Data	
Saturations	S_w = 35% avg
API Gravity	27°
Gas-Oil Ratio	100:1
Other	—
Production Data	
Oil and/or Gas in Place	~130 million BO
Ultimate Recovery	26 million BO; URE = 20%
Cumulative Production	13.6 million BO (2.04 million m^3) through November 1979

Remarks: IP 205 BOPD, IP mechanism water drive. Discovered 1958.

11
Porosity Development in the Mississippian Pisolitic Limestones of the Mission Canyon Formation, Glenburn Field, Williston Basin, North Dakota

Lee C. Gerhard

Location and Discovery

The Glenburn Field is located in parts of Renville and Bottineau counties, North Dakota, approximately 100 miles (160 km) north of Bismarck, in the eastern part of the Williston Basin (Fig. 11-1). The field was discovered in February 1958 on a seismically defined structural high, by the Anschutz #1 Einar Christiansen well, which had initial production of 205 barrels of oil per day with 27° API gravity oil and a gas-oil ratio of about 100:1. Glenburn produces oil and associated gas from limestones of the Frobisher-Alida interval of the Mission Canyon (Mississippian). The nature of these limestone reservoirs has been a subject of differing interpretations and is the principal topic of this case study.

General Features of Williston Basin Mississippian Fields

The Frobisher-Alida is one of a series of informally named zones in the Mission Canyon that produce in the North Dakota, Montana, and Canadian portions of the Williston Basin. Certain of these zones are shown in Figure 11-2. Glenburn is one of a long cluster of Mississippian oil fields distributed along an arc-shaped trend extending from southern Saskatchewan and Manitoba southeastward into north-central North Dakota.

Oil production in the North Dakota part of the basin was first discovered in 1951, in Devonian carbonates along the Nesson Anticline, a major north-south structure in the northwestern part of the state. Mississippian production was established soon after the original Devonian discoveries.

Unlike the Nesson Anticline fields, the fields arranged along the northwest–southeast trend previously mentioned have little structural closure and seem to involve very little fracturing. They are located in a series of arcs in which the lowest zones of the Mississippian are productive in the eastward arcs and progressively younger zones in the arcs to the west. This arrangement of productive zones, as well as the distribution and shapes of the fields, suggests a strong influence on reservoir distribution of facies belts arranged generally parallel to depositional strike of the sedimentary basin. Drilling in flush areas soon demonstrated that fields are largely limited to small structural noses with updip evaporite barriers. Evaporites lateral to the reservoirs assist trapping where structures are not sufficient to form effective traps. The Mission Canyon reservoirs include a variety of limestone and dolomitic limestone facies.

Glenburn is the southeasternmost of the Frobisher-Alida fields along the arc-shaped trend in North Dakota. Another field producing from the same zone, Wiley, lies 15 miles (24 km) to the northwest and has similar reservoir lithologies

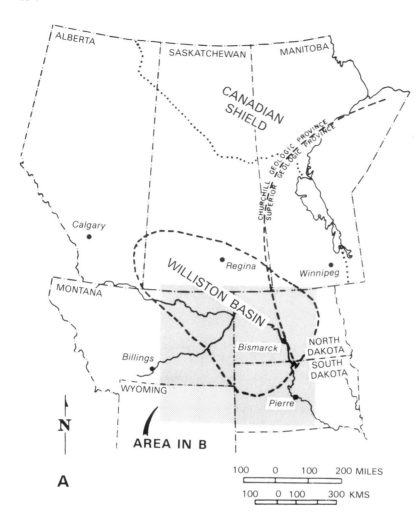

Fig. 11-1. Index map showing location of Glenburn Field. *A.* Regional setting of Williston Basin.

and trapping mechanisms. Glenburn was selected for investigation in this study because of its geographic position just north of extensive untested areas in Ward and McHenry counties, North Dakota, and because relatively good core control was available for study.

Glenburn Reserves and Production

Since its discovery in 1958, Glenburn has been on continuous production from the Frobisher-Alida interval of the Mission Canyon. In 1979 it produced 31,000 barrels of oil from 54 pumping wells and 160,000 barrels of water per month. Cumulative production to December 1, 1979, was 13,597,191 barrels of oil and 33,350,092 barrels of water. Ultimate recoverable reserves were originally estimated at 20 million barrels of oil and have since been revised to 26 million barrels.

Since that time, recognition of the stacked porosity zones deriving from the porosity development processes suggested here and in Gerhard *et al.* (1978) has resulted in a major redrilling and recompletion of the field. Originally spaced on one well per 80 acres, the field now sustains two wells per spacing unit, effectively decreasing the spacing to 40 acres. Production in June 1982 was about 62,000 barrels of oil from 93 wells. Nearly original reservoir pressures have been encountered in porosity zones unconnected to the original completed zone. Stacking of porosity zones implies that oil-producing zones may underlie water-saturated zones because of lack of continuity of the weathering-derived porous intervals. A similar recompletion of the Wiley Field to the north has further substantiated the concepts of porosity development outlined here.

Fig. 11-1B. Structural elements in USA portion of Williston Basin.

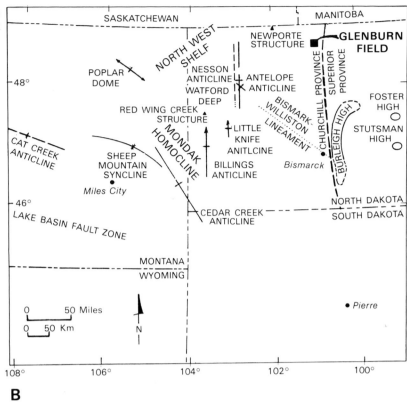

B

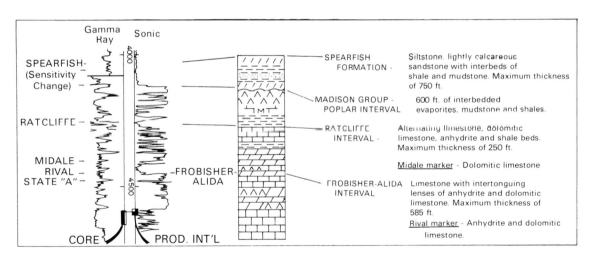

Fig. 11-2. Gamma-ray/laterolog response with a generalized lithologic column of the Frobisher-Alida zone in the Texota Oil Weber #1 well (NDGS #3630), Section 14-T158N-R82W, Renville County.

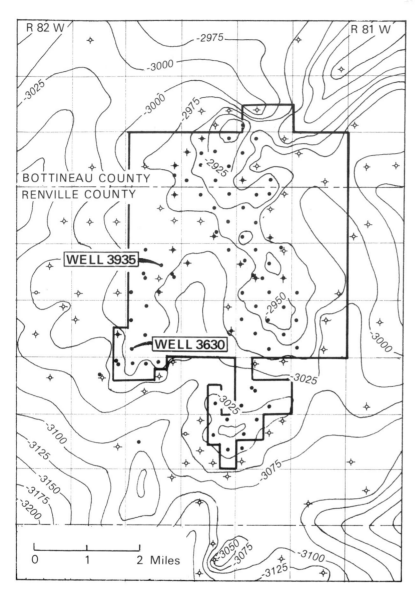

Fig. 11-3. Map of structure on the Frobisher-Alida carbonate, Glenburn Field, North Dakota. Contour interval 25 feet. Texota #1 Weber well (NDGS #3630) and Vaughn #1 Weber well (NDGS #3935), Section 11-T158N-R82W. Note the northeast and northwest trends. The northeast trend is interpreted to be the result of salt solution below the carbonates (e.g., northeast gradient of map); the less distinct northwest trend is interpreted to be parallel to the paleostrandline and reflects the shape of the eastern flank of the Williston Basin.

Reservoir and Trap

The top of the Frobisher-Alida reservoir lies at an average depth of 4464 feet (1361 m) below surface, with a maximum oil column approximately 100 feet (30 m) thick. The field is irregular in shape (Fig. 11-3) covering about 11,840 acres (47.4 km²). Structure as defined by the top of the Frobisher-Alida has relatively low relief of about 75 feet (23 m). Entrapment is accomplished by updip facies change from porous limestones to anhydrites with some structural assistance. The vertical seal is provided by thin terrigenous shales and massive anhydrites.

Frobisher-Alida reservoir rocks, with a net pay thickness averaging 10 feet (3 m) as originally drilled, are principally pisolitic limestones (grainstones and packstones) with vuggy fenestral and vuggy interparticle porosity. These facies and their fabrics and porosity are chiefly the result of subaerial exposure and a complex of vadose-diagenetic processes. The porous intervals tend to be relatively extensive and continuous areally. Their fabrics vary considerably in detail.

Porosity is in the form of vugs, in both matrix (muddy rocks) and between particles, and range up to 4 millimeters in longest dimension. No detailed quantitative determination of porosity is possible from the well logs available. Porosities by visual estimates range up to 15 to 20 percent, but more typically are lower. For example, in the Texota #1 Weber well, which was cored extensively, the most porous zone is a 5-foot (1.5-m) interval with 8 to 10 percent porosity.

Initial production was assumed to be a strong water drive, but some solution-gas drive is now evident; water saturations in the reservoir are high. Typical completions were open-hole, and Frobisher-Alida rocks were penetrated only a few feet so as to avoid encountering water. Typically, little acid and no fracturing treatment were used. Only in a few wells was the entire Frobisher-Alida section penetrated.

Previous Studies

Early studies of the Mississippian oil fields in North Dakota identified algal pisolites as a major lithology of the Mission Canyon, with updip permeability barriers of anhydrite as major seals for stratigraphic traps (Johnson, 1956). Other studies published around the same time described and illustrated "algal" and "oolitic" structures and lithologies in the MC-3 and MC-5 pay zones of the Mission Canyon, and interpreted these reservoir rocks as marine strandline facies (Fuller, 1956), generally near-shore marine deposits (George, 1962; Kinard, 1964; Moore, 1964), or "bank" facies (Harris et al., 1966). In an earlier study, the producing sequence at Glenburn itself was described as "algal and oolitic vuggy limestone" (George, 1962).

The present study of cores, logs, and samples from Glenburn Field indicates that subaerial processes rather than marine algal growth or car-

bonate "bank" environments were mainly responsible for creating the limestone fabrics and porosity in the Frobisher-Alida reservoirs of Glenburn. It is probable that Mission Canyon limestones in other oil fields of the Williston Basin had similar origins. Hydrocarbon entrapment in the Glenburn Field is largely accomplished by a structurally assisted updip permeability pinchout into evaporites and evaporite-cemented limestones.

Geologic Characteristics

Upper Mission Canyon rocks of the Glenburn Field include limestone, dolomite, and anhydrite, in varying ratios. Each of these rock types has several fabrics. Cores and core chips from many wells were examined, but core from one well, the Texota #1 Weber (NDGS #3630) was especially important to this study because of its length and the central position of the well in the field (Figs. 11-3 and 11-4). In addition, much lithologic information was obtained from the Vaughn #1 Weber well (NDGS #3935), which is located on the west flank of the field.

Dolomite and Anhydrite

Dolomite and anhydrite fabrics appear to be relatively simple. Dolomite is present in two distinct forms: (1) primary or "supratidal" dolomite, characterized by very small crystals (25 μ) in thin beds and associated with desiccation structures (Gerhard et al., 1978); and (2) dolomite that occurs as distinct rhombs in limestone (and primary dolomite), replacing an earlier carbonate. Dolomitization does not appear to be pervasive; most of the secondary dolomite-rhombs appear in dolomitic limestone.

"Chickenwire" anhydrite, massive bedded anhydrite, and clear, vug-filling anhydrite are present in varying amounts. At Glenburn, the "chickenwire" form is considered to be of sabkha origin, and is commonly associated with primary dolomites and desiccation features. Massive anhydrite is characterized by a subhedral crystalline fabric and is distinctly bedded and interfingered by thin carbonate beds, many of which are fine-grained dolomite beds. The clear anhydrite that fills fenestrae in carbonate beds occurs as large subhedral to anhedral crystals. There are sharp

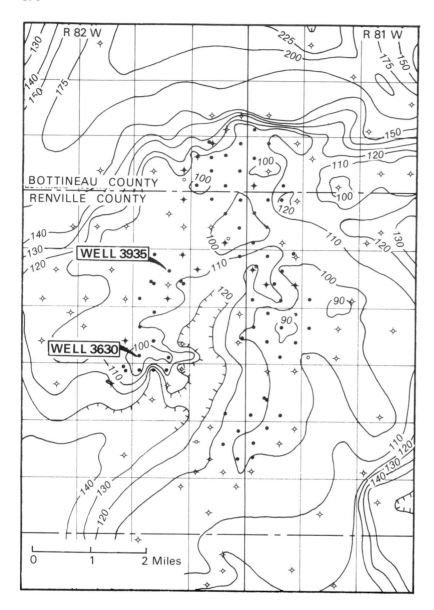

Fig. 11-4. Isopach map of the Frobisher-Alida evaporite, Glenburn Field. Contour interval 10 feet (3.3 m).

contacts between the anhydrite crystals and the walls of the fenestrae.

Limestones

Limestones are more complex texturally than either of the foregoing lithologies. The most obvious features of the limestones are rounded, commonly spherical, concentric bodies averaging a centimeter in diameter. In places these occur randomly distributed in matrix, in other places packed in grain-to-grain contact with high interparticle porosity. These are the bodies that were described as oolitic, pisolitic, and oncolitic by

earlier workers, and were attributed to construction by blue-green algae. Porous fine-grained limestone beds, thin nonporous limestone beds, and shale partings are interbedded with rocks bearing these round grains (Fig. 11-5).

A variety of primary sedimentary structures and fabrics can be seen in thin sections of the limestones, and are used as the basis for recognizing three major facies: (1) skeletal pisolitic wackestone, (2) mudstone, and (3) pisolitic packstone or grainstone. It is probable that the pisolitic fabrics are diagenetic because primary fabrics are difficult to recognize. These structures and fabrics suggest a strandline origin for the

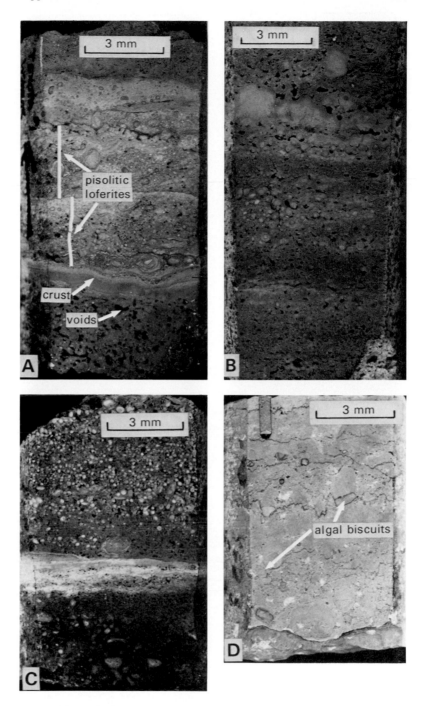

Fig. 11-5. Photograph of core slabs of Frobisher-Alida limestone. *A.* Löferite-type shrinkage or desiccation voids overlain by subaerial laminated crust, in turn overlain by pisolitic löferites. Texota #1 Weber well, 4600 feet (1402 m). From Gerhard *et al.*, 1982. Published by permission, American Association of Petroleum Geologists. *B.* Weathering pisolite in löferite. Texota #1 Weber well, 4596.5 feet (1401 m). *C.* Weathering pisolite overlain by laminated crust, in turn overlain by mixture of coated grains, löferite and weathering pisolite. Texota #1 Weber well. 4614 feet (1406 m). *D.* Algal biscuits with stylolites. Algal biscuits are faint caps on micritic intraclasts. Allochems are hypersaline oolites. Note the abundant stylolites. Texota #1 Weber well, 4623 feet (1409 m).

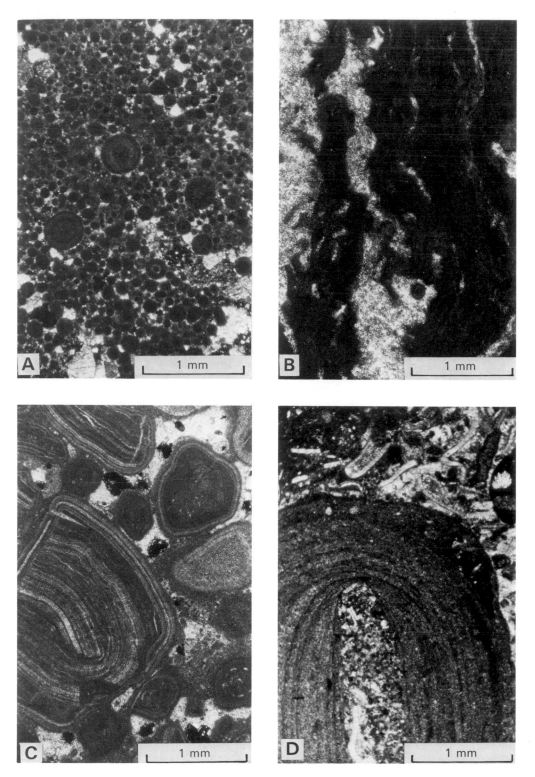

Fig. 11-6. Photomicrographs, cross-polarized light. *A.* Hypersaline oolite. Note the radial fibrous habit of the larger grains, distinct from normal-marine oolites (Bahamian) that are concentrically laminated without the radial structure. Texota #1 Weber well, 4623 feet (1409 m). *B.* Blue-green algal stromatolite interbedded with anhydrite. View is oriented 90° from horizontal. Vaughn #1 Weber well, 4557.3 feet (1389 m). *C.* Vadose pisolite in the Mississippian Frobisher-Alida zone. Note the distinct radial fibrous habit and interrupted laminates. Texota #1 Weber well, 4624 feet (1409 m). *D.* Oncolite in skeletal packstone from the Meagher Limestone (Cambrian), Montana, demonstrating discontinuous and uneven lamination.

Table 11.1. Classification of Spheroidal Sedimentary Structures (Based on average whole rock characteristics, or on n = 10)

I. Positive evidence for algal origin
   Oncolite
II. No evidence for algal origin
   A. Larger than 2 mm diameter
      1. Coated
         a. *vadose pisolite*—radial fibrous w/concentric coats
         b. *incipient weathering pisolite*—grain nucleus w/amorphous or micritic coatings
      2. Uncoated, micritic
         a. *weathering pisolite* (other evidence for weathering present)
         b. *peloid*
   B. Smaller than or equal to 2 mm diameter
      1. Concentric coating
         a. *micropisolite* (occurs w/vadose pisolite)
         b. oolite
            (1) Fully concentric coating w/pseudouniaxial "figure," or radial fibrous habit within concentric coatings; *common oolite*
            (2) Major aspect is radial fibrous, concentricity subtle; *hypersaline oolite*
      2. No coating
         a. Micritic: *peloid*
         b. Allochem: *abraded grain* or *primary grain*

Table 11.2. Characteristics of Oncolites and Subaerial Pisolites.[1]

Subaerial
   Vadose Pisolite
   Fully concentric banding
   Radial fibrous fabric
   Multiple generation (nuclei of earlier pisolite)
   Gravity cements
   Grain-to-grain contacts
   Reverse grading common

   Weathering Pisolite
   No banding or faint banding
   No observable nuclei
   Mud-supported fabric
   Often associated with crusts

Oncolites (Algal)
   Anastomosing laminae
   Skeletal or intraclastic nuclei
   Grain-to-grain contact common
   Algal filaments sometimes present
   Associated "community" or other organisms
   Muddy laminae, *not* radial fibrous

[1] Several of these characteristics are illustrated in Figures 11.5 through 11.7.

rocks, with hypersalinity occurring episodically during deposition. Oolites and stromatolites are present in the Vaughn #1 Weber core. The oolites have radial fibrous structure, however, and are similar to Holocene oolites of the Great Salt Lake, Utah (Eardley, 1938) rather than to the concentric oolites with pseudo-uniaxial optics in marine oolite sands of the Bahamas (Fig. 11-6). It is suggested that the radial fibrous habit is characteristic of hypersalinity (Table 11-1). Stromatolite structures, attributed here to blue-green algae, are interbedded with anhydrite (Fig. 11-6B).

Mudstones and wackestones are virtually devoid of skeletal intraclasts, except for occasional replaced fragments, suggesting that faunas were restricted. Desiccation structures and fenestral porosity also suggest subaerial exposure or supratidal deposition. Pisolites are of several types (Table 11-1).

Vadose Fabrics and Oncolites

Dunham (1969) first demonstrated the similarity in structure and fabric between the pisolite in the Yates Formation (Permian) of the Capitan reef complex, West Texas and New Mexico, and modern caliche. The Yates pisolite had been considered to be oncolitic—that is, the concentrically banded spherical structures formed by blue-green algae accreting mud particles in an agitated aqueous environment. Since Dunham's first paper, more examples have been identified in the geologic record. These include: (1) Holocene and Mississippian exposure surfaces, described by Harrison and Steinen (1978); (2) vadose pisolites (Dunham, 1969); and (3) crusts (Kahle, 1977) and micritized fabrics. These origins are topics of current investigation.

Different environmental interpretations and exploration strategies could result if oncolites were confused with subaerial fabrics. For instance, in the Glenburn Field, part of the trapping mechanism is the filling of interpisolite porosity by anhydrite cement in laterally adjacent rock. Algal pisolites (oncolites) would require an agitated environment that is not compatible with evaporite pore fillings. If a skeletal or other grain nucleus can be clearly demonstrated for these pisolitic bodies, or primary organic structures (filaments) are preserved in lineations, then an algal or at least a subaqueous origin may be inferred (Table 11-2 and Fig. 11-5). However, in most of the Frobisher-Alida zone, these criteria are not met.

Comparison With Other Carbonates

A direct comparison of Frobisher-Alida pisolites with pisolites described by Dunham (1969) clearly demonstrates the subaerial origin of coated multi-generation pisolites if Dunham is correct (Fig. 11-6). However, examples from the literature do not present enough data for comparison to establish an origin for micritic pisolites.

A series of samples of subaerially exposed Holocene (or late Pleistocene) skeletal sands from St. Croix, U.S. Virgin Islands, illustrates one possible sequence of generation of weathering pisolite (Fig. 11-7). Micritic and aphanitic rims are initially formed around skeletal nuclei. Further weathering progressively micritizes grains until only featureless pseudo-pisolites or pseudo-oolites are left. Apparently, this process can also take place in mud-rich fabrics, but the sequence of pisolite formation cannot be demonstrated as clearly.

Porosity

Two major porosity fabrics are present in the Glenburn reservoir. Fenestral porosity is commonly associated with micrites, some of which have isolated micritic pisolites "floating" in micrite. This vuggy fenestral porosity was described as löferites by Fischer (1964) and suggests analogy with desiccation porosity in modern muds, since vugs are elongate parallel to the assumed depositional surface and may have secondary vertical elongation (Fig. 11-5). Vuggy fenestral porosity such as this occurs in sediments on partly desiccated tidal flats in Florida Bay mud islands.

The second major porosity type is interparticle vug porosity, associated with weathering (mic-

Table 11.3. Outline of Frobisher-Alida Paragenesis, Glenburn Field, North Dakota

1. Salt solution in buried Prairie Evaporite created surface structures
2. Transgressive episode: Deposition of shallow-marine mud-based carbonate sediments
3. Stable episode: Oscillatory sea stand within small limits. Created marginal evaporite facies, distal sabkha, minor weathering pisolite. Some evaporite filling of pores
4. Regressive episode: Created major weathering pisolite, vadose pisolite, massive evaporite (off-structure), primary dolomite, clastic mud layers, major crusts. Pisolitic porosity developed. Proximal sabkha. (Episodes II, III, and IV are cyclical)
5. Offlap of sabkha environment
6. Trap sealed by lateral evaporites
7. Development of stylolites
8. Oil migration into trap

ritic) pisolites (Figs. 11-5, 11-7). If the interpretation regarding pisolite origin is correct, the porosity is diagenetic and formed by subaerial, near-surface vadose processes. It is this porosity type that is most common in the reservoir.

Origin and Paragenesis

Other observations pertinent to further interpretation:

1. Anhydrite occurs as vug fillings in carbonates, and as massive beds in apparent stratigraphic equivalence to carbonates.
2. Production has been established along topographic (structural) highs without regard for structural closure (Fig. 11-3).
3. Structural noses on carbonates, plunging basinward, are separated by evaporites at the same structural position.

A low depositional slope, arid climate, a slightly unstable sea level, and presence of slight structural deformation either penecontemporaneous with or pre-Frobisher-Alida deposition are assumed. Selective salt solution in older evaporites could easily have accomplished the deformation (Anderson and Hunt, 1964).

In the area of the Glenburn Field during Frobisher-Alida time, carbonate muds and a few muddy skeletal sands were deposited during a

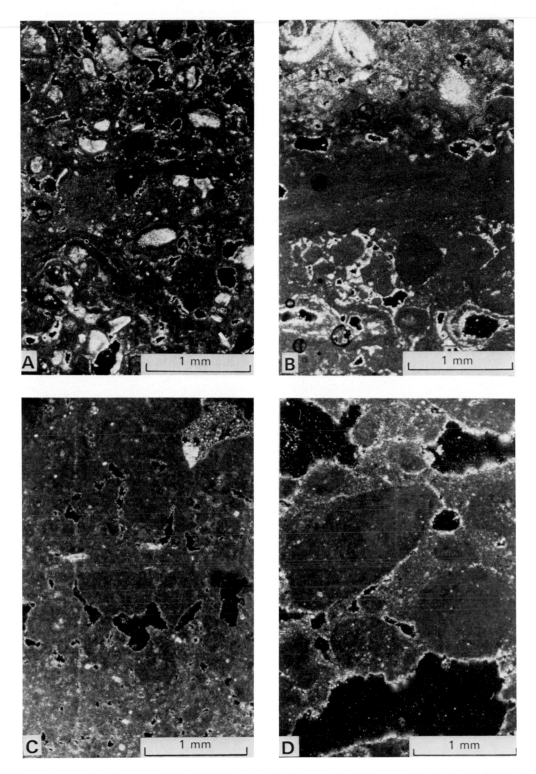

Fig. 11-7. Photomicrographs (in cross-polarized light) comparing some Holocene porosity and pisolites with Mississippian samples. Porosity is black. *A*. Incipient pisolites with coated and partially micritized grains and well-developed porosity, St. Croix, U.S. Virgin Islands. *B*. Weathering pisolites with crust and intrapisolite porosity, Holocene, St. Croix, U.S. Virgin Islands. *C*. Löferite showing micritic pisolites and vertically oriented laminated cracks. Texota #1 Weber well, 4617 feet (1407 m). *D*. Weathering pisolites with interparticle porosity. Texota #1 Weber well, 4595.5 feet (1391.6 m).

transgressive episode. Water depths were probably never more than a few feet. It is possible that a low depositional slope coupled with tidal phenomena could have caused environments to vary rapidly from intertidal to very shallow lagoonal in depth with only minor fluctuations in sea level. Supratidal desiccation may have occurred, creating primary fenestral porosity (löferite).

Regression, with minor fluctuations, exposed the carbonates. Lithification probably occurred by both early hypersaline and later vadose cementation. Flushing of hypersaline lenses by fresh water appears to be a common modern cementation process, as exemplified by case hardening of beachrock (Moore, 1973; Hanor, 1978). Exposed sediment, inferred to have been partly or poorly lithified, was present on structural highs with intervening hypersaline pans (Fig. 11-8). As weathering of the carbonates progressed, evaporite pumping may have occurred along the periphery of the limestone, leading to dolomitization. Partial plugging of primary fenestral porosity by anhydrite occurred.

Porosity enhancement, resulting from the formation of vadose and weathering pisolites, was particularly pronounced in the uppermost parts of the limestone high. Topographic relief was probably less than six or seven feet (2 m) between the top of the limestone mounds and the bottom of the evaporite pans.

Updip to the east, the limestones interfingered with sabkha evaporites, which created impermeable barriers. On the flanks of the limestone mounds, anhydrite that formed by penecontemporaneous evaporitic deposition occludes porosity, creating barriers to lateral oil migration. As mentioned previously, the areas of evaporite pans may be sites of earlier salt solution, either prior to or during Mission Canyon time.

The final seal at the top of the Frobisher-Alida porous zone originated in either of two ways. One was by deposition of a thin shale cap of terrigenous muds, which washed over the evaporite-filled pans and covered the porous limestones. These thin shales are only a fraction of an inch thick, but appear to be effective seals. In conjunction with or independent of this area are calcareous caliche crusts of low permeability, which may form seals as well (Fig. 11-5). The other major type of seal is massive anhydrite, perhaps representing salt-pan deposition during the next

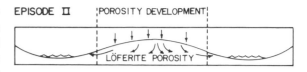

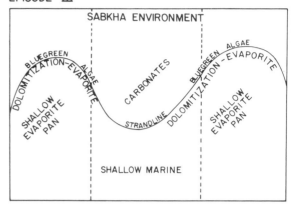

Fig. 11-8. Diagrammatic representation of episodes II to IV of the sequence of porosity development in the Glenburn Field Frobisher-Alida rocks. Refer to Table 11-3 for episode descriptions. From Gerhard *et al.*, 1982. Published by permission, American Association of Petroleum Geologists.

minor transgressive interval. Pre-existing permeability was unaffected by the massive anhydrite development, due to the presence of hypersaline fluid in the pores or earlier anhydrite. This fluid inhibited the mixing with external, sulfate-rich fluids needed for anhydrite formation; early anhydrite has been remobilized.

Post-trap intrastratal solution is indicated by widely distributed stylolites that appear to predate oil migration. The stylolites are inferred to have created lenticular zones of permeability.

Summary and Conclusions

Limestones of the Frobisher-Alida zone of the upper Mission Canyon produce hydrocarbons in the Glenburn Field from porosity largely developed through subaerial diagenetic processes.

Vadose pisolites, carbonate crusts, and solution features are major fabrics in cores of the pay zone. The development of porosity was in part coeval with anhydrite deposition in pans or supratidal sabkhas on the flanks of a very low topographic arch. Flank salt solution or ancestral structural development are plausible mechanisms for the construction of the limestone prominences. Dolomitization along the flanks, coupled with minute interbeds of anhydrite, support contemporaneous porosity development and evaporite deposition.

Early cementation was followed by the formation of vadose pisolite and other solution/desiccation features. The vadose pisolite may have been in part contemporaneous with early cementtation. The last major diagenetic phase was intrastratal solution. Porosity is lenticular in part. The oil-water contact is probably an irregular surface because of varying pore dimensions.

Acknowledgments This study was supported by the North Dakota Geological Survey, the University of North Dakota Carbonate Studies Program funded by industry, and a U.S. Geological Survey grant (#14-08-0001-G-597) for studies of subaerial diagenesis. S. B. Anderson and James Berg were instrumental in producing the earliest report of the project (Gerhard *et al.*, 1978). Published by permission of the North Dakota Geological Survey.

References

ANDERSON, S.B., and J.B. HUNT, 1964, Devonian salt solution in north central North Dakota: Third Internat. Williston Basin Symp., p. 93–104.

DUNHAM, R.J., 1969, Vadose pisolite in the Capitan reef (Permian), New Mexico and Texas: *in* Friedman, G.M., ed., Depositional Environments in Carbonate Rocks—a symposium: Soc. Econ. Paleontolotists and Mineralogists, Spec. Publ. 14, p. 182–191.

EARDLEY, A.J., 1938, Sediments of Great Salt Lake, Utah: Amer. Assoc. Petroleum Geologists Bull., v. 50, p. 1305–1412.

FISCHER, A.G., 1964, The löfer cyclothems of the alpine Triassic: Kansas Geol. Survey Bull. 169, p. 107–149.

FULLER, J.G.C.M., 1956, Mississippian rocks in the Saskatchewan portion of the Williston Basin: First Internat. Williston Basin Symp., p. 29–35.

GEORGE, R.S., 1962, Glenburn Field: *in* Oil and Gas Fields of North Dakota: North Dakota Geol. Soc., p. 109–110.

GERHARD, L.C., S.B. ANDERSON, and J. BERG, 1978, Mission Canyon porosity development, Glenburn Field, North Dakota Williston Basin: *in* Economic Geology of the Williston Basin: Williston Basin Symp., Montana Geol. Soc., p. 177–188.

HANOR, J.S., 1978, Precipitation of beachrock cements: mixing of marine and meteoric waters vs. CO_2-degassing: Jour. Sedimentary Petrology, v. 48, p. 489–501.

HARRIS, S.B., C.B. LAND, JR., and J.H. MCKEEVER, 1966, Mission Canyon stratigraphy, North Dakota: Amer. Assoc. Petroleum Geologists Bull., v. 50, p. 2269–2276.

HARRISON, R.S., and R.P. STEINEN, 1978, Subaerial crusts, caliche profiles, and breccia horizons: Comparison of some Holocene and Mississippian exposure surfaces, Barbados and Kentucky: Geol. Soc. America Bull., v. 89, p. 385–396.

JOHNSON, W., 1956, Mississippian oil fields of northeastern Williston Basin: First Internat. Williston Basin Symp., p. 52–53.

KAHLE, C.F., 1977, Origin of subaerial Holocene calcareous crusts: Role of algae, fungi and sparmicritisation: Sedimentology, v. 24, p. 413–435.

KINARD, J.C., 1964, Recent Mission Canyon discoveries in the Roth-N.E. Landa area, Bottineau County, North Dakota: Third Internat. Williston Basin Symp., p. 241–245.

MOORE, C.H., JR., 1973, Intertidal carbonate cementation Grand Cayman, West Indies: Jour. Sedimentary Petrology, v. 43, p. 591–602.

MOORE, W.L., 1964, Mechanical log and phase relationships in the carbonates of the Mississippian of the Williston Basin: Third Internat. Williston Basin Symp., p. 263–265.

NORTH DAKOTA GEOLOGICAL SOCIETY, 1967, Glenburn Field: *in* Oil and gas fields of North Dakota, 1967 Supplement: North Dakota Geol. Soc., p. 20.

ROEHL, P.O., 1967, Stony Mountain (Ordovician) and Interlake (Silurian) facies analogs of Recent low-energy marine and subaerial carbonates, Bahamas: Amer. Assoc. Petroleum Geologists Bull., v. 51, no. 10, p. 1979–2032.

12
North Bridgeport Field
Philip W. Choquette and Randolph P. Steinen

RESERVOIR SUMMARY

Location & Geologic Setting   Lawrence County, T3-4N, R12-13W, SE Illinois, Illinois Basin, USA

Tectonics   Field on crest of LaSalle Anticline, a NNE-trending narrow anticline of possible wrench-fault origin

Regional Paleosetting   Cratonic basin, shallow-marine portion

Nature of Trap   Structural/stratigraphic (depositional and diagenetic)

Reservoir Rocks
  Age   Mississippian (Meramecian)
  Stratigraphic Units(s)   St. Genevieve Formation (upper)
  Lithology(s)   Limestone and dolomite
  Dep. Environment(s)   Shelf or shallow-ramp lagoon with oolite shoals
  Productive Facies   Ooid grainstone and microcrystalline dolomite (mudstone-wackestone)
  Entrapping Facies   Lagoonal lime mudstone and wackestone
  Diagenesis/Porosity   Partial cementation of oolite (mostly phreatic); localized early diagenetic dolomitization of porous lime mudstone-wackestone

Petrophysics
  Pore Type(s)   Primary cement-reduced in ooid grainstone; intercrystalline and moldic dolomites
  Porosity   3.8–28.8%, avg 13.7%[1]; 0.1–40.0%, avg 27.0%[2]
  Permeability   0.1–9500+ md, avg 250 md[1]; 0.2–241 md, avg 12 md[2]
  Fractures   Moderate in dolomite, negligible in oolite

Source Rocks
  Age   NA
  Lithology(s)   NA
  Migration Time   NA

Reservoir Dimensions
  Depth   1600 ft (490 m) avg
  Thickness   16 ft (5 m)[1], 39 ft (12 m)[2]—maximum gross
  Areal Dimensions   [1,2] 0.3–1.5 mi (0.5–2.5 km) × 0.3–2.7 mi (0.5–4 km)
  Productive Area   1100 acres (4.4 km²) for 2 waterflood areas

Fluid Data
  Saturations   NA; field being waterflooded
  API Gravity   39°
  Gas-Oil Ratio   NA
  Other   —

Production Data
  Oil and/or Gas in Place   NA
  Ultimate Recovery   ~4.0 million BO[3]
  Cumulative Production   3.16 million BO[3] through March 1983

Remarks: [1] Ooid grainstone reservoirs, [2] microcrystalline dolomite reservoirs. [3] Data are for the 2 waterflood areas noted above. IP mechanism solution-gas drive. Discovered 1907.

12
Mississippian Oolite and Non-Supratidal Dolomite Reservoirs in the Ste. Genevieve Formation, North Bridgeport Field, Illinois Basin

Philip W. Choquette and Randolph P. Steinen

Location and Geologic Setting

North Bridgeport is a multiple-pay field producing oil from carbonate and sandstone reservoirs of Middle to Late Mississippian age. The field is located in southeastern Illinois approximately 50 miles (80 km) northeast of the structural center of the Illinois Basin, along a major trans-basin structure, the LaSalle Anticline (Fig. 12-1). The principal carbonate reservoirs in North Bridgeport Field occur in the upper part of the Ste. Genevieve Formation, a relatively thin (70–170 ft; 20–55 m) sequence of shallow-marine carbonates, Meramec in age, which extends over a five-state area of the basin and crops out along all but its northern perimeter. The Ste. Genevieve carbonates were deposited on a gently southwest-sloping intracratonic shelf (Swann, 1963; Sedimentation Seminar, 1966, 1972), as summarized more recently by the authors (Choquette and Steinen, 1980).

Reservoir Trap Characteristics

Ste. Genevieve production in North Bridgeport comes from a series of elongate, lenticular reservoirs situated at an average depth of about 1600 feet (490 m), along a north-northwest-trending sector of the LaSalle Anticline (Figs. 12-1 and 12-2). The reservoirs are of two main lithologic and facies types: ooid grainstone (also known as McClosky Oolite in Illinois Basin oil-field terminology), and microcrystalline dolomite. These reservoir facies occur as combination stratigraphic and structural traps. Entrapment in the oolite reservoirs is due to structural closure coincident with changes in depositional facies both laterally and upward into low-permeability lime mudstone and wackestone. Entrapment in the dolomite reservoirs is due to structural closure and lateral lithofacies transition into undolomitized, essentially nonporous and impermeable lime mudstone and wackestone. Some dolomite reservoirs are contiguous with overlying oolite reservoirs; others are overlain, in general gradationally, by lime mudstone/wackestone (Figs. 12-3, 12-4, 12-5).

This study utilized information from about 150 wells drilled into and completed in the Ste. Genevieve, including 47 wells cored in the upper Ste. Genevieve that were studied in detail. Dimensions and geometry of the reservoirs are known from well control, which in parts of the field provide unusually detailed data (e.g., cross-section A–A' in Fig. 12-3). Cross-sections show the oolite lenses to be essentially flat-bottomed, convex-upward features, whereas the dolomite reservoirs have more varied forms (Fig. 12-3). The maps in Figures 12-6 through 12-8 reveal a strong preferential northeast-southwest orientation of both oolite and dolomite facies patterns, and an even more striking elongation of narrow prisms of sandy, ooid-pellet pack-

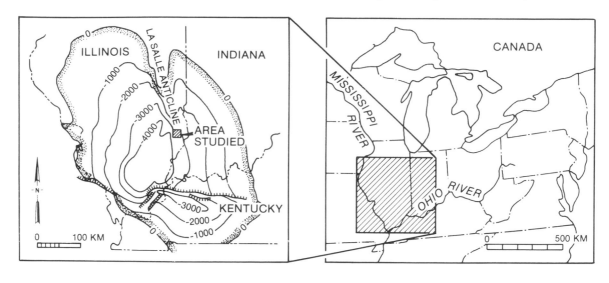

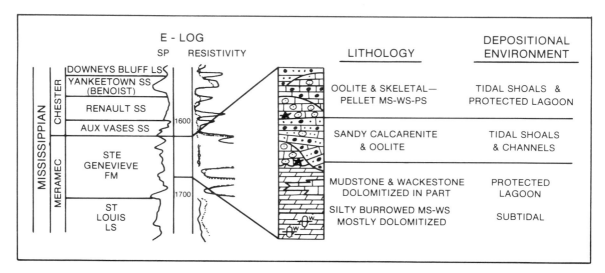

Fig. 12-1. Location maps and generalized Ste. Gene-vieve lithologic column. Map at left shows regional structure on top of the New Albany Shale (Upper Devonian-Lower Mississippian). Lithologic column is a conceptualized version of sequence cored in well 12, Section 19-T4N-R12W. Depths and contours in feet.*

stone interpreted as marine shelf channel fills (Fig. 12-6) (Choquette and Steinen, 1980).

Production and Reserves

The discovery well in North Bridgeport was drilled in 1909 by International Oil and Gas Company with a cable-tool rig. Production for a half century was mainly by solution-gas drive, but has been supplanted since 1959 by a series of

secondary waterflood recovery projects. Ste. Genevieve crude oil has negligible sulfur content and 39° API gravity. Production data and es-timates of ultimate recovery are available for three waterflood units in the North Bridgeport Field. These waterflood areas comprise 987 acres (3.94 km²), 70 active production wells, and 50 water injection wells, and have estimated ultimate

*Figures 12-1–12-3, 12-6–12-8, and 12-12 from Choquette and Steinen (1980), published by permission of Society of Economic Geologists and Paleontologists.

Fig. 12-2. Structure map of the Ste. Gene-vieve Formation, North Bridgeport Field. Contours are in feet and were drawn using the well control shown in Figures 12-6–12-8.

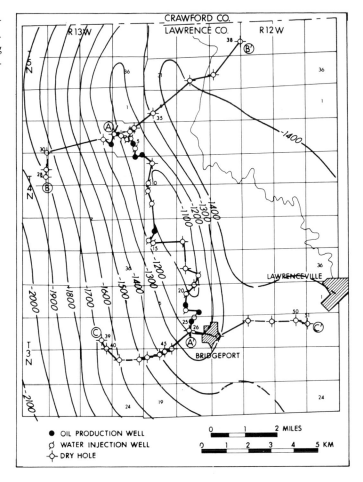

recoverable reserves by both primary and secondary mechanisms of about 4.0 million barrels. Approximately 3.16 million barrels, or 93 percent, of the recoverable reserves had been produced through March 31, 1983, by the oolite and dolomite reservoirs combined in these three areas (W. F. Bandy, personal communication).

Petrophysical Properties of Reservoir Rocks

Although short, near-vertical fractures can be seen in many cores, production behavior of the reservoirs in waterflooding suggests that fractures play a minor part, if any, in the production. Fractures are generally thin extension or hairline fractures, unenlarged by dissolution and healed by blocky calcite and/or dolomite spar cement. Rarely do more than two fractures appear in a given length of core. Dolomite intervals contain patches of irregular breccia fractures, apparently of collapse origin, but these too are filled by coarse carbonate cements.

As shown in Table 12-1, the two kinds of rock types comprising these reservoirs have quite different pore systems, resulting in different porosity and permeability characteristics. A semi-log plot of porosity and permeability values of the ooid grainstones displays a trend characteristic of many carbonate "sands" containing well-connected systems of interparticle pores (Fig. 12-9). By comparison, the microcrystalline dolomites are composed mainly of rhombic dolomite crystals which, although an order of magnitude smaller than the ooids, are equally "well-sorted." The dolomites have mixed pore systems in which

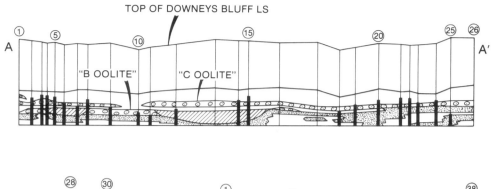

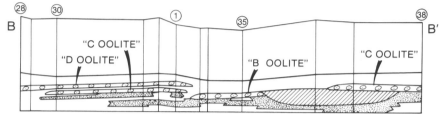

Fig. 12-3. Stratigraphic cross sections of upper Ste. Genevieve Formation. Locations of sections are shown in Figure 12-2. Datum is a widespread, thin unit of silty burrowed mudstone-wackestone (Fig. 12-1). Lime mudstone or wackestone facies (unpatterned) underlying oolite lenses was dolomitized in patches, generally to greatest

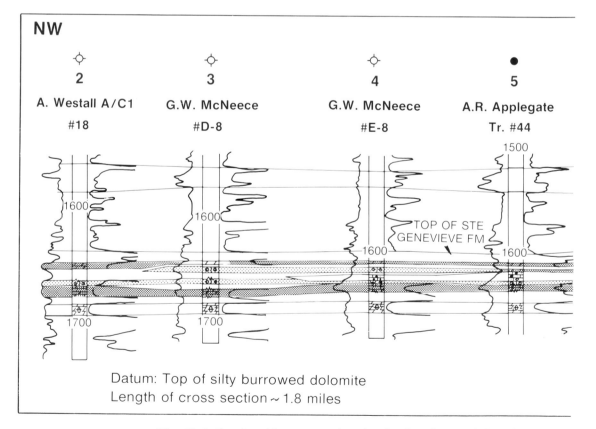

Fig. 12-4. Stratigraphic cross section showing log characteristics of oolite and microdolomite reservoir intervals. Cross-section is part of A–A' (Figs. 12-2 and

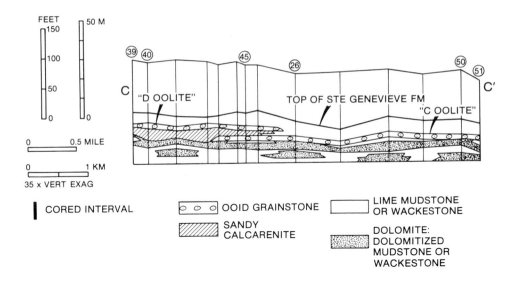

depths beneath calcarenite channel fills. Interval between Ste. Genevieve and a unit above, the Downeys Bluff Limestone (Fig. 12-1), thins reciprocally over the anticline, suggesting that the structure had sea-floor expression (Choquette and Steinen, 1980). Cores were available from wells on cross-section A–A'.

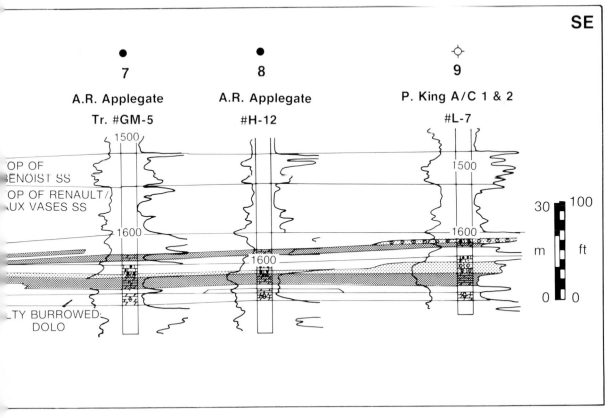

2-3). Resistivity contrast between reservoir lithologies is due to a combination of higher porosity and higher water saturation in the microdolomites.

Fig. 12-5. Core sequence from part of upper Ste. Genevieve sequence in well 20 of Figure 12-2. Base of oolite is gradational over 2–3 centimeters. Differences between lime mudstone (MS) and dolomite mudstone (DO) are subtle, but dolomite has less well-defined thin lamination, fewer skeletal fragments, and no stylolites whatsoever. Dolomitization in Ste. Genevieve lime-mud sediments is interpreted to have preceded the development of stylolites. Depths in feet.

Fig. 12-6. Thickness map of lower mudstone (both limestone and dolomite) beneath B and C oolites, and distribution of channel-fill calcarenite (stippled areas). Well control used to construct this map was also used to construct the maps in Figures 12-2, 12-7 and 12-8.

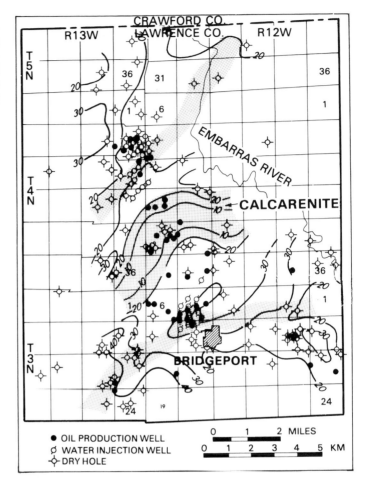

Table 12.1. Comparison of Some Reservoir Properties in Oolite and Dolomite Reservoirs[1]

	Oolite	Dolomite
Porosity types	Interparticle (BP)	Microintercrystal (mBC)
	Intraparticle (WP)	Moldic (MO)
Inferred origin	Primary, reduced by cementation	Dissolution of $CaCO_3$
Pore size range (apparent)	10^{-2}–10^{0} mm	10^{-4}–10^{-2} mm (BC)
		10^{-1}–10^{0} mm (MO)
Porosity average	12%	27%
range	2–22%	13–40%
n	117	90
Permeability average	250 md	12 md
range	0.1–9500 md	0.7–130 md
n	115	87

[1] Porosity types follow Choquette and Pray (1970). Porosity and single-point gas permeability were determined by standard core analysis of 3/4-inch (1.9-cm) diameter plugs drilled from cores.

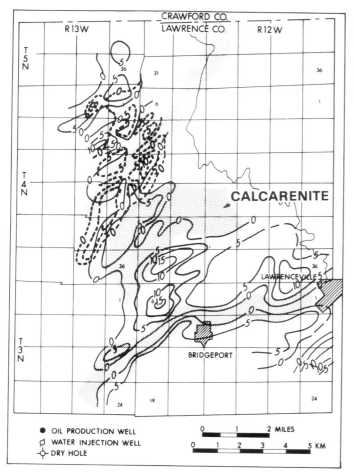

Fig. 12-7. Thickness map of two oolite sequences (B and C) in the upper Ste. Genevieve. Oolitic-sand lenses appear to be localized along the LaSalle Anticline (Fig. 12-2). Contours in feet. Well control omitted due to complexity of contours.

intercrystal voids predominate but small molds of ooids, pellets, and fine fossil fragments are common in places. Figure 12-9 shows a "flatter" porosity-permeability trend for these microdolomites, reflecting their greater pore volume but smaller-sized pores and pore-throat interconnections.

Typically, the microdolomite reservoirs are distinguishable by strongly positive SP-log expression and very low resistivity compared with the oolites (Fig. 12-4). The low resistivity is attributable to unusually high porosity and to high water saturations that result in part from very fine-scale pore systems. Commonly, the electric logs differentiate between reservoir-quality dolomite and the thin unit of less resistive, burrowed silty

dolomite that we have used to mark informally the base of the upper Ste. Genevieve sequence (Figs. 12-1, 12-2, 12-4). This thin dolomite generally has more clay, higher water saturations, and little or no oil saturation compared with the reservoir dolomites.

Nature and Origin of Reservoir Rocks

The geology and origin of Ste. Genevieve microdolomites have been discussed in detail elsewhere (Choquette and Steinen, 1980) and are only outlined here. Characteristics of the oolite reservoir rocks in North Bridgeport Field are

Fig. 12-8. Distribution map of dolomite. This is a normalized map expressing dolomite thickness as a percentage of mudstone thickness (cf. Fig. 12-4); much the same pattern appears when thicknesses rather than percentage distributions are mapped. Note the strong resemblance between this diagenetic pattern and the depositional pattern seen in Figures 12-4 and 12-5.

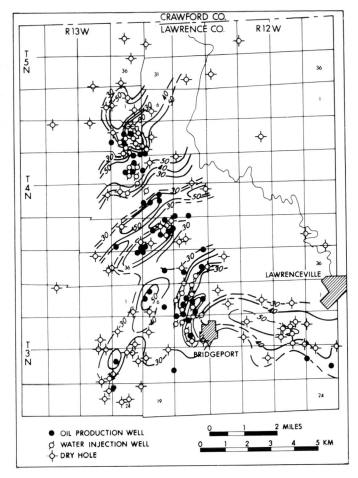

Ooid Grainstone Reservoirs

Oolites in North Bridgeport occur in a series of elongate lenses at three stratigraphic levels designated simply B, C, and D as illustrated in Figure 12-3. Of the two main groups of oolites, the "B oolite" is restricted to the northern part of the field, whereas the "C oolite" is found all along the crest of the structure (Fig. 12-7). Available well control suggests that both groups of ooid-sand bodies were restricted more or less to the area of the present anticline axis. From thickness and facies relationships, we have interpreted this part

hitherto undescribed, however, and are reported in some detail in this case study.

of the LaSalle anticline as a paleoshoal along which oolitic sands were localized (Choquette and Steinen, 1980).

Geometry of Oolites

In form and extent, the oolites closely resemble modern tide-dominated, oolitic-sand shoals in the Bahamas, particularly those along the northwestern perimeter of Great Bahama Bank south of Bimini (*e.g.*, Newell *et al.*, 1960; Purdy, 1963). As a group, the oolitic-sand bodies are arranged in a north-northwest-trending linear shoal but individually are oriented at high angles to this trend (Fig. 12-7). Like oolites known in the outcrop belt (Carr, 1973), these are typically flat-bottomed and convex upward. Oolite lenses are 0.3 to 0.9

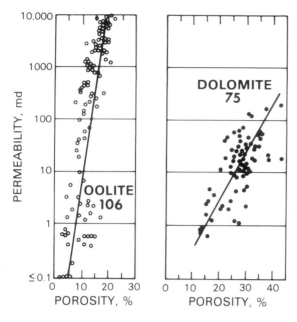

Fig. 12-9. Semi-log plots of percent porosity vs. air permeability in an oolite reservoir (left) and a dolomite reservoir (right). Oolite data are from four cored wells in Sections 12 and 13-T4N-R13W, northern part of field; dolomite data are from three cored wells in Section 30-T4N-R12W, central part of field (Fig. 12-2).

mile (0.5–1.5 km) wide, 0.3 to 2.4 miles (0.5–4 km) or more in length, and up to 16 feet (5 m) thick. In cores, the oolitic grainstone lenses are cross-stratified in sets up to a meter or so in thickness, with planar to trough-shaped bases and avalanche-face dips up to 24 to 26 degrees. Bimodal dips are common both on outcrop and in North Bridgeport Field cores, suggesting transport by ebb and flood tides as inferred by Carr (1973) for oolites exposed to the east in Indiana.

Texture

Oolite lenses are almost entirely grainstone, except for intervals of bioturbated ooid packstone and wackestone found at the base of some lenses. These "muddy" zones have been dolomitized, and consist now of grain-sized molds, either empty or filled by dolomite spar, in a groundmass of microcrystalline dolomite. Oolitic grainstones, by contrast, rarely contain dolomite, except at the bases of a few oolite lenses where dolomite occurs as both a cement and a replacement of allochems.

Grainstones range from fine-grained and well-sorted where the grains are well-developed ooids, to medium and coarse or very coarse-grained where the grains are mostly bioclastic fragments. These fragments commonly include the assemblage of echinoderm ossicles and plates, punctate brachiopods and fenestrate bryzoans so common in Mississippian shallow-marine carbonates. Pellets and intraclasts of grapestone and lime mudstone occur in places. Ooids with multiple cortices comprise anywhere from 30 to 95+ percent of the grains. An "average" oolite would have a grain composition of about 55 percent ooids, 40 percent coated and uncoated or micrite-rimmed skeletal fragments, and 5 percent pellets and intraclasts.

Cements and Their Origins

Reservoir quality of the ooid grainstone lenses is determined mainly by the amounts and distribution of carbonate cements. The principal types of cements are illustrated by photomicrographs in Figure 12-10, and their inferred relative time relationships and origins are outlined in Figure 12-11.

The upper parts of oolite lenses generally have an initial stage of either isopachous, fringing calcite cement interpreted as synsedimentary marine in origin (Fig. 12-10A; stage 1 of Fig. 12-11) or of fringing calcite cement, which in larger pores is pendant from grains roofing the pores (Fig. 12-10B; stage 2 of Fig. 12-11). The pendant or "stalactitic" cement is interpreted as marine-vadose or beachrock cement, following the interpretations of Purser (1971) and Scholle (1978, p. 161). Both of these early cements are minor volumetrically and occluded about 10 percent at most of the original depositional pore space.

Meniscus cement, precipitated at grain contacts as limpid microspar crystals, occurs in the upper parts of some of the oolite lenses. This cement is shown as stage 2 in Figure 12-11. Although it is minor volumetrically, following Dunham (1971) it is interpreted here as a product of precipitation in a water-air system, probably in meteoric-vadose diagenetic conditions. Thus it indicates at least one episode of emergence for the oolite lenses involved.

The question of emergence is important to an understanding of both diagenesis in the oolites and dolomitization of the underlying lime mudstone/wackestone sequence. Although vadose cement does occur, other considerations suggest

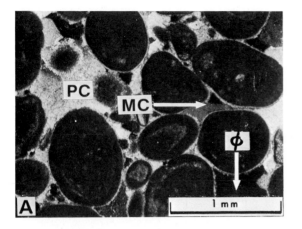

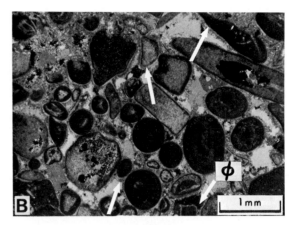

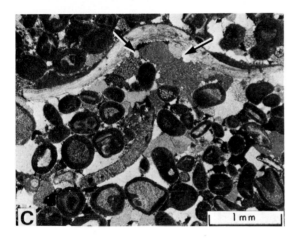

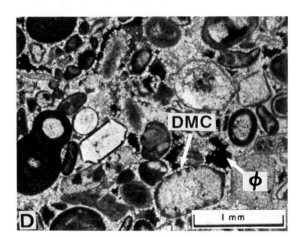

Fig. 12-10. Photomicrographs of cements in oolitic grainstone reservoirs. All photos are in cross-polarized light; porosity is black. *A.* Isopachous fringes of calcite interpreted as synsedimentary marine cement of stage 1 (Fig. 12-9) lightly bond the grains; they appear as off-white fringes against the dark grains. Porosity is occluded mainly by blocky calcite spar interpreted as phreatic (PC), stages 3 and 4. About 7% porosity remains between the grains. From well 2, depth 1677 feet (511.1 m) *B.* Slightly pendular rinds of calcite cement (arrows) interpreted as marine phreatic (beachrock, stage 1) in origin. Remaining interparticle porosity almost completely filled by blocky phreatic calcite (stages 3 and 4). From well 3, depth 1654.4 feet (504.3 m). *C.* Early phreatic cement (stage 3; arrows) and later coarse phreatic cement (stage 4) beneath small brachiopod valve. Grains were lightly bonded previously, by meniscus cement (stage 2), which is barely visible at this magnification. From well 3, depth 1658 feet (505.4m). *D.* Dolomitized grainstone. Earlier isopachous fringes of marine cement have been dolomitized (DMC, stage 3D) along with the grains. Coarse ferroan dolomite (stage 4D) subsequently filled some of the interparticle porosity. About 12% interparticle porosity remains. From well 7, depth 1639.5 feet (499.7 m).

that emergence of the oolite shoals may have been minor and relatively brief. The ooid grainstones do not contain oomoldic or other solution porosity and only rarely exhibit solution enlargement of primary interparticle pores. None of the lenses has capping soil profiles or calcretes or other features typical of well-developed exposure surfaces in Carboniferous carbonates (*e.g.,* Walkden, 1974; Walls *et al.,* 1975; Riding and Wright, 1981), and even meniscus cements are rare. All of these observations, although negative, suggest that whether the paleoclimate was humid or arid, freshwater lenses probably did not reside long in the oolites.

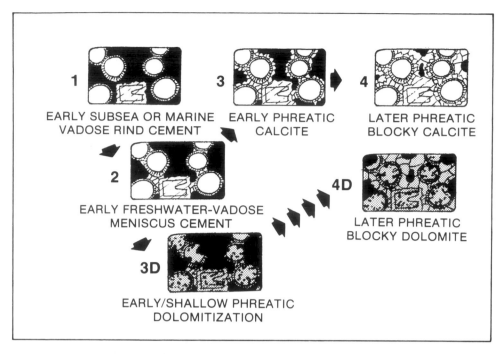

Fig. 12-11. Stages in oolitic sand diagenesis. The usual course of diagenesis is via stages 1–2–3–4. Dolomitized oolite, found only at bases of oolite lenses, is interpreted to have followed the course 1–2–3D–4D. Examples of cements are shown in Figure 12-8.

Coarse, blocky calcite spar is the most abundant cement volumetrically. This cement, which ranges in crystal size from 30 to 50 microns to about a millimeter in some of the larger interparticle and intrabrachiopod pores (Figs. 12-10A,B,C), is shown as stage 4 in Figure 12-11. It is commonly associated with equally coarse ferroan dolomite, which seems to be essentially contemporaneous with, or slightly later than, the calcite spar. Both of these coarse cements generally were preceded by finer calcite spar, shown as stage 3 in Figure 12-11. The stage 3 cement is probably best interpreted as early phreatic in origin and may be analogous to cements described by Halley and Harris (1979) in Holocene oolitic sands of Joulters Cay in the Bahamas.

The influence of grain types on the extent and style of cementation is evident from the relatively coarse syntaxial calcite cement overgrowths on echinoderm fragments. Grainstones with an abundance of crinoid ossicles tend to be cemented more extensively than echinoderm-poor grainstones, and reservoir quality is generally poor.

The isotopic compositions of the coarse spar cements have marine-like $\delta^{13}C$'s of around +1 to +3‰ *vs.* PDB, but have $\delta^{18}O$ values of about −5 to −8‰. These values suggest precipitation in

relatively hot formation water, perhaps somewhere in the range of 50° to 130° C (Choquette, 1971). An alternative interpretation, that the oxygen-isotope compositions are the result of precipitation from highly ^{18}O-depleted, meteoric ground water in the shallow subsurface, is unlikely in view of plate reconstructions that locate the area somewhere between the paleoequator and 20° S latitude during the Early Carboniferous (Drewry *et al.*, 1974; Scotese *et al.*, 1979).

Microcrystalline Dolomite Reservoirs

The other reservoir lithology in the Ste. Genevieve Formation at North Bridgeport Field, microcrystalline dolomite, occurs variably distributed within a unit of carbonate mudstone and wackestone (Figs. 12-2 and 12-3). This unit is generally 20 to 30 feet (6–9 m) thick except where it either thickens off-structure or is partly eroded beneath calcarenite channel fills (Figs. 12-3 and 12-6).

Geometry of Dolomite
Dolomite tends to be most abundant in the upper part of the mudstone-wackestone unit. It occurs in short tabular to lenticular bodies that are best

developed beneath prisms of sandy calcarenite or the thicker lenses of oolite. The dolomite bodies range from 0 to 38 feet (0–12 m) in aggregate thickness, comprising 50 percent or more of the mudstone unit in places (Fig. 12-8). In plan, they range from about 0.3 to 1.5 miles (0.5–2.5 km) in width and 0.7 to 4.5 miles (1.2–7.7 km) or more in length.

Detailed correlations between closely spaced wells show that the dolomite reservoirs are generally elongate, with flat bases and flat to concave-upward tops (Fig. 12-3). Dolomite bodies interfinger laterally with undolomitized lime mudstone or wackestone. Contacts between dolomite and lime mudstone/wackestone in a given well are gradational over as much as two meters of vertical section.

Petrography and Porosity

Details of the petrography of Ste. Genevieve microdolomites in North Bridgeport have been reported earlier (Choquette and Steinen, 1980). The features of significance in interpreting the dolomite reservoirs and their pore systems may be summarized as follows:

1. *Lithology and texture*—In general, the dolomites are poorly stratified, locally burrow-mottled, dolomitized mudstones and wackestones containing small amounts (generally <20%) of dissolved brachiopods, crinoid ossicles, bryozoans, and sparse reworked ooids that occur in the form of empty or cement-filled molds. Most of the microdolomites are quite uniform in texture and composed of 5- to 20 micron dolomite rhombs (Fig. 12-12). At first glance, the microcrystalline dolomites closely resemble the lime mudstones and wackestones with which they intergrade, although they commonly contain fewer fossil fragments (as molds) and are less well-laminated. Also, oil saturation gives the microdolomites a medium to light yellowish brown color instead of the light grayish tan or light gray color typical of the limestones. The two lithologies are well represented by the core photographs in Figure 12-5.

2. *Associated limestones*—The lime mudstones and wackestones interbedded with dolomites are also slightly dolomitic, containing around 5 percent dolomite on the average. However, they have no productive porosity (<3%) and are impermeable (<0.05 md) and unstained

by oil, except along rare fractures. These limestones are sparsely fossiliferous, nearly pure carbonate, essentially lithographic limestones, with only traces of clay and only a few percent of silt-size quartz. By analogy with Holocene lime-mud counterparts, their initial porosities may have been in the vicinity of 70 percent (Enos and Sawatsky, 1981), but were reduced to near zero during burial and diagenesis (Figs. 12-13, 12-14).

3. *Porosity*—The dolomites are characterized by very high porosities of up to 40 percent (Table 12-1). Their pore systems are dominantly of microintercrystal type, with very small pores of dolomite-rhomb size and associated small moldic pores in some dolomites. Porosity and permeability are directly proportional to volumetric ratios of dolomite to calcite as determined by detailed X-ray diffraction, petrographic, and core analysis of samples from dolomite to limestone transition zones (Figs. 12-12 and 12-13A,B). Vertical transitions in percent porosity from essentially undolomitized limestone with less than 5 percent dolomite to calcite-free dolomite are quite consistent and follow a pattern that has been noted in other dolomitized sequences (*e.g.*, Murray, 1960; Lucia, 1962; Wardlaw, 1979).

4. *Limestone-to-dolomite transitions*—The spatial changes in porosity, permeability, and dolomite-calcite ratios through transitions from lime mudstone-wackestone to calcite-free microdolomite are assumed to represent sequential stages in the dolomitization process as well. If so, the evolution of porosity in these microdolomites has probably taken place in distinct stages, as discussed next.

Dolomite Development and Porosity Evolution

Porosity evolution in the microdolomites seems to have taken place in two main stages: a dolomite-rhomb addition stage and a calcite-dissolution stage. Figure 12-12 is meant to illustrate these stages as sampled in one limestone-dolomite transition zone, and Figure 12-13 is a representation of their effects on permeability and pore volume based on data from representative transition zones in four wells. The *dolomite-addition stage* was characterized by the appearance of new dolomite-rhombs until dolomite composed (or now composes) 50 to 55 percent of the carbonate fraction and effectively formed a load-supporting

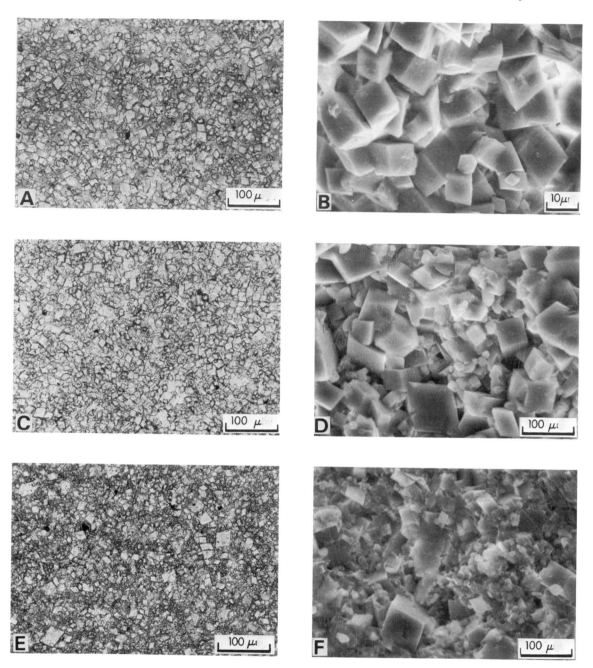

Fig. 12-12. Photomicrographs of stages in a limestone-dolomite transition. Three samples are represented, from well 22, depth 1653.0 to 1658.0 feet (503.8–505.4 m), as observed in thin sections with a light microscope (*A, C, E*) and on broken rock surfaces with a scanning electron microscope at (*B, D, F*). *A, B.* Microcrystalline dolomite with no calcite. Notice the plane-sided and sharply terminated rhombs, the lack of corrosion or pressure solution, and the high porosity (gray in photo). Porosity 39%. Depth 1653.0 feet (503.8 m). *C, D.* Calcitic microdolomite with dolomite to calcite ratio of 79:21 and 25% porosity. Small loaf-shaped crystals of calcite microspar occur between rhombs. Depth 1653.7 feet (504.0 m). *E, F.* Dolomite lime mudstone with dolomite to calcite ratio of 25:74 and 3% porosity. Depth 1658.0 feet (505.4 m).

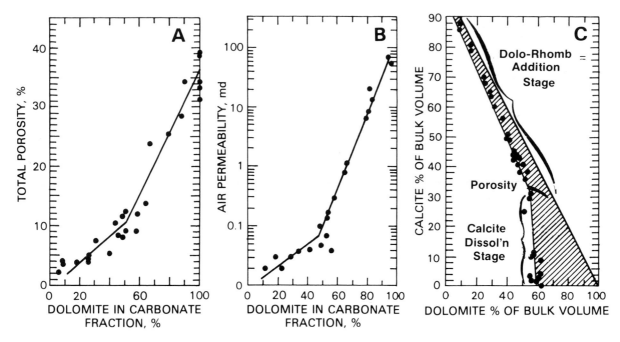

Fig. 12-13. Graphs relating pore volume (*A,C*) and permeability (*B*) to variations in dolomite content. Samples represented in these graphs are from lime mudstone to microdolomite transition zones in four wells (2, 18, 20, and 22 in section A–A', Fig. 12-3), at average depths of about 1650 feet (503 m). Porosity and permeability were determined by standard core analysis methods on 1.90-cm-diameter cylindrical core plugs; percent dolomite and calcite were estimated from d_{104} peak-height ratios determined by X-ray diffraction.

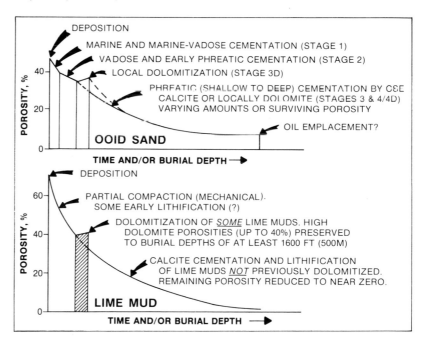

Fig. 12-14. Inferred sequence of diagenetic events and effects on porosity in oolitic sand and lime mud.

crystal framework. At that composition, physical compaction of the sediment would have slowed or stopped, as proposed by Weyl (1960). A "local source" could have provided the Ca^{++} and CO_3^{--} ions needed for new dolomite, in a chemically closed or partly closed system (Weyl, 1960; Sibley, 1982). Any $CaCO_3$ (presumably, porous lime mud) that had not been replaced by that point later lost its residual porosity (Figs. 12-13, 12-14) by some combination of mechanical compaction, $CaCO_3$-mineral neomorphism, and pore infill by microcrystalline calcite cements.

The *CaCO₃-dissolution stage* then followed in which relatively little new dolomite was added, supplied with Ca^{++} and CO_3^{--} by hitherto unreplaced calcite or aragonite remaining within the framework of new dolomite-rhombs. The overgrowths of dolomite on these earlier-formed rhombs can be seen in cathodoluminescence (Choquette and Steinen, 1980) and may have formed during this stage. The stage ended with the complete removal of $CaCO_3$ from between dolomite-rhombs, as illustrated by a microdolomite with 40 percent porosity (Fig. 12-12A, B). The high porosity, once it was established, survived subsequent burial to a minimum of around 1600 feet (480 m), the present average depth of the reservoirs. Possible processes involved in the dolomitization and sources of the magnesium required were suggested in an earlier paper (Choquette and Steinen, 1980).

Conclusions

Lenticular oil reservoirs occur in two contrasting carbonate facies in the Ste. Genevieve Formation at North Bridgeport Field. The two reservoir lithologies, oolite and microdolomite, have contrasting porosity-permeability characteristics (Fig. 12-9, Table 12-1) that were acquired in quite different ways. Porosity in the oolite reservoirs is residual primary interparticle pore space, which survived progressive cementation mainly during deep-phreatic burial diagenesis. Pore space in the dolomite reservoirs is mainly of microintercrystal type supplemented by moldic pores. The favorable porosity and permeability of these dolomites resulted directly from early diagenetic dolomitization of lime-mud-rich sediments and were then preserved in subsequent burial diagenesis. The barrier and sealing beds are undolomitized mudstones/wackestones that were

compacted and their pores cemented to essentially zero porosity and permeability. Thus the reservoirs produce from partly stratigraphic traps of both depositional and purely diagenetic origins.

Acknowledgments   Discussions with our present or former Marathon colleagues, D. H. Craig, D. B. MacKenzie, J. C. Harms, and L. C. Pray, and with D. H. Zenger of Pomona College, were helpful in the development of our ideas. R. E. Grove and W. F. Bandy supplied production and reserves data. The late J. H. Buehner made available subsurface structural information shown in Figures 12-1 and 12-2. Published by permission of Marathon Oil Company.

References

CARR, D.D., 1973, Geometry and origin of oolite bodies in the Ste. Genevieve Limestone (Mississippian) in the Illinois Basin: Indiana Dept. Nat. Resources, Geol. Survey Bull. 48, 81 p.

CHOQUETTE, P.W., 1971, Late ferroan dolomite cement, Mississippian carbonates, Illinois Basin, USA: *in* Bricker, O.P., ed., Carbonate Cements: Johns Hopkins Univ. Studies in Geology, No. 19, p. 339–348.

CHOQUETTE, P.W., and L.C. PRAY, 1970, Geologic nomenclature and classification of porosity in sedimentary carbonates: Amer. Assoc. Petroleum Geologists Bull., v. 54, p. 207–250.

CHOQUETTE, P.W., and R.P. STEINEN, 1980, Mississippian non-supratidal dolomite, Ste. Genevieve Limestone, Illinois Basin: evidence for mixed-water dolomitization: *in* Zenger, D.H., J.B. Dunham, and R.L. Ethington, eds., Concepts and Models of Dolomitization—a symposium: Soc. Econ. Paleontologists and Mineralogists, Spec. Publ. 28, p. 163–196.

DREWRY, G.E., A.T.S. RAMSAY, and A.G. SMITH, 1974, Climatically controlled sediments, the geomagnetic field, and trade wind belts in Phanerozoic time: Jour. Geology, v. 82, no. 5, p. 531–553.

DUNHAM, R.J., 1971, Meniscus cement: *in* Bricker, O.P., ed., Carbonate Cements: Johns Hopkins Univ. Studies in Geology, No. 19, p. 297–301.

ENOS, P., and L.H. SAWATSKY, 1981, Pore networks in Holocene carbonate sediments: Jour. Sedimentary Petrology, v. 51, p. 961–985.

HALLEY, R.B., and P.M. HARRIS, 1979, Freshwater cementation of a 1000-year-old oolite: Jour. Sedimentary Petrology, v. 49, p. 969–988.

LUCIA, F.J., 1962, Diagenesis of a crinoidal sediment: Jour. Sedimentary Petrology, v. 32, p. 848–865.

MURRAY, R.C., 1960, Origin of porosity in carbonate rocks: Jour. Sedimentary Petrology, v. 30, p. 59–84.

NEWELL, N.D., E.G. PURDY, and J. IMBRIE, 1960, Bahamian oolitic sand: Jour. Geology, v. 68, p. 481–497.

PURDY, E.G., 1963, Recent calcium carbonate facies of the Great Bahama Bank—2. Sedimentary facies: Jour. Geology, v. 71, p. 472–497.

PURSER, B.F., 1971, Middle Jurassic synsedimentary marine cements from the Paris Basin, France: *in* Bricker, O.P., ed., Carbonate Cements: Johns Hopkins Univ. Studies in Geology, No. 19, p. 178–184.

RIDING, R., and J.P. WRIGHT, 1981, Paleosols and tidal flat/lagoon sequences on a Carboniferous carbonate shelf—sedimentary associations of triple disconformities: Jour. Sedimentary Petrology, v. 51, p. 1323–1340.

SCHOLLE, P.A., 1978, A color-illustrated guide to carbonate rock constituents, textures, cements, and porosities: Amer. Assoc. Petroleum Geologists Mem. 27, 241 p.

SCOTESE, C.R., R.K. BAMBACH, C. BARTON, R. VAN DER VOO, and A.M. ZIEGLER, 1979, Paleozoic base maps: Jour. Geology, v. 87, no. 3, p. 217–277.

SEDIMENTATION SEMINAR, 1966, Cross-bedding in the Salem Limestone of central Indiana: Sedimentology, v. 6, p. 95–114.

SEDIMENTATION SEMINAR, 1972, Bethel Sandstone (Mississippian) of western Kentucky and south-central Indiana, a submarine channel fill: Kentucky Geol. Survey, Ser. 10, 24 p.

SIBLEY, D.F., 1980, Dolomitization of Plio-Pleistocene carbonates, Bonaire, N. A.: *in* Zenger, D.H., J. B. Dunham, and R.L. Ethington, eds., Concepts and Models of Dolomitization—a symposium: Soc. Econ. Paleontologists and Mineralogists, Spec. Publ. 28, 320 p.

SIBLEY, D.F., 1982, The origin of common dolomite fabrics—clues from the Pliocene: Jour. Sedimentary Petrology, v. 52, p. 1087–1100.

SWANN, D.H., 1963, Classification of Genevievian and Chesterian (Late Mississippian) rocks of Illinois: Ill. Geol. Survey Rept. Invest. 216, 91 p.

WALKDEN, G.M., 1974, Paleokarstic surfaces in Upper Visean (carboniferous) limestones of the Derbyshire Block, England: Jour. Sedimentary Petrology, v. 44, p. 1232–1247.

WALLS, R.A., W.B. HARRIS, and W.E. NUNAN, 1975, Calcareous crust (caliche) profiles and early subaerial exposure of Carboniferous carbonates, northeastern Kentucky: Sedimentology, v. 22, p. 417–440.

WARDLAW, N.D., 1979, Pore systems in carbonate rocks and their influence on hydrocarbon recovery efficiency: *in* Geology of Carbonate Porosity: Amer. Assoc. Petroleum Geologists Contin. Ed. Course Note Series 11, p. E1-E24.

WEYL, P.K., 1960, Porosity through dolomitization—conservation-of-mass requirements: Jour. Sedimentary Petrology, v. 30, p. 85–90.

13
Seminole Southeast Field
S. J. Mazzullo

RESERVOIR SUMMARY

Location & Geologic Setting	Gaines Co., west Texas, part of Permian Basin province, USA
Tectonics	Post-Strawn folding and faulting of Strawn shelf
Regional Paleosetting	East edge of Central Basin Platform
Nature of Trap	Stratigraphic; updip facies change and porosity pinchout into lime mudstone-wackestone
Reservoir Rocks	
Age	Middle Pennsylvanian
Stratigraphic Units(s)	Strawn Limestone
Dep. Environment(s)	Shelf-margin, shallow-marine coralgal and other reef-mounds
Productive Facies	*Chaetetes*-algal-ooid-bioclastic grainstone; phylloid-algal wackestone
Entrapping Facies	Shale and lime mudstone-wackestone
Diagenesis/Porosity	Early vadose and phreatic dissolution associated with subaerial exposure
Petrophysics	
Pore Type(s)	Intercrystal, vug and channel
Porosity	3–18%, avg 13%
Permeability	0.1–80 md, avg 29 md
Fractures	Common vertical fractures, hairline to solution-enlarged
Source Rocks	
Age	Coeval Pennsylvanian and overlying Wolfcampian
Lithology(s)	Basinal shales
Migration Time	Late Early to Middle Permian(?)
Reservoir Dimensions	
Depth	10,800 ft (~3300 m)
Thickness	12 ft (3.7 m) & 5 ft (1.5 m) upper & lower reservoirs
Areal Dimensions	1.5 × 1 mi (2.4 × 1.6 km)
Productive Area	1500 acres (6 km^2)
Fluid Data	
Saturations	S_w = 20–80% avg for reservoirs
API Gravity	43–48.2
Gas-Oil Ratio	450–2992:1
Production Data	
Oil and/or Gas in Place	7.0 million BO
Ultimate Recovery	2.0 million BO; URE = 28.6%
Cumulative Production	1.53 million BO and 2.2 billion CFG (0.3 billion m^3) through 1981

Remarks: IP 240 BOPD. Discovered 1973.

13
Pennsylvanian Facies-Diagenetic Reservoir, Lower Strawn Formation, Seminole Southeast Field, Midland Basin, West Texas

S. J. Mazzullo

Introduction

Carbonate rocks of Desmoinesian (Middle Pennsylvanian Strawn) age form hydrocarbon reservoirs in many oil and gas fields in the Permian Basin of west Texas (Fig. 13-1A) and southeastern New Mexico. The reservoirs occur in stratigraphic and stratigraphic-structural combination traps, and are predominantly algal and coralgal buildups ("reefs") and associated carbonate grainstones of shallow-marine origin. A modest literature exists on the sedimentology and diagenesis of these reservoirs (*e.g.*, Stafford, 1954; Myers, *et al.*, 1956; Thornton and Gaston, 1968; Thomas, 1970; Vest, 1970; Toomey and Winland, 1973; Randolph, 1974).

A representative example of many such Strawn fields in the Permian Basin is the Seminole Southeast (SE) oil field, west Texas (Fig. 13-1B). This field is a stratigraphic-structural combination trap involving shelf-limestone reservoirs, in which variations in porosity and permeability resulted mainly from contrasting processes and products of early vadose and phreatic diagenesis. A later overprint related to burial diagenesis also can be seen. This paper describes the depositional facies and diagenetic evolution of this lower Strawn reservoir as determined by analysis of slabbed cores and cutting samples. Although specific to Seminole SE Field, the depositional-diagenetic models described herein apply more widely to Strawn carbonate reservoirs throughout the Permian Basin.

Location and Geologic Summary

Seminole SE Field is located in sections 160 (Blk. G WTRR), 15, 16, and 17 (Blk. C-44 PSL) of Gaines County, west Texas (Fig. 13-1B). The lower Strawn limestones of the field were deposited on a shallow shelf adjoining the ancestral Central Basin Platform and shallow Midland Basin (Fig. 13-1A), as a thin sequence of prograding, shoal-water carbonates. The upper part of this sequence includes two stacked, separate reservoirs in cyclic limestones and an overlying coralgal bioherm ("reef") grainstone shoal complex. Uplift of the Central Basin Platform immediately adjoining the field areas in Middle Pennsylvanian (Desmoinesian) time, and simultaneous subsidence of the Midland Basin, resulted in the present structural configuration of the field, a small fold or dome. Seminole SE Field is situated southeast of the larger Seminole Field (Fig. 13-1A). Although presently unsubstantiated, it is believed that (overlying) basinal Wolfcampian and/or coeval Strawn shales supplied the hydrocarbons for Seminole SE Field, during the late Early to Middle Permian (Wolfcampian-Leonardian?), subsequent to the completion of vadose and phreatic diagenesis.

Discovery and Field Data

Oil production at Seminole SE Field was established in May 1973 with the completion of the

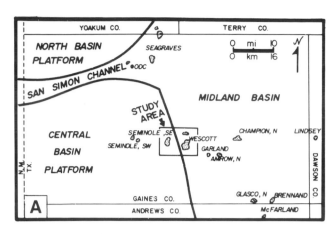 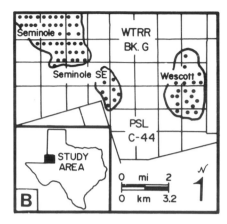

Fig. 13-1. A. Index and location map of Gaines and surrounding counties in west Texas showing major paleogeographic provinces and location of Strawn fields in the vicinity of the study area. *B.* Detailed location map of Seminole SE Field in Gaines County.

Union Texas Petroleum (UTP) No. 1 Wyatt (Fig. 13-2). The well was drilled as a lower Strawn reef test on the basis of subsurface geology and seismic data, subsequent to the discovery of Strawn production in the nearby Wescott Field (Fig. 13-1B). The discovery well reported an initial daily flowing potential of 240 barrels of oil and 26 barrels of water from 1/2-inch (1.28-cm) choke-through perforations at 10,780–10,792 feet (3285.7–3289.4 m); subsequent wells were perforated and acidized with 500 to 2000 gallons (~1900-7600 l) of dilute hydrochloric acid.

Development drilling up to 1977 by UTP defined the limits of the field and established the reservoirs as comprising a slightly folded, biohermal-shaped buildup originally situated on and flanking a low-relief, structural dome mapped on the top of the lower Strawn (Fig. 13-2). Production comes from two limestone reservoir facies, *coralgal bioherm* and grainstone shoal, the distributions of which closely reflect the lower Strawn structure (Figs. 13-2 and 13-3). Formation evaluation suggests that the Union California *et al.* No. 1 Stanley well (Sec. 160), which was plugged and abandoned in 1954, should also be productive from the lower Strawn.

The structural attitude of the lower Strawn, which is nearly harmonious with that of a deeper, Devonian datum, reflects the shelf-to-"basin" topography of the Strawn bank complex, modified by late or immediate post-Desmoinesian deformation in the study area. The absence by nondeposition of the two reservoir facies in areas peripheral to the field and inferred conceptually to

the west of the field (Fig. 13-3), further illustrates the essentially stratigraphic nature of the field.

Proven productive area of the ten-well field, which is about 0.5 × 1.0 mile (0.8 × 1.6 km) in extent, is approximately 1500 acres (5.9 km²). The upper and lower reservoirs average 12 feet and 5 feet (3.7 and 1.5 m) in thickness. Cumulative production to March 1983 was 1.53 million barrels of oil and 2.2 billion cubic feet (0.3 billion m³) of gas, and the estimated oil reserves are 7.0 million barrels originally in place and 2.0 million barrels recoverable. The field is presently on secondary-recovery, waterflood production. Gasoil ratios in the field ranged initially from 450 to 2992 and oil gravities from 43 to 48.2° API, respectively; original water saturations varied from approximately 20 to 80 percent, depending on position of individual wells with respect to the oil-water contact (Fig. 13-2).

Lithofacies and Depositional Framework

The entire Strawn section in the field area averages 233 feet (71 m) in thickness, and unconformably(?) overlies a thin, regressive sequence of marginal-marine siliciclastic rocks and thin shallow-marine limestones of Atokan (Bend) age (Fig. 13-2). Formal stratigraphic names have not been proposed for this sequence. Deposition of siliciclastics, Unit A, was continuous across the lower Strawn-Atoka time boundary, and was succeeded by rapid inundation and deposition of

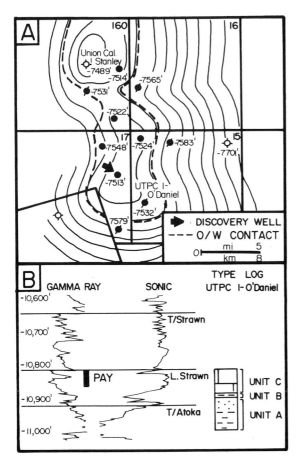

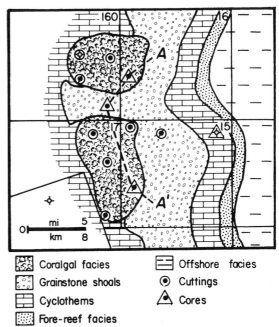

Coralgal facies   Offshore facies

Grainstone shoals   ⊙ Cuttings

Cyclothems   ⊿ Cores

Fore-reef facies

Fig. 13-3. Generalized map illustrating distribution of paleofacies toward the close of lower Strawn time (upper Unit C). The cyclothem facies occurs to the east and west of the field proper because it is a transgressive unit on top of the fore-"reef" and offshore facies. The western edge of coralgal and grainstone facies is conceptual (no well data available). Cross-section A–A' is shown in Figure 13-5.

Fig. 13-2. A. Structure map of the top of lower Strawn, showing location of discovery well and original oil-water contact. Contour interval 25 feet. *B.* Type log in Seminole SE Field. Location of upper and lower reservoir zones in the field indicated in Figure 13-5.

offshore marine shales, Unit B, in a local depocenter between Seminole and Wescott fields. The following discussion focuses on the upper part of the lower Strawn lithofacies, Unit C, 72 to 107 feet (22–33 m) in thickness, which was identified on the basis of megascopic and petrographic examination of cores from four wells, and samples from eight other wells in the field area (Fig. 13-3).

The sedimentological history of the Unit C sequence involved:

1. Shoaling atop the underlying offshore-marine shale facies, Unit B, and subsequent deposition, in regressive order, of forereef and then cyclic-shelf facies (lower Unit C).
2. Establishment and minor progradation of a coralgal bioherm and grainstone shoal com-

plex (upper Unit C) over the higher portions of the field.
3. Onlap of offshore-marine deposits (upper Strawn) over the field area, providing the vertical seal on the reservoir.

The inferred facies recognized within the sequence deposited in stages 1 and 2 are grainstone shoals, coralgal biolithites, and a cyclic sequence that includes: low-energy shelf limestones, and phylloid algal mounds and associated crestal grainstones. The facies deposited in stages 1 and 3 include forereef limestones and offshore-marine shales.

Grainstone and Coralgal Facies

The high-energy grainstone shoal facies of the middle and upper part of Unit C (Fig. 13-4A) consists of horizontally laminated and/or cross-stratified, oolitic and bioclastic limestones. These limestones contain crinoids, fusulinids, foraminifers, ooids, and abundant abraded fragments

Fig. 13-4. Photographs of core slabs demonstrating lithofacies: *A*. Cross-stratified skeletal ooid grainstone; 10,983 feet (3348 m), Amerada Hess 1-O'Daniel "G". Length of scale 5 cm. *B*. Vertical sequence of Chaetetes biolithites in matrix of brecciated and porous grainstone, stratigraphic top at upper left; 10,835–10,843 feet (3303–3305 m), UTPC 1-Fleeman (pay), Length of scale 6 cm. *C*. Bioturbated, low-energy shelf facies; typical lower member of cyclothemic limestones; 10,865 feet (3312 m), UTPC 1-Fleeman. Length of scale 2 cm. *D*. Core slab of phylloid algal wackestone facies, typical middle member of cyclothemic limestones; 10,883 feet (3317 m), UTPC 1-Fleeman. Length of scale 7 cm. *E*. Vertical sequence of graded fore-"reef" deposits, argillaceous mudstone-wackestone with displaced shelf corals (1) and internally graded to cross-stratified biopackstones (2) top at upper left; 11,034–11,038 feet (3363–3364 m) Amerada Hess 1-O'Daniel "G". Length of scale 7 cm.

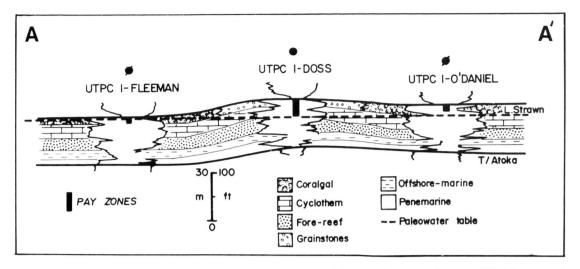

Fig. 13-5. Cross-section A–A', illustrating facies and pay zones and their relation to inferred paleowater table at −7575 feet (2309 m) subsea. The occurrence of inferred tidal-inlet grainstones at the highest structural position on this section (in the UTPC1-Doss well) reflects post-depositional folding. No horizontal scale, all logs are gamma-ray/sonics.

of *Chaetetes* and *Komia* in the vicinity of the coralgal facies. This facies is porous. The pronounced westward grainstone re-entrant in the SE/4 of section 160 (Fig. 13-3) probably represents tidal inlet deposits.

The coralgal facies of Unit C occurs on the higher portions of the field as massively bedded, bioherm-shaped masses of limestone up to 16 feet (4.8 m) in thickness. *In situ* colonies of the tabulate coral (?) *Chaetetes* sp. (few cm–0.5 m in height) occur in a porous matrix of skeletal packstone-grainstone (Fig. 13-4B) with crinoids, (encrusting and benthonic) foraminifers, fusulinids, bryozoans, *Komia* sp., *Chaetetes* fragments, coated grains, and scattered superficial ooids. The coral buildups appear to have been deposited in somewhat sheltered areas along the leeward margins (west) of the grainstone facies tract in a fashion suggested by Rich (1969). The robust *Komia*, an organism variously regarded as a dendroid stromatoporoid (Wilson *et al.*, 1963) or red alga (Johnson, 1963), probably inhabited a narrow zone transitional between the grainstone and coralgal facies, such as along a tidal inlet. The cumulative thickness of the grainstone-coralgal reservoir facies varies within the field from 12 to 20 feet (3.7–6.0 m).

Cyclically Repeated Facies

In the field area proper, the facies just described overlie a shoaling-upward sequence (Figs. 13-3, 13-5) that comprises the lower part of Unit C and includes, in ascending area: (1) black, calcareous shales, the offshore-marine facies of Unit B; (2) graded, argillaceous limestones (forereef facies); and (3) a relatively thin (15–30 ft, 4.6–9.2 m) sequence of stacked, shallow-marine cyclic limestone facies deposited across the field area (Figs. 13-3, 13-5). Locally, the argillaceous and cyclic limestone facies (2 and 3, above) are in lateral juxtaposition with the grainstones and coralgal biolithites (upper Unit C) of the field proper. A complete profile through a typical cyclic limestone section averages 6.6 feet (2.0 m) in thickness, and consists of three superimposed members: (1) a substratum of bioturbated lime mudstones and wackestones, interpreted as a low-energy shelf facies (Fig. 13-4C); (2) phylloid algal wackestone-packstones of biohermal to biostromal geometry, deposited in low-energy environments (Fig. 13-4D); and (3) bioclastic-oolitic grainstones (*e.g.*, Fig. 13-4A), comprising the high-energy crestal deposits that commonly overlie such algal mounds (Wilson, 1975). The forereef facies are dark-colored, argillaceous mudstone-wackestones of bioturbated to homogeneous texture (Fig. 13-4E). Thin (1-ft, 0.3-m) interbeds of graded biowackestone and packstone with displaced shelf fossils and poorly developed Bouma C-D turbidite sequences occur within this final section and are interpreted to represent debris-flow deposits derived from the wastage of the coralgal biolithite-grainstone section (upper

Unit C) in the field area proper. These rocks pass rapidly downdip (to the east) into the calcareous and spiculitic, black shales (Unit B) interpreted as deeper-water, offshore marine facies.

Petrophysics: Early Diagenesis and Porosity Evolution

Comparative Vadose-Phreatic Effects

Hydrocarbon production in the field is mainly from the upper coralgal and grainstone facies, with subordinate production in the field area proper above 7575 feet (2310 m) subsea from thin, discontinuous lenses of porous grainstone and phylloid-algal limestone within the cyclothemic sequence (Fig. 13-5). Despite the local occurrence of porous limestone lenses below 7575 feet subsea, these facies are not productive because of their low porosities and permeabilities (less than 5% and 0.1 md).

Petrographic analysis of the grainstones and biolithites throughout the lower Strawn section suggests a causal relationship between degree of reservoir development and environment of early diagenesis. The excellent, upper coralgal and grainstone reservoir averages 12 feet (3.7 m) in thickness, with average whole-core porosity of 13 percent and core-plug permeabilities of 29 millidarcys (maximum 94 md); these rocks appear to have been altered in the subaerial vadose environment. The impermeable limestones below 7575 feet subsea appear to have been affected solely by freshwater, phreatic diagenetic processes. The intervening limestone section exhibits diagenetic features and reservoir characteristics that are intermediate between the vadose and phreatic-altered rocks; average thickness, porosities, and permeabilities of the included lower reservoir zones are 5 feet (1.5 m), 5.8 percent, and 11.2 millidarcys, respectively. These rocks appear to have been stabilized mineralogically within a zone defined by a fluctuating water table. Inasmuch as the inferred paleo-water table is a horizontal datum (viewed parallel to strike: Fig. 13-5), emplacement of the freshwater lens occurred during or after uplift and folding in immediate post-Desmoinesian (pre-Permian) time. The comparative effects of vadose and phreatic

diagenesis observed in these rocks (Table 13-1, Fig. 13-6) are similar to those observed elsewhere by many other workers.[1]

Despite the emphasis in this discussion on diagenesis as an influence on reservoir quality, it is clear that fracturing also has a significant control on productivity by enhancing vertical permeabilities. Open, vertical hairline and solution-enlarged fractures are ubiquitous in the field cores examined, and many are stained with residual hydrocarbons. The formation of these fractures is probably related to post-Strawn (Upper Pennsylvanian-Lower Permian) deformation of the study area.

Features of Burial-Diagenetic Origin

The petrographic products of deep-burial diagenesis in the mesogenetic realm are ubiquitous in the lower Strawn limestones examined. However, the effects of such diagenesis are much more limited both volumetrically and in relation to their role in reservoir quality compared with development of early diagenesis and fracturing. Burial-diagenetic features include:

1. Partial occlusion of scattered vugs, solution channels, and inter- and intraparticle pores by medium to coarsely crystalline dolomite (Fig. 13-7A) or baroque dolomite. Some of the calcite that cements together layers of ruptured ooid cortex is also of burial-diagenetic origin (Fig. 13-7E).

2. Minor porosity enhancement resulting from partial dissolution of pore-filling cements, characteristically originating along intercrystal boundaries (Fig. 13-7B). Such a process has also been inferred in some Atokan carbonates from the Delaware Basin of Texas (Mazzullo, 1981).

3. Pervasive replacement of particles, cements, and micrite matrix by coarsely crystalline baroque dolomite (Fig. 13-7C); partial replacement of skeletal fragments by normal and length-slow chalcedony (Fig. 13-7D); partial pseudomorphic replacement by mosaic quartz of intraskeletal calcite cements within *Chaetetes* masses (Fig. 13-6A).

[1] Roehl (1967), Dunham (1969, 1971), Land (1970), Muller (1971), Oldershaw (1971), Steinen and Matthews (1973), Steinen (1974), Jacka and Brand (1977), Halley and Harris (1979).

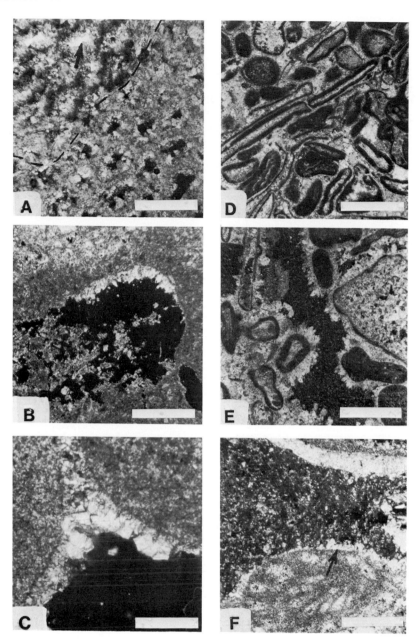

Fig. 13-6. Photomicrographs of examples showing comparative vadose and phreatic diagenesis. Cross-polarized light. *A–C* in vadose zone, samples above −7575 feet (2309 m) subsea. *D–F* in phreatic zone, samples below −7575 feet (2309 m) subsea. *A.* Chaetetes head with empty and partially occluded corallites (black) with small calcite cement crystals (below dashed line) and patches of completely occluded pores (above dashed line) with megaquartz-replaced cement (arrow). Length of scale 400. *B.* Non-fabric-selective vug (black) with gravitational (pendant) calcite cement and crystal silt. Length of scale 500. *C.* Meniscus calcite cement between adjacent peloids (pore is black). Length of scale 300. *D.* Pervasive cementation in a grainstone. Note fabric-selective dissolution of bioclasts (now occluded). Length of scale 600. *E.* Two generations of calcite cementation within a grainstone. Isopachous rim of scalenohedral crystals preceded coarse, pore-filling crystal (black). Length of scale 600. *F.* Competition between incipient crystal druse (arrow) and overgrowth (dark) on crinoid. Length of scale 500.

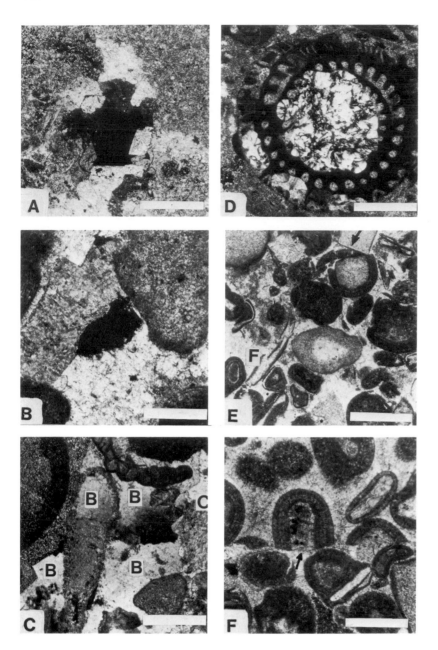

Fig. 13-7. Photomicrographs of diagenetic features in the deep-burial environment. Cross-polarized light. *A.* Dolomite crystals partially filling vug (black). Length of scale 150. *B.* Secondary porosity enhancement via dissolution of calcite cement along an intercrystalline contact. Length of scale 150. *C.* Baroque dolomite (b) replacing skeletal fragments and calcite cement (c). Length of scale 500. *D.* Length-slow chalcedony replacing fusulinid. Length of scale 400. *E.* Squashed and ruptured ooids, grain penetration (arrow), and broken skeletal fragments (f). Length of scale 500. *F.* Microsolution seam in ooid grainstone (arrow). Length of scale 500.

Table 13.1. Comparative Effects of Vadose and Phreatic Diagenesis, Seminole SE Field[1]

Vadose	Phreatic
Porosity in grainstones is primary inter- and intraparticle, with only minor amounts of puntal cements; porosity is primary intraparticle in corals	Inter- and intraparticle pores almost completely occluded by calcite
Porosity types in muddy rocks include non-fabric-selective vugs and solution channels, and micrite-selective solution	Selective dissolution of allochems (now almost completely occluded)
Brecciation and internal sediments common	Brecciation and internal sediments absent
Patchy development of partially and more completely cemented areas in grainstones and within corals	Uniform cementation within rocks
Occurrence of meniscus and gravitational cements	Absence of meniscus and gravitational cements
Calcite cements are non-ferroan, finely crystalline blades, rhombs, or, less commonly, scalenohedra	Calcite cements are ferroan, coarsely crystalline scalenohedra and equant to rhombic crystals
Absence of isopachous cement distributions	Presence of isopachous and multigeneration cements with drusy mosaics; competitive rim cementation
High porosities and permeabilities	Low porosities and permeabilities

[1] Refer to Figures 12-6 and 12-7 for photomicrographic documentation of these features.

4. An unusually high incidence of squashed and broken ooids and skeletal fragments (Fig. 13-7E); some of the latter are coated with similarly broken, isopachous cement rinds, the fragments later healed by coarse, baroque dolomite or calcite (Fig. 13-7E).

5. The common occurrence of micro-solution seams (Fig. 13-7F) and stylolites, including the "horsetail" variety described and illustrated by Shinn (1979).

Summary

Porosity evolution in the coralgal, phylloid-algal-grainstone facies in Seminole SE Field is related directly to contrasting processes and products of freshwater, eogenetic vadose and phreatic diagenesis. A vug and solution-channel pore system with excellent permeability evolved primarily as a result of pervasive, non-fabric-selective dissolution in the vadose zone. Porosity preservation was a consequence of slow rates of cementation and resultant incomplete occlusion of voids. In contrast, those rocks altered in the paleophreatic zone, although initially quite porous and permeable, were more rapidly stabilized and cemented and thus are poor or commercially unproductive reservoirs. Diagenesis in the mesogenetic (burial) environment had no appreciable effect on reservoir development because of the counterbalance between secondary porosity enhancement and occlusion via dolomite precipitation. However, permeability enhancement via fracturing is an integral aspect of the field reservoirs.

Acknowledgments   This study was done while at Union Texas Petroleum Corporation, with whose authorization this paper is published. Tom Nelson (UTPC) kindly provided background information on the history and development of the field, and read an earlier version of this manuscript. Miles W. Gray (UTPC) assisted in the generation of the production data cited in the text. Log-calculated porosities were determined by W. Markgraf (UTPC). The manuscript was typed by Fawn Meek.

References

DUNHAM, R.J., 1969, Early vadose silt in Townsend mound (reef), New Mexico: *in* Friedman, G.M., ed., Depositional Environments in Carbonate Rocks—a symposium: Soc. Econ. Paleontologists and Mineralogists, Spec. Publ. 14, p. 139–181.

DUNHAM, R.J., 1971, Meniscus cement: *in* Bricker, O.P., ed., Carbonate Cements: Johns Hopkins Univ. Studies in Geology, No. 19, p. 297–301.

HALLEY, R.B., and P.M. HARRIS, 1979, Freshwater cementation of a 1000-year-old oolite: Jour. Sedimentary Petrology, v. 49, p. 969–98.

JACKA, A.D., and J.P. BRAND, 1977, Biofacies and development and differential occlusion of porosity in a Lower Cretaceous (Edwards) reef: Jour. Sedimentary Petrology, v. 47, p. 366–381.

JOHNSON, J.H., 1963, Pennsylvanian and Permian algae: Quarterly, Colorado School of Mines, v. 58, 211 p.

LAND, L.S., 1970, Phreatic vs. vadose meteoric diagenesis of limestones—evidence from a fossil water table: Sedimentology, v. 14, p. 175–185.

MAZZULLO, S.J., 1981, Facies and burial diagenesis of a carbonate reservoir: Chapman Deep (Atoka) field, Delaware Basin, Texas: Amer. Assoc. Petroleum Geologists Bull., v. 65, p. 850–865.

MULLER, G., 1971, Gravitational cement: an indicator for the vadose zone of the subaerial diagenetic environment: *in* Bricker, O.P., ed., Carbonate Cements: Johns Hopkins Univ. Studies in Geology, No. 19, p. 301–302.

MYERS, D.A., P.T. STAFFORD, and R.J. BURNSIDE, 1956, Geology of the late Paleozoic Horseshoe Atoll in west Texas: Texas Bur. Econ. Geology Bull. 5607, 113 p.

OLDERSHAW, A.E., 1971, The significance of ferroan and non-ferroan calcite cements in the Halkin and Wenlock Limestones (Great Britain): *in* Bricker, O.P., ed., Carbonate Cements: Johns Hopkins Univ. Studies in Geology, No. 19, p. 225–229.

RANDOLPH, E., 1974, Lithostratigraphy and subsurface study of the *Chaetetes*-bearing Lower Strawn Formation (Pennsylvanian), Gaines County, Texas: M.S. Thesis, Univ. Oklahoma, Norman, 51 p.

RICH, M., 1969, Petrographic analyses of Atokan carbonate rocks in central and southern Great Basin: Amer. Assoc. Petroleum Geologists Bull., v. 53, p. 340–366.

ROEHL, P.O., 1967, Stony Mountain (Ordovician) and Interlake (Silurian) facies analogs of Recent low-energy marine and subaerial carbonates, Bahamas: Amer. Assoc. Petroleum Geologists Bull., v. 51, no. 10, p. 1979–2032.

SHINN, E.A., 1979, Conference on deep-burial diagenesis in carbonates, Tulsa, OK.

STAFFORD, P.T., 1954, Scurry field: *in* Occurrence of Oil and Gas in West Texas: Texas Bur. Econ. Geology Publ. 5716, p. 295–302.

STEINEN, R.P., 1974, Phreatic and vadose diagenetic modification of Pleistocene limestone—petrographic observations from subsurface of Barbados, West Indies: Amer. Assoc. Petroleum Geologists Bull., v. 58, p. 1008–1024.

STEINEN, R.P., and R.K. MATTHEWS, 1973, Phreatic vs. vadose diagenesis: stratigraphy and mineralogy of a cored borehole on Barbados, WI: Jour. Sedimentary Petrology, v. 43, p. 1012–1021.

THOMAS, C.M., 1970, Petrology of Pennsylvanian carbonate bank and associated environments, Azalea Field, Midland County, Texas (abst.): Amer. Assoc. Petroleum Geologists Bull., v. 54, p. 872–873.

THORNTON, D.E., and H.H. GASTON, 1968, Geology and development of Lusk Strawn Field, Eddy and Lea Counties, New Mexico: Amer. Assoc. Petroleum Geologists Bull., v. 52, p. 66–81.

TOOMEY, D.F., and H.D. WINLAND, 1973, Rock and biotic facies associated with Middle Pennsylvanian (Desmoinesian) algal buildup, Nena Lucia Field, Nolan County, Texas: Amer. Assoc. Petroleum Geologists Bull., v. 57, p. 1053–1074.

VEST, E.L., 1970, Oil fields of Pennsylvanian-Permian Horseshoe Atoll, west Texas: *in* Halbouty, M.T., ed., Geology of Giant Petroleum Fields: Amer. Assoc. Petroleum Geologists Mem. 14, p. 185–203.

WILSON, E.C., R.H. WAINES, and A.H. COOGAN, 1963, A new species of *Komia Korde* and the systematic position of the genus: Paleontology, v. 6, p. 246–253.

WILSON, J.L., 1975, Carbonate Facies in Geologic History: Springer-Verlag Inc., Berlin, 471 p.

14
Happy and Seberger Fields
William J. Ebanks, Jr. and W. L. Watney

RESERVOIR SUMMARY

Location & Geologic Setting	Rawlins Co., NW Kansas, USA
Tectonics	SE flank of Las Animas Arch
Regional Paleosetting	NW Kansas paleoshelf of the Hugoton Embayment
Nature of Trap	Stratigraphic/structural; small structural anomalies on regional dip, in part due to phylloid-algal reef-mounds
Reservoir Rocks	
Age	Late Pennsylvanian (Missourian)
Stratigraphic Units(s)	Lansing-Kansas City Groups
Lithology(s)	Limestone
Dep. Environment(s)	Shallow-marine
Productive Facies	Phylloid algal wackestone-packstone & associated grainstone
Entrapping Facies	Shale
Diagenesis/Porosity	Enhancement of original porosity by early dissolution; later cementation by calcite
Petrophysics	
Pore Type(s)	Skeletal moldic, vuggy
Porosity	2–12%, 4% avg
Permeability	0.1–900 md, 1 md avg
Fractures	Mild brecciation due to solution-collapse
Source Rocks	
Age	Late Pennsylvanian
Lithology(s)	NA
Migration Time	NA
Reservoir Dimensions	
Depth	4250 ft (1295 m)
Thickness	12–18 ft (3.7–5.5 m)
Areal Dimensions	Seberger 0.5 × 0.8 mi (0.8 × 0.4 km), Happy 0.8 × 0.3 mi (1.2 × 0.4 km)
Productive Area	40–80 acres (0.16–0.32 km^2)
Fluid Data	
Saturations	NA
API Gravity	35°
Gas-Oil Ratio	NA
Other	paraffinic
Production Data	
Oil and/or Gas in Place	NA
Ultimate Recovery	<100,000 BO both fields
Cumulative Production	~98,100 BO both fields through 1981

Remarks: [1]Fields 1.5 mi (2.4 km) apart; IP 17 BOPH, IP mechanism solution-gas drive. 3 producing wells. Discovered 1973 (Happy) and 1976 (Seberger).

14
Geology of Upper Pennsylvanian Carbonate Oil Reservoirs, Happy and Seberger Fields, Northwestern Kansas

William J. Ebanks, Jr. and W. L. Watney

Location and Discovery

Happy and Seberger fields are small fields located in Rawlins County, Kansas, approximately 30 miles (50 km) north of Colby and 40 miles (64 km) northeast of Goodland (Fig. 14-1). Oil and gas in these fields are produced from limestone formations of the Lansing and Kansas City groups (Missourian). Rawlins County is situated at the northeastern end of the Las Animas arch, near the intersection of this low-relief, subsurface structure with the trend of the Cambridge arch. This part of the state, the northwest Kansas Shelf, is a northward extension of the larger Hugoton Embayment, a sedimentary basin deepening southward during the late Paleozoic.

Happy Field was discovered in 1973 and consists of one producing well surrounded by dry holes (Fig. 14-2). Nearby is Seberger Field, which was found in 1976. It consists of two producing wells and also is delimited by dry holes.

Most oil and gas traps in this area are found by mapping subsurface marker beds from well data or by seismic reflection surveys. Happy and Seberger fields appear as small, positive structures that are anomalies on the regional dip (Fig. 14-2). Because of fairly sparse well control in parts of this region, the presence of such small features can easily be overlooked.

Reservoir Trap Characteristics

The productive formations in Happy and Seberger fields occur at depths that vary from about 4250 to 4300 feet (1295–1310 m). Each field comprises between 40 and 80 productive acres (0.16–0.32 sq. km), which emphasizes their small size. Other similar fields in the area, such as Cahoj (Fig. 14-1), include many more wells and are several hundred acres in size.

Reserves and Production

Most of the small fields in Rawlins County contain less than 100,000 barrels of recoverable reserves, but one field, Cahoj, has produced more than 6.5 million barrels of oil since its discovery in 1959. Ultimate recovery of oil from Cahoj Field will probably be about 8 million barrels.

Accurate estimates of reserves in Happy and Seberger fields are not available, but through 1981 the three wells in these two fields had produced about 98,100 barrels of oil. This oil comes from multiple completions in thin limestones of the Lansing and Kansas City groups. Other fields in this part of the state produce from the same units or from other, lithologically similar, units above and below. Initially, the discovery well of Seberger Field produced at a rate of 17 barrels of

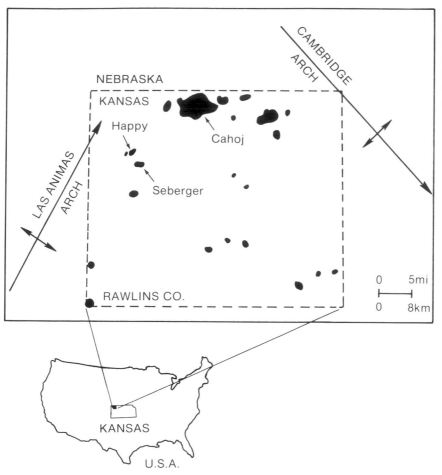

Fig. 14-1. Location map of Happy and Seberger fields in Rawlins County, northwestern Kansas, showing important paleotectonic features in this area.

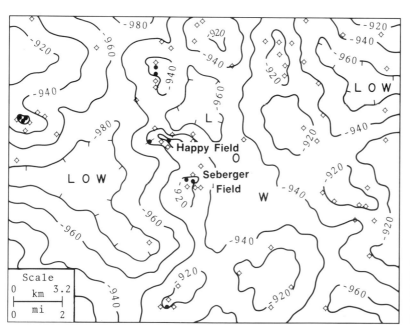

Fig. 14-2. Structure map on top of the Lansing group in Rawlins County, Kansas, near Happy and Seberger fields. Contour interval 20 ft.

oil per hour, by swabbing the well. The oil has a gravity of approximately 35° API and is thought to be paraffinic in composition. A rapid initial decline in rate of oil production of the wells and the low water-oil ratio during their early life indicate a solution-gas-drive producing mechanism.

Geological Characteristics of Reservoirs

Areal continuity of the alternating limestone and shale formations of the Lansing and Kansas City groups is good, and is typical of most of the middle and upper Pennsylvanian sequences in the American midcontinent. Regional dip of the Missourian (Watney, 1980) section in northwestern Kansas is eastward or southeastward. Very small departures from regional dip, such as low-relief tectonic folds or "noses," or closure caused by differential compaction, are sufficient to trap oil and gas locally.

Porosity in these reservoirs is seemingly unpredictable, but oil traps are found where porosity is present in areas that appear slightly higher than surrounding areas on maps of subsurface structure. An interesting aspect of these areally limited reservoirs is the fact that their porosity, and in part their "structure," are products of sedimentologic and diagenetic factors (Watney and Ebanks, 1978; Watney, 1980).

Formations of the Lansing and Kansas City groups in northwestern Kansas are informally named by industry as lettered zones (Fig. 14-3), rather than by their equivalent formation names in surface exposures (Morgan, 1952). In Happy Field, production of oil and gas comes from the D and J zones, whereas in Seberger Field, the D and E zones produce. The D and E zones are probably equivalent stratigraphically to the Spring Hill Limestone Member and Merriam Limestone Member, respectively, of the Plattsburg Formation; and the J zone is equivalent to the Winterset Limestone Member of the Dennis Formation in surface outcrops. Oil-field terminology is used in discussing these units here, and the J zone will not be discussed further.

Beneath the E zone is reddish-brown, green-mottled, slightly calcareous, fractured shale (Fig. 14-4). This is a typical "outside" shale in the scheme of Heckel (1977), by which he discussed the significance of midcontinent cyclothemic sequences.

The E zone, the next highest unit (Fig. 14-4), is a "transgressive," or "middle," limestone in Heckel's terminology. Transgressive limestones are not commonly reservoirs for oil and gas because of their usual dense, nonporous texture, which is attributed to their deposition in relatively deep, quiet, marine environments. In Seberger Field, the E zone has unusual thickness and a fabric somewhat like the D zone above. There is also diagenetic evidence of subaerial exposure of the E zone.

The E and D zones are separated by 5 feet (1.5 m) of green to gray or brownish-black, fossiliferous shale (Fig. 14-4). This is the "offshore" or "core" shale of Heckel (1977), who interprets it to have been deposited at the height of marine inundation during formation of this cyclothem. The upper part of this marine shale is transitional upward into 2.5 feet (0.8 m) of tan, brachiopod-foram wackestone, the lowest part of the D zone.

The next higher 10 feet (3.0 m) of the D zone comprises less well-bedded, slightly dolomitic, oil-stained, phylloid algal-foraminiferal wackestone. The top 2 feet (0.7 m) is fusulinid-foraminiferal, quartz-silt packstone (Fig. 14-4). Porosity in the D zone is also mostly vuggy and moldic, but includes some interparticle and intercrystal matrix porosity. The D-zone limestone is a "regressive," or "upper," limestone (Heckel, 1977). Watney (1980) has found that this and other regressive limestones most frequently are the reservoir rocks for oil pools in the study area, because they comprise the most porous primary facies and most commonly have had primary porosity enhanced by subaerial diagenesis. Because of its natural position in the sequence, this diagenesis probably occurred soon after deposition during regression of the shallow, midcontinental sea.

The D zone is overlain by several feet of reddish-brown, green-mottled, slightly calcareous, fractured shale (Fig. 14-4) that is very similar to the shale underlying the E zone. This outside shale was probably deposited in a continental environment, during marine regression, and it is part of the evidence for early subaerial exposure of the underlying limestone.

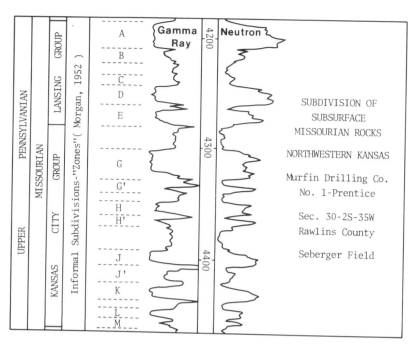

Fig. 14-3. Gamma-ray/neutron log of the Murfin Drilling Co. No. 1 Prentice well showing stratigraphic subdivisions of the Lansing and Kansas City groups that are the subject of this paper.

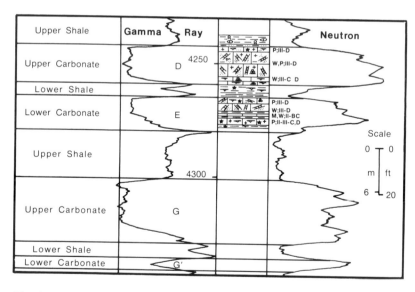

Fig. 14-4. Well log and core description from the Murfin Drilling Co. No. 1 Prentice in Seberger Field, showing cyclically recurring carbonate and clastic facies in northwestern Kansas. Well log character is, in most cases, diagnostic of these sequences of rock types. Dots in lithologic log of cored interval indicate terrigenous silt partially or completely filling vuggy or shelter porosity. Location of well: Section 30-T2S-R35W, Rawlins Co., Kansas. Letters and numerals to right of core log indicate classification of types of porosity (Archie, 1952), and of types of rock fabric (Dunham, 1962).

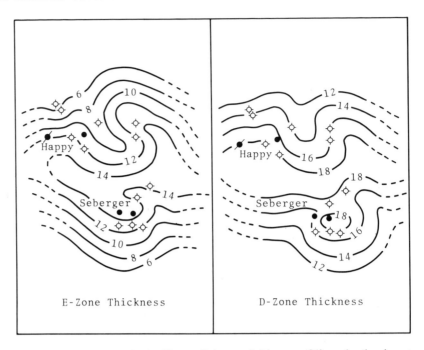

E-Zone Thickness   D-Zone Thickness

Fig. 14-5. Thickness maps of E- and D-zone carbonates in the Happy-Seberger fields area. Oil production is not solely related to trends of reservoir thickness, but is controlled by the occurrence of suitable porous and permeable rock. Contour interval 2 feet.

Alternations of limestone and shale such as these comprise the entire Lansing and Kansas City groups in this region. Relative thicknesses of the limestones and shales vary, and the relative abundances of different fossil constituents of the limestones may change, but the pattern is remarkably similar throughout a large area.

Thus, E- and D-zone limestones were deposited in a shallow-marine shelf setting, near an area that was intermittently a source of fine-grained terrigenous detritus. During Lansing-Kansas City time, frequent changes in sea level occurred (Irwin, 1965; Crowell, 1978; Watney, 1977, 1980; Heckel *et al.*, 1979; Heckel, 1980; Dubois, 1980). Small differences in thickness may have been important in determining which areas ultimately were exposed to subaerial diagenesis for the longest periods, during the time following deposition when slight withdrawals of the sea exposed broad areas of the Middle Pennsylvanian depositional shelf (Watney, 1980; Heckel *et al.*, 1979; Heckel, 1980).

Isopach maps of the porous E- and D-zone limestones (Fig. 14-5) illustrate the slight changes in thickness of these units in the area of Happy

and Seberger fields. Locally, the E zone increases in thickness from 4 or 5 feet (1.2–1.5 m) nearby to 12 or 14 feet (3.7–4.3 m) in the area of the fields, with thickening of the phylloid algal wackestone facies, and a similar change of facies in D zone accompanies thickening from 13 feet (4.0 m)

Table 14.1. Core Plug Analysis of Oil-Producing Formations in Murfin #1 Prentice well, Seberger Field (Fig. 14-4).[1]

Zone	Depth (ft.)	Porosity (%)	Permeability (md.)
D	4244	5.6	0.1
D	4245	3.0	0.1
D	4247	2.4	0.1
D	4250	2.9	0.1
D	4251	2.9	0.1
D	4254	5.3	30.4
D	4257	1.6	0.1
E	4268	11.6	897.0
E	4270	5.6	8.0
E	4275	2.6	0.1

[1]Porosity is 3–4% less than values indicated by log analysis. Sample 4268 has horizontal, solution-related fractures.

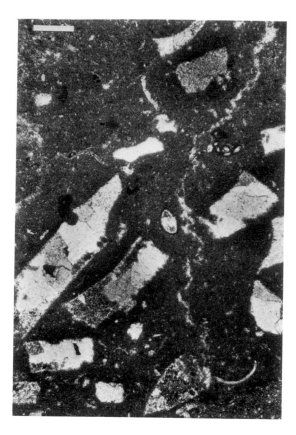

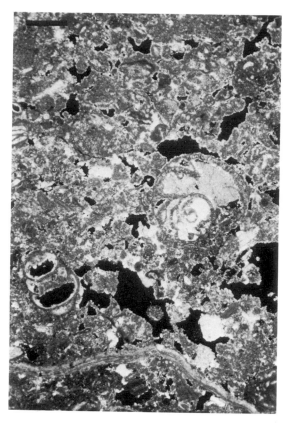

Fig. 14-6. Photomicrograph of molds of phylloid-algal blades that have been infilled by blocky calcite. Cross-polarized light. Bar scale is 0.5 mm. Murfin Drilling Co. No. 1 Prentice, 4265 feet (1300 m).

Fig. 14-7. Photomicrograph of vuggy pores in a mixed-skeletal packstone, with an overall "moth-eaten" appearance caused by irregular dissolution of matrix, grains, and cement. Cross-polarized light. Bar scale is 0.5 mm. Murfin Drilling Co. No. 1 Prentice, 4243 feet (1293 m).

nearby to 16 or 18 feet (4.9–5.5 m) in the producing fields. Throughout Rawlins County, Kansas, thickening of E- and D-zone limestones also correlates with thickening of the phylloid algal wackestone portion of each unit (Watney, 1980).

Porosity of E- and D-zone limestones is variable but quite low in Happy and Seberger fields (Table 14-1). Average values of core plug porosity are not as crucial to evaluating reservoir potential as is the total thickness of limestone having a porosity, by log analysis, greater than some predetermined cut-off value. In the experience of oil operators in this area, this minimum porosity value is about 8 percent. Porosity measured in small core plugs is consistently lower than that inferred from neutron logs (Table 14-1). Measurements of permeability of core plugs from these limestones in Happy and Seberger fields are also given in Table 14-1. These very low values of matrix permeability probably are not representative of effective permeability of the total reservoirs, which is enhanced by diagenetic fractures and fissures.

Alteration of the E- and D-zone limestones by processes related to subaerial exposure has been the most significant diagenetic change that affected their quality as oil reservoir rocks (Watney and Ebanks, 1978). These processes have both increased and reduced the porosity at different stages of alteration. The net effect, however, is

Fig. 14-8. Photomicrograph of "grains" composed of pellet-wackestone that had been the "matrix" cemented to phylloid algal blades until the algal carbonate was dissolved, leaving only a ghost of the original wall with a line of mud-filled marginal tubules to mark its former presence. Microbrecciation and recementation of the "grains" followed the earlier dissolution. Plane light. Bar scale is 0.25 mm. Murfin Drilling Co. No. 1 Prentice, 4271 feet (1302 m).

Fig. 14-9. Photomicrograph of terrigenous silt (grainy, white), filling a clay-lined vuggy pore in skeletal wackestone. Wisps of clay (gray) form partings that suggest the silt was sedimented into an open cavity after non-selective dissolution had truncated matrix and the large fusulinid test. Plane light. Bar scale is 0.5 mm. Murfin Drilling Co. No. 1 Prentice, 4250 feet (1295 m).

probably enhancement of porosity and permeability and of the ability of these rocks to produce oil and gas.

Dissolution of selected skeletal constituents, especially phylloid algal blades, formation of vuggy porosity that transects grains and finer matrix, and incipient solution brecciation are the most evident effects of percolation of undersaturated solutions through these rocks (Figs. 14-6, 14-7). The moldic and vuggy porosity formed in this manner has been partially occluded in parts of these reservoir rocks by cementation with sparry calcite (Figs. 14-6, 14-8) and by infiltration of terrigenous silt and clay (Fig. 14-9), probably from the shale overlying D-zone limestone.

Porosity is less affected by silt infiltration in the E zone than in the D zone.

Interpretation of these diagenetic effects in relation to subaerial exposure requires information, such as that presented by Watney and Ebanks (1978), from an area larger than that of the fields discussed here. In contiguous areas of northwestern Kansas and southwestern Nebraska, "regressive" limestones, such as the D zone, include laminar subaerial carbonate crusts, carbonate solution breccias, mixed-clast conglomerates, dissolution fissures and vugs, vadose-type internal sediment in pores and vugs, and paleosoil structures. Some of these features are illustrated in Figures 14-10A–D.

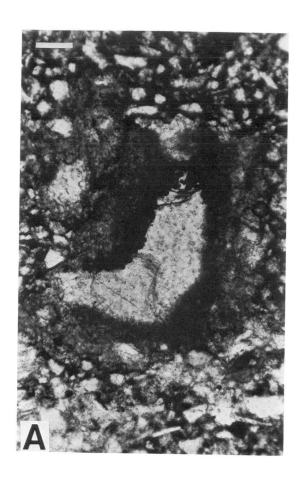

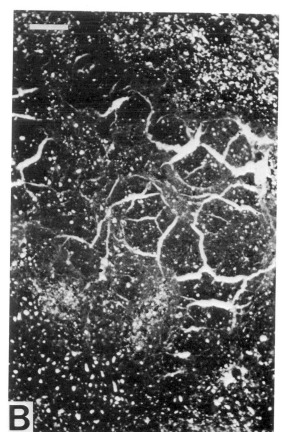

In some units, the subaerial crusts are gradational with subjacent-supratidal facies, and include desiccation-cracked laminae and pedotubules (Figs. 14-10A, B). Similarly, the nonmarine, "upper" shales that occur above the regressive limestones contain caliche nodules and calcrete cement, and are commonly perforated by networks of pedotubules. Most of these criteria have been cited by Perkins (1977) as evidence of recurrent subaerial exposure of Pleistocene limestone in south Florida. Evidence of subaerial processes diminish, and marine aspects become more prevalent, from north to south in western Kansas. Thus, the intensity of alteration of Missourian limestones by subaerial exposure decreases in the same direction (Rascoe, 1962; Watney, 1980). Limestones become relatively thicker and more numerous southward, terrigenous units become less silty and include more marine shales in that direction, and the entire sequence thickens toward a shelf margin in Oklahoma (Moore, 1979).

Discussion and Summary

The E- and D-zone limestones that are oil reservoirs in Happy and Seberger fields were de-

---

◄ *Fig. 14-10. A.* Photomicrograph of a pedotubule in silty limestone, with calcite (gray) filling and dark, stained wall. Plane light. Bar scale is 0.5 mm. Skelly 1 Bartosovsky, Section 9-T1S-R34W, Rawlins Co., Kansas, 4166 feet (1270 m). *B.* Photomicrograph of shrinkage (nontectonic) cracks in matrix of a ferruginous, calcareous siltstone. Plane light. Bar scale is 0.5 mm. Gore 1 Wertz, Section 6-T2N-R32W, Hitchcock Co., Nebraska, 3648 feet (1112 m). *C.* Photograph of laminated, brown crust capping regressive limestone (shaly, sparsely fossiliferous, vuggy wackestone); dark mottling in lower half is oil stain. Bar scale is 1 cm. Murfin Drilling Co. No. 1 Prentice, Section 30-T2S-R35W, Rawlins Co., Kansas, 4301 feet (1311 m). *D.* Photograph of mixed lithoclast conglomerate (chert, dolomite, limestone, siltstone) in a red, silty matrix. Scale is in cm. Empire 1 Rathe, section 9-T3N-R30W, Red Willow Co., Nebraska, 3663 feet (1116 m.) *C* and *D* from Watney, 1980. Published by permission of Kansas Geological Survey.

posited on a shallow-marine shelf, near an area that was an intermittent source of fine-grained terrigenous detritus, and at a time when frequent changes in sea level were occurring (Watney, 1980; and others). These and other reservoir limestones of the Lansing and Kansas City groups have facies that vary in composition laterally and vertically. In some areas they are shaly, mixed-skeletal wackestones and skeletal lime mudstones that probably represent deposition in a quiet, sheltered marine environment. In others they are phylloid algal wackestone and bioclastic, oolitic, or lithoclast-pellet grainstone or packstone that represent more shallow and current-agitated environments with varying amounts of local bathymetric relief. This variability contributes to the difficulty of predicting porous facies ahead of drilling, but it also offers the prospect of more than one porous zone being present at any particular location.

Diagenetic alteration of original rock fabric has been important in determining the final quality of these limestones as potential oil reservoirs. Fabric-selective and non selective dissolution have enhanced original porosity, but cementation by calcite and infiltration of pores by terrigenous silt and clay have subsequently reduced this porosity. The net effect is preservation of fairly good moldic and vuggy porosity in some areas and elimination of porosity in others. These alterations prevent the prediction of porosity occurrence at any particular locality.

Understanding the repetitive nature of the events that led to the deposition and subsequent diagenesis of the reservoir limestones is pivotal to future discovery of other similar, or larger, oil pools in this area and adjacent areas with similar geologic history, such as southern Nebraska (Dubois, 1980), eastern Colorado (Rascoe, 1962), northwestern Missouri-southwestern Iowa (Heckel, 1975), and northern Texas (Brown, 1979). New oil pools may well be discovered by combining regional maps of lithofacies with maps of subtle variations in subsurface structure of key stratigraphic markers.

Acknowledgments   Appreciation is expressed to Murfin Drilling Company, Wichita, Kansas, for providing materials necessary to this study and for granting permission to publish. The Director, Kansas Geological Survey, also granted permission to publish this paper.

References

ARCHIE, G.E., 1952, Classification of carbonate reservoir rocks and petrophysical considerations: Amer. Assoc. Petroleum Geologists Bull., v. 36, no. 2, p. 278–298.

BROWN, A.A., 1979, Evidence for subaerial exposure supports eustatic control of the growth of Missourian carbonate mounds, eastern shelf of the Midland Basin, Texas (abst.): Ninth Internat. Congr. of Carboniferous Stratigraphy and Geology, Prog., p. 25.

CROWELL, J.C., 1978, Gondwanan glaciation, cyclothems, continental positioning, and climatic change: Amer. Jour. Sci., v. 278, p. 1345–1372.

DUBOIS, M.K., 1980, Factors controlling the development and distribution of porosity in the Lansing-Kansas City "E" Zone, Hitchcock County, Nebraska: Unpubl. M.S. Thesis, Univ. Kansas, Lawrence, KN 100 p.

DUNHAM, R.J., 1962, Classification of carbonate rocks according to depositional texture: *in* Ham, W.E., ed., Classification of Carbonate Rocks—a symposium: Amer. Assoc. Petroleum Geologists Mem. 1, p. 108–121.

HECKEL, P.H., 1975, Field guide to Stanton Formation (Upper Pennsylvanian) in southeastern Kansas: Kansas Geol. Soc. 31st Reg. Field Conf. Guidebook, p. 41, Fig. 5.

HECKEL, P.H., 1977, Origin of phosphatic black shale facies in Pennsylvanian cyclothems of Mid-continent North America: Amer. Assoc. Petroleum Geologists Bull., v. 61, p. 1045–1068.

HECKEL, P.H., 1980, Paleogeography of eustatic model for deposition of mid-continent upper Pennsylvanian cyclothems: *in* Fouch, T.D. and E. R. Magathan, eds., Paleozoic Paleogeography of West-Central United States: Soc. Econ. Paleontologists and Mineralogists, Rocky Mtn. Sect., Symp. 1.

HECKEL, P.H., L.L. BRADY, W.J. EBANKS, JR., and R.K. PABIAN, 1979, Field guide to Pennsylvanian cyclic deposits in Kansas and Nebraska: Kansas Geol. Survey Guidebook, Ser. 5, p. 1–60.

IRWIN, M.L., 1965, General theory of epeiric clear water sedimentation: Amer. Assoc. Petroleum Geologists Bull., v. 49, p. 445–459.

MOORE, G.E., 1979, Pennsylvanian paleogeography of the southern Mid-Continent: Tulsa Geol. Soc., Spec. Publ. 1, p. 1–12.

MORGAN, J.V., JR., 1952, Correlation of radioactive logs of the Lansing and Kansas City groups in central Kansas: Amer. Inst. Mining and Metallurgy, Petrol. Engrs. Petrol. Trans., v. 195, p. 111–118.

PERKINS, R.D., 1977, Depositional framework of Pleistocene rocks in south Florida: Geol. Soc. America Mem. 147, pt. II, p. 131–198.

RASCOE, B., JR., 1962, Regional stratigraphic analysis of Pennsylvanian and Permian rocks in western mid-continent, Colorado, Kansas, Oklahoma, and Texas: Amer. Assoc. Petroleum Geologists Bull., v. 46, p. 1345–1370.

WATNEY, W.L., 1977, Lithofacies of upper Pennsylvanian Lansing-Kansas City groups in northwest Kansas-southwest Nebraska (abst.): Amer. Assoc. Petroleum Geologists Bull., v. 61, p. 839–840.

WATNEY, W.L., 1980, Cyclic sedimentation of the Lansing-Kansas City groups in northwestern Kansas and southwestern Nebraska: A guide for petroleum exploration: Kansas Geol. Survey Bull. 220, 72 p.

WATNEY, W.L., and W.J. EBANKS, JR., 1978, Early subaerial exposure and freshwater diagenesis of upper Pennsylvanian cyclic sediments in northern Kansas and Southern Nebraska (abst.): Amer. Assoc. Petroleum Geologists Bull., v. 62, p. 570–571.

15
Tarchaly, Rybaki, and Sulecin Fields
S. Depowski and T. M. Peryt

RESERVOIR SUMMARY

Location & Geologic Setting	Fore-Sudetic area, western Poland		
Tectonics	Central European Permian Basin		
Regional Paleosetting	Margin of stable cratonic basin		
Nature of Trap	Structural: brachyanticline-type structures complicated by faults		
Reservoir Rocks			
Age	Early Permian		
Stratigraphic Units(s)	Zechstein Main Dolomite		
Lithology(s)	Dolomite		
Dep. Environment(s)	Peritidal		
Productive Facies	Dolomitized oolitic grainstone, dolomite mudstone†. Dolomitized oolitic grainstone and mudstone*; NA**.		
Entrapping Facies	Zechstein anhydrite cap		
Diagenesis/Porosity	NA		
Petrophysics			
Pore Type(s)	Interparticle		
Porosity (avg)	8.2%†	3.5%*	10.6%**
Permeability (avg)	0.3 mg†	NA*	5.7 md**
Fractures	NA		
Source Rocks			
Age	Main Dolomite Zone, Zechstein. Tarchaly gas in part from Carboniferous		
Lithology(s)	Basinal and deep-shelf lime mudstone-wackestone facies of Main Dolomite		
Migration Time	NA		
Reservoir Dimensions			
Depth	4500 ft (1375 m)†	5330 ft (1630 m)*	9100 ft (2750 m)**
Thickness (gross pay)	180 ft (55 m)†	50 ft (18 m)*	50 ft (30 m)**
Areal Dimensions	NA†	1.9 × 0.9 mi (3 × 1.4 km)*	NA**
Productive Area	1875 acres (7.5 km²)†	990 acres (4 km²)*	630 acres (2.5 km²)**
Fluid Data			
Saturations	NA	NA	NA
API Gravity	—	34°	37°
Gas-Oil Ratio	—	100:1 avg	100:1 avg
Other	—	(¹)	(¹)
Production Data			
Oil and/or Gas in Place	NA	NA	NA
Ultimate Recovery	NA	NA	NA
Cumulative Production	NA	882,000BO*²	314,000 BO**²

Remarks: [1]IP mechanism solution-gas drive. † = Tarchaly, * = Rybaki, ** = Sulecin
[2]to May 1983. Discovered 1961 (Rybaki), 1965 (Tarchaly), and 1973 (Sulecin).

15
Carbonate Petroleum Reservoirs in the Permian Dolomites of the Zechstein, Fore-Sudetic Area, Western Poland

S. Depowski and T. M. Peryt

Introduction

In the Polish part of the Central European Permian basin, 6 oil and 13 gas fields have been discovered in the Zechstein Main Dolomite (Upper Permian) in the Fore-Sudetic area, western Poland (Figs. 15-1, 15-2). The Main Dolomite reservoirs are of stratigraphic-structural type and are principally dolomitized oolitic and oncolitic grainstones and fractured, vuggy, dolomitized mudstones.

Two oil fields discussed briefly in this paper, Rybaki and Sulecin, were discovered in 1961 and 1973, and a gas field, Tarchaly, in 1965 (Fig. 15-1). The discovery wells were located on the basis of reflection seismic data. Initial potentials for gas production ranged from 25.2 million to 0.74 billion cubic feet (0.72–21 million m^3) per day at Tarchaly Field, and for oil at Rybaki and Sulecin fields, from 190 to 630 barrels per day through 0.12 to 0.2 inch (3–5 mm) chokes.

Depositional Facies and Environments in the Main Dolomite

In this part of the Central European Basin, two broad areas of shallow intracratonic shelf and deep intracratonic shelf environments can be distinguished in the Main Dolomite (Fig. 15-1). Carbonate facies of the shallow shelf originated in shallow/subtidal to supratidal environments. They are generally dolomites now, containing minor insoluble residues (<1%) and significant amounts of anhydrite (10–30%). The main depositional rock types, now completely dolomitized, are oolitic and vadose pisolitic grainstones and mudstones (see Peryt, 1983). The carbonate facies of the deep shelf originated in deep-water environments and are lime mudstones interbedded with terrigeneous shales. Our interpretations of the regional paleobathymetry and facies relationships are shown in cross-section form in Figures 15-3 and 15-4.

The grainstone lithofacies (Figs. 15-5A–D and 15-6A,B,D) is generally composed of ooids that have undergone vadose diagenesis (*e.g.*, Fig. 15-5C,D). Previously, grains that were modified in a vadose environment were commonly interpreted, especially by petroleum geologists, as being oncoids (Peryt, 1983). In many cases, ooids in the Main Dolomite are accompanied by intraclasts (Fig. 15-5A,B,D), peloids, and mollusk shells. Most of the oolitic grainstones seem to be structureless. The oolitic grainstone lithofacies is in places accompanied by packstones composed either of ooids and peloids (Figs. 15-5E, 15-6C) or of peloids alone (Fig. 15-7B,E,F), and in places by stromatolite boundstone (Fig. 15-5A).

Mudstones are massive (Figs. 15-7D, 15-8B) or bedded (Figs. 15-5C–F, 15-7A) and locally they contain such isolated fossils as mollusks, os-

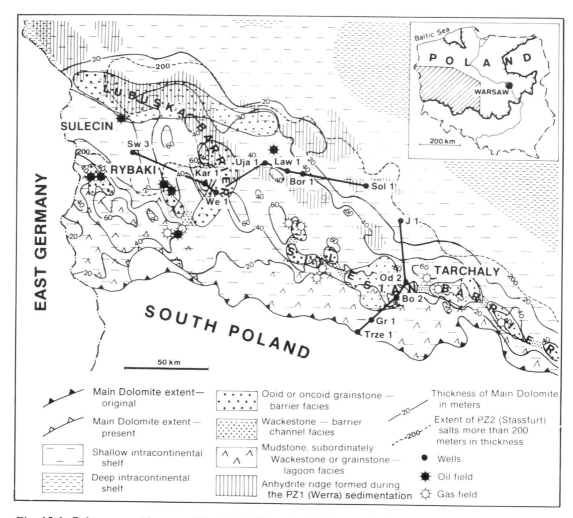

Fig. 15-1. Paleogeographic map of the Main Dolomite (Ca2) zone in the Fore-Sudetic area, western Poland. After Depowski (1975). Cross-section lines refer to Figures 15-3 and 15-4.

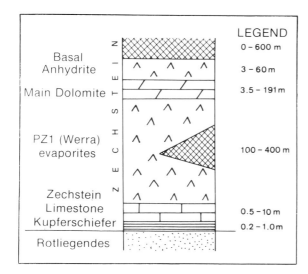

Fig. 15-2. Simplified stratigraphic column of the lower Zechstein. Cross-hatch pattern is salt.

tracodes and foraminifers. These may occur in abundance (Fig. 15-8A), and the mollusk wackestones form thin layers in thick mudstone units. The absence of stenohaline organisms in the Main Dolomite deposits seems to indicate sedimentation in environments of higher than normal salinity.

The oolitic facies occur in the shallow-shelf province (Fig. 15-1). They form broad belts, interpreted as barriers, which are commonly superimposed on ridges of Werra anhydrite (PZ1) below the Main Dolomite (Fig. 15-2), and they also occur as isolated shoals in different areas of the shallow shelf. In addition to banks built by grainstones, mud mounds composed of carbonate mud stabilized by non-stromatolitic blue-green algae are also found in the shallow-shelf province, especially east of the Lubuska Barrier (Fig. 15-1).

Fig. 15-3. Regional paleofacies cross-section of Main Dolomite, west to east from well Swiebodzin 3 to well Solec 1 (Fig. 15-1), showing inferred bathymetric relationships of facies at the end of Main Dolomite deposition.

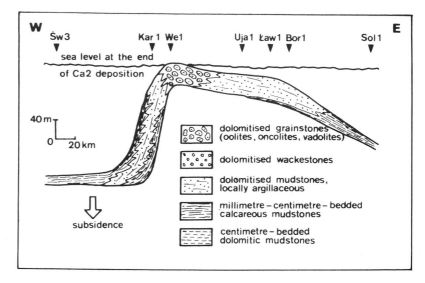

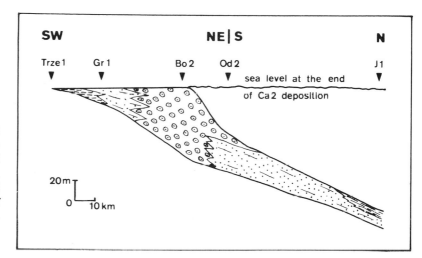

Fig. 15-4. Regional paleofacies cross-section of Main Dolomite, south to north from well Trzebnica 1 to well Jarocin 1 (Fig. 15-1), showing inferred bathymetric relationship of facies at the end of Main Dolomite deposition.

The barriers are located near the edge of the Main Dolomite platform.

In general, the bank forms were initiated during deposition of the lower part of the Main Dolomite, especially in the northwest-trending barrier zone, and also continued to develop later. In places, within the barrier zone in the lowermost part of the Main Dolomite, mudstones interbedded or interlaminated with shales occur, overlain successively by mudstones and oolites. This sequence may result from the existence of relatively shallow depressions, perhaps in intrabarrier areas, which were quickly filled up; or, the sequence may result from basinward migration of oolitic facies. The latter would constitute a generally regressive sequence of lithofacies in the area of shallow shelf.

The Main Dolomite is generally thicker in the barrier zone than in adjacent areas, exceeding 130 feet (40 m) there. Locally it is quite diminished, as thin as 11.5 feet (3.5 m) on paleohighs inferred from the development and thickness of underlying and overlying strata, and exhibits a breccia development. The strongly increased thickness, up to 620 feet (191 m) near the Lubuska Barrier (Figs. 15-1, 15-3), probably results from deposition on slopes.

Diagenesis and Porosity Characteristics and History

The deposits of the Main Dolomite have been subjected to rather intensive diagenetic processes that occurred primarily in the early stages of

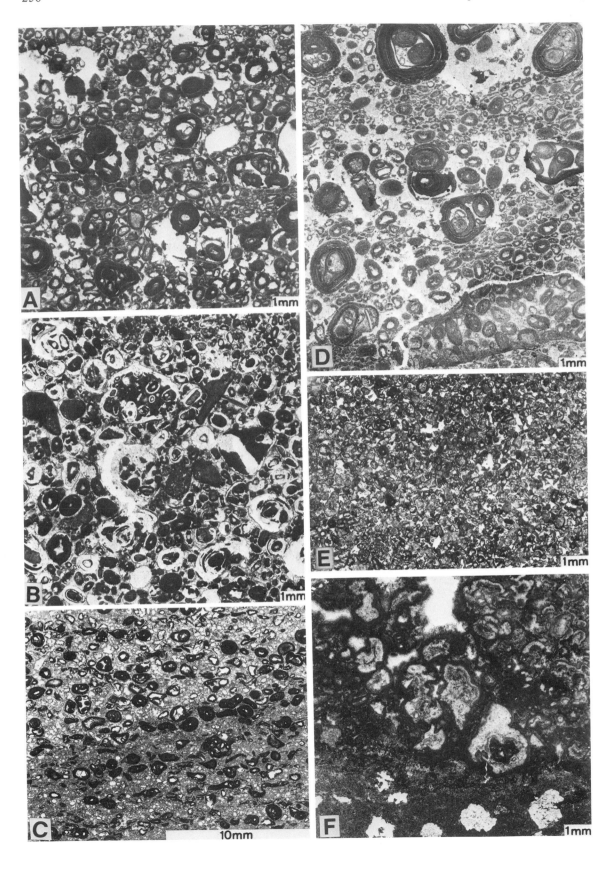

diagenesis. The dominant processes were dolomitization and anhydritization; other processes include dissolution, cementation, and vadose compaction. During later stages of diagenesis, the following processes were important: anhydritization, halitization, dedolomitization, calcitization of anhydrite, aggradational neomorphism, and pressure solution (Peryt, 1978).

Soon after deposition and penecontemporaneous dolomitization, early leaching occurred under vadose conditions that caused the removal of the centers of many ooids, oncoids, and shells and thus created excellent moldic porosity (Fig. 15-5B,E,F). Later, this porosity was in places destroyed by anhydrite cement. Late dissolution, believed to have been related to generation of CO_2 by thermal degradation of organic matter (Clark, 1980), created new pores that in turn were filled by anhydrite and halite. The varied sequences of pore-forming and pore-destroying processes result in wide variations of reservoir properties both vertically and laterally.

Because of diagenesis related to the vadose environment, and especially because of non-uniformly bedded cementation (characteristic of vadose environments), the ooids underwent early vadose compaction. This is of importance for reservoir properties in that the compacted layers generally have very low permeabilities and appear to act as barriers to vertical fluid movement; thus, they commonly divide a given sequence into several parts having different reservoir properties. Although grainstone sequences were especially strongly affected by vadose diagenesis and later cementation, the best reservoir properties are found in areas of thick grainstone. The oncolitic-oolitic rocks exhibit porosities up to 25 percent and permeabilities up to a hundred-plus millidarcys. Mudstones on average have porosities of several percent and permeabilities of several millidarcys. Finally, mudstones interlaminated with shales generally have porosities below 1 percent and permeabilities below 1 millidarcy.

Hydrocarbon Accumulations

The reservoirs have both fabric-selective and non-fabric-selective types of porosity. Fabric-selective types include both moldic (Figs. 15-5A–D,F; 15-6B,F; and 15-8A) and intercrystal pores; non-fabric-selective pore spaces are generally vug

---

◄ *Fig. 15-5.* Photomicrographs of oolitic grainstone and related rocks, all completely dolomitized. All photos in plane light. *A.* Oolitic grainstone. Cortices of ooids are smoothly laminated, and usually the individual envelopes are difficult to recognize except for bigger grains (mainly compound ooids). Nuclei are generally peloids and commonly are sparitic or anhydritic. Anhydrite (white areas) filled the molds of ooids, produced by partial dissolution of aragonite, and also partly filled other voids of irregular shapes. There are rare intraclasts consisting of ooids, as at right center edge. Depth 2985 m, Duszniki 1 well, 85 km NE of well Sw3 in Figure 15-1. *B.* Oolitic grainstone. The rock differs from that in A in that cortices of many larger ooids and some intraclasts, as well as some of the matrix, are replaced by anhydrite (white). Nuclei of small ooids and the central parts of peloids also are anhydritic. Little porosity is preserved. Duszniki 1 well, Depth 2977.6 m, 85 km NE of well Sw3 in Figure 15-1. *C.* Oolitic grainstone that underwent vadose compaction in the lower part and little or none in the upper part. Flattened and deformed grains like these are common in the Main Dolomite in western Poland and else-where in the Main Dolomite Basin (Fuchtbauer, 1964; Clark, 1980) Clark (1980) relates the origin of overpacked texture to vadose compaction. The textures seen here evidently testify to two different histories. The loosely packed layer was cemented earlier in the depositional environment or during subaerial exposure, and this cementation inhibited the compaction that affected uncemented layers. Variability of compaction intensity is probably related to percolating meteoric water that dissolved aragonite or high-Mg calcite grains at their contacts, leading to compacted texture. Depth 1548.7 m, Smilowo 2 well. *D.* Oolitic grainstone. Note loosely compacted intraclasts in compacted host, suggesting that the intraclasts were cemented before their incorporation in the sediment that later underwent vadose compaction (Clark, 1980). Depth 2985.6 m, Duszniki 1 well. *E.* Oolitic-pelletoidal packstone, with common solution-interparticle pores. Depth 2999.2 m, Duszniki 1 well. *F.* Vadose zone example from a site overlying the clotted-textured rock of a caliche profile. Some solution vugs are partly filled by calcite cement. Depth 1459.5 m, Radziadz 4 well.

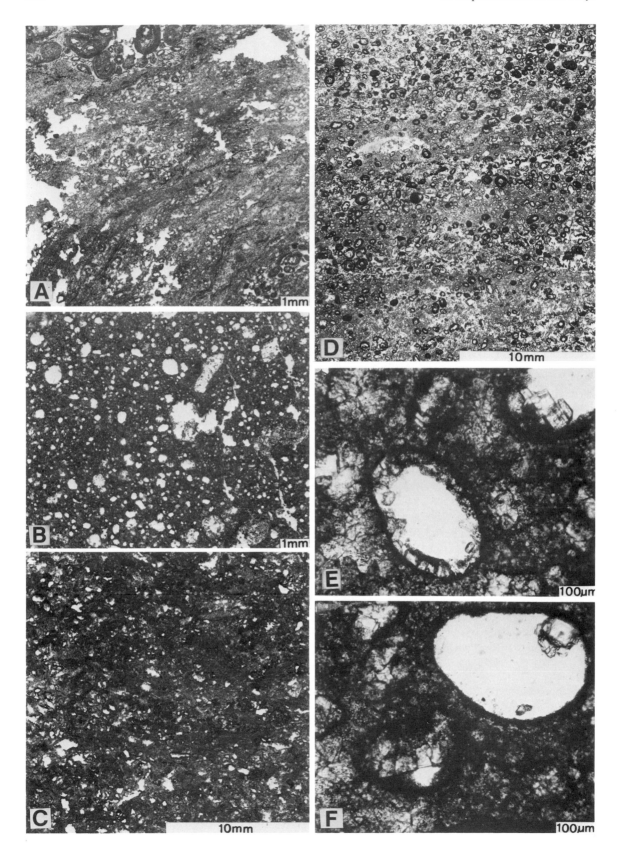

and/or fracture types (Figs. 15-5F, 15-7B,E). In fault zones, systems of fractures and microfractures result in increased flows of natural gas and oil compared with parts of fields that are not tectonically disturbed. Because the oncolite and oolite grainstones are of prime importance as reservoirs in the Main Dolomite, hydrocarbon accumulations are usually found in barrier areas (Figs. 15-1, 15-9) or adjacent areas (Figs. 15-1, 15-10, 15-11), and rarely in local mounds and banks in the shallow-shelf province (Fig. 15-11). The reservoirs usually occur within brachyanticline-type structures that are complicated as a direct result of faulting (Fig. 15-9), or in local uplifts near faults (Depowski, 1975).

Rybaki Oil Field

The Rybaki Field was discovered by the Rybaki 1 well on a structure located during seismic reflection studies. Thereafter, 17 wells were drilled, 5 of which were productive from the Main Dolomite at initial rates of 6 to 13 barrels per day through 0.2-inch (5-mm) chokes. The producing interval at the crest of the structure, a faulted dome, lies at a depth of about 5330 feet (1630 m). The Main Dolomite (Ca^2 zone) is about 185 feet (50 m) thick with a net thickness averaging 59 feet (18 m). The field has a productive area of about 990 acres (4 km²), and the two small productive structures are about 1.9 miles (3 km) by 0.9 miles (1.4 km) in combined length and width, respectively (Fig. 15-9). The Main Dolomite reservoir consists of mudstones in its lower part, overlain by mudstones with varying thicknesses of interbedded oolitic grainstones. Average porosity is 3.5 percent, and the permeability as determined on small samples averages a few millidarcys. Cores commonly show fractures. The oil is 3.05% paraffinic, with a density of 0.857 gm/cm³

(API gravity 34°) at 20° C. Gas-oil ratios averaged 100:1. The gas fraction has been found to contain 0.20 percent CO_2, 15.5 percent N_2, 0.05 percent H_2, and 0.01 percent H_2S by volume. The production mechanism is solution-gas drive. To May 1983, the Main Dolomite reservoir had produced a little over 882,000 barrels.

Sulecin Oil Field

TheSulecin Field was discovered by the Sulecin 1 well on a structure located during seismic reflection studies. Of 11 wells drilled in the field, four produced oil at initial rates of 190 to 315 barrels per day through 0.12-to 0.2-inch (3–5-mm) chokes. The Main Dolomite reservoir lies at a depth of about 9100 feet (2750 m) at its shallowest point in the field (Fig. 15-10), and has average gross and net thicknesses of 56 feet (17 m) and 33 feet (10 m), respectively. The average porosity is 10.6 percent, and the average permeability 6.67 millidarcys. The oil is of paraffinic type (5.23% paraffins) with a density of 0.838 gm/cm³ (API gravity 37°). Gas-oil ratio averaged 100:1 initially. The gas fraction contained on average 2.67 percent CO_2, 22.23 percent N_2, and 0.2 percent H_2S. The production mechanism also is similar to that at Rybaki Field: solution-gas drive. Initial reservoir pressure was about 30 percent higher than hydrostatic, as at Rybaki. The productive field area is 630 acres (2.5 km²), and oil production to May 1983 was about 314,000 barrels.

Tarchaly Gas Field

The Tarchaly Field was discovered by the Tarchaly 1 well on a brachyanticline structure located previously in seismic-reflection studies. The discovery well found reservoirs in the

---

Fig. 15-6. Photomicrographs of dolomitized oolitic grainstone. All photos in plane light. *A.* Oolitic grainstone (upper left corner) and oolitic packstone with weakly developed stromatolitic incrustations. Abundant solution vugs (white). Depth 2984.5 m, Duszniki 1 well. *B.* Leached and then dolomitized oolitic grainstone in which porosity is mostly occluded by anhydrite cements (white) that filled former molds of

ooids and other grains. Depth 1676.7 m, Janowo 3 well. *C.* Oolitic-pelletoidal packstone. Pores are oomolds and solution vugs (white). Sulecin 8 well, Sulecin Field. *D.* Oolitic grainstone. Smilowo 1 well, 35 km NW of well J1 in Figure 15-1. *E,F.* Portions of oolitic grainstone shown in *D.* Oomolds are filled partially by dolomite cement.

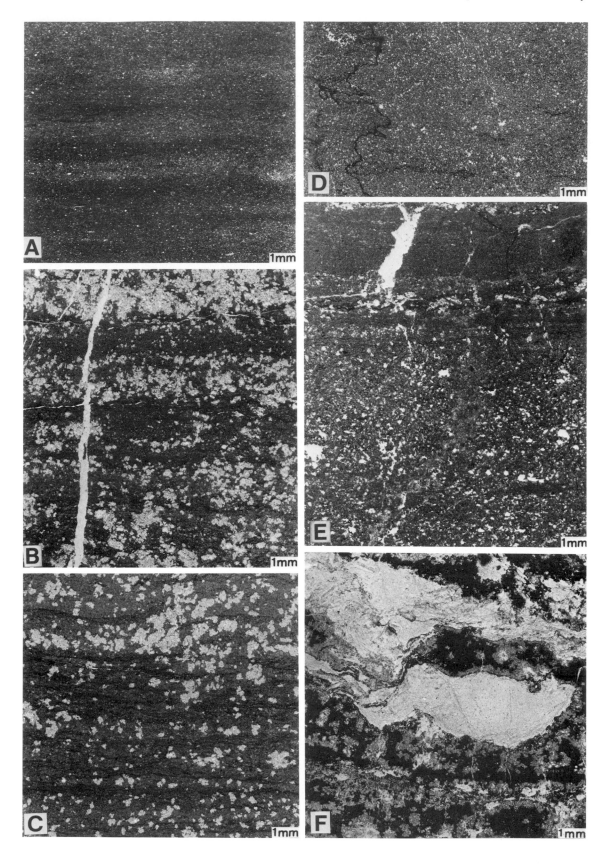

Rotliegendes and the Zechstein Limestone. Subsequent wells delineated reservoirs also in the Main Dolomite. Of 40 wells drilled in Tarchaly Field, 25 are productive from the Rotliegendes and the Zechstein Limestone, and 5 from the Main Dolomite.

The gas reservoir in the Main Dolomite at the highest point in Tarchaly Field occurs at a depth of 4500 feet (1375 m). The Main Dolomite has an average gross thickness within the gas field of about 180 feet (55 m), and is composed of a lower sequence of dolomite mudstones and an upper sequence of dolomitized oolitic grainstone. The productive area of the Main Dolomite reservoir is 1 75 acres (7.5 km^2) and that of the Rotliegendes and the Zechstein Limestone is 2750 acres (11 km^2). The average porosity of the Main Dolomite reservoir is 8.2 percent, and the average permeability (as determined using small core samples) is 0.3 millidarcys.

Although the initial potential for gas was 590 million cubic feet (20.9 million m^3) per day in the discovery well, potentials were 60.9 million cubic feet (2.2 million m^3) per day or less in subsequent wells, indicating that fractures are probably important in this reservoir. The initial reservoir pressure was about 20 percent higher than hydrostatic. Analysis of the gas gave the following compositions by percentage of volume: 59.1 CH$_4$, 1.8 C$_2$H$_6$, 0.7 C$_3$H$_8$, 0.3 C$_4$H$_{10}$, 0.2 C$_5$H$_{12}$, 0.2 CO$_2$, 37.2 N$_2$, and 0.4 He. Gases from the Rotliegendes and the Zechstein Limestones have similar compositions in general, but contain less methane, 55.7 percent, and more nitrogen, 43.5 percent.

Petroleum Source Rocks

According to geochemical studies by Calikowski and Glogoczowski (1976), the source rocks for crude oils in the Main Dolomite are believed to be the basinal facies, the deep-shelf facies, and adjacent parts of the shallow shelf facies of the Main Dolomite itself, which as a group are characterized by organic matter contents up to 1 percent (organic carbon) of sapropel-humus type. The same studies suggest that hydrocarbon generation in the Main Dolomite took place at depths of 6500 feet (2000 m) and more, and suggest a genetic relationship between crude oils and organic bitumens in this stratigraphic unit based on the assemblage of n-paraffin and aromatic hydrocarbons in these oils (Calikowski and Glogoczowski, 1976).

The gas seems to have had a twofold origin according to these same workers. In the Lubuska Barrier area (Fig. 15-1), the gas contains condensates and seems to derive, like the oil, from the Main Dolomite itself. It should be stressed that in the Lubuska Barrier, no hydrocarbon accumulations have been found below the Main Dolomite, and only small accumulations of nitrogen gas have been recorded in the Rotliegendes. On the other hand, in the Silesian Barrier area to the southeast (Fig. 15-1), the gas in Main Dolomite reservoirs (*e.g.*, at Tarchaly Field) contains small amounts of heavier hydrocarbons and is believed to have derived, at least partly, from the underlying Carboniferous deposits (Calikowski and Glogoczowski, 1976). In that region there are local basins filled by upper Car-

Fig. 15-7. Photomicrographs of dolomitized mudstones. All photos in plane light. *A.* Mudstone with weakly developed thin lamination. Depth 3714.0 m, Florentyna IG2 well, 70 km NW of well J1 in Figure 15-1. *B.* Pelletoidal packstone extensively replaced by anhydrite (pale gray). Fracture is open, however. Depth 1503.5 m, Czeszow 9 well, 30 km south of well J1 in Figure 15-1. *C.* Anhydritized, mm-laminated mudstone. Depth 1503.3 m, Czeszow 9 well. *D.* Massive mudstone; note vertically oriented stylolites. Depth 2536.3 m, Ujazd 1 well (Uja-1 in Fig. 15-1). *E.* Pelletoidal packstone with abundant small solution vugs, overlain by micritic crust. White areas including fractures are open pores. Depth 1452.0 m, Radziadz 4 well. *F.* Pelletoidal packstone extensively replaced by anhydrite. Note solution vug filled by anhydrite. Depth 1503.2 m, Czeszow 9 well.

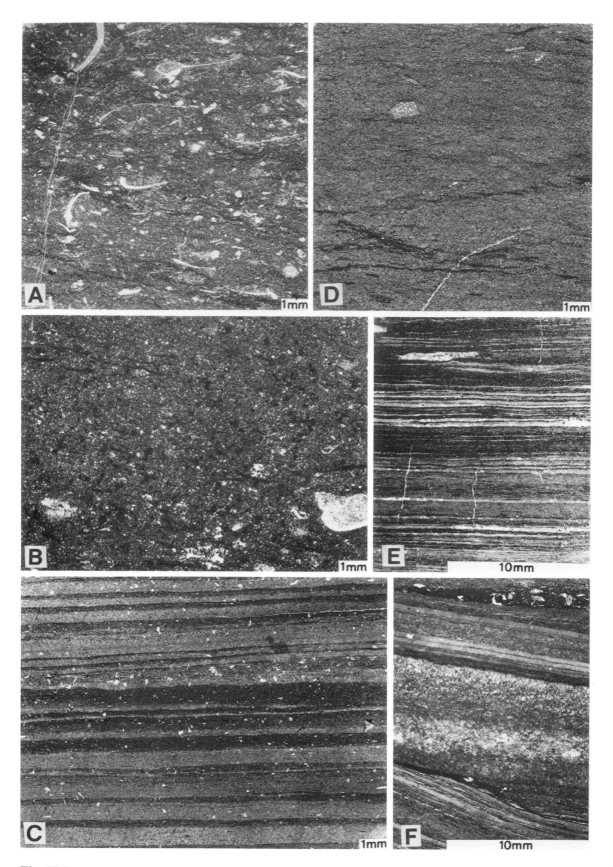

Fig. 15-8

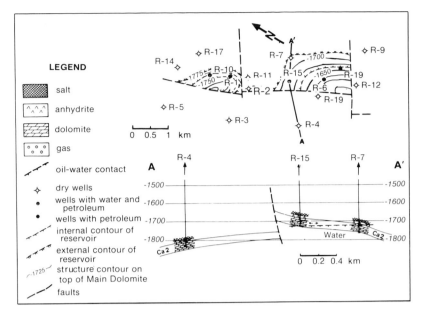

Fig. 15-9

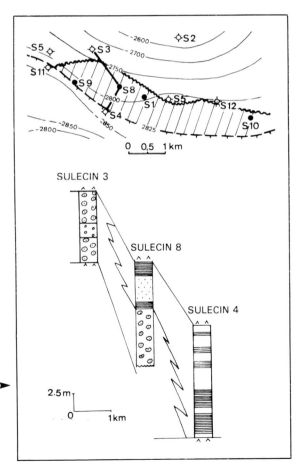

◄ *Fig. 15-8.* Photomicrographs of dolomitized mudstones and wackestone. All photos in plane light. *A.* Indistinctly bedded mollusk wackestone. Mollusks now occur as molds, some of which are filled by anhydrite cement. Depth 2549.7 m, Ujazd 1 well. *B.* Massive mudstone with abundant organic matter. Depth 2535.3 m, Ujazd 1 well. *C.* Mudstone with mm lamination showing numerous fining-upward cycles. Depth 2184.0 m, Sycowice 1 well, 12. 5 mi., (20 km) south of well Sw3 in Figure 15-1. *D.* Indistinctly bedded mudstone with bryozoan fragment. Depth 2536.6 m, Ujazd 1 well. *E.* Mudstone with mm-lamination. Organic layers have been compacted, resulting in bituminous laminae between carbonate laminae. Depth 2912.3 m, Sulecin 8 well. *F.* Graded wackestone (center) and mudstone. Depth 2913.1 m, Sulecin 8 well.

Fig. 15-9. Rybaki Field as represented by structure map and cross-section. Structure-contour map drawn on top of Ca2 zone. Overlying anhydrite and salt and a system of faults form the seals. Depths and contours are in meters subsea.

Fig. 15-10. Sulecin Field as represented by structure ► map and cross-section. The difference in bathymetry at the end of Main Dolomite deposition is believed to have been over 330 feet (100 m) between the Sulecin 3 and Sulecin 4 wells. Field area is shown by hatching. In cross-section, the lined pattern is dolomite mudstone. Light stipple pattern is wackestone, and other patterns are oolitic and pelletoidal grainstone.

Fig. 15-10

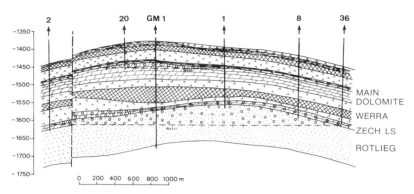

Fig. 15-11. Geologic cross-section through Tarchaly gas field. Lithologic patterns are the same as Figures 15-9 and 15-10. Open circles define the two gas zones in the Main Dolomite, both overlain by anhydrite seals.

boniferous shales that contain 1 to 3 percent of humus-type organic matter. The maturation level of this organic matter corresponds in general to that of semi-anthracite coals. The gases in the Rotliegendes Sandstone and the Zechstein Limestone also appear to have derived from the upper Carboniferous.

Conclusions

In the Upper Permian Main Dolomite, the best reservoir facies are those of the upper carbonate-platform slope and adjoining parts of the platform itself. It should be stressed that most of the oil and gas fields so far discovered occur in this general facies belt both in western Poland and in other parts of the Main Dolomite Basin (Fig. 15-1; Depowski *et al.*, 1981; Sannemann *et al.*, 1978). Intensive subaerial cementation within the carbonate-platform facies during early diagenesis essentially eliminated porosity and seals for lateral migration of hydrocarbons, except along a narrow zone adjoining the platform slope.

Acknowledgments The authors are greatly indebted to the editors, P. W. Choquette and P. O. Roehl, for numerous suggestions to improve the text.

References

CALIKOWSKI, J., and J.J. GLOGOCZOWSKI, 1976, Geochemical prospection of bitumens in Poland (in Polish, with English summary): Nafta, v. 32, p. 227–232.

CLARK, D.N., 1980, The sedimentology of the Zechstein-2 Carbonate Formation of eastern Drenthe, the Netherlands: Contr. Sedimentology, no. 9, p. 131–165.

DEPOWSKI, S., 1975, Hydrocarbon occurrences in the Permian Basin, Polish Lowlands: Biuletyn Instytut Geologicznego, v. 252, p. 175–184.

DEPOWSKI, S., T.M. PERYT, T.S. PIATKOWSKI, and R. WAGNER, 1981, Paleogeography versus oil and gas potential of the Zechstein Main Dolomite in the Polish Lowlands: Internat. Symp. Central European Permian Proc., p. 587–595. Geological Institute, Warsaw.

FÜCHTBAUER, H., 1964, Fazies, Porosität und Gasinhalt der Karbonatgesteine des norddeutschen Zechsteins: Z. dt. geol. Ges., v. 114, p. 484–531.

PERYT, T.M., 1978, Microfacies of the carbonate sediments of the Zechstein Werra and Stassfurt cyclothems in the Fore-Sudetic Monocline (in Polish, with English summary): Studia Geologica Polonica, v. 54, 88 p.

PERYT, T.M., 1983, Coated grains from the Zechstein Limestone/Upper Permian/of western Poland: *in* Peryt, T., ed., Coated Grains: Springer-Verlag Inc., Berlin, p. 587–598.

SANNEMANN, D., J. ZIMDARS, and E. PLEIN, 1978, Der basale Zechstein (A2–T1) zwishen Weser und Ems: Z. dt. geol. Ges., v. 129, p. 33–69.

16
North Anderson Ranch Field
M. Malek-Aslani

RESERVOIR SUMMARY

Location & Geologic Setting	Lea Co., T15S, R32E, Lea Co., SE New Mexico, USA
Tectonics	Subtle basement fault-block highs
Regional Paleosetting	NW shelf of cratonic Delaware Basin
Nature of Trap	Stratigraphic; updip facies change into lime mudstone, plus capping shale

Reservoir Rocks
Age	Permian (Wolfcampian)
Stratigraphic Units(s)	Bursum Formation
Lithology(s)	Limestone
Dep. Environment(s)	Patch reefs deposited in shoaling-upward cycles
Productive Facies	*Tubiphytes*-phylloid algal boundstone; *Tubiphytes*-skeletal grainstone and packstone-wackestone
Entrapping Facies	Shale and lime mudstone
Diagenesis/Porosity	Solution enlargement of primary porosity, local solution collapse

Petrophysics
Pore Type(s)	Interparticle, shelter, and moldic-vug
Porosity	1.2–12.5%, 9.6% avg
Permeability	0.1–1000+ md, 124 md avg
Fractures	Open, near-vertical fractures common in cores

Source Rocks
Age	Pennsylvanian to Permian (Wolfcampian)
Lithology(s)	Basinal black shales (Pennsylvanian) and lagoonal lime mudstone-wackestone (Wolfcampian)
Migration Time	NA

Reservoir Dimensions
Depth	9825–9990 ft (2994–3044 m)
Areal Dimensions	35 ft (11 m) net pay
Productive Area	920 acres (3.68 km^2)

Fluid Data
Saturations	S_w = 20% avg
API Gravity	42°
Gas-Oil Ratio	1550:1
Other	NA

Production Data
Oil and/or Gas in Place	12.9 million BO*
Ultimate Recovery	NA
Cumulative Production	3.34 million BO produced through 1981*

Remarks: IP 200 BOPD. Discovered in 1960. *From Petroleum Data Systems, Information System Program, University of Oklahoma, Norman, OK, data retrieval November 1983.

16
Permian Patch-Reef Reservoir, North Anderson Ranch Field, Southeastern New Mexico

M. Malek-Aslani

Introduction

North Anderson Ranch Field, located in Lea County, southeastern New Mexico (Fig. 16-1), is a stratigraphic trap that is productive from Lower Wolfcampian (Permian, Sakmarian) limestones. The field was discovered in 1960 on a subsurface and seismic prospect. At the end of 1981, it had produced 3.340 million barrels of oil from 16 wells. The field covers 920 acres (3.68 km^2), its gross pay is 100 feet (31 m), and the net pay is 35 feet (11 m) in thickness (Barnhill, 1966).

The depth to the top of the pay ranges from 9825 to 9990 feet (2994–3045 m). The average porosity of the pay is 9.4 percent, the average permeability is 124 millidarcys, and the average water saturation is 20 percent.

Regional Setting and Structure

The North Anderson Ranch Field is situated on a Permian physiographic feature known as the Northwest Shelf of the Delaware Basin (Fig. 16-1). Figure 16-2A is the structural configuration of the North Anderson Ranch Field contoured on top of the pay. North Anderson Ranch Field is interpreted as an updip porosity pinch-out on a homocline.

Figure 16-2B is an isopach of the interval between the overlying shale marker and the top of the Lower Wolfcamp pay. Thinning of the interval within the field suggests that the North Anderson Ranch Field was a bathymetric high during deposition. The constructed facies patterns for the main Wolfcampian cycles studied in the area support this interpretation.

Environmental Stratigraphy

The Wolfcampian sediments over the Northwest Shelf of the Delaware Basin consist of numerous upward-shoaling cycles deposited during a regressive phase that started in Wolfcampian (Lower Permian) and continued through Ochoan (Late Permian) time.

During Lower Wolfcamp deposition the Northwest Shelf was fringed by a more or less continuous barrier-type reef along the north and northwest margins of the Delaware Basin (Malek-Aslani, 1970). Figure 16-3 is a paleoenvironmental map of the western part of Lea County, New Mexico, in Lower Wolfcamp time. The Anderson Ranch and North Anderson Ranch fields are phylloid-algal patch reefs whose locations are controlled by structures formed as a consequence of differential vertical movements along basement fault blocks (Stipp, 1960). Figure 16-3 also shows locations of oil fields that produce from generally similar phylloid-algal reefs of Lower Wolfcamp age. These reefs are informally named the Bough member of the Hueco Formation and are correlative with the Kemnitz-Townsend reef

Fig. 16-1. Location map showing Wolfcampian paleo-geography.

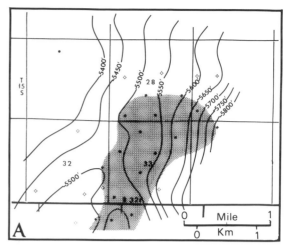

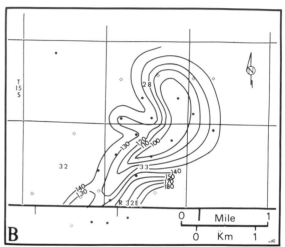

and North Anderson patch reef. Morton Field, shown on Figure 16-3, is one of the Bough fields which is described elsewhere in this volume by Cys and Mazzullo.

In North Anderson Ranch Field, the patch-reef reservoirs occur within two cycles that lie near the base of the Wolfcamp. Within the limits of the field, each cycle shows marked thickness variations (Fig. 16-4). The thickness of Cycle 1, the younger of the two, varies from 30 feet (9 m) to a maximum of 140 feet (43 m), (Fig. 16-4). The thickness of Cycle 2 varies from an average of less than 100 feet (31 m) to a maximum of 220 feet (67 m). Each cycle is thickest along the productive fairway.

Shale markers define the top and bottom of each cycle and can be correlated from well to well with the aid of mechanical logs (gamma-ray/sonic and SP-induction). Figure 16-5 is a composite facies and mechanical-log cross-section of the field.

Petrology and Petrography

A total of 927 feet (282.5 m) of cored intervals from seven wells in the North Anderson Ranch area were used for the petrographic analyses of the Lower Wolfcampian reservoirs. Following is a list of wells used in the study:

Fig. 16-2A. Structur-contour map on top of Cycle 1 in North Anderson Ranch Field. Contour interval 50 feet (15.2 m). B. Isopach map of the interval from datum A (Fig. 16-5) to the top of Cycle 1. Contour interval 10 feet (3 m).

	Location Numbers on Figure 16-5
Union Oil of California State 1-33	3
Union Oil of California State A-1-33	—
Union Oil of California State 2-33	—
Union Oil of California State 3-33	4
Union Oil of California State B. 1-28	—
Union Oil of California 4-32	2
Tenneco Gulf State B-1	—

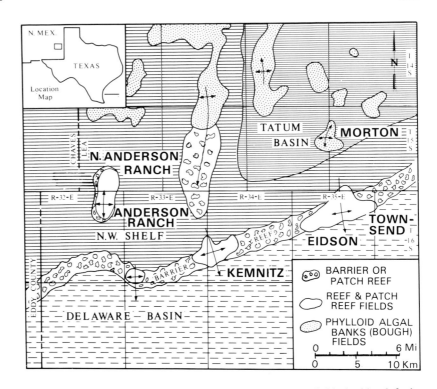

Fig. 16-3. Map of Lower Wolfcamp paleoenvironments in Lea County, New Mexico. The Lower Wolfcampian Hueco Formation includes barrier-reef, patch-reef, and phylloid-algal bank facies, all of which form hydrocarbon traps. The phylloid-algal bank facies is informally referred to as the Bough member of the Hueco Formation.

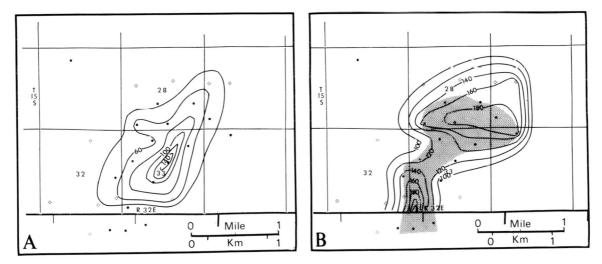

Fig. 16-4. Isopach maps of (A) Cycle 1 and (B) Cycle 2 in North Anderson Ranch Field. *A.* Contour interval 20 feet (6 m). *B.* Contour interval 10 feet (3 m).

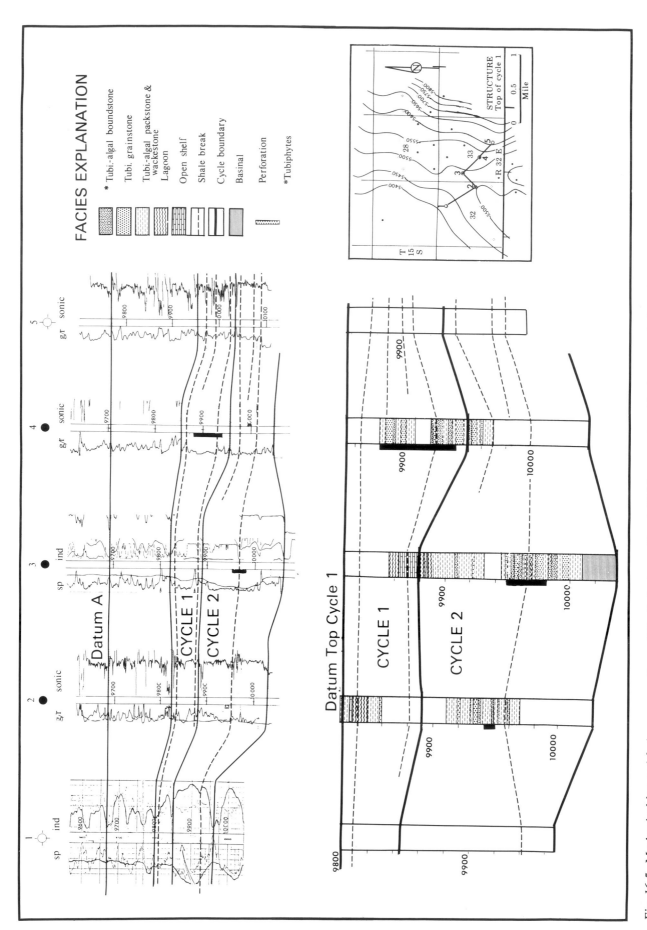

Fig. 16-5. Mechanical log and facies cross-sections of North Anderson Ranch Field, on stratigraphic datums as indicated.

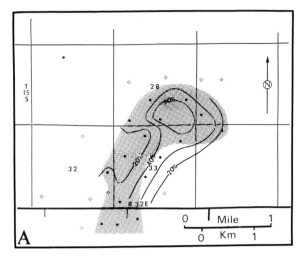
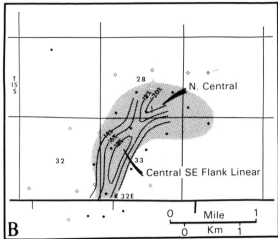

Fig. 16-6. Isopach maps. *A.* Percentage of grainstone per interval of Cycle 2. *B.* Percentage of Tubiphytes per interval of Cycle 2 in North Anderson Ranch Field. Stippled area shows limits of production for Cycle 2.

The available cores do not provide uniform coverage of the two productive depositional cycles. Petrologic observations were made on polished, etched, and stained core slabs. Petrographic observations were made on stained thin sections. To assess the areal variations of various core attributes, the data had to be normalized so as to provide comparable information for all the wells. This was achieved by calculating the average percentage of each attribute per interval described for each cycle.

Twenty six maps showing the areal variations of depositional fabrics, principal grain types, and diagenetic fabrics were prepared and used to interpret the depositional and diagenetic patterns within the North Anderson Ranch Field. Figure 16-6 illustrates two such maps for Cycle 2. Map 16-6A shows that grainstones constitute up to 60 percent of the interval and are best-developed on the northeast, east, and southeast side of the field. Map 16-6B shows that remains of *Tubiphytes*, an encrusting reef-building alga, account for less than 20 percent of the volume per interval and are concentrated along the axis of the isopach thick within the field (*cf.* Fig. 16-6). Similar analyses of other attributes helped provide the basis for an environmental model of North Anderson Ranch Field.

Figure 16-7 shows a number of east-west profiles that depict variations of various depositional and diagenetic fabrics and grain types for Cycle 2. For the location of the east-west transect, see

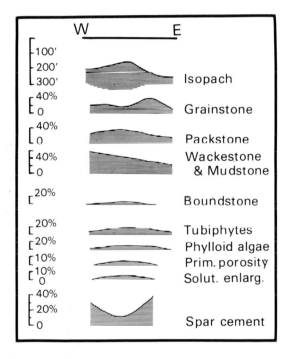

Fig. 16-7. West-to-east profiles of variations in depositional fabrics, principal grain types, and diagenetic attributes in North Anderson Ranch Field. For location of profile, see section W–E in Figure 16-8.

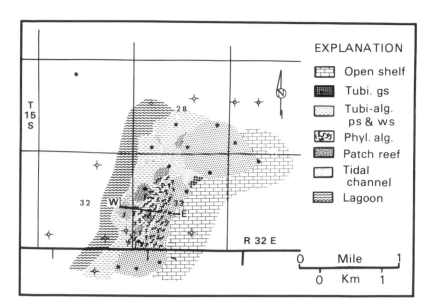

Fig. 16-8. Generalized paleo-environmental map of Cycle 2 in North Anderson Ranch Field.

Figure 16-8. The thickness profile at the top shows configurations of the top and the bottom of Cycle 2 using Marker A (Fig. 16-5) as datum. Grainstones are prevalent on the east side, packstones show a high concentration near the crest of the buildup, and wackestones and mudstones are the predominant fabric on the west part. The maximum occurrence of boundstone is limited to the crestal position of buildup. *Tubiphytes* and phylloid algae are concentrated along the crest of the buildup. Primary porosity and solution-enlargement porosity have maximum values along the crest, decreasing to the east and west. The spar cement shows a minimum value along the crest but increases eastward and westward.

Depositional History

Each cycle in the North Anderson Ranch area started with a transgressive facies composed of shaly spiculitic limestone. This facies was deposited during a rapid rise in sea level and followed by a period of relative stability during which a regressive upward-shoaling cycle was deposited. Once the upward-shoaling facies reached sea level, offshore progradation began east of the North Anderson Ranch area. Over the bathymetric high, small patch reefs formed in which *Tubiphytes* stabilized the bottom and phylloid algae trapped the sediments. Some of the patch reefs appear to have been partly eroded, perhaps by storm generated wave action. Detrital frag-

ments of *Tubiphytes*, mixed with other skeletal and nonskeletal grains, were transported eastward to form submarine banks of grainstone. To the west of the patch reefs, within a protected lagoon, packstone and wackestone were deposited, and they contain a higher percentage of fusulinids than the more proximal lagoon facies. Percentages of *Tubiphytes* and phylloid algae decrease with distance from the patch reef. In addition to fusulinids, dasycladacean algae, worm tubes, ostracodes, and sponge spicules also occur in the lagoonal facies.

The contrasting pattern of grainstone distribution between the barrier reef (Fig. 16-3) and the North Anderson Ranch patch reef suggests a differing mode of sediment transport from the two reef sites. Wave energy, propelled by the prevailing south wind, transported grainstones to the north of the barrier reefs and dissipated over the shallow shelf (Malek-Aslani, 1970). Distribution of the grainstones in the North Anderson Ranch area was controlled by seasonal weather fronts that blew in from the Pedernal landmass to the west and produced storm-generated waves that transported skeletal grains to the east of the patch reef.

Figure 16-8 is a generalized paleoenvironmental map of Cycle 2 in the North Anderson Ranch area. The interpretation of a tidal channel is based on the occurrence of an unusual concentration of ooids and coated grains in one well. Such grains are uncommon in other wells.

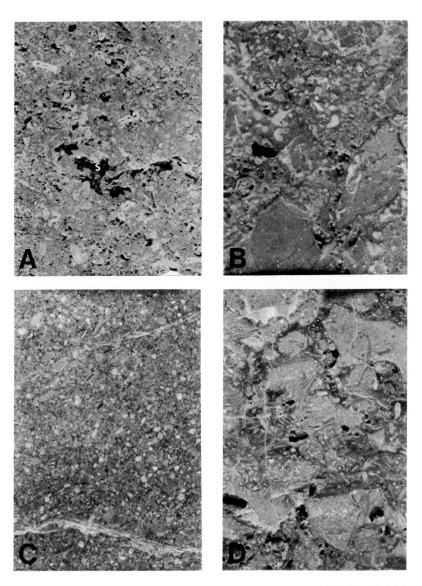

Fig. 16-9. Photographs of polished core slabs. *A.* Tubiphytes boundstone. Note Tubiphytes (T) and solution channels (S). Union State A well 1-33, 9873–9874 feet (3009.3–3009.6 m). *B.* Phylloid-algal boundstone. Note sheltered and solution enlarged pore spaces. Union State A well 1-33, 9872–9873 feet (3009.0–3009.3 m). *C.* Tubiphytes. Skeletal and non-skeletal grainstone. Union State well 2-23, 10,014–10,015 feet (3052.2–3052.6 m). *D.* Solution-collapse in Tubiphytes, phylloid-algal and skeletal wackestone. Note primary inter-lithoclast and secondary solution pores. Union State A well 1-33, 9871–9872 feet (3008.6–3008.9 m).

Porosity Types

Primary—Interparticle porosity occurs in the *Tubiphytes* grainstones, and shelter porosity is common in the *Tubiphytes* boundstones and phylloid-algal boundstones (Fig. 16-9B).

Secondary Porosity—Solution-enlarged primary porosity, vugs, and solution-interparticle pores formed by dissolution of mud matrix, and moldic porosity are all common in the *Tubiphytes* packstone and wackestones (Fig. 16-9A).

Solution-Collapse Breccia—During subaerial exposure subterranean dissolution produced caverns into which the overlying reef collapsed, providing loosely packed lithoclasts with interparticle porosity (Fig. 16-9D).

Table 16.1. Summary of Depositional Environments, Diagenetic Fabrics and Environments, and Reservoir Properties, Permian Carbonates, North Anderson Ranch Field

Reservoir Facies	Depositional Environment	Diagenetic Fabrics	Diagenetic Environment	Porosity Type	Porosity and Permeability Range
Tubiphytes-phylloid algal boundstone	Patch reef	1. Isopachous calcspar 2. Equant calcspar 3. Dolospar 4. Late calcspar	FW phreatic and vadose subsurface (mesogenetic)	1. Prim. sheltered 2. Secondary A. Solution B. Channel C. Vugs	Porosity: 1.8–4.1% Perm: 0.1–43 md
Skeletal-*Tubiphytes* grainstone	Submarine bars east of patch reef	1. Palisade calcspar 2. Equant calcspar	Marine phreatic FW phreatic and/or vadose	Between particles	Porosity: 1.4–11.9% Perm: 0.1–118 md
Tubiphytes-algal-skeletal packstones and wackestones	Patch reef complex	Solution	FW phreatic and vadose zones (eogenetic)	Secondary A. Solution B. Channels C. Vugs	Porosity: 1.2–12.5% Perm: 0.1–1000 md

Note: Tectonic fracturing is evident in all the reservoir facies.
  FW = freshwater

Fractures in the patch-reef complex, produced either by penecontemporaneous spalling off of partly lithified deposits, or by tectonic fracturing, is very common in North Anderson Ranch Field. Fracture porosity undoubtedly is an important factor in the productivity of the field.

Cement Types

Sediments derived from the original patch-reef complex were cemented in a variety of diagenetic environments, each producing a particular cement type (Table 16-1) as outlined below:

Palisade Cement—This type appears as fibro-radial blades that circumscribe the grains, and is most common in coated-grain grainstones. It is interpreted as the result of paramorphic replacement of marine aragonitic cement (Folk, 1977).

Blocky Cement—Equant calcite crystals range in size (10–100 μ) and are common in *Tubiphytes* grainstones. This cement type is probably the product of freshwater vadose or phreatic environments (above the meteoric water table or within the interval filled with the meteoric waters).

Coarse Dolomite—This is a late cement, which in paragenetic sequence followed those above. It is characterized by well-defined euhedral dolomite rhombs (100–250 μ) that line the large, original vugs. Coarse dolomite cement was probably formed in the mesogenetic zone of Choquette and Pray (1970).

Coarse Late Calcite—The latest cement in the paragenetic sequence is a coarse calcite spar, which occupies pores not filled with coarse dolomite. In large original pores, particularly within the boundstones, coarse dolomite surrounds the coarse calcite.

Diagenetic History

During the deposition of each cycle, available evidence suggests the principal diagenetic process was submarine cementation by aragonite and possibly high-magnesium calcite. At the end of each cycle, the top of the buildup was subaerially exposed and underwent freshwater diagenesis. The aragonite cement was replaced by low-magnesium calcite palisade cement, the high-magnesium calcite inverted to low-magnesium calcite, and dissolution of various depositional constituents took place. This episode was fol-

lowed by a relative rise in sea level which started the next cycle. During advanced stages of burial, the late cements were precipitated.

The principal reservoir facies in the North Anderson Ranch Field include: *Tubiphytes*-phylloid algal boundstones, and *Tubiphytes*-skeletal-non-skeletal grainstones, packstones, and wackestones. Figure 16-10 is a semi-log plot of porosity versus permeability for the three principal reservoir facies in the North Anderson Ranch Field. Linear regression lines through the data for packstone, wackestone, and boundstone facies show steeper slopes than those for grainstones. Thus, it follows that the increased porosity in packstone, wackestones, and boundstones results in higher

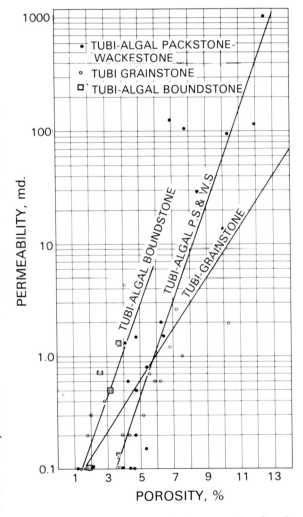

Fig. 16-10. Porosity-permeability cross-plot for the three principal reservoir facies in North Anderson Ranch Field.

permeability than in the grainstones. Since solution-enlarged porosity is the principal type in boundstone, packstone, and wackestone in North Anderson Ranch Field, the above statistical data show that these depositional fabrics were more likely to attain higher permeabilities with increased porosity than the grainstones in which interparticle primary porosity predominated. The presence of diagenetic cements must have retarded development of channeling in grainstones.

Source Rocks

Geochemical analysis and crude-extract correlation were not included in the present study. The following, therefore, is a speculation based on the stratigraphic relations between possible source rocks and the Lower Wolfcamp reservoirs.

Based on lithologic character, the low-energy, spiculitic limestones and black shales that occur below the reservoir facies are good candidates for the source of petroleum in the North Anderson Ranch Wolfcamp Field. These beds are several hundred feet thick and occur at the base of the Wolfcamp and in the Upper Cisco (Pennsylvanian). The location of the North Anderson Ranch Field, well back on the Northwest Shelf, suggests that the lagoonal facies are the likely source rocks, as in the case of other similarly located fields (Jones and Smith, 1965).

After thermal maturation, both the low-energy carbonates and shales could have released hydrocarbons for vertical migration upward into the Lower Wolfcamp reservoirs. An extensive fracture system observed in the cores provided avenues for vertical migration.

Summary

North Anderson Ranch Field produces from Lower Wolfcamp patch-reef facies that constitute two upward-shoaling cycles. The reservoir facies include *Tubiphytes* and phylloid-algal boundstones, *Tubiphytes* grainstones, and *Tubiphytes* and phylloid-algal packstones and wackestones. Primary porosity is limited to grainstones and to a minor extent, boundstones. Secondary solution porosities are fabric-selective and occur primarily in boundstones, packstones, and wackestones. The low-energy limestones and shales that occur below the patch-reef complex are possible source rocks.

References

BARNHILL, W.B., 1966, North Anderson Ranch Field: *in* Roswell Geol. Soc. Symp., p. 62–63.

CHOQUETTE, P.W., and L.C. PRAY, 1970, Geologic nomenclature and classification of porosity in sedimentary carbonates: Amer. Assoc. Petroleum Geologists Bull., v. 54, p. 207–250.

JONES, T.S., and H.M. SMITH, 1965, Relation of oil composition and stratigraphy in the Permian Basin of west Texas and New Mexico: *in* Fluids in Subsurface Environments: Amer. Assoc. Petroleum Geologists Mem. 4, p. 101–224.

FOLK, R.L., 1977, Oral presentation at Rice University, Houston, TX.

MALEK-ASLANI, M., 1970, Lower Wolfcampian reef in Kemnitz Field, Lea County, New Mexico: Amer. Assoc. Petroleum Geologists Bull., v. 54, p. 2317–2335.

STIPP, T.F., 1960, Major geological features and geologic history of southeastern New Mexico: *in* Oil and Gas Fields of Southeastern New Mexico: Roswell Geol. Soc., p. 27–30.

17
Morton Field
John M. Cys and S. J. Mazzullo

RESERVOIR SUMMARY

Location & Geologic Setting	Lea Co., New Mexico; South Tatum Basin, NW shelf of Delaware Basin
Tectonics	Intrabasin uplift
Regional Paleosetting	Shelf edge cratonic basin
Nature of Trap	Structural/stratigraphic
Reservoir Rocks	
Age	Early Permian (Wolfcampian)
Stratigraphic Units(s)	Hueco Formation
Lithology(s)	Limestone
Dep. Environment(s)	Shallow-shelf biohermal
Productive Facies	Phylloid algal bafflestone and bioclastic grainstone-packstone
Entrapping Facies	Shale
Diagenesis/Porosity	Dissolution of some aragonite grains, some occlusion by phreatic cement
Petrophysics	
Pore Type(s)	Moldic, solution-enlarged interparticle, growth-framework
Porosity	3–20%, 7% avg
Permeability	67 md avg
Fractures	Minor
Source Rocks	
Age	Pennsylvanian (late) and Permian (early)
Lithology(s)	Shale & lime mudstone
Migration Time	Mesozoic
Reservoir Dimensions	
Depth	10,400 ft (3170 m)
Thickness	100 ft gross pay (30.5 m)
Areal Dimensions	1.25 × 0.5 mi (2.0 × 0.8 km)
Productive Area	720 acres (2.9 km^2)
Fluid Data	
Saturations	$S_o = 10\%$, $S_w = 25\%$
API Gravity	44°
Gas-Oil Ratio	2200:1
Other	—
Production Data	
Oil and/or Gas in Place	NA
Ultimate Recovery	1.82 million BO
Cumulative Production	1.813 million BO through 1982

Remarks: IP mechanism solution-gas. Discovery method subsurface mapping. Discovered 1964.

17
Depositional and Diagenetic History of a Lower Permian (Wolfcamp) Phylloid-Algal Reservoir, Hueco Formation, Morton Field, Southeastern New Mexico

John M. Cys and S. J. Mazzullo

Location and Setting

Morton Field is a small structural-stratigraphic reservoir located in the Tatum Basin of southeastern New Mexico, on the Northwest Shelf of the very prolific Permian Basin region (Fig. 17-1). The Tatum Basin is a local depocenter that appeared in early Pennsylvanian time, with maximum development in the Middle Pennsylvanian. By late Wolfcampian (early Permian) time, the basin had lost its integrity as a result of depositional infilling. The field is located in the southern portion of the basin (Fig. 17-1) and produces from a phylloid-algal bioherm complex in the informal "Lower Wolfcamp" and Bough members of the Hueco Formation, a unit of Wolfcampian age. This paper describes the depositional and diagenetic facies in the Hueco Formation of Morton Field, as identified in cores from five wells.

Discovery and Reservoir Data

Hydrocarbon production from Lower Permian limestones at Morton Field was established in 1964 with the completion of the Union Oil of California No. 1 State 7 in Section 7, T-15-S, R-35-E, Lea County, New Mexico (Fig. 17-2). The discovery method was subsurface mapping. The operator reported an initial potential of 220 barrels of oil per day from perforations at 10,383 to 10,391 feet (3164.7–3167.2 m). The average

depth of the reservoir below surface is 10,400 feet (3170 m). The reservoir is lenticular in shape and is approximately 1.25 miles (2 km) in length and 0.5 mile (0.8 km) in width. Subsequent drilling on 80-acre spacing until 1968 defined the limits of the field, which now covers approximately 720 acres. However, because the 80-acre spacing was not a mandatory requirement for field development, the Huber Corp. Well No. 2 Stoltz-Federal, located in the SW 1/4 of SE 1/4 of Section 12, was drilled in 1979 on 40-acre spacing (Fig. 17-2). This well was subsequently completed, and reported an initial flowing potential of 289 barrels of oil per day plus 486 barrels of water per day. Water encroachment has resulted in abandonment of 6 of the 11 producing wells in the field (Fig. 17-2). Cumulative oil production at the end of 1982 was 1.813 million barrels, in excess of the ultimate recovery estimated by API in 1976 (1.8 million barrels) for this solution-gas drive reservoir.

Jordan (1967) reported an average of 100 feet (30.5 m) of gross pay, 40 feet (12.2 m) of net pay, and 67 millidarcys permeability in the field. Total reservoir thickness, including the "Lower Wolfcamp" to base of the Bough "B" (Fig. 17-3), varies from 267 to 304 feet (81.4–92.7 m) in the productive area of the field. Including areas peripheral to the field, the thickness varies from 257 feet (78.3 m) in the Union of California No. 1-7 Stansel (SW 1/4 of NE 1/4, Sec. 7) to 364 feet (110.9 m) in the Texaco No. 1 Arreguy (SW

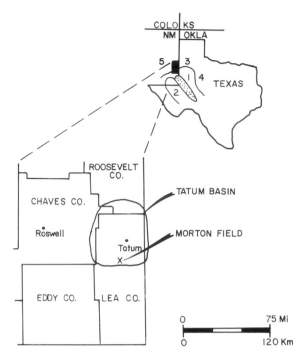

Fig. 17-1. Location map of Morton Field and the Tatum Basin in southeastern New Mexico. (1) Midland Basin; (2) Delaware Basin; (3) North Basin Platform; (4) Eastern Shelf; (5) Northwest Shelf.

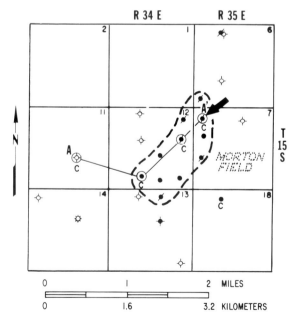

Fig. 17-2. Map of Morton Field (enclosed by dashed outline) showing location of cross section A–A' (Figs. 17-9 and 17-10), the five cored wells studied for this report (indicated by a "c"), and the field discovery well (arrow).

1/4 of NE 1/4, Sec. 13) (Fig. 17-2). Field production is from the "Lower Wolfcamp" and Bough "A" members (Fig. 17-3); however, the Bough "B" also is commonly included as part of the reservoir because it is porous and yields significant amounts of sulfur water on drill-stem tests. Field-average porosities measured from wireline logs are 5 to 9 percent, with a productive range between 3 and 20 percent. Porosity distribution in the two pay zones of the field is shown in Figures 17-4 and 17-5.

A structure-contour map drawn on the top of the "Lower Wolfcamp" member (Fig. 17-6) illustrates the essential structural aspect of the field, with its approximately 170 feet (51.8 m) of closure. However, seismic data and a limited amount of deep subsurface control indicate thinning of Lower and Middle Pennsylvanian strata over a paleostructural high in the Morton Field area. This high formed in early and middle Pennsylvanian time and provided a favorable substratum for optimum biohermal growth in the field during Bough through "Lower Wolfcamp" time. Regional mapping illustrates that the thickest reservoir zones are developed within the field area. The presence of dry holes located in areas structurally higher than producing field wells results from a depositional and diagenetic facies control on distribution of porosity. Morton Field is therefore an example of a combination stratigraphic-structural trap with depositional "reef-constructional" topography, modified slightly by later tectonic deformation.

Geologic Characteristics

Stratigraphy

Morton Field is developed in phylloid-algal and associated limestone facies of the informal "Lower Wolfcamp" and Bough members of the Lower Permian Hueco Formation (Fig. 17-3). This formation occurs in both the surface and the subsurface of the Permian Basin, and consists of limestones with variable amounts of interbedded shales and conglomerates. The "Lower Wolfcamp" member is confined to the southern and central parts of the Tatum Basin, and includes oil-productive phylloid-algal bioherms and bio-

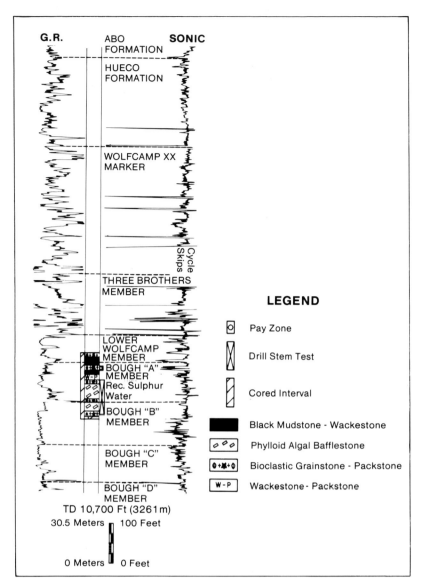

Fig. 17-3. Gamma-ray/sonic log of the discovery well of Morton Field, showing basic stratigraphic sequence, lithologies, well tests, and pay zones.

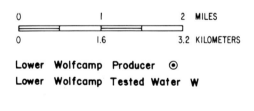

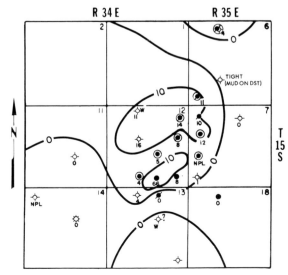

Fig. 17-4. Isopach map of net porosity (>5%), Lower Wolfcamp member, Morton Field area. Contour interval is 10 feet (3 m). NPL means "no porosity log." For legibility, the 20- to 60-foot contours have been omitted.

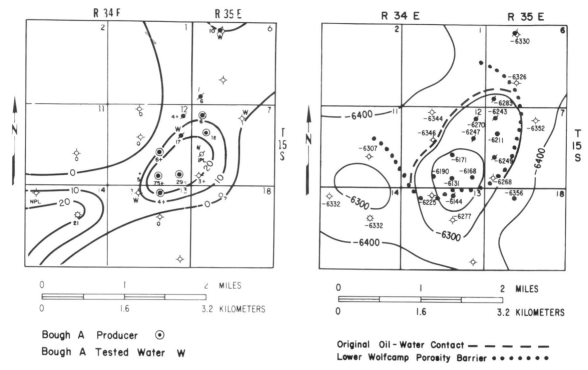

Bough A Producer ⊙
Bough A Tested Water W

Original Oil - Water Contact — — — — —
Lower Wolfcamp Porosity Barrier • • • • • •

Fig. 17-5. Isopach map of net porosity (>5%), Bough "A" member, Morton Field area. Contour interval is 10 feet (3 m). NPL means "no porosity log." For legibility, the 30- to 70-foot contours have been omitted.

Fig. 17-6. Structural map, top of informal "Lower Wolfcamp" member identified on Figure 17-3. Contour interval is 100 feet (30.5 m). Original oil-water contact is at a subsea datum of −6318 feet (−1926 m).

stromes and associated grainstone shoals in eight local fields. The Bough members are the most important upper Paleozoic producing sequences in the Tatum Basin. The Bough "C" is a prolific oil producer in several fields in the northern Tatum Basin, where it includes phylloid-algal biostromes and associated grainstones. The Bough "A," "B," and "C" all produce oil from a number of fields in the southern and central Tatum Basin.

The exact age of the Bough members is controversial, because they apparently straddle the Pennsylvanian-Permian time boundary. Fusulinids recovered from cores in the upper Bough "A" in the Union of California No. 1 State 7 well were identified by Garner L. Wilde (personal communication, 1980) as forms transitional between *Triticites* and *Schwagerina*; however, despite their smaller size, they are probably more closely related to *Schwagerina grandensis* Thompson than to any other described species. Such taxonomic assignment would place these

fossils and strata in the Lower Wolfcampian, equivalent to the Neal Ranch Formation faunizone type section of the Wolfcamp located in the Glass Mountains. In the McAlester No. 1 State EL well (NW 1/4 SW 1/4 Sec. 7), "Bursum" (Early Wolfcampian) fusulinids were reported from 10,400 to 10,700 feet (3170–3261 m) and Cisco (Virgilian) fusulinids from 10,720 to 10,780 feet (3267.5–3285.7 m) (Paleontological Laboratory commercial report). The depths 10,400 to 10,700 feet coincide with the Bough 'A," "B," and "C" members, and the depths 10,720 to 10,780 feet are in the upper Bough "D" member. Thus, in Morton Field, the Permian-Pennsylvanian boundary appears to be at the top of the Bough "D" member.

Depositional Facies

The reservoir in Morton Field consists of a complex of "stacked" phylloid-algal bioherms and grainstone shoals that accumulated along a local shelf edge within the Tatum Basin. The "Lower Wolfcamp" and the Bough "A–C" bioherm members each consist of shoaling-upward carbonate cycles separated by intervals of shale and/or argillaceous limestone (Fig. 17-3). An idealized cycle within the main bioherm facies (Fig. 17-3) consists of three superposed units:

1. A basal transgressive marine facies, 4 to 20 feet (1.2–6.1 m) thick, of black shale and interbedded dark gray, argillaceous wackestone with crinoids and fusulinids; beds of wispy, argillaceous, fusulinid packstone and wispy lime mudstones with chert nodules are common.
2. An open-marine shelf facies of phylloid-algal bafflestones with abundant encrusting tubular foraminifers and subordinate *Tubiphytes*, crinoids, bryozoans, fusulinids, gastropods, and thin-shelled brachiopods (Fig. 17-7A); also common are interbeds of gray crinoidal and fusulinid wackestones with tubular foraminifers and subordinate mollusks, brachiopods, intraclasts, and oncolites (Fig. 17-7B).
3. Crestal high-energy bioclastic grainstones and packstones comprise the upper facies and include tubular foraminifers, fusulinids along with minor amounts of crinoids, *Tubiphytes*, ooids, and intraclasts (Figs. 17-8A,B).

The phylloid algae are primarily *Eugonophyllum*. Local patches of *Tubiphytes*-tubular foraminifer boundstone occur within the phylloid-algal bafflestone facies.

Away from the main locus of algal bioherm development, the tripartite cyclic sequence just described is replaced by a bipartite cycle that consists of a basal, transgressive shale and argillaceous wackestone facies, overlain by an open-marine, crinoid-fusulinid wackestone facies (Figs. 17-9, 17-10). In the Union of California No. 1 Scott 11 well (Sec. 11), this sequence also includes units of black, very argillaceous mudstones and wackestones with scattered crinoids and fusulinids (Fig. 17-7C), and intervals of black, calcareous shale. The limestones are massively bedded, nodular (Fig. 17-7D), or graded. These rocks are believed to represent the relatively deeper-water basinal facies deposited seaward of the shelf-edge bioherm complex.

The updip lateral barrier of the reservoir is provided by the gradation from porous phylloid-algal bafflestone and crestal bioclastic grainstones of the bioherm complex into nonporous crinoid-fusulinid wackestones of the open-marine shelf environment leeward of the bioherm. The nonporous character of the open-marine wackestones probably results from a lack of subaerial and vadose exposure and subsequent dissolution. This wackestone facies is porous where it is interbedded with the bioherm complex and was subjected to subaerial exposure. The vertical seal of the reservoir is the thick basal shale of the Three Brothers Member immediately overlying the reservoir.

Diagenesis and Reservoir Development

The effects of syndepositional marine and subsequent meteoric diagenesis are observed in each of the cyclic reservoir sequences represented by the Bough "A–C" and "Lower Wolfcamp" members. Reservoir porosity is most abundant in the algal bafflestone members of each cycle, but occurs in the associated open-marine crinoid wackestones and crestal grainstones and packstones as well. Marine diagenesis involved the precipitation of botryoidal and isopachous, acicular cements in shelter cavities within the bafflestone facies. Such mosaics, including radiaxial-fibrous and radial-fibrous calcites, commonly are encrusted by marine organisms and are similar to Holocene and ancient marine aragonite and high-Mg calcite cements (*e.g.*, Mazzullo and Cys, 1979; Mazzullo, 1980). Tubular foraminifers, *Tubiphytes*, and ooids within the crestal grainstone facies locally are isopachously coated by radially oriented, stubby crystals of possible marine origin. Of significance to reservoir development is the observation that the inferred marine cements were not dissolved by fresh waters during subsequent meteoric exposure. Occlusion of primary porosity by pervasive marine cementation, and subsequent wholesale retention of cements by paramorphism before any dissolution occurred, could conceivably account for the lack of porosity in some algal bioherms encountered in the subsurface. An example of this apparent inhibition of solution-porosity in surface exposures of Wolfcampian bioherms has been described by Mazzullo and Cys (1979).

The subaerial exposure of individual bioherm

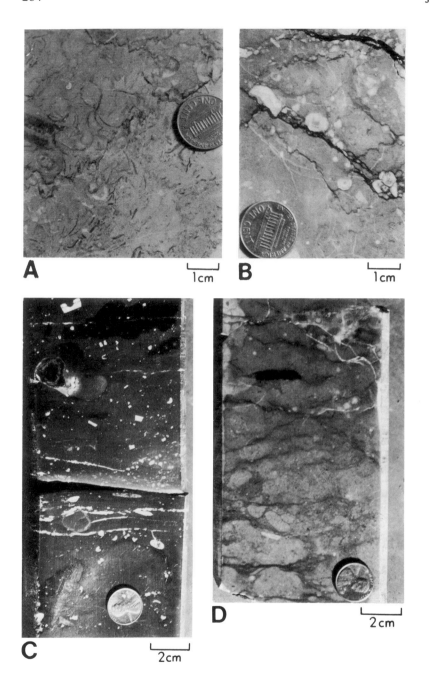

Fig. 17-7. Photographs of polished core slabs of depositional facies; diameter of coin is 1.85 cm. All samples stratigraphically oriented. *A.* Phylloid-algal bafflestone with scattered crinoids. Note stylolites and pressure-solution of crinoids in upper center. Bioherm core facies. Union of California No. 1 State 7 well. Depth 10,451 feet (3185.5 m). *B.* Fusulinid-crinoid wackestone of the open-marine shelf environment. Stylolites and pressure-solution of crinoids are prominent. Union of California No. 1 State 7 well. Depth 10,395 feet (3168.4 m). *C, D.* Basinal facies from the Union No. 1 Scott 11 well. *C.* Black, argillaceous, crinoid mudstone and sparse wackestone at a depth of 10,402 feet (3170.5 m). *D.* Nodular bedding in a mudstone-wackestone at a depth of 10,402 feet (3170.5 m).

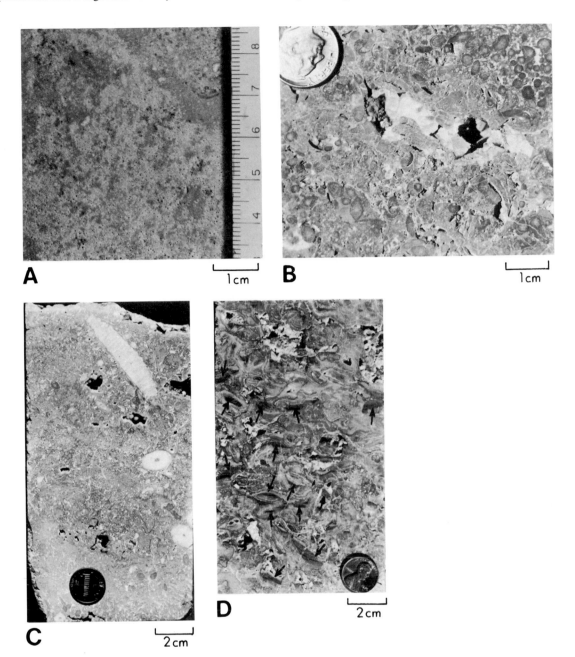

Fig. 17-8. Photographs of polished core slabs showing porosity development, Morton Field. Stratigraphic top is up in all photos. *A.* Fine-grained bioclast-oolitic grainstone with abundant "pinpoint" porosity of oomoldic, primary intraskeletal, and primary interparticle types. Union of California No. 1 State 7 well. Depth 10,392 feet (3167 m). Scale is in centimeters. *B.* Fusulinid-intraclast packstone-grainstone with partial occlusion of both solution vugs and solution-enlarged and simple fossil molds. Union of California No. 1-12 Gulf-Federal well. Depth 10,337 feet (3150.7 m). Diameter of coin is 1.7 cm. *C.* Crinoid wackestone facies of open-marine shelf with partial occlusion of: fossil molds, solution-enlarged fossil molds, and reduced vugs. Union No. 1 Reed well. Depth 10,411 feet (3171.3 m). Diameter of coin is 1.85 cm. *D.* Phylloid-algal bafflestone facies of bioherm core with primary shelter cavities, solution vugs, and phylloid-algal molds. Most of the porosity has been partially reduced by sparry calcite cement and/or internal vadose silt (indicated by arrows). Union No. 1 Reed well. Depth 10,434 feet (3180.3 m). Diameter of coin is 1.85 cm.

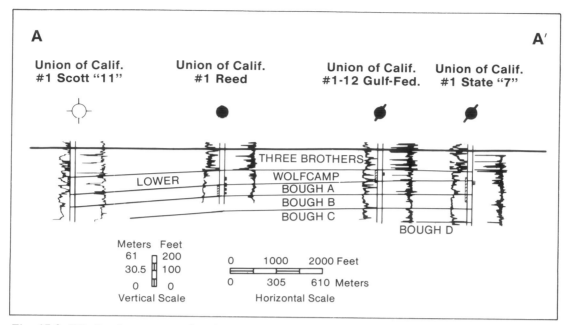

Fig. 17-9. Wireline-log cross-section A–A' illustrating stratigraphic relationships in the Morton Field area; datum is top of the Three Brothers Member. Symbols are the same as in Figure 17-3. Line of section shown in Figure 17-2.

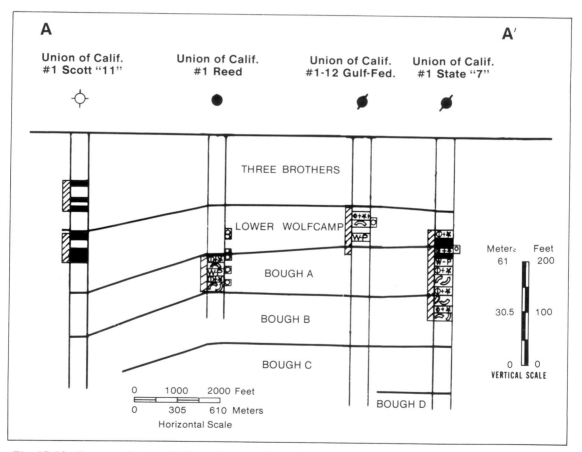

Fig. 17-10. Cross-section A–A' illustrating core-determined lithofacies. Datum same as in Figure 17-9. Line of section shown in Figure 17-2.

complexes resulted in concomitant vadose and phreatic diagenesis of these rocks. In general, limestones that underwent diagenesis in the phreatic environment are more thoroughly cemented, and of poorer reservoir quality, than rocks exposed to vadose alteration.

Porosity types in the bioherm core and associated open-marine crinoid wackestone facies that were exposed to vadose diagenesis include: (1) moldic porosity derived by fabric-selective dissolution of phylloid-algal plates and other aragonitic skeletal fragments and ooids; (2) solution vugs and channels created by pervasive dissolution of micrite (Fig. 17-8C), including pockets of solution breccia; and (3) primary shelter cavities beneath phylloid-algal plates (Fig. 17-8D).

Within the crestal grainstones, the following porosity types are observed: (1) primary intraskeletal porosity within fusulinids and tubular foraminifers; (2) incompletely cemented interparticle pores; (3) oomoldic and biomoldic (*e.g.*, Fig. 17-8A); and (4) micrite-matrix, dissolutional vugs and biomolds, common where packstone lenses occur in grainstones (Fig. 17-8B). In general, most of the porosity in all these rocks is biomoldic and vuggy, and was subsequently solution-enlarged and then somewhat occluded by a later generation of sparry calcite cements and/or internal silt (Choquette and Pray, 1970) (Fig. 17-8D). No vadose cementation features, such as gravitational or meniscus cements, were observed in these rocks.

In contrast, those rocks exposed to phreatic diagenesis are characterized by nearly complete occlusion of porosity by calcite cement. Such contrasts between vadose and phreatic diagenesis are common (see Mazzullo, this volume). From the top downward in the bioherm complex, the transition from the porous zone of vadose diagenesis to the low-porosity zone of phreatic diagenesis is marked by the appearance of pores occluded to increasing degrees by coarsely crystalline, equant to blocky, non-ferroan, sparry calcite. Cementation dependent on successive paleowater tables is the preferred mechanism of alteration, but it is possible that some of these cements, as well as those that partially occlude solution vugs, are of late-stage mesogenetic origin (Figs. 17-8B,C). Features of probable burial-diagenetic origin in these rocks include stylolites and associated partially dissolved allochems (Figs. 17-7A,B).

Summary

The stratigraphic-structural carbonate reservoirs at Morton Field are stacked phylloid-algal bioherm complexes deposited along an intrabasinal shelf edge within the Tatum sub-basin of the Northwest Shelf, Permian Basin, New Mexico. Porosity in these limestones developed as a result of subaerial exposure terminating upward-shoaling cycles. Porosity in the bioherm complexes is mainly the result of fabric-selective and pervasive dissolution of allochems and micrite matrix; the most porous reservoir rocks are those that resided in vadose zones.

Paleontological data and regional correlations indicate that the buildup cycles in Morton Field are approximately coeval with the Kemnitz-Townsend shelf-edge cycles to the south, described in detail by Malek-Aslani (1970 and this volume) and Dunham (1969).

Acknowledgments The authors express appreciation to Union Oil of California, Midland, Texas, for permission to examine cores from their wells in the field area and to publish this paper. Mapco Production Company and Union Texas Petroleum supported this study and permitted its publication. Special thanks go to Garner L. Wilde, Exxon Company, Midland, Texas, for his identification of fusulinids from the Union of California No. 1 State 7 well core. R. V. Hollingsworth, Midland, Texas, kindly permitted us to use certain fusulinid data from the commercial report of the Paleontological Laboratory on the McAlester No. 1 State EL well.

References

CHOQUETTE, P.W. and L.C. PRAY, 1970, Geologic nomenclature and classification of porosity in sedimentary carbonates: Amer. Assoc. Petroleum Geologists Bull., v. 54, p. 207–250.

DUNHAM, R.J., 1969, Early vadose silt in Townsend mound (reef), New Mexico: *in* Friedman, G.M., ed., Depositional environments in carbonate rocks—a symposium: Soc. Econ. Paleontologists and Mineralogists, Spec. Pub. 14, p. 139–181.

JORDAN, J.B., 1967, Morton Lower Wolfcamp field: *in* Kinney, E.E., ed., 1966 supplement to symposium of oil and gas fields of southeastern New Mexico: Roswell Geol. Soc., p. 136–137.

MALEK-ASLANI, M., 1970, Lower Wolfcampian reef in Kemnitz Field, Lea County, New Mexico: Amer. Assoc. Petroleum Geologists Bull., v. 54, p. 2317–2335.

MAZZULLO, S.J., 1980, Calcite pseudospar replacive of marine acicular aragonite, and implications for aragonite cement diagenesis: Jour. Sedimentary Petrology, v. 50, p. 409–422.

MAZZULLO, S.J., and J.M. CYS, 1979, Marine aragonite seafloor growths and cements in Permian phylloid algal mounds, Sacramento Mountains, New Mexico: Jour. Sedimentary Petrology, v. 49, p. 917–936.

18
Reeves Field

Stewart Chuber and Walter C. Pusey

RESERVOIR SUMMARY

Location & Geologic Setting SE Yoakum Co., 75 mi (120 km) north of Midland, west Texas, NW shelf of Midland Basin

Tectonics Basement block upthrusts

Regional Paleosetting Stable NW shelf of cratonic basin

Nature of Trap Structural/stratigraphic, on high-standing basement block

Reservoir Rocks
  Age Permian
  Stratigraphic Units(s) San Andres Formation
  Lithology(s) Dolomite and some limestone
  Dep. Environment(s) Nearshore-marine to evaporative sabkha
  Productive Facies Dolomitized skeletal wackestone and mudstone
  Entrapping Facies Lime mudstone and cement-filled grainstone
  Diagenesis/Porosity Dolomitization and dissolution of unreplaced $CaCO_3$
  Petrophysics
    Pore Type(s) Intercrystalline, moldic, vuggy
    Porosity 7.8–17.6%, avg 10.4%
    Permeability 0.01–230 md, avg 2.2 md
    Fractures Present trend dominantly NE–SW parallel to paleoshelf edge

Source Rocks
  Age Basinal equivalents of San Andres Formation
  Lithology(s) Interbedded fine-grained gray clastics and gray lime mudstone
  Migration Time Post-dolomitization

Reservoir Dimensions
  Depth 5500–5600 ft (1675–1705 m)
  Thickness 10–30 ft (3–9 m)
  Areal Dimensions 6 × 2.5 mi (9.6 × 4 km)
  Productive Area 5480 acres (22.2 km²)

Fluid Data
  Saturations S_o = 13%, S_w = 42% (avg)
  API Gravity 33°
  Gas-Oil Ratio 361:1
  Other —

Production Data
  Oil and/or Gas in Place 106.8 million BO*
  Ultimate Recovery 30.0 million BO; URE = 28.1%
  Cumulative Production 22.1 million BO through 1982

Remarks: IP 275 BOPD (43.7 m³), IP mechanism solution-gas drive. Discovered 1957. From Petroleum Data System, Information Systems Program, University of Oklahoma, Norman, OK; data retrieval November 1983.

18
Productive Permian Carbonate Cycles, San Andres Formation, Reeves Field, West Texas

Stewart Chuber and Walter C. Pusey

Introduction

Until recently (*e.g.*, Longacre, 1980), little has been written about the sedimentary and diagenetic history of the San Andres Formation and the reasons for its production, notwithstanding the fact that it is a substantial hydrocarbon producer in the Permian Basin. This paper is primarily a detailed subsurface-petrographic study of the Reeves Field, Yoakum County, Texas (see also Chuber and Sipes, 1966).

The San Andres Formation currently produces 20 percent of the annual oil and accounts for 15 percent of the cumulative oil produced in the Permian Basin. Because of its shallow depth and response to secondary recovery stimuli, the San Andres is undergoing redevelopment by additional infill drilling. Thus, the need to understand its depositional patterns is as great now for this redevelopment as it has always been for exploration.

Location, Discovery, Production and Reservoir Parameters

Reeves Field is situated at the north end of the Midland basin in the extreme southeast corner of Yoakum County, Texas, roughly 75 miles (120 km) north of Midland and 330 miles (530 km) west of Dallas (Fig. 18-1). Originally drilled on a Devonian seismic anomaly, the discovery well, J. S. Abercrombie *et al.* No. 1 J. C. Rogers, was plugged back from its total depth of 13,561 feet (4133 m). The producer had 7-inch (17.8-cm) casing set at 5719 feet (1743 m) and was perforated from 5544 to 5574 feet (1690–99 m). After an acid treatment of 750 gallons and a 10,000-gallon frac, the initial flowing potential was 274.5 barrels of oil per day (43.6 m^3) through a 1/2-inch (1.27-cm) choke on April 1, 1957. The original reservoir pressure is estimated at 2000 psig.

Reeves Field was fully developed by 1962, covering 5480 acres (21.9 km^2) in a vase-shaped area 6 miles (9.7 km) long and 2-1/2 miles (4 km) at its widest (Fig. 18-2). Of the original 144 producing wells, about 80 percent were diamond-cored and analyzed for permeability, porosity, total water saturation, and residual oil saturation. Waterflooding began in 1963, and infill drilling started in the mid-1970s. In 1983 there were 73 producing wells; cumulative production through April 1983 reached 22,805,226 barrels (3.6 million m^3), and 1982 production was 588,984 barrels. The field has estimated total recoverable reserves of 30 million barrels.

From analyses of 4147 core samples, permeabilities range from 0.01 to 230 millidarcys (avg. 2.2 md) and porosities from 7.8 to 17.6 percent (avg. 10.5%). However, 61 percent of these samples had less than 1 millidarcy permeability and 11.4 percent porosity. Average oil saturation

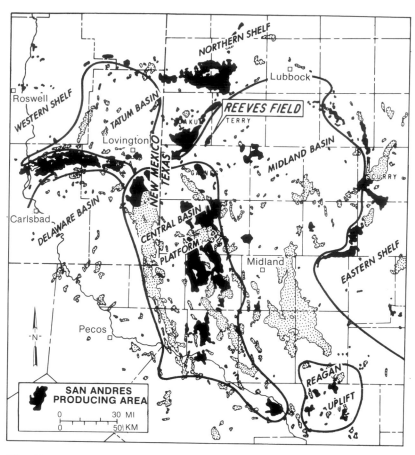

Fig. 18-1. Regional location map showing Permian Basin province boundaries, San Andres production, and the Reeves Field. Stippled areas are fields producing from Pennsylvanian and other (non-San Andres) Permian carbonates in the basin.

is 13 percent, and average water saturation is 42 percent. Oil gravity is 33° API, and the gas-oil ratio is 361:1. Reeves Field is a solution-gas drive reservoir.

Method of Study

Numerous electric log sections (scale 1″ = 100′) were correlated to show bed equivalence and define sedimentary and productive cycles within Reeves Field. Log correlations in the field are reliable, and numerous markers can be recognized. Wireline logs of the field wells invariably include gamma-ray-neutron or gamma-ray-laterolog curves. Correlations are made primarily using the gamma-ray character. The most continuous picks were thin (1–3 ft., 0.3–1 m) shale beds with high gamma readings that could be traced throughout the field area. Some of the markers appear to be correlatable for 30 miles (50 km) or more. Small-scale regional log sections were constructed perpendicular and parallel to the long axis of Reeves Field to demonstrate cross-correlations and depositional topography.

Nine wells in the center of the field were selected for petrographic study because the wells were extensively cored for engineering data; however, only core chips were saved. The core chips, about 1″ × 1″ × 1/2″, were broken from each foot of core. Approximately 1000 core chips were examined in detail, and 52 thin sections were made from selected lithologies observed in the chips. One slabbed core was obtained from the Buttes Gas & Oil Co. No. 1 Smith Brownfield at the extreme north end of the field (Fig. 18-2).

Figure 18-3 is an examination flow sheet that

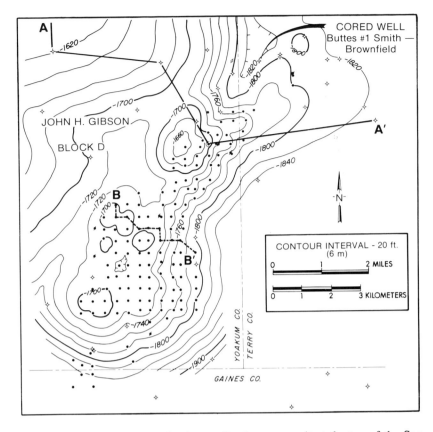

Fig. 18-2. Structure map of the Reeves Field, contoured on Lovington Sandstone, a unit at the top of the San Andres Formation.

evolved during the study and helped standardize the authors' hand specimen and microscope descriptions. Five steps make up the procedure: (1) Preparation: on a mass-production basis, the core chips are ground and polished with 100 and 300 corundum. Half the chip is then etched in hydrochloric acid for about five seconds and washed. (2) Hand-specimen description: the color and sedimentary structures are described and a Dunham (1962) rock-texture name applied to the chip (see Appendix of this volume). (3) Low-power examination: under the binocular microscope, the chips, coated with an alcohol-water mixture, are studied to determine grain proportion, size, and type. (4) High-power examination: the type and proportion of porosity, dolomite-rhomb size, and kind and amount of anhydrite are estimated under high binocular-microscope magnification. (5) Recording: fractures, pyrite, dead or live oil, earthy appearance, and stylolites were recorded. No chemical tests to determine carbonate type were made on the San Andres chips, but the slow reactions with concentrated acid suggest that the rocks are entirely dolomite.

Regional Geology

Reeves Field is situated on the northwest margin of the Midland Basin. During Guadalupe time, this area was a stable shelf, but the deep tectonic style is that of vertical basement block upthrusts. About 14,000 feet (4270 m) of sedimentary rocks overlie the Devonian in the field as shown by east-west cross-section A–A' (Fig. 18-4). Most of these rocks are Permian in age.

Geology of the Reeves Field

Reeves is typical of many Permian Basin San Andres fields that result from stratigraphic porosity loss on or flanking a structural or topographic

1. Sample Preparation: grind, etch (30% HCl), wash
2. Hand Specimen Description
   A. Color
   B. Sedimentary Structures
      1. Bedded   (bd)
      2. Laminated   (1)
      3. Mottled   (m)
      4. Homogeneous   (h)
      5. Stromatolitic   (s)
      6. "Birdseye"   (b)
   C. Dunham Rock Name: G, P, W, M (with sedimentary structure in lower case)

3. Low Power Examination (9X, water & alcohol)
   A. Grain Percentage
   B. Grain Size
   C. Grain Types: pellets, lumps, intraclasts, oolites, pisolites, skeletal (specify)
   D. Carbonaceous Material
4. High Power Examination (40X, water & alcohol)
   A. Porosity
      1. ϕ_l — Leached
      2. ϕ_i — Intergranular
      3. Intercrystalline (log rhomb size)

B. Anhydrite
   1. An_n — Nodular
   2. Void filling
      a. An_c — Castic
      b. An_i — Intergranular (poikolitic)
   3. An_r — Replacement
   4. An_e — Euhedral (rare)
5. Remarks
   A. Shale, pyrite, silica
   B. Dead oil
   C. Stylolites
   D. Earthy-chalky appearance ("soil")
   E. Fractures

Logging Sheet (Composite Example)

Depth	FAB-STRX	Grains	Porosity Percent	Rhomb Size	An_c	An_i	Total	An_r	An_n	Remarks
5590	G_{bd}	75 70 ool (200 μ) / 5 pell.	ϕ_i 5	N/A	1	19	25	2	0	well sorted An_i
5591	P_m	70 60 pell—300-600 μ / 10 skel. (Br., Mol. 500 μ)	ϕ_i 8 pell	10 μ	20	0	28	10	< 1	ϕ isolated by An_r
5592	W_m	30 10 spic 50 μ / 5 other skel. 400 μ / 15 pell 200 μ	ϕ_i 10 spic	35 μ	2	0	12	1	2	effective ϕ
5593	M_s	5 pell—100μ	ϕ_i 1	15 μ	5	0	6	1	4	An_n w/shaly lam 10% S_iO_2 in An_n
5594	Shale	< 1 qtz ≈ 20 μ	0	N/A	0	0	0	0	10	90% red brown carbonaceous (plant?) frags.

Fig. 18-3. Examination procedure for Reeves Field core chips.

feature. Closure in such fields is a combination of downdip contact with single or multiple water levels and updip termination of porosity in single or multiple cycles, causing hydrocarbon entrapment. A result is that oil columns are considerably in excess of the structural closures. For instance, the Wasson Field, located 10 miles (16 km) west of Reeves, has only 140 feet (43 m) of structural closure but about 400 feet (120 m) of vertical oil and gas column (Schneider, 1943, p. 498, 508, 509).

Structure

Figure 18-2 is a structure contour map on the base of the Lovington sandstone member of the San Andres Formation. This marker is about 100 feet (30 m) below the top of the San Andres and 400 feet (120 m) above the Reeves zone (Fig. 18-4). Maximum closure is about 40 feet (12 m); however, a productive stratigraphic closure of at least 100 feet (30 m) exist. Reeves Field is elongate northeast–southwest and dips steeply southeast into the Midland Basin. Essentially, Reeves is a structural terrace and is attributed primarily to drape and compaction over a pre-San Andres shelf edge.

Vertical fractures were observed in cores from many wells. The highest concentration is in a northeast–southwest trending belt extending from the south-central to the northeast corner of the field. The fractures resulted from the forces of overburden compaction and appear to have concentrated San Andres breakage along an underlying carbonate shelf-edge trend.

Stratigraphy

The "Reeves" zone of the San Andres Formation has been subdivided into four cycles based on lithologic and wireline log character. The cycles are numbered stratigraphically from 1 to 4 (Fig.

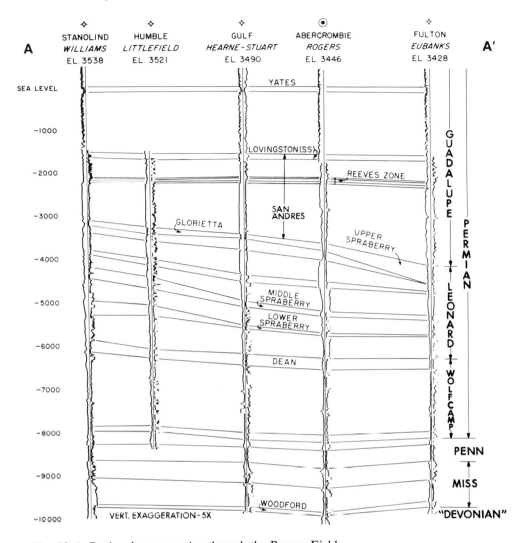

Fig. 18-4. Regional cross-section through the Reeves Field.

18-5). Cycle 1 varies from 27 to 32 feet (8.3–9.8 m), cycle 2 from 14 to 23 feet (4.3–7.0 m), cycle 3 from 12 to 21 feet (3.7–6.4 m), and cycle 4 from 15 to 23 feet (4.6–7.0 m). Each cycle has a similar lateral lithologic facies gradation across the field. The facies reflect the crossing of a transitional energy interface, which began with low-energy, moderate-depth marine conditions succeeded by high-agitation, shallow-marine conditions that finally evolved into low-energy shallow-lagoon, tidal flat or sabkha conditions. The carbonates are muddy on the southeast or basinal side, and they become grainier and more winnowed near the apex of the field. On the northwest or shelf side, the rocks once again grade into carbonate mudstone facies.

Petrography

A diverse assemblage of particulate grain types was identified in thin sections of the Reeves Field dolomites. In order of relative abundance, the following skeletal grain types were noted: brachiopod-mollusks, echinoderms, ostracodes, bryozoans (rhomboporid and fenestrate), sponge spicules, calcareous algae, forams, and plant fragments (Fig. 18-6). All calcareous-type organisms were preserved as molds or anhydrite casts. This type of preservation negated a more specific breakdown of skeletal grain types. Fish scales and conodonts were noted in the black shales. In addition to these organisms, there was a gray worm-like tube that was usually found in dolomitized

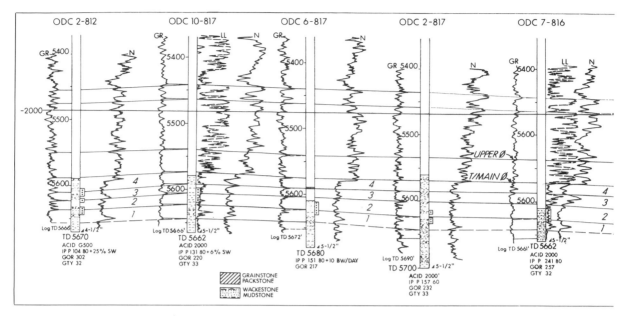

Fig. 18-5. Detailed electric-log cross-section showing lithology and production data. Location of section shown in Figure 18-2.

lime mud. Remarkably, the tubes survived the mud recrystallization and served as an indicator of mud, even when it was recrystallized to microspar.

The San Andres seas are often portrayed as extremely saline; consequently, the observed diversity and abundance of skeletal types was unexpected. However, fusulinids and corals were conspicuously absent, a fact that may reflect special environmental conditions, possibly insufficient nutrients and poorly oxygenated water at the shallow edge of a lagoon with restricted circulation.

Non-skeletal grains were far more abundant than skeletal grains. In order of relative abundance, the following types of nonskeletal grains were logged: pellets, ooliths, lumps, and intraclasts (Fig. 18-6). This observation tends to confirm a shallow, hypersaline lagoonal margin environment.

Dolomitized carbonate mud was by far the most abundant constituent of the Reeves Field suite. A large variation in dolomite rhomb size (from 5 to 40 microns) made recognition of mud difficult, but the presence of gray organic tubes provided a convenient indicator.

The dolomite mud occurred in several types of sedimentary fabric as follows: homogenous, mottled, stromatolitic, and birdseye (Fig. 18-7). In addition to carbonate mud, dark-gray shale was

present as thin beds and laminae. This shale is nearly opaque in thin section and appears to be mostly carbonaceous plant material (Fig. 18-7).

Depositional Cycles and Depositional Environment

The rock components and sedimentary structures occur in a definite vertical sequence that is repeated, *i.e.*, cyclic. San Andres cycles in Reeves Field are asymmetric (hemicycles), for they record a progressive shallowing of the water followed by a disconformity, but no obvious record of deepening water. The return to relatively deep water is, in effect, "instantaneous." The cycles can be divided into two types: (1) shelf-edge cycles, which culminate with well-developed oolites, and (2) back-shelf cycles, which culminate with stromatolitic mud.

Back-Shelf Depositional Cycle

Figure 18-8 shows two complete cycles from the Buttes No. 1 Smith Brownfield just west (*i.e.* shelfward) of the field. The base of each cycle rests on a subaerially exposed surface. This surface is irregular and contains vertical (desiccation) cracks. Each cycle begins with organic shale

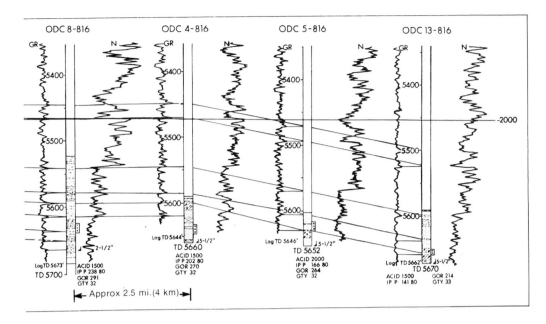

laminae overlain by dark gray-brown dolomitized mudstone. The mudstone contains rare fossils, but occasionally it grades laterally to a spiculitic wackestone. It grades upward into a medium brown mottled dolomitized wackestone that contains fossils and pellets. The wackestone grades upward and is interbedded with pinkish weathered stromatolites. The stromatolites contain vertical shrinkage cracks and birdseye structures, which clearly indicate supratidal deposition (Shinn, 1968). These are exemplified where stromatolitic laminae are curved upward, as if the core penetrated large desiccation polygons (e.g. at 5953.5 ft., 1814.6 m). Finally, the cycle ends abruptly with a gray weathering zone about an inch thick. The cycle described above typifies the sequence shelfward of the oil field. It is interpreted as the record of an initial transgression (dark mudstone) followed by a relative shallowing of the water until the upper part of the cycle is supratidal. Despite evidence that the depth of water decreased, there is little evidence of more grainy rocks indicative of increased wave energy.

Shelf-Edge Depositional Cycles

In contrast to the back shelf, cycles near the shelf edge on the east side of Reeves Field contain oolites, which indicate the presence of high-energy events in the upper part of each cycle (Fig. 18-9). This type of cycle characterizes the apex of the field.

The vertical facies in each cycle is more difficult to establish because of incomplete preservation and/or removal of component parts. An ideal cycle is shown in Figure 18-9 and consists of the following segments from the base upward:

1. A brown, mottled, burrowed organic shale approximately 2 feet (0.66 m) thick, with gradational upper and sharp lower contact.
2. Dark-gray mudstone becoming lighter in color vertically and increasing in grains and fossils vertically; about 9 feet (2.7 m) thick, with gradational upper and lower contact.
3. Gray fossiliferous wackestone about 6 feet (2 m) thick, with gradational contacts.
4. Gray or brown fossiliferous packstone about 3 feet (1 m) thick, with gradational contacts.
5. Gray or brown oolitic and fossiliferous grainstone with a gradational lower contact and a sharp, probably erosional, upper contact.

This vertical sequence is a transgressive marine cycle that begins and ends in the basal shale. The shale is probably a soil zone inundated by a rapid transgression to a water depth approximately equal to the thickness of the overlying cycle (uncorrected for compaction). Dark carbonate mudstones were deposited above and mixed into the organic shale by bottom burrowers. As sediment

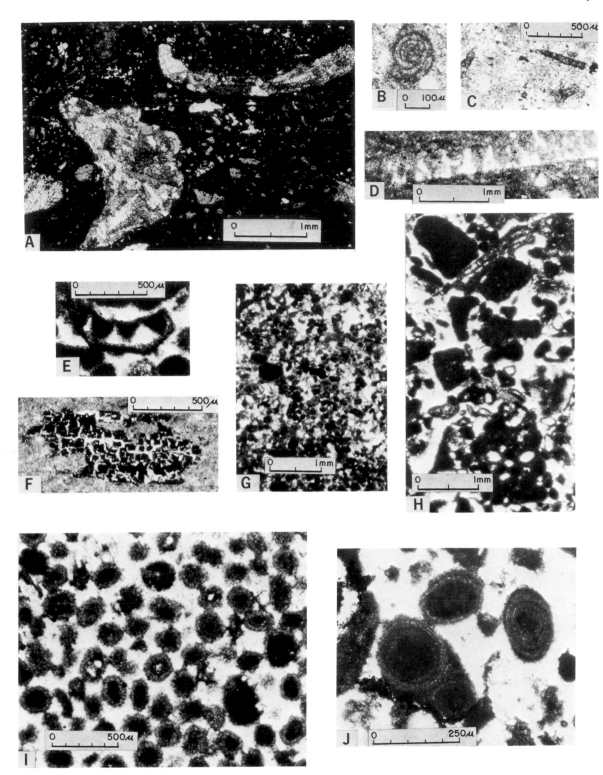

Fig. 18-6. Photomicrographs of grain types in the San Andres at Reeves Field. *A.* Elongate grain is brachiopod or mollusk. Grains to the left of the field are probably echinoderm plates. All grains are anhydrite-filled molds. Oil Development Co. Well No. 4-876A, 5653.0 feet (1723.0 m). Cross-polarized light. *B.* Possible foraminifer, Apterinella sp. Oil Development Co. Well No. 4-876A, 5628.6 feet (1715.6 m). Plane light.

accreted in the area, an upward increase in fossil fragments, grains, and lighter color reflects the shoaling water and increased energy. The mudstones grade vertically to wackestones, packstones, and grainstones while shoaling continues. Ultimately the marine environment is pierced, and periodic subaerial exposure of the grainstones takes place. Total exposure initiates soil development, which completes the cyclothem and sets the stage for renewed subsidence or transgression. The cycle is easily misinterpreted as gradual seaward progradation (regression?), but can be distinguished by the multiple succession of similar cycles. It is in fact evidence of cyclic stillstand.

Grain-supported rocks, which contain interparticle porosity, occur in the upper third of the shelf-edge cycles. Figure 18-10A illustrates depositional porosity in an oolite. The depositional porosity in this grain-supported rock is approximately 25 percent. However, 80 percent of the well-completion perforations are in mud-supported carbonates, which by core analysis are permeable and porous. The reason for this apparent discrepancy is that the depositional porosity in the oolite has been totally infilled by anhydrite (Fig. 18-10B), whereas the dolomitized mud-supported rocks contain leached porosity and are not infilled with anhydrite (Fig. 18-

10C,D). Consequently, the post-depositional processes reversed the porosity pattern expected from depositional processes.

Diagenesis and Anhydrite

Three major related modifications of Reeves Field carbonates occurred: dissolution of skeletal grains, complete dolomitization and interstitial precipitation of gypsum.[1]

The carbonate cycles of Reeves Field closely resemble contemporary sequences of modern-day sabkha sediments of the Persian Gulf. For instance, the Reeves back-shelf sequence appears to include all the facies seen by McKenzie *et al.* (1980) and shown in their idealized vertical profile (Fig. 3; p. 13). It is logical to assume, therefore, that the hydrologic evaporative pumping model proposed by them accounts for both the carbonate diagenesis and the anhydrite emplacement in Reeves Field. In San Andres time, Reeves was located at the seaward edge of a supratidal flat. The diagenetic fabric and mineralogy described by Shearman (1966) and Kinsman (1966) in the Persian Gulf are similar to the diagenetic features of San Andres dolomite in Reeves Field.

Anhydrite occurs in four basic forms in Reeves Field: tabular crystals, nodules, void fillings, and replacement. Beds of laminated anhydrite were not present.

C. Organic tubes. Oil Development Co. Well No. 2-812, 5621.0 feet (1713.3 m). Plane light. *D.* Possible algal blade. Oil Development Co. Well No. 6-817, 5673.0 feet (1729.1 m). Plane light. *E.* Possible algal fragment. Coline Well No. 2-881 Willard, 5630.0 feet (1716.0 m). Plane light. *F.* Carbonized leaf fragment. Oil Development Co. Well No. 2-812, 5621.0 feet (1713.3 m). Plane light. *G.* Pellets and lumps. Coline Well No. 2-881 Willard, 5592.9 feet (1704.7 m). Plane light. *H.* Lumps and lithoclast (near bottom). Coline Well No. 2-881 Willard, 5630.0 feet (1716.0 m). Plane light. *I.* Ooliths (note uniform size). Oil Development Co. Well No. 4-876A, 5590.0 feet (1703.8 m). Plane light. *J.* Detail of ooliths showing preservation of laminated structure. Oil Development Co. Well No. 4-876A, 5590 feet (1703.8 m). Plane light.

Tabular Crystals

Long, extremely thin (200 μ) crystals occur in sections in several wells. The laths are generally 2×20 millimeters in section, although some up to 100 millimeters long occur. They are dark brown and can easily be mistaken for plant fragments. This type of anhydrite is rare relative to other types of anhydrite. Possibly the tabular

---

[1]It is assumed that the gypsum was converted to anhydrite after burial at several thousand feet. See Murray (1964) and Kinsman (1966) for pros and cons of primary gypsum versus primary anhydrite.

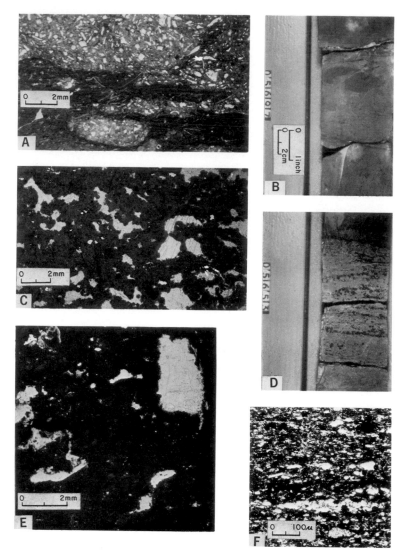

Fig. 18-7. Photomicrographs of sedimentary structures in the San Andres, Reeves Field. *A*. Skeletal packstone (light) mottled into shaly wackestone (dark). Oil Development Company Well No. 6-817, 5671 feet (1728.5 m). Plane and reflected light. *B*. Polished core, mottled wackestone and shale. Buttes Gas and Oil Well No. 1 Brownfield, 5697–5698 feet (1754.7–1754.9 m). *C*. Birdseye pellet mudstone with birdseye structure. Birdseyes are infilled with clear crystalline anhydrite. Coline No. 2-881 Willard, 5568.9 feet (1797.4 m). Plane and reflected light. *D*. Polished core, stromatolitic mudstone. Dark blebs are birdseyes infilled with anhydrite. Dark shale and shaly mudstone, near top of core, mark base of overlying cycle. Buttes Gas and Oil Well No. 1 Brownfield, 5653.5 feet (1723.2 m). *E*. Birdseye mudstone. Large vug in upper right contains geopetal filling in large burrow. Oil Development Co. Well No. 2-812, 5595.0 feet (1705.4 m). Plane and reflected light. *F*. Gray organic shale. Dark grains are reddish-brown plant materials; white grains are quartz silt. Oil Development Co. Well No. 4-876A, 5649.0 feet (1721.8 m). Plane light.

Fig. 18-8. Back-shelf cycles of deposition in the San ► Andres Formation, Buttes Gas and Oil Well No. 1 Brownfield, slabbed cores (depths indicate bottom foot of each piece of core). Each core slab is approximately 3 inches (7.6 cm) across. Two complete cycles are in this illustration. The older cycle (5655.8–5660.5 ft.; 1723.9–1725.3 m) begins with dark gray mudstone resting on an irregular gray surface (5660.6 ft.). Three feet of dolomitized and mottled wackestone dominate the cycle. At 5657.1 feet (1724.3 m) and above, the mudstone is stromatolitic and contains nodular anhydrite. The younger cycle begins with shaly gray, dolomite wackestone resting on an irregular surface (5655.8 ft.; 1723.9 m). Within a foot it is succeeded by 3 feet of stromatolitic dolomite. At 5653.4 to 5653.7 feet (1723.2–1723.3 m) the stromatolite laminae are curved upward as if dessicated (core fragment at 5653.7 ft. was inserted upside down). The cycle ends with a weathered birdseye mud at 5652.2 feet (1722.8 m).

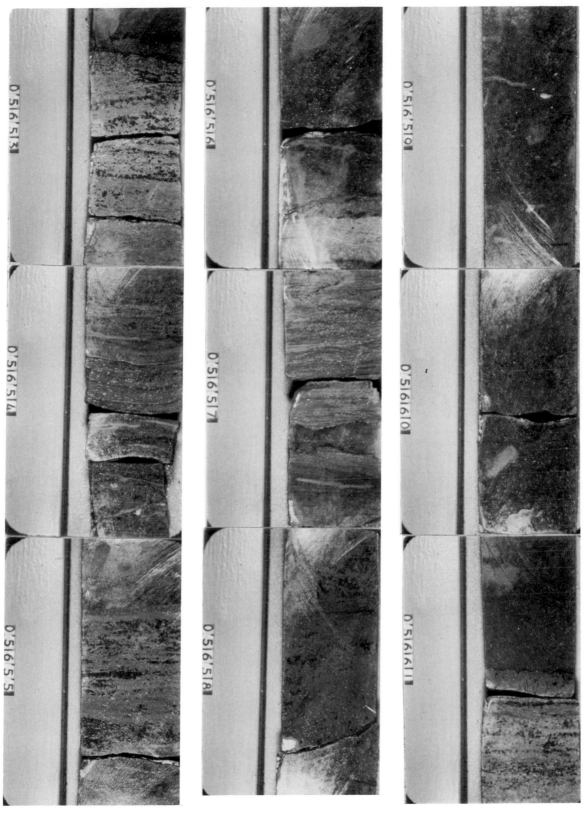

0   1 inch

0   2 cm

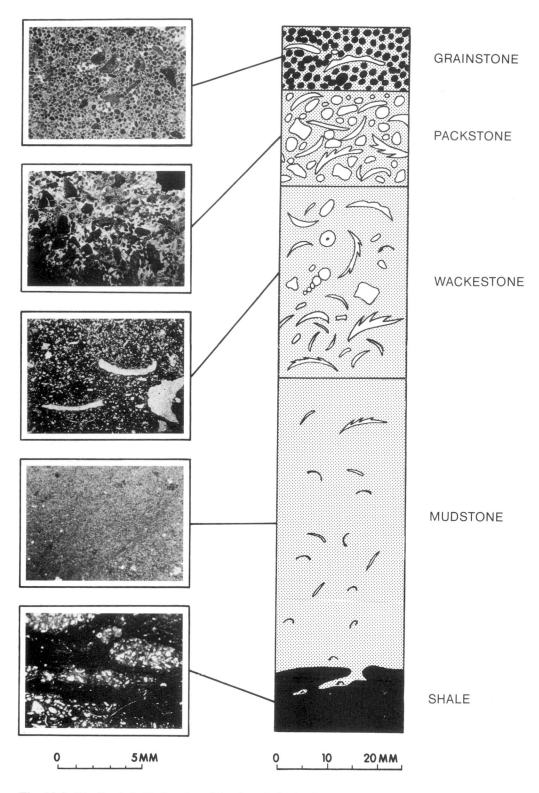

GRAINSTONE

PACKSTONE

WACKESTONE

MUDSTONE

SHALE

0   5MM

0   10   20 MM

Fig. 18-9. Idealized shelf-edge depositional cycle in the San Andres Formation, Reeves Field.

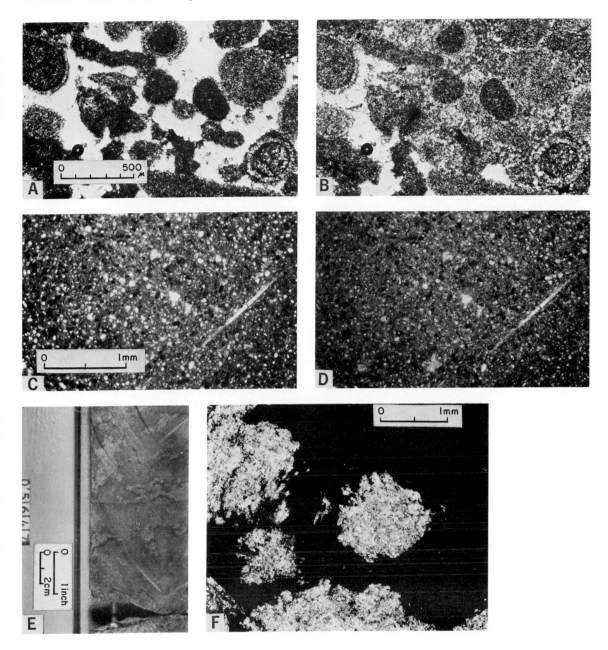

Fig. 18-10. San Andres lithofacies. *A.* Photomicrograph of oolitic packstone containing apparent depositional porosity. Oil Development Co. Well No. 4-876A, 5608.3 feet (1709.4 m). Plane light. *B.* Same as *A*, in cross-polarized light. Poikilitic anhydrite can be seen to infill all porosity. *C.* Photomicrograph of spicular mudstone containing 22% leached and intercrystalline porosity. Oil Development Co. Well No. 4-876A, 5628.6 feet (1715.6 m). Plane light. *D.* Same as *C*, in cross-polarized light. Less than 1% anhydrite in spicule and blebs. *E.* Photograph of polished core, nodular anhydrite in mudstone. Buttes Gas and Oil Well No. 1 Brownfield, 5646–5647 feet (1720.9–1721.1 m). *F.* Photomicrograph of nodular anhydrite in dark (opaque) mudstone. Note fibrous anhydrite laths. Oil Development Co. Well No. 2-812, 5606.0 feet (1708.7 m).

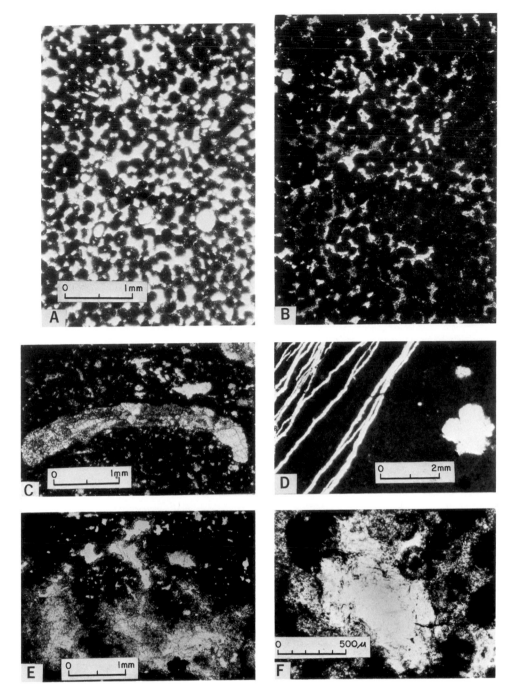

Fig. 18-11. Photomicrographs of San Andres rocks. *A.* Pellet packstone. Transparent anhydrite cement (white) gives a false appearance of good porosity. Oil Development Co. Well No. 2-876, 5590 feet (1703.8 m). *B.* Same as *A*, in cross-polarized light, showing large poikilitic crystals infilling interparticle porosity. *C.* Clear pore-filling anhydrite in mold of brachiopod or mollusk shell. Oil Development Co. Well No. 4-876A, 5653 feet (1723.0 m). Cross-polarized light. *D.* Dolomitized mud with porphyroblastic anhydrite replacing mud. Fractures at left are infilled with clear anhydrite. Coline Well 2-882 Willard, 5642.5 feet (1719.8 m). Cross-polarized light. *E.* Birdseye infilled with clear anhydrite. The anhydrite has begun to replace dolomite mud beyond the birdseye wall, and replacement anhydrite gives a "dusty rim" to the birdseye. In the bottom half, replacement anhydrite almost equals void-filling anhydrite. Oil Development Co. Well No. 6-817, 5617 feet (1712.1 m). Plane light. *F.* Poikilitic anhydrite replacing ooliths and pellets. Note the relict grain structure in the large clear anhydrite crystal. Oil Development Co. Well No. 6-817, 5643 feet (1720.0 m). Plane light.

crystals represent an early single crystal stage of nodular anhydrite.

Nodular Anhydrite

Nodules of anhydrite occur most commonly in lime mud (Fig. 18-10E) or shale (Fig. 18-10F). The nodules are thought to form relatively early because they push the mud aside. Anhydrite nodules have little direct effect on porosity destruction. They are thought to have formed from gypsum rosettes such as those illustrated by Kerr and Thomson (1963) and Masson (1955). With burial, the crystals become crushed and dewatered to form anhydrite nodules. Kerr and Thomson (1963; Figs. 12, 13, and 14) offer convincing proof of this by illustrating anhydrite nodules from the Permian that still maintain their gypsum crystal form.

Void-Filling Anhydrite

Clear, coarsely crystalline anhydrite is the most abundant form present. It is present in depositional void spaces such as intergranular porosity (Fig. 18-11B), early diagenetic voids such as burrows and birdseyes (Fig. 18-11B), post-leaching voids such as fossil molds (Fig. 18-11C), and post-lithification fractures (Fig. 18-11D).

Intergranular anhydrite tends to occur in very large single crystals of irregular geometry that engulf one or several carbonate grains. This anhydrite probably grew as gypsum sand crystals near the sediment surface in a manner similar to sand crystals reported by Illing et al. (1965) and Kinsman (1966) in Persian Gulf sabkhas. The carbonate sand grains associated with these crystals have not been leached, suggesting that this type of gypsum occurs early, perhaps contemporaneous with nodular gypsum. The grain-encompassing anhydrite versus the grain-excluding type appears to be controlled by the grain size of the host sediment and does not reflect a fundamental difference in environment. It should be noted, however, that the encompassing crystals are usually reported from very shallow depths (1–2 ft.), whereas nodules are commonly reported at depths as great as 20 feet (6 m) (Masson, 1955).

As noted previously, the skeletal morphology of dissolved calcareous grains is preserved as anhydrite-filled molds. Fine dolomite crystals often line the molds and indicate that the shells were not only leached, but also were partially infilled before gypsum deposition.

The destruction of primary porosity by void-filling anhydrite is complete in Reeves Field. The few feet of porous oolite in which porosity was logged contained relict anhydrite, which suggests that the porosity is produced secondarily by dissolution of other void-filling cements or anhydrite (Fig. 18-12E,F).

Replacement Anhydrite

Anhydrite that replaces dolomite is commonly associated with all anhydrite types. In grainstones it appears as dusty rims (Fig. 18-11E,F) and commonly completely replaces the carbonate. Replacement anhydrite is the main destroyer of intercrystal porosity in dolomitized muds and also the "muds" in packstones (Figs. 18-12E,F). Minor amounts on the order of 1 to 2 percent do not affect the permeability (Fig. 18-12B), but amounts of 10 percent and more (Fig. 18-12C) are invariably associated with low porosities and low permeabilities.

The precipitation of multiple gypsum generations effectively destroyed the depositional porosity of the grainstone facies. Partial anhydrite replacement and plugging in the less permeable wackestones and mudstones allowed subsequent dolomitization and leaching to convert these rocks into the relatively permeable reservoir beds producing petroleum today.

Thus, diagenesis has reversed the reservoir pattern expected from depositional facies. The reservoir is largely in the muds, and the seal is formed by oolitic sands.

Summary and Conclusions

Reeves Field contains the record of an upper Permian shelf-edge. Depositional cycles begin with an initial low-energy phase representing relatively deep water (20–30 ft; 6–9 m). Sedimentation aggraded the sea floor into relatively shallow water. On the seaward and crestal parts of the field area, oolitic grainstones predominated; on the back-shelf area, stromatolitic mudstones predominated. This depositional pattern resulted in repeated subaerial exposure, which, coupled with a high rate of evaporation, established a hydro-

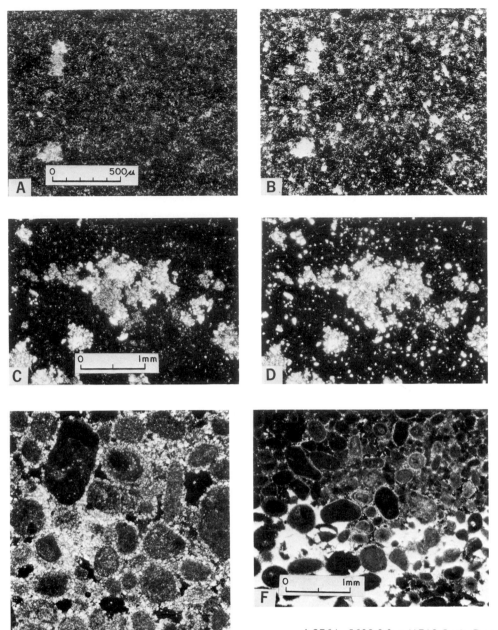

Fig. 18-12. Photomicrographs of San Andres lithofacies. *A*. Dolomite wackestone with about 1% replacement anhydrite. Oil Development Co. Well No. 4-876A, 5631.2 feet (1716.4 m), permeability 11 md, porosity 18% by whole-core analysis. Cross-polarized light. *B*. Same as *A*, in plane light. Note the porosity (white). *C*. Dolomite wackestone with about 20% replacement anhydrite. Oil Development Co. Well No. 4-876A, 5638.2 feet (1718.5 m). Cross-polarized light. *D*. Same as in *C*, plane light, showing that some porosity remains but is greatly reduced compared to B. Oil Development Co. Well No. 4-876A, 5638.2 feet (1718.5 m) permeability 10 md, porosity 11% by whole-core analysis. *E*. Pellet packstone with interparticle porosity remaining after partial cementation. The small patch of white poikilitic anhydrite near the bottom of the slide suggests the porosity is the result of leached poikilitic anhydrite. Oil Development Co. Well No. 4-876A, 5596.8 feet (1705.9 m). Cross-polarized light. *F*. Pellet packstone, boundary between open porosity (at top of slide) and porosity plugged by poikilitic anhydrite cement. Oil Development Co. Well No. 4-876A, 5600.0 feet (1706.9 m). Cross-polarized light.

logic evaporative pump system. As a result, there occurred post-depositional precipitation of gypsum and the simultaneous dissolution, dolomitization, and precipitation of carbonates. Although depositional porosity in the grainstones was destroyed by gypsum, porosity and permeability developed in the dolomitized mudstones.

After lithification, the weight of the overburden produced a fracture system that may have provided access routes for oil migration.

Acknowledgments The authors acknowledge with thanks the permission given by the West Texas Geological Society to republish the text and figures, which are slightly revised and updated from the original (Chuber and Pusey, 1969).

References

CHUBER, S., and W.C. PUSEY, 1969, Cyclic San Andres facies and their relationship to diagenesis, porosity and permeability in the Reeves Field, Yoakum County, Texas: *in* Elam, J.G. and S. Chuber, eds., Cyclic sedimentation in the Permian Basin—a symposium: West Texas Geol. Soc. Publ. 69-56, p. 136–151.

CHUBER, S., and L.D. SIPES, JR., 1966, Reeves Field: *in* Oil and gas fields in west Texas a symposium: West Texas Geol. Soc. Publ. 66-52, p. 308–312.

DUNHAM, R.J., 1962, Classification of carbonate rocks according to depositional texture, *in* Ham, W.E., ed., Classification of Carbonate Rocks—a symposium: Amer. Assoc. Petroleum Geologists Mem.1, p. 108-121.

ILLING, L.V., A.J. WELLS, and J.C.M. TAYLOR, 1965, Penecontemporaneous dolomite in the Persian Gulf: *in* Pray, L.C. and R.C. Murray, eds., Dolomitization and Limestone Diagenesis—a symposium: Soc. Econ. Paleontologists and Mineralogists, Spec. Publ. 13, p. 89–111.

KERR, S.D., JR., and A. THOMSON, 1963, Origin of nodular and bedded anhydrite in Permian shelf sediments, Texas and New Mexico: Amer. Assoc. Petroleum Geologists Bull., v. 47, no. 9, p. 1726–1732.

KINSMAN, D.J.J., 1966, Gypsum and anhydrite of recent age, Trucial Coast, Persian Gulf: Permian Basin Sec., Soc. Econ. Paleontologists and Mineralogists, 11th Annual Meeting, Midland, Texas, 34 p.

LONGACRE, S.A., 1980, Dolomite reservoirs from Permian biomicrites: *in* Halley, R.B. and R.G. Loucks, eds., Carbonate Reservoir Rocks: notes for SEPM Core Workshop No. 1, Denver, Colorado, June 1980: Soc. Econ. Paleontologists and Mineralogists, p. 105–117.

McKENZIE, J.A., K.J. HSÜ, and J.F. SCHNEIDER, 1980, Movement of subsurface waters under the sabkha, Abu Dhabi, UAE, and its relationship to evaporative dolomite genesis: *in* Zenger, D.H., J.B. Dunham, and R.L. Ethington, eds., Concepts and Models of Dolomitization—a symposium: Soc. Econ. Paleontologists and Mineralogists, Spec. Publ. 28, p. 11–30.

MASSON, P.N., 1955, An occurrence of gypsum in southwest Texas: Jour. Sedimentary Petrology, v. 25, no. 1, p. 72–77.

MURRAY, R.C., 1964, Origin and diagenesis of gypsum and anhydrite: Jour. Sedimentary Petrology, v. 34, p. 512–523.

SCHNEIDER, W.T., 1943, Wasson Field, Yoakum and Gaines counties, Texas: Amer. Assoc. Petroleum Geologists Bull., v. 27, no. 4, p. 479–523.

SHEARMAN, D.J., 1966, Origin of marine evaporites by diagenesis: Inst. Mining and Metallurgy Trans. (B), v. 75, p. 208–215.

SHINN, E.A., 1968, Practical significance of birdseye structures in carbonate rocks: Jour. Sedimentary Petrology, v. 38, no. 1, p. 215–223.

19
Blalock Lake East Field
George B. Asquith and John F. Drake

RESERVOIR SUMMARY

Location & Geologic Setting Glasscock Co., west Texas; margin of eastern shelf of Midland Basin, USA

Tectonics Westward regional dip into cratonic Midland Basin

Regional Paleosetting Shallow marine shelf

Nature of Trap Stratigraphic; porous skeletal reef-mounds and oolite bars in homoclinally west-dipping limestone-shale sequence

Reservoir Rocks
  Age Early Permian (Wolfcampian)
  Stratigraphic Units(s) Wolfcamp-A Zone
  Lithology(s) Limestone
  Dep. Environment(s) Reef-mound (bank), forebank, oolite shoal
  Productive Facies Phylloid-algal mudstone and and bafflestone, *Fistulipora-Tubiphytes* boundstone and grainstone, ooid grainstone
  Entrapping Facies Dark calcareous shales and nonporous grainstone
  Diagenesis/Porosity Dissolution in meteoric water associated with subaerial exposure; later partial phreatic cementation

  Petrophysics
    Pore Type(s) Moldic, vuggy and solution-enlarged interparticle
    Porosity 8–10%, 9% avg
    Permeability NA
    Fractures Minor open vertical fractures

Source Rocks
  Age NA
  Lithology(s) NA
  Migration Time NA

Reservoir Dimensions
  Depth 7850 ft (2390 m) avg
  Thickness ~10 ft (3 m)
  Areal Dimensions NA
  Productive Area 890 acres (3.6 km^2)

Fluid Data
  Saturations S_o = 60–75%, S_w = 25–40%
  API Gravity 41°
  Gas-Oil Ratio 385:1
  Other —

Production Data
  Oil and/or Gas in Place 4.65 million BO
  Ultimate Recovery 930,000 BO; URE = 20.0%
  Cumulative Production 711,600 BO through 1983

Remarks: IP 21 BOPD pumping to 284 BOPD flowing. IP mechanism solution-gas drive. Discovered 1971.

19

Depositional History and Reservoir Development of a Permian *Fistulipora–Tubiphytes* Bank Complex, Blalock Lake East Field, West Texas

George B. Asquith and John F. Drake

Location and Discovery

Blalock Lake East Field is located on the eastern shelf of the Midland Basin in west Texas approximately 30 miles (48 km) east-southeast of the city of Midland (Fig. 19-1). Specific geographic location of the field is identified by Sections 11 through 14 of Block 35, T-3-S, T & P Railroad Survey, Glasscock County, Texas.

Subsurface mapping and interpretation of CDP seismic record sections led to discovery of the field in October 1971. The discovery well, the Belco Petroleum Company No. 1 Powell in Section 11, was perforated from 7823 to 7830 feet (2384–2387 m) and 7904 to 7914 feet (2409–2412 m) in the Permian Wolfcamp-A zone. Initial production from the No. 1 Powell well was 30 barrels of oil per day with API gravity of 40°C and gas-oil ratio of 1,800-1, and 17 barrels of salt water.

Reservoir Trap and Production

Hydrocarbon accumulation at Blalock Lake East Field is stratigraphically controlled by an eastward (updip) decrease in porosity resulting from facies change and by depositional thinning onto the eastern shelf of the Midland Basin (Fig. 19-2). The field is 890 acres (3.6 km^2) in productive areal extent (Fig. 19-3) and contains eight wells that produce from thin (~10 ft, 3 m) discontinuous porosity zones at depths ranging from 7814 to 7914 feet (2382–2412 m). From 1971 through 1979, the eight wells produced 667,596 barrels of oil by the mechanism of solution-gas drive. Ultimate recovery from the field should approximate 930,000 barrels of oil.

Petrography

The Lower Permian Wolfcamp-A zone at Blalock Lake East Field is a *Fistulipora-Tubiphytes* bank complex that grew along the eastern shelf of the Midland Basin (Fig. 19-1) and attained a maximum thickness of 188 feet (57 m). Five of the seven facies that Wilson (1975) associates with Late Paleozoic carbonate buildups can be identified in the Wolfcamp-A zone bank complex. These facies are: (1) phylloid algal mudstone core, (2) *Fistulipora-Tubiphytes* crestal boundstone, (3) phylloid-algal foraminiferal grainstone and packstone flank beds, (4) bank detritus, and (5) *Mizzia*-oolite capping grainstone. An additional basinal facies, seaward of the bank complex, is a sequence of argillaceous lime mudstones interbedded with dark spicule-bearing calcareous shales. These dark calcareous shales are believed to be probable source beds. Similar argillaceous mudstones and dark calcareous shales overlie the bank complex and form the reservoir seal.

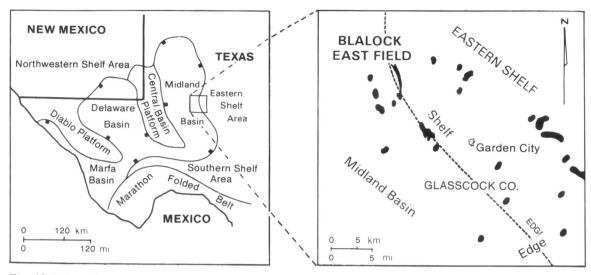

Fig. 19-1. Location map of Blalock Lake East Field in the Midland Basin, Glasscock County, West Texas.

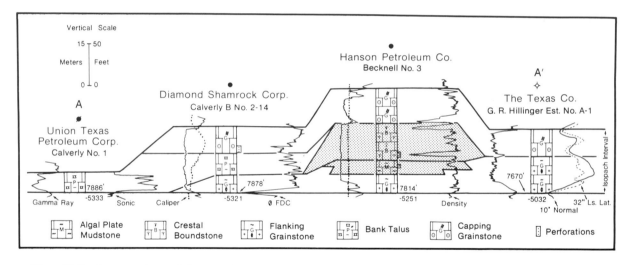

Fig. 19-2. Cross-section of Lower Permian Wolfcamp-A zone, Blalock Lake East Field. The line of cross-section is illustrated in Figure 19-3. Cross-section datum is on the base of the Wolfcamp-A zone. Perforated intervals are indicated on The Diamond Shamrock Corp. Calverly B No. 2-14 and the Hanson Petroleum Co. Becknell No. 3.

Phylloid Algal Plate Mudstone

This facies consists chiefly of plates of phylloid algae in a dark lime mud matrix (Fig. 19-4) commonly encrusted with tubular foraminifers. The algal mudstones represent the lime-mud core facies. Vuggy and fossil-moldic porosity is developed because of the leaching of lime mud matrix and phylloid-algal plates.

Fistulipora-Tubiphytes Boundstone

Prominent in this facies is skeletal debris composed of crinoids, bryozoans, trilobites, brachiopods, and phylloid-algal plates. The skeletal debris is bound together (Figs. 19-5, 19-6) by *Tubiphytes*, an encrusting problematic alga (Horowitz and Potter, 1971) or hydrozoan (Rigby, 1958); by *Fistulipora*, an encrusting

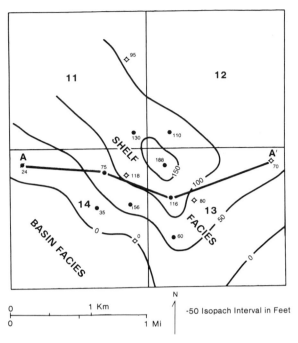

Fig. 19-3. Isopach map of Lower Permian Wolfcamp-A zone, Blalock Lake East Field, Glasscock County, Texas.

bryozoan (Horowitz and Potter, 1971); and by tubular foraminifers. Vuggy porosity, created by dissolution of the lime mud matrix, is commonly present (Fig. 19-6). The *Fistulipora-Tubiphytes* boundstones make up a crestal boundstone facies and overlie the lime mud core (Wilson, 1975).

Phylloid Algal Foraminiferal Grainstone and Packstone

This facies is made up of fusulinids, small foraminifers, and peloids, together with fragments of brachiopods, phylloid algae, crinoids, and bryozoans; all have some intergranular lime mud (packstone) or equant calcite spar as cement (grainstone) (Fig. 19-7). The flank facies is nonporous because of the presence of equant calcite-spar cement or lime mud (micrite) matrix. The skeletal fragments in these grainstones and packstones are commonly encrusted with *Tubiphytes* (Fig. 19-7). Grainstones and minor associated packstones represent the flank facies.

Bank Detritus

Detritus in this facies consists of large, poorly sorted, angular to subrounded fragments, up to 15 cm, of *Fistulipora-Tubiphytes* boundstone and algal foraminiferal grainstone-packstone (Fig. 19-8). Detritus occurs with an intergranular calcareous mud fraction containing spicules. The result is a dark-colored rock with little or no porosity or permeability.

Mizzia-Oolite Grainstone

Limestones in this facies are composed of fragments of dasycladacean algae (*Mizzia*) and

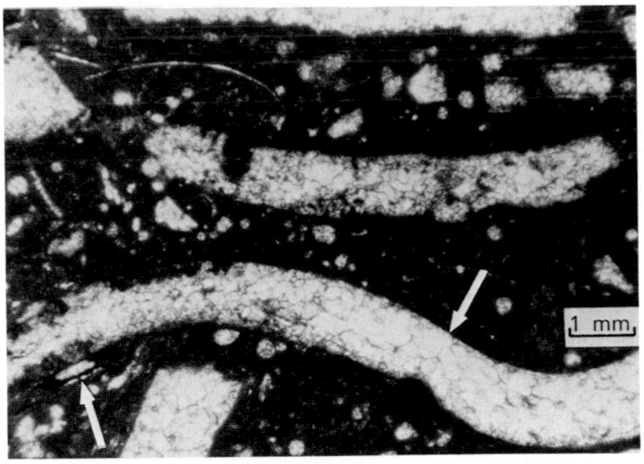

Fig. 19-4. Photomicrograph of phylloid-algal mudstone (algal-plate mudstone core). Long arrow points to phylloid algae and short arrow to tubular foraminifer. Plane light.

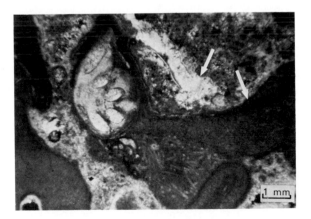

Fig. 19-5. Photomicrograph of *Tubiphytes* and tubular foraminifers encrusting a bryozoan fragment (*Fistulipora-Tubiphytes* crestal boundstone). Long arrow points to *Tubiphytes* and short arrow to vuggy porosity. Plane light.

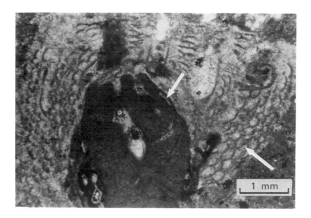

Fig. 19-6. Photomicrograph of *Fistulipora* (arrow) encrusting *Tubiphytes* and tubular foraminifers (*Fistulipora-Tubiphytes* crestal boundstone). Plane light.

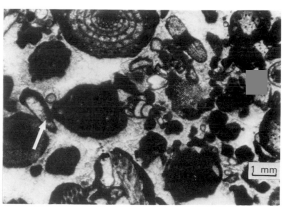

Fig. 19-7. Photomicrograph of algal-plate-foram grainstone (flanking grainstone) with *Tubiphytes* encrusting some of the fossil debris. Arrow points to fragment of phylloid algae encrusted with *Tubiphytes*. Plane light.

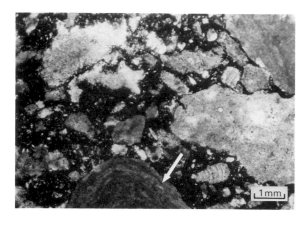

Fig. 19-8. Photomicrograph of bank detritus with poorly sorted fragments of bank debris in a dark, calcareous mudstone matrix. Arrow points to *Tubiphytes* fragment. Plane light.

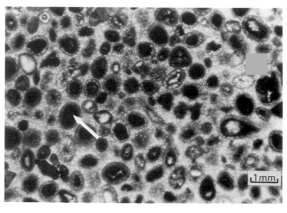

Fig. 19-9. Photomicrograph of *Mizzia*-oolite grainstone with oomoldic porosity (capping oolite grainstone). Arrow points to oolmoldic porosity. Cross-polarized light.

oolites cemented with equant calcite-spar cement (Fig. 19-9). Fossil-moldic, oomoldic, and minor interparticle porosity is commonly present. The *Mizzia*-oolite grainstones appear to represent the capping lime-sand facies of the bank complex (Wilson, 1975).

Geologic Characteristics of the Reservoir

With the exception of minor interparticle porosity preserved in the *Mizzia*-oolite capping grainstones, all of the porosity in the bank complex is secondarily developed by dissolution of skeletal debris, oolites, and/or lime mud matrix in each of the facies just described, except the bank detritus. The porosity zones are thin (<10 ft, ~3 m) and discontinuous and have an average porosity of 8 to 10 percent, calculated from logs, with water saturations ranging from 25 to 40 percent. The development of oomoldic and fossil-moldic porosity in the capping grainstones and the presence of vuggy and fossil-moldic porosity in the *Fistulipora-Tubiphytes* crestal boundstone and phylloid-algal mudstones indicate that, after deposition of the capping grainstones, parts of the bank complex were exposed and subjected to freshwater dissolution. The presence of equant calcite-spar cement that partly fills moldic, vuggy, and interparticle pore space in the reservoir facies, and completely fills interparticle pores in the topographically lower flanking grainstones (Fig. 19-2), suggests a vertical downward transition from vadose to phreatic diagenesis (*i.e.*, a paleo-water table). In addition to the moldic, vuggy and minor interparticle porosity, open vertical fractures are present in places. A summary of the paragenetic sequence is presented in Table 19-1.

Depositional History

A model for Permian Wolfcamp banks in the Permian Basin of west Texas and New Mexico was proposed by Dunham (1969) and Malek-Aslani (1970) as a linear mud bank with a "cap" of *Tubiphytes*-tubular foram boundstone. From detailed studies of Late Paleozoic buildups, Wilson (1975) proposed a general bank model composed of seven depositional facies: (1) bioclastic pile, (2) phylloid-algal mudstone core, (3) *Tubiphytes*-tubular foram crestal boundstone, (4) organic veneer, (5) foraminiferal grainstone-packstone flank beds, (6) bank talus, and (7) dasycladacean-oolitic capping grainstone. Wilson (1975) noted that a single bank complex commonly will not contain all the different facies. Only two of the seven expected facies proposed by Wilson (1975) for Late Paleozoic carbonate buildups are absent from the Wolfcamp-A zone at Blalock Lake East Field. These two facies are the bioclastic pile and organic veneer.

The Wolfcamp-A zone at Blalock Lake East Field is an elongate carbonate buildup that varies in thickness from about 24 to 188 feet ($\sim$7–57 m) (Figs. 19-2, 19-3). Similar Wolfcamp banks are present in the Townsend-Kemnitz Field in Lea County, New Mexico (Dunham, 1969; Malek-

Aslani, 1970), on the margin of the northwestern shelf of the Delaware Basin (Fig. 19-1).

At Blalock Lake East Field, bank growth began with the deposition of phylloid-algal lime muds. These sediments originated in a quiet-water subtidal environment as accumulations probably trapped or baffled by phylloid algae (Elias, 1963; Asquith, 1979). Entrapment of lime mud by phylloid algae has also been suggested as the sediment-accumulation mechanism for the Pennsylvanian banks of north central Texas (Wermund, 1975).

As the phylloid algal mudstone bank grew upward, it entered wave base, at which time the *Fistulipora-Tubiphytes* facies accumulated as wave-resistant crestal boundstones along with flanking facies of algal-foraminiferal grainstones and minor packstones (Fig. 19-2). Unlike the *Tubiphytes*-tubular foram boundstones of Townsend-Kemnitz Field, the encrusting bryozoan *Fistulipora* is common in the crestal boundstones at Blalock Lake East Field (Fig. 19-6). The abundance of *Fistulipora-Tubiphytes* crestal boundstones relative to the phylloid-algal mudstone core suggests that the bank grew into wave base early in its development. The flanking grainstones and minor packstones of the bank represent bank detritus winnowed and redistributed by waves and currents both basinward and shelfward of the bank crest (Fig. 19-2).

Bank detritus (talus) overlies the flanking grainstones and packstones (Fig. 19-2), and consists of poorly sorted debris eroded from the crestal boundstone and flank facies. Deposits of detritus are localized on the basinward side of the bank and extend approximately a mile (0.6 km) into the basin from the bank crest, grading seaward into the basinal facies of thin-bedded, argillaceous, lime mudstone and dark, spicule-bearing, calcareous shales (Fig. 19-3).

Table 19-1. Paragenetic Sequence of Post-Depositional Events and Features

| Deposition | Diagenesis | | |
	Eogenetic	Mesogenetic	Telogenetic
	Vadose leaching and minor equant calcite cementation	Fracturing	Not recognized
	Phreatic equant calcite cementation		
	Minor recrystallization of lime mud matrix to microspar		
Interparticle porosity	Moldic and vuggy porosity	Fracture porosity	

Steeper depositional gradients on the seaward side of the bank (Figs. 19-2, 19-3) probably resulted in slumping or partial slumping of the forebank and bank facies, as suggested for other Permian forebank deposits (Pray and Stehli, 1963; McDaniel and Pray, 1967; and Elam, 1972). Flanking grainstones and packstones accumulated in greater thicknesses shelfward from the bank crest than did the basinward (forebank) flanking grainstones and packstones and the bank detritus facies (Figs. 19-2, 19-3). The bank's center in Blalock Lake East Field, as inferred from thickness relationships, is located in the northwest corner of Section 13 (Fig. 19-3). Progradation of the mudstone core and boundstone facies from the bank center resulted in flank deposits underlying the mudstone core and crestal boundstone facies in the Hanson Production Company Becknell No. 3 well located just southeast of the bank center (Figs. 19-2, 19-3).

The entire bank complex is overlain by *Mizzia*-oolite grainstones that cap the complex and represent shallow-marine deposits (Fig. 19-2). Similar shoaling resulted in capping oolite grainstones overlying the Pennsylvanian phylloid-algal banks of the Aneth Field in the Four Corners region of the Paradox Basin (Peterson and Ohlen, 1963).

Summary

The Wolfcamp-A zone (Permian) is the major reservoir in Blalock Lake East Field, west Texas. This sequence represents a *Fistulipora-Tubiphytes* bank complex that accumulated at the eastern-shelf margin of the Midland Basin. The bank complex, which has a maximum thickness of nearly 190 feet (~58 m), contains most of the facies typically found in late Paleozoic buildups of this region.

Reservoirs are developed in three facies of the Wolfcamp-A zone: the algal-mudstone core facies, the *Fistulipora-Tubiphytes* crestal-boundstone facies, and the *Mizzia*-oolite capping grainstone. Porosity developed mainly by freshwater dissolution of skeletal debris, oolites and/or lime mud during exposure of the bank complex. The hydrocarbon accumulation is stratigraphically controlled and results from an updip (shelfward) loss of porosity due to facies change and shelfward depositional thinning. Dark spicule-bearing lime mudstones and calcareous shales form the overlying seal.

Reservoir development in Permian bank complexes, such as this one in the southwestern United States, is dependent mainly on subtle facies differences, topographic buildup, and exposure to freshwater diagenesis.

References

ASQUITH, G.B., 1979, Subsurface Carbonate Depositional Models—a Concise Review: The Petroleum Publishing Corp., Tulsa, OK, 121 p.

DUNHAM, R.J., 1969, Early vadose silt in Townsend mound (reef), New Mexico: *in* Friedman, G.M., ed., Depositional Environments in Carbonate Rocks—a symposium: Soc. Econ. Paleontologists and Mineralogists, Spec. Publ. 14, p. 139–181.

ELAM, J.G., 1972, The tectonic style in the Permian Basin and its relationship to cyclicity: *in* Elam, J.G. and S. Chuber, eds., Cyclic sedimentation in the Permian Basin—a symposium: West Texas Geol. Soc. Publ. 69-56, p. 65–76.

ELIAS, G.D., 1963, Habitat of Pennsylvanian algal bioherms, Four Corners area: *in* Bass, R.O. and S.L. Sharps, eds., Shelf Carbonates of the Paradox Basin—a symposium: Four Corners Geol. Soc., 4th Field Conf., p. 185–203.

HOROWITZ, A.S., and P.E. POTTER, 1971, Introductory Petrography of Fossils: Springer-Verlag, Inc., New York, 302 p.

MALEK-ASLANI, M., 1970, Lower Wolfcampian reef in Kemnitz Field, Lea County, New Mexico: Amer. Assoc. Petroleum Geologists Bull., v. 54, p. 2317–2335.

MCDANIEL, P.N., and L.C. PRAY, 1967, Bank-to-basin transition in Permian (Leonardian) carbonates, Guadalupe Mountains, Texas (abst.): Amer. Assoc. Petroleum Geologists Bull., v. 51, no. 9, p. 1903.

PETERSON, J.A., and H.R. OHLEN, 1963, Pennsylvanian shelf carbonates, Paradox Basin: *in* Bass, R.O. and S.L. Sharps, eds., Shelf Carbonates of the Paradox Basin—a symposium: Four Corners Geol. Soc., 4th Field Conf., p. 65–79.

PRAY, L.C., and F.G. STEHLI, 1963, Allochthonous origin, Bone Springs "patch reefs" West Texas (abst.): Geol. Soc. America Spec. Paper 73, p. 218–219.

RIGBY, J.K., 1958, Two new upper Paleozoic hydrozoans: Jour. Paleontology, v. 32, p. 583–586.

WERMUND, E.G., 1975, Upper Pennsylvanian limestone banks, North Central Texas: Univ. Texas, Austin, Bur. Econ. Geology Circ. 75-3, 34 p.

WILSON, J.L., 1975, Carbonate Facies in Geologic History: Springer-Verlag, Inc., New York, 439 p.

Mesozoic Reservoirs

20
Qatif Field
Augustus O. Wilson

RESERVOIR SUMMARY

Location & Geologic Setting Coast of Saudi Arabia 100 mi (160 km) north of Qatar, on NE flank of Arabian Plate

Tectonics Low-amplitude anticline related to right-lateral slip along basement fault, in part during Jurassic

Regional Paleosetting Shallow shelf to sabkha

Nature of Trap Essentially structural

Reservoir Rocks
  Age Late Jurassic (Tithonian and Kimmeridgian)
  Stratigraphic unit(s) Arab Formation, C and D members (zones)
  Lithology(s) Limestone and some dolomite
  Dep. Environment(s) Offshore bar, lagoonal and supratidal
  Productive Facies Dasyclad-peloidal packstone and grainstone, ooid-mollusk grainstone (Arab-C); peloidal grainstone and dolomitized grainstone (Arab-D)
  Entrapping Facies Nodular anhydrite and lime mudstone
  Diagenesis/Porosity Extensive dissolution; local marine cements, sabkha(?)
  Petrophysics
    Pore Type(s) Primary interparticle; moldic and intercrystal
    Porosity 12–25%*, 25–31%**, 21–31%#, 5–25%+, 15–26%++
    Permeability 12–100 md*, 80–100 md**, 250–5000 md#, 4–500 md+, 50–500 md#
    Fractures NA

Source Rocks
  Age Jurassic (Oxfordian)
  Lithology(s) Laminated peloidal limestone in Hanifa and Tuwaiq Mtn formations
  Migration Time Late Cretaceous through Tertiary

Reservoir Dimensions
  Depth Arab-C 6900 ft (2100 m) along crest of anticline
  Thickness Arab-C 95–130 ft (29–40 m). Arab-D 70–190 ft (21–58 m)
  Areal Dimensions 27 mi × 3–4.4 mi (44 × 5–7 km)
  Productive Area ~55,000 acres (220 km^2)

Fluid Data
  Saturations NA
  API Gravity 31° Arab-C, 38° Arab-D
  Gas-Oil Ratio 53:1 Arab-C; 138:1 Arab-D
  Other —

Production Data
  Oil and/or Gas in Place NA
  Ultimate Recovery 9.0 million BO (Halbouty *et al.*, 1970)
  Cumulative Production 684 million BO through 1980

Remarks: Discovered in 1946. IP mechanism water drive, 18 wells producing. *Dasyclad-peloid limestone, dolomitized, Arab-C. **Same, undolomitized. #ooid-mollusk grainstone, Arab-C. +fine grainstone, Arab-D. ++Same, dolomitized.

20
Depositional and Diagenetic Facies in the Jurassic Arab-C and -D Reservoirs, Qatif Field, Saudi Arabia

Augustus O. Wilson

Location and Discovery

Qatif Field is a giant oil field that straddles the northeastern Persian Gulf coast of Saudi Arabia (Fig. 20-1). ARAMCO discovered Qatif in early 1946, after surface geology and gravity surveys had indicated the presence of a major north–south anticlinal structure. Qatif is interpreted as a low-amplitude, "banana-shaped" shear fold, convex westward (Fig. 20-2), which was probably formed by right-lateral displacement along basement faults, and possibly enhanced by deep movement of salt from the Cambrian Hormuz Salt. The structure lies on the eastern side of the Arabian Plate, but well to the west of the intense Zagros Mountains fold belt near the plate boundary in Iran.

General Geology and Production Data

Approximately 27 miles long and 3 to 4.4 miles wide (44 × 5–7 km), Qatif contains several Jurassic carbonate reservoirs of which two, the Arab-C and Arab-D members of Upper Jurassic (Tithonian-Kimmeridgian) age, are the most important (Fig. 20-3). The Arab Formation is divided into Arab-A, -B, -C, and -D reservoirs (from top down); the B, C, and D members consist of carbonate reservoir and anhydrite units (Fig. 20-3). The Arab Formation is underlain by the Jubaila Formation and overlain by the Hith Anhydrite. The Arab-C reservoir is about 95 to 130 feet (29–40 m) thick, and the Arab-D reservoir about 70 to 190 feet (21–58 m) thick; the two reservoirs lie, respectively, at depths of about 6900 feet (2100 m) and 7100 feet (2165 m) subsea at the crest of the structure.

As of July 1980, 18 wells were producing 115,000 barrels of oil per day. The average API gravity of produced oil is 39° for the Arab-C and 38° for Arab-D. Cumulative oil production through 1980 was 684 million barrels. The initial and continuing production mechanism is a water drive.

The Qatif Field is an enormous structural trap. Arab-C and -D seals are widespread units of bedded to nodular anhydrite and, in the Arab-C member, also lime mudstone in the middle of the reservoir that separates productive intervals over part of the field. The reservoir rocks in the Arab-C are *Clypeina*-peloidal lime grainstones and packstones and oolitic-molluscan lime grainstones, whereas the principal reservoir rocks of the Arab-D are fine-to-medium grainstones and dolomitized grainstones composed of varying mixtures of foraminifers, peloids, and superficial ooids. Diagenetic features are primarily early, and in many cases, closely controlled by depositional environment.

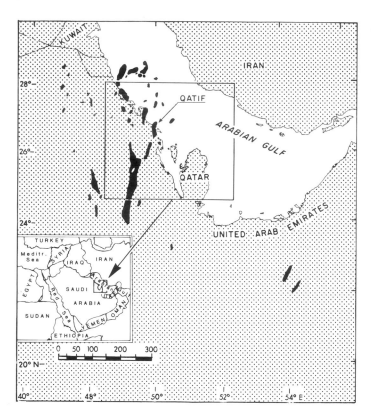

Fig. 20-1. Map showing location of Qatif Field and other fields in Saudi Arabia. Qatif is a north–south-trending anticline straddling the Persian Gulf shoreline.

Pre-Arab Geologic Setting

The Jurassic carbonates contain virtually no terrigeneous material, for the nearest land lay hundreds of miles to the west on the site of the present-day Arabian Shield. Prior to deposition of the Arab Formation, a rather restricted shelf basin existed, which was confined largely to eastern Saudi Arabia; Qatif Field is near the eastern boundary of this basin. In the Qatif area, the pre-Arab sequence in this basin consists of the Jubaila, Hanifa, Tuwaiq Mountain, and part of the Dhrumas Formation. The Fadhili, Hadriya, and Hanifa reservoirs are porous facies that capped the accumulations of tight, deep subtidal basin facies in the Dhruma, Tuwaiq Mountain, and Hanifa formations, respectively. The most likely sources for all the Arab oil are laminated, peloidal carbonates in the upper Tuwaiq Mountain and lower Hanifa, which have up to 12 percent total organic carbon content by weight (Ayres *et al.*, 1982). This sequence is the most likely source for much of the Jurassic oil, for the other intervals are much less organic-rich.

Fig. 20-2. Outline map of Qatif Field showing line of cross-section in Figure 20-5.

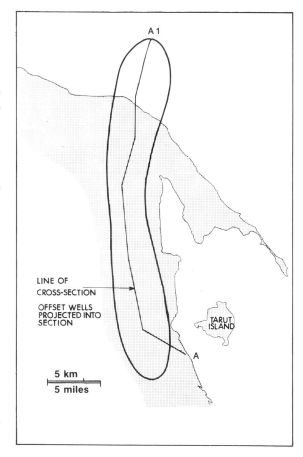

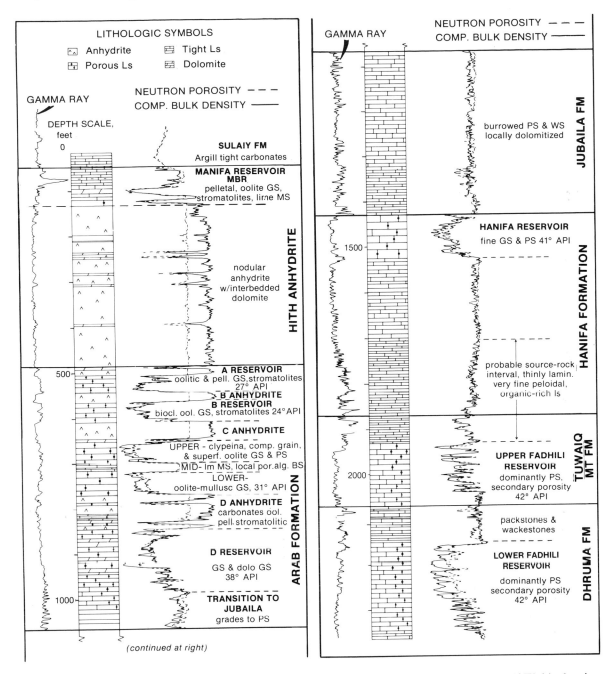

Fig. 20-3. Typical log and lithologic column in the Hith, Arab, and Jubaila formations in Qatif Field, showing porous carbonates facies underlain by tight, restricted-shelf basin Jubaila carbonates or interbedded and overlain by anhydrites (Arab-Hith).

Arab-Hith Sequence

The Arab carbonate-anhydrite cycles and overlying Hith Anhydrite represent the most restricted phases of Jurassic sedimentation. They are the result of progressive basin infilling that culminated in a regressive sabkha complex punctuated by brief episodes of transgressive carbonate cycles. This record reflects the predominance of sedimentation relative to subsidence and eustatics. The Arab-D reservoir facies of high-energy grainstones was the final upward-shoaling, "basin-filling" phase, whereas the Arab-A, -B, and -C reservoirs are transgressive phases.

The depositional environments for each Arab reservoir facies from the Arab-D upward are progressively more restricted, with increasing proportions of lagoonal and intertidal sediments in the upper reservoirs. Wilson (1975, p. 290) noted this to be true regionally for the Arab sequence. Each reservoir facies terminated with the deposition of evaporites (anhydrites), which are inferred to have sealed the underlying carbonates at an early stage of burial, forming a closed system protected from extensive late diagenetic cementation.

Arab-D Reservoir Rocks

Lower Transition

The Arab-D grainstone facies grades upward from the underlying burrowed wackestones and packstones of the Jubaila Formation (Fig. 20-4P). Stylolites, rare in most of the Arab Formation, are common in the transition intervals between cycles. Jubaila fossils are of limited diversity (Fig. 20-5F), which together with the lime-mud-rich character of the sediments indicate that the Jubaila sediments were deposited in deep subtidal environments into which Arab-D grainstones prograded.

Fine Grainstones

The fine grainstones dominant in the Arab-D at Qatif were deposited in current- and tide-swept environments, as indicated by superficial oolites, other rounded grains, good sorting, and cross-bedding (Figs. 20-5C–E). In general, the grains are fine-to-medium size and in the aggregate are commonly so micritized and tightly packed that only careful observation in thin sections at high magnification can distinguish them from packstones and wackestones. The grainstones are thicker along the crest of the Qatif structure, where early structural growth provided shallower, higher-energy environments (Figs. 20-6 and 20-7A). Regionally, the grainstones thicken towards the north end of the field, and to the south are completely replaced by very finely crystalline dolomite, but "ghost" sedimentary structures (laminations, cross-beds, and hardgrounds) similar to those in the undolomitized grainstones suggest that grainstones originally dominated the sequence in the south as well. In well J, shown in Figure 20-6, are lime mudstones that may correspond to a middle lime mudstone interval noted by Powers (1962) in Ghawar and some other Saudi Arabian fields.

Upper Coarse Grainstones

The fine grainstones of the Arab-D shoal upward to coarser grainstones (Fig. 20-5B) 10 to 20 feet (3–6 m) thick, which have bored hardgrounds with intraclasts (Figs. 20-4J and 20-5A) and are only slightly dolomitized. These are shallow subtidal and lagoonal grainstones, as indicated by superficial oolites and dasycladacean algae. At the top of the reservoir, stromatolites and associated lime mudstones occur locally (Fig. 20-4I). The coarse lagoonal grainstone facies are similar to facies comprising the bulk of the Arab-D reservoir and aquifer to the south and west of Qatif (Powers, 1962) and towards Abu Sa'Fah Field to the east.

Arab-D Diagenesis

Cementation

Calcite cements in the Arab-D grainstones are dominantly very thin fringes of isopachous cement with lesser amounts of late blocky calcite spar (small crystals 15–300 μ). The echinoderm fragments have syntaxial calcite overgrowths. Most of the fine grainstones are tightly packed, indicating that the cements provided little resistance to mechanical compaction. The isopachous

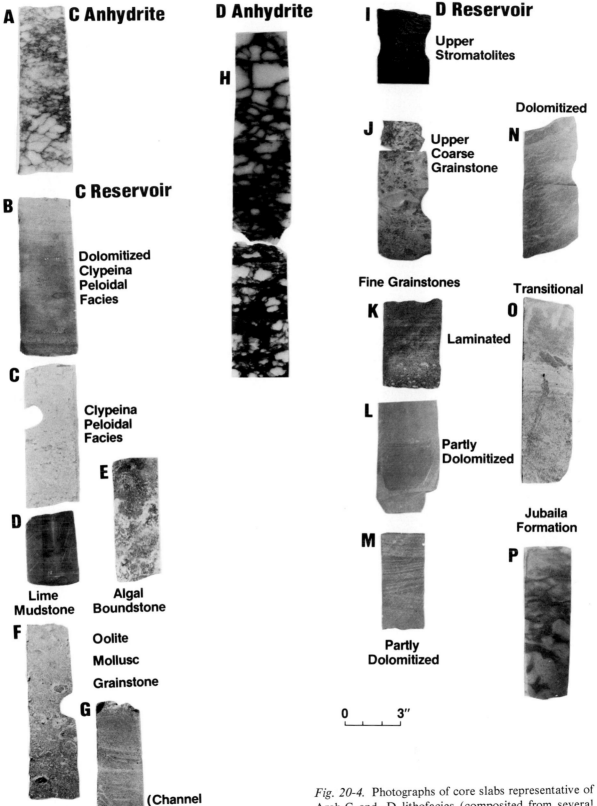

A **C Anhydrite**

D Anhydrite

I **D Reservoir**

Upper
Stromatolites

Dolomitized

C Reservoir

B

Dolomitized
Clypeina
Peloidal
Facies

H

J

Upper
Coarse
Grainstone

N

C

Clypeina
Peloidal
Facies

Fine Grainstones

Transitional

K

O

Laminated

E

D

L

Partly
Dolomitized

Lime
Mudstone

Algal
Boundstone

Jubaila
Formation

M

P

F

Oolite

Mollusc

Grainstone

Partly
Dolomitized

0   3"

G

(Channel
Fill)

Fig. 20-4. Photographs of core slabs representative of Arab-C and -D lithofacies (composited from several wells).

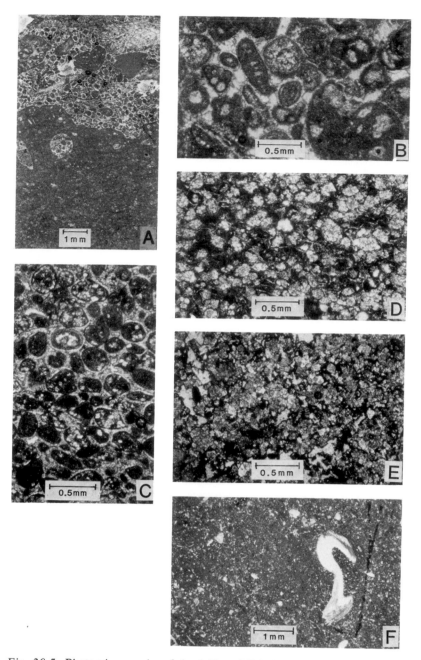

Fig. 20-5. Photomicrographs of Arab-D and Jubaila carbonate rock cores (all photos in cross-polarized light, except *E*). *A*. Bored surface, Arab-D upper coarse grainstone facies. Bored surface is overlain by intraclasts and rounded bioclastics; borings are filled with grains. Close inspection reveals micritized grain texture of underlying rock. Porosity 24%, permeability 107 md. *B*. Typical Arab-D coarse grainstone. Forams, superficial oolites, and mollusks are cemented by thin, isopachous cement. Grains are micritized and some are partly dolomitized. Porosity 26%, permeability 173 md. *C, D,* and *E*. Arab-D fine grainstones, partly to completely dolomitized. Rounded grains, mostly superficial oolites and forams were cemented initially by isopachous cement. Photo D shows less cement than photo C and reveals greater compaction, as evidenced by more grain contacts. Grains in photo D are predominantly dolomitized; those in E are completely dolomitized. Both have high porosity, but there is a progressive increase in permeability with dolomitization: (*C*) Porosity 29%, permeability 4 md. (*D*) Porosity 20%, permeability 10 md. (*E*) Porosity 20%, permeability 252 md. *F*. Jubaila packstone. Brachiopod fragment and fine bioclastic debris (rich in echinoderms) typical of the Jubaila.

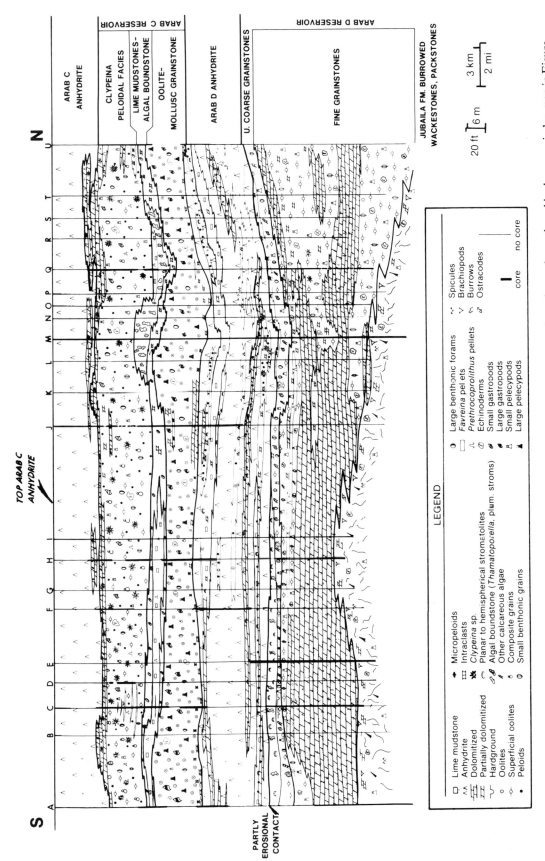

Fig. 20-6. Cross-section showing the Arab-C and -D facies sequence. Line of section is approximately north-south and is shown on index map in Figure 20-2.

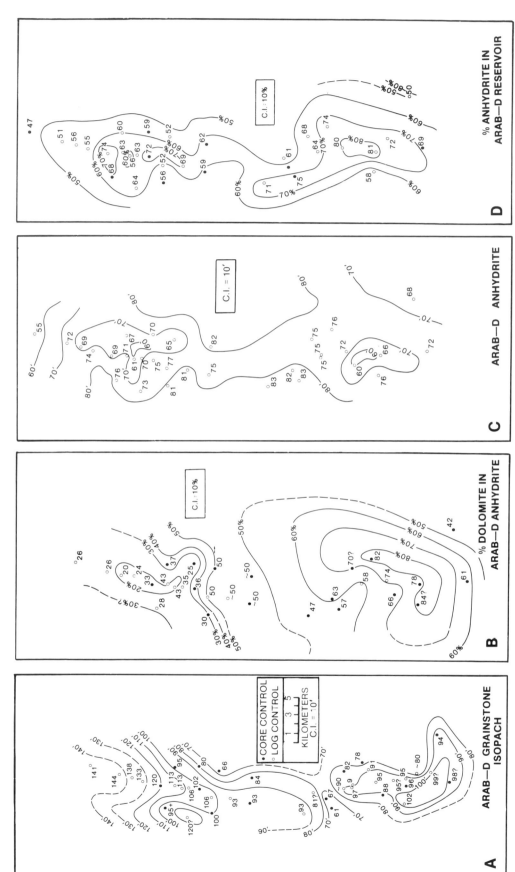

Fig. 20-7. Isopach and percentage maps of the lithologies comprising Arab-D reservoir facies. *A.* Isopach map of the Arab-D grainstone reservoir facies, showing isopach on-structure thickening in the south. This indicates that early structural growth influenced deposition. The facies also thickens northward in the field. *B.* Percentage of completely dolomitized rock in the Arab-D reservoir. Dolomite also increases on-structure in the south but decreases northward. The percentage of anhydrite beds in the Arab-D anhydrite has a similar pattern (as in Fig. 20-7D). *C.* Isopach map of the total Arab-D anhydrite interval, showing on-structure thinning. This may be due to erosion and/or structural growth during deposition. Compare to Fig. 20-7D. *D.* Percentage of "pure" anhydrite beds in the Arab-D anhydrite, showing on-structure increase, which indicates that early structural growth provided more intense sabkha conditions on-structure.

cements were precipitated early, because they do not fill molds of leached mollusk shells and are probably marine-phreatic.

Dolomite: A Diagenetic Facies

The Arab-D reservoir contains a spectrum of partly to completely dolomitized grainstones. The dolomites are a diagenetic overprint on a depositional facies and formed preferentially in compacted fine grainstones in which the grains were mostly micritized. A further control was proximity to evaporitic conditions in the overlying anhydrite sabkhas. Both the percentage of completely dolomitized rock (25–95%) in the Arab-D reservoir and the percentage of anhydrite (47–81%) in the Arab-D anhydrite increase on-structure, but decrease northward in the Qatif Field (Figs. 20-6, 20-7B, D).

A sabkha-derived brine origin for these Arab-D dolomites is supported by geochemical data. Two samples of these dolomitized fine grainstones yielded $\delta^{18}O$ of -3.09 and $-1.87‰$, and $\delta^{13}C$ values of $+2.95$ and $+3.01‰$ (PDB standard). These values are similar to those in two samples of dolomite taken from sites between sabkha anhydrite nodules in the Arab-C anhydrite, which yielded $\delta^{18}O$ values of -2.53 and $-2.13‰$ and $\delta^{13}C$ values of $+2.76$ and $+2.61‰$ (PDB standard). The two dolomitized fine grainstones analyzed consist of stoichiometric dolomite with 50.7 and 49.7 mole percent $CaCO_3$, further supporting the inference of an evaporite-derived brine origin (Lumsden and Chimahusky, 1980). (These data are preliminary results of an unpublished study by J. R. Allan, Chevron Oil Field Research Co.)

The dolomitizing brines apparently soaked through the grainstones and formed brine lenses in the fine grainstones atop the tight Jubaila sediments along porosity "channels" in the reservoir. These brines were trapped in the sediment as deposition of the Arab sequence continued; hence, the dolomitization probably continued long after burial.

Grain-selective dolomitization in Qatif resulted in the best permeability, for it retained and enhanced interparticle porosity. As illustrated in Figures 20-5C–E, dolomitization increased permeability in three Arab-D grainstones with similar porosities. Most dolomites with less than 10 millidarcys permeability occur lower in the reservoir and have relict sedimentary structures similar to those of laterally equivalent rocks with low porosity; dolomite replaced these rocks without significantly increasing the porosity.

Anhydrite

Tabular crystals of anhydrite and less commonly anhydrite nodules formed during early and late diagenesis throughout the Arab-D reservoir as well as all the Arab carbonates. Crystals replace the rock fabric or fill pore space and increase in abundance near beds of nodular anhydrite.

Arab-D Anhydrite

The Arab-D anhydrite, overlying the reservoir, grades to marine carbonates east and north of the Qatif Field, and reflecting this trend, the anhydrite interval within the field itself has 20 to 50 percent interbedded carbonate (Figs. 20-6, 20-7D). The total Arab-D anhydrite interval thins to the north, and also on-structure due to contemporaneous structural growth. The percentage of anhydrite within the interval also decreases northward, but increases on-structure, suggesting that structural growth caused more evaporitic conditions. Regionally, the top of the Arab-D anhydrite is a time surface, but locally it is at least partly erosional, for example in wells O and P (Fig. 20-6).

The anhydrites in both the Arab-D and Arab-C members represent in fact a diagenetic facies that originated beneath flat sabkha surfaces behind prograding shorelines. These Jurassic sabkhas apparently were similar to those containing anhydrite along the coast of the present-day Gulf, but were different in that they lacked terrigenous sediment and consisted entirely of prograding carbonates and evaporites. Figure 20-8 depicts late Arab-D reservoir and early Arab-D anhydrite sedimentation.

The sulfate nodules, perhaps gypsum initially, grew from interstitial brines formed by influx of sea water across and through the sabkha sediments. The nodules both displaced and replaced pre-existing carbonate sediment, which com-

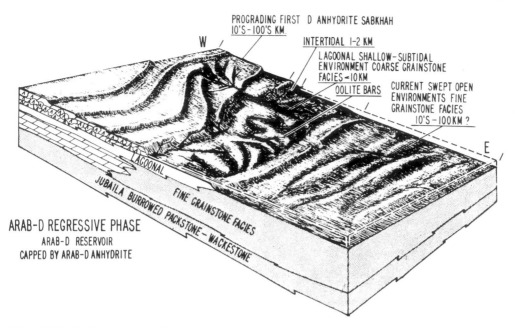

Fig. 20-8. Block diagram of Arab-D regressive phase. Near the close of Arab-D reservoir deposition, lagoonal carbonate facies and sabkha anhydrites prograded over former current-swept environments of the fine grainstone facies.

monly is preserved between nodules. Complete sequences of this process exist, as in the Arab-D anhydrite core in Figure 20-4.

Carbonates interbedded with the nodular anhydrite are lagoonal, "strandline", and/or intertidal deposits. They are commonly dolomitized. Those with *Prethrocoprolithus* pellets and *Clypeina* dasycladacean algae are interpreted to be lagoonal, whereas *Favreina* pellets are probably characteristic of the shorefaces of low intertidal flats (Fig. 20-9A). Some are stromatolitic and are probably peritidal. Others are oolitic, lagoon-mouth channel deposits (Fig. 20-9B). Intertidal ostracod-bearing lime mudstones also are common (Fig. 20-9C).

Arab-C Reservoir Rocks

The Arab-C reservoir is subdivided into three depositional facies, which comprise a transgressive-regressive-transgressive sequence capped by a regressive anhydrite (Arab-C). The lower facies is oolitic-molluscan grainstone, which is the best reservoir lithology in the Arab-C and -D. The middle facies consists of tight lime mudstones and porous algal boundstones; the tight mudstones may be a permeability barrier over much of the field, except where fractured. The upper facies is composed of *Clypeina* algal and peloidal packstone and grainstone. Each of these facies also has unique diagenetic characteristics that are a

Fig. 20-9. Photomicrographs of oolitic-molluscan grainstone from the lower Arab-C reservoir and carbonates from the Arab-D Anhydrite Member. All photos in cross-polarized light. A. Carbonate grainstone within the Arab-D anhydrite, comprised of *Favreina* pellets, oolites, superficial oolites, coated intraclast, and leached mollusks, all with isopachous cement. Porosity 19%, permeability 177 md. B. Oolite grainstone bed in Arab-D anhydrite. Well-preserved oolites and mollusk fragments are bonded together by isopachous cement. Spalling of cement and interpenetration of grains occurred during compaction. Porosity 14%, permeability 7 md. (Porosity and permeability low due to anhydrite in the core plug, which is not shown in the photo.) C. A lime mudstone preserved between felty anhydrite nodules and composed in part of ostracodes. The nodules are bordered by blocky anhydrite crystals. Location: Nodular anhydrite interval in Arab-D anhydrite. Porosity 4.5%, permeability 0 md. D. An oolitic-molluscan grainstone that has

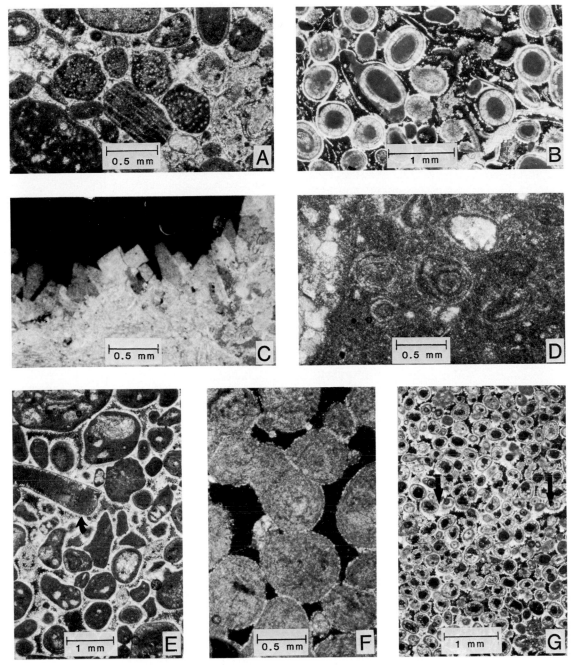

been partly micritized from the upper transition of the lower Arab-C reservoir. Blocky anhydrite crystals are on the left. Porosity 25%, permeability 103 md. *E*. Arab-C oolitic-molluscan grainstone exhibiting isopachous cements. The facies contains oolites, forams, coated intraclasts, and leached mollusk fragments. There is no isopachous cement in the leached shell (arrow), indicating early cementation, before leaching of the aragonitic shell. Porosity 21%, permeability 200 md. *F*. Arab-C oolitic-molluscan grainstone facies: sparsely cemented zone. Micritization obscured oolitic laminations. Oolites are held together more by compaction and grain interpenetration than by the very sparse cement. A few crystals of late, equant spar cement are present. Porosity 20%, permeability 921 md. *G*. Oolitic-molluscan grainstone from the basal transition zone of the lower Arab-C reservoir. Partly leached oolites are seen with meniscus and microstalactitic cements (arrows) of possible vadose origin, suggesting that this grainstone is beachrock. Porosity 12%, permeability 0.1 md. Core analysis sample probably anhydrite plugged.

product of processes closely associated with their different depositional environments.

Lower Reservoir—Oolite-Mollusk Grainstone

Depositional Phases

The facies of the lower Arab-C accumulated during three major phases: a rapid initial transgression, a main transgressive phase, and a regressive phase (Fig. 20-10A–C).

1. *Initial Trangression*: Oolitic beachrocks and intertidal sediments were formed on an eroded and reworked sabkha surface and related channels. Cut-and-fill structures (Fig. 20-4G), intraclastic pebbles, and vadose-cemented grainstone with leached oolites (Fig. 20-9G) are typical of these initial transgressive sediments. Locally, they are dolomitized.

2. *Main Transgressive Phase*: This phase is represented by a thin oolite with few shells progressing upward and "offshore" into oolite with increasing numbers of large gastropods and pelecypods. Echinoderms, foraminifers, and dasycladacean algae are also present. A

sequence similar to this one is forming today along the beaches of the Gulf.

3. *Regression*: Oolites continue to predominate in the early stage of the regressive phase.

Depositional strike in the Qatif area was roughly north–south, parallel to present-day structure, and consequently the oolitic-molluscan facies thickens eastward (Fig. 20-11A). The cross section in Figure 20-6 shows that the oolitic-molluscan grainstone interval is fairly uniform in thickness, except where its top interfingers with the overlying lime mudstone (wells Q and J). The east–west-trending isopach thicks produced by this interfingering are probably "tidal deltas" that may have formed in channels between regressive tongues (Fig. 20-10B). The early Qatif structure may have partly blocked the more restricted western environments at this time from the more marine environments eastward, and thus provided a "barrier" across which these channels developed.

Diagenesis

Four diagenetic zones can be identified in the oolitic-molluscan grainstone. These zones extend across the field, although they require core control for identification and correlation.

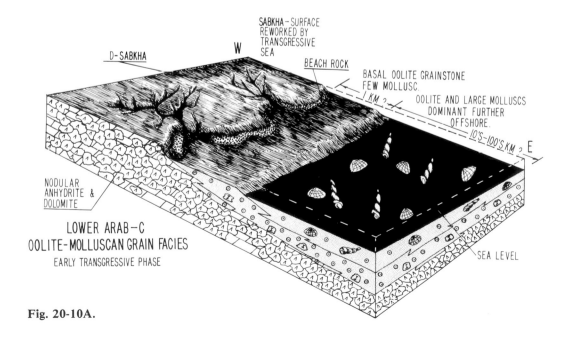

Fig. 20-10A.

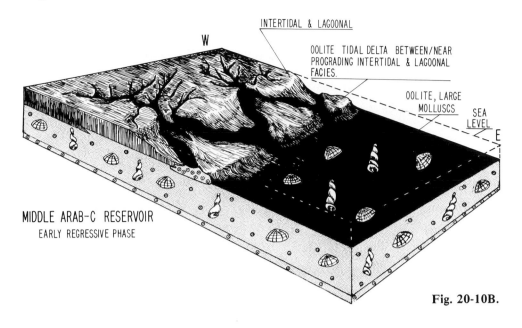

MIDDLE ARAB-C RESERVOIR
EARLY REGRESSIVE PHASE

Fig. 20-10B.

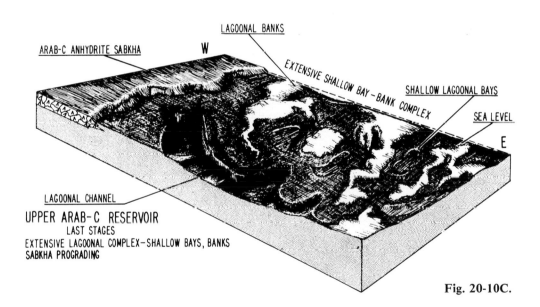

UPPER ARAB-C RESERVOIR
LAST STAGES
EXTENSIVE LAGOONAL COMPLEX—SHALLOW BAYS, BANKS
SABKHA PROGRADING

Fig. 20-10C.

Fig. 20-10. Block Diagrams—Arab-C Developmental Sequence *A*. The lower Arab-C oolitic-molluscan grainstone facies transgressed over the Arab-D anhydrite sabkhas. *B*. The middle Arab-C reservoir regressive phase formed oolite tidal deltas between tongues of prograding intertidal lime mud. *C*. The upper Arab-C Clypeina-peloidal facies was deposited in a lagoonal bay-bank complex, in which sharp lateral and vertical facies changes plus local exposure formed varied lithologies and diagenetic types.

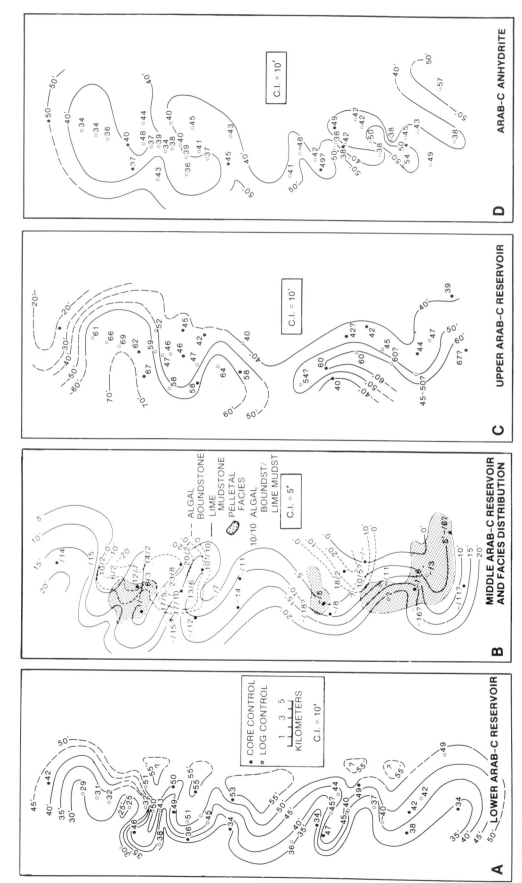

Fig. 20-11. Isopach maps of Arab-C reservoir and facies distribution. A. Isopach map of oolitic-molluscan grainstone facies (lower Arab-C), showing east-west-oriented thins and thicks, which reflect development of oolite tidal deltas during late stages of deposition of this facies. B. Isopach map of the middle Arab-D reservoir facies. The impermeable lime mudstone (solid lines) thickens westward. Algal boundstones formed in two shallow subtidal embayments (dashed lines) and locally, there were *Favreina* and *Prethrocoprolithus* pelletal facies (shades). C. *Clypeina*-peloidal facies isopach (upper Arab-C reservoir) showing a slight tendency towards east–west-oriented thins and thicks. D. Isopach map of the Arab-C anhydrite, showing a pattern like that in C.

1. *Basal Zone*: CaCO$_3$ cement was deposited in beachrocks, and dissolution developed some oomoldic porosity (Fig. 20-9G). This is interpreted as a vadose event of the early transgressive phase. Locally, the grainstones are dolomitized.

2. *Sparsely Cemented Zone*: Within the oolite-dominated beds in the lower parts of the reservoir, there is very little cement and the rock is quite friable (Fig. 20-9F). Packing is moderate, and interparticle porosity reaches 28 percent and permeability up to 5000 millidarcys.The grains are micritized, and the oolitic laminae are barely visible. A hydrographic regime of nearly constant agitation is required to prohibit early cementation, as noted in modern analogs by Shinn (1969) and Gebelein (1978). Some early syntaxial calcite cement was precipitated on echinoid plates, and some later blocky calcite cement was precipitated (as in other zones), but the effects of these local cements on reservoir properties appear small.

3. *Isopachous Cement Zone*: This zone occurs in the portion of the main transgressive phase, where large shells are mixed with oolites. Isopachous cements occur as uniform layers around grains but are not found inside leached-molluscan molds (Fig. 20-9E). The degree of isopachous cementation varies, ranging from only a thin lining of pore spaces to a coalescive cement entirely filling some pore spaces. The isopachous cements are interpreted as marine-phreatic, and appear to have been formed in closely spaced submarine hardgrounds similar to those described in the Bahamas (Dravis, 1979). A lower-energy hydrographic regime with slower sedimentation prevailed in these environments than in the sparsely cemented zone below.

4. *Upper Transitional Zone*: Grainstones with isopachous calcite cements are typical of this zone, but were locally micritized (Fig. 20-9D) or dolomitized. Some beds were leached until little structural competence remained, and subsequently were compacted with fabrics like those in the upper Arab C. The dolomitization and dissolution were later events controlled by local emergence associated with the overlying regressive facies.

The combination of high interparticle and molluscan-moldic porosities and sparse cementation makes this oolitic-molluscan grainstone facies the best reservoir. The highest porosities tend to occur in the isopachous cement zones, but the highest permeabilities are found in the sparsely cemented zone, even though packing is tighter there. Permeabilities are so much higher in the sparsely cemented zone that in some producing wells most of the flow comes from this zone.

Middle Reservoir—Lime Mudstone and Algal Boundstone

Burrowed, ostracode-bearing lime mudstones, *Favreina* and *Prethrocoprolithus* pelletal grainstones and packstones, and algal boundstones characterize this facies. The algal boundstones are a combination of *Thamatoporella* dasycladaceans and an unidentified alga, both of which occur in growth position, and stromatolitic binding, which resulted in plumose stromatolites (Figs. 20-4E, 20-12A). In the vicinity of well M (Fig. 20-6), these algal structures formed a small "mound". Regressive lagoonal and intertidal environments are envisioned.

Differential cementation is typical within the algal boundstones. As a result, stromatolitic portions are relatively tight, whereas areas between stromatolite "fingers" and within *Thamatoporella* growth frameworks, porosity and permeability are commonly quite high.

In contrast, the lime mudstones are partially dolomitized, have low-amplitude stylolites, and are tight. The algal boundstones formed in two lagoonal embayments on the east side of the field, whereas the lime mudstones were more prevalent to the west (Fig. 20-11B). The pelletal facies are more local. Laterally, these middle Arab-C facies grade to sabkha anhydrites south of the Qatif Field. Where present, the lime mudstones probably serve as a vertical permeability barrier separating the upper and lower Arab-C reservoirs.

Upper Reservoir—*Clypeina*-Peloidal Carbonate Sands

Depositional Facies

The upper Arab-C reservoir is a complex of carbonate sands rich in fragments of the dasyclada-

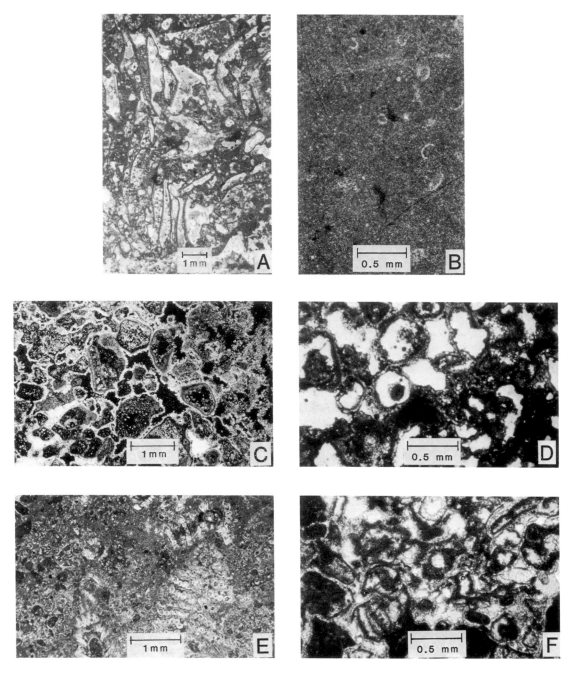

Fig. 20-12. Photomicrographs of middle and upper Arab-C carbonate rock cores. B, C, E, and F in cross-polarized light. *A. Thamatoporella* algal boundstone, middle Arab-C reservoir. The *Thamatoporella* algae are in growth position with "fingers" infilled by fine pelletal and bioclastic grains. Porosity 16%, permeability 459 md. *B.* Ostracod-bearing lime mudstone, middle Arab-C reservoir. Porosity 1.4%, permeability 0 md. *C.* Dolomitized peloids, top upper Arab-C reservoir. Some are oolitically coated and have dolomitized isopachous cement. Porosity 15%, permeability 7 md. *D.* Dolomitized and leached, poorly sorted grainstone, top of upper Arab-C reservoir. The dolomitized cement is all that remains in much of this thin section. Porosity 23%, permeability 100 md. *E.* Dasyclad-*Clypeina* sp. in low mud packstone "matrix" of very fine to medium peloids. upper Arab-C reservoir. Porosity 25%, permeability 83 md. *F.* Leached peloids, algal, and molluscan debris in upper Arab-C reservoir. Porosity 32%, permeability 424 md.

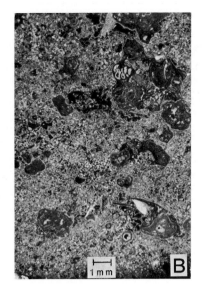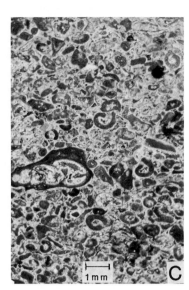

Fig. 20-13. Photomicrographs of upper Arab-C *Clypeina*-peloidal facies in carbonate rock cores. All photos in cross-polarized light. *A*. Squashed peloidal fabric, showing compacted grains and isopachous cement. Prior to leaching and compaction, grains were cemented by isopachous cement. Were these leached grains or pellets still soft when compacted? Porosity 20%, permeability 19 md. *B*. Composite grains of forams and mollusks in a highly altered, "chalkified" matrix. It is not known whether the rock was originally a packstone or poorly sorted grainstone. Thin section was impregnated with blue epoxy, which reveals a high matrix microporosity. Porosity 27%, permeability 108 md. *C*. High-spired small gastropod (with "composite-grain" coating) in *Prethrocoprolithus* pelletoidal grainstone. Such small gastropods are typical of the upper Arab-C reservoir. Porosity 26%, permeability 26 md.

cean alga *Clypeina*. These sands were deposited in a final transgressive phase before progradation of the Arab-C anhydrite sabkhas. The transgressive phase sequence is interpreted as having been initiated by shallow lagoonal deposition. The rocks of this facies are dominantly poorly sorted grainstones, but also include "low-mud" packstones and thin ostracod-bearing lime mudstones, particularly near well C (Fig. 20-6). Variations in micrite content, grain sorting, roundness, and the lagoonal aspects of the grains suggest a depositional setting of shallow bay and bank environments (Figs. 20-10C, 20-12, and 20-13).

This facies is thickest on the west side of the field (Fig. 20-11C). Just as with the oolitic-molluscan facies, an east–west trend is apparent, reflecting variations in thickness of underlying facies and thus depositional topography.

Diagenesis

Diagenetic facies zones in the upper Arab-C reservoir cannot be separated, like those in the lower Arab-C reservoir. The upper reservoir underwent more extensive early diagenesis than the lower. The diagenesis includes isopachous cementation (Fig. 20-13C, D), dissolution, micritization and "chalkification" of grains, compaction, dolomitization, and replacement and/or cementation by anhydrite. Early marine cementation, local emergence, and circulation of brines from the encroaching Arab-C anhydrite sabkha controlled the diagenetic changes.

Much of the upper Arab-C reservoir is highly leached (Fig. 20-12F). Micrite in packstones has been recrystallized and finely leached, and the fine grains in poorly sorted grainstones have been micritized, making distinction between packstones and grainstones difficult. The resulting "matrix" fabric (Fig. 20-13B) has a very fine, relatively ineffective "chalky" microporosity, which is visible with intense impregnation by blue epoxy.

Figure 20-13A shows a squashed peloidal fabric in which isopachous-cemented, peloidal grains have been crushed by compaction. The grains either were soft pellets initially or were leached until the structural integrity of the fabric was destroyed, allowing compaction.

The top 3 to 10 feet (1–3 m) of the upper Arab-C reservoir was replaced by very finely crystalline dolomite, which delicately preserved pre-existing depositional and diagenetic textures (Figs. 20-12C, D). Dolomitizing solutions derived from the Arab-C anhydrite sabkhas were apparently confined to the top of the reservoir by a micritic phase that occurs just below the dolomite over much of the field.

Blocky anhydrite crystals of both void-filling and replacement origins occur throughout the upper Arab-C reservoir, but are more common near the overlying nodular anhydrites.

Arab-C Anhydrite

The Arab-C anhydrite is similar to the Arab-D anhydrite, but is a purer anhydrite sequence. The isopach map in Fig. 20-11D shows northwest–southwest-oriented thins and thicks, which may reflect original depositional trends or may be due to erosion and dissolution during the Arab-B reservoir transgression.

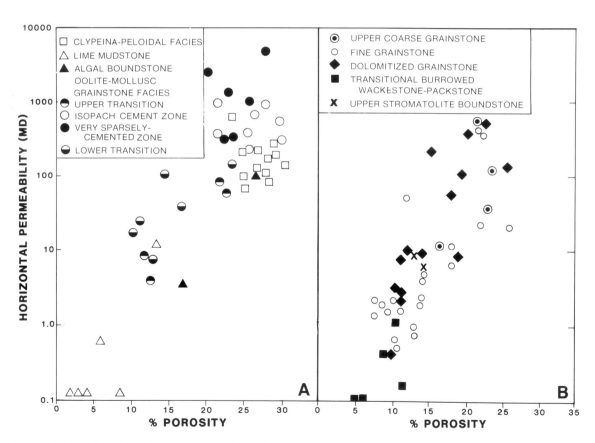

Fig. 20-14. Semilog plots of average porosities and permeabilities. *A*. Arab-C reservoir facies in several wells, showing the isopachously cemented and sparsely cemented zones of the oolitic-molluscan grainstone facies to have the best reservoir quality. *B*. Arab-D reservoir facies in several wells. Grain-selective dolomitization in the fine grainstones comprising much of the Arab-C reservoir improves permeability. Most dolomitized fine grainstones are more permeable than undolomitized grainstones with similar porosities. Grainstone facies intervals (both undolomitized and dolomitized) with lower average porosities and permeabilities were more compacted or are lower in the reservoir and near the transition.

Comparison of Porosity and Permeability in Arab-C and Arab-D Facies

Plots in Figure 20-14A of average core plug porosities and permeabilities clearly show the oolitic-molluscan facies of the Arab-C, with isopachous and very sparse cements to be the most porous and permeable. Porosities in the upper Arab-C *lagoonal facies* tend to be slightly higher, but are associated with substantially lower permeabilities because the upper reservoir has more ineffective secondary porosity, in contrast to the effective primary porosities of the lower oolitic-molluscan grainstones. This ineffective secondary porosity is usually oil-saturated, but is relatively less productive.

In the *oolitic-molluscan facies*, porosities tend to be higher in the isopachous cement zone that resisted compaction, but permeabilities are higher in the barely cemented zone because the interparticle porosity is not significantly reduced or complicated by cement. The upper and lower transitions in this facies have lower porosities and permeabilities, mainly due to diagenetic reduction of porosity by greater cementation, by micritization of the grainstone fabric, by partial dolomitization, and by void-filling with anhydrite. The Arab-C *middle lime mudstones* are quite tight, whereas the *algal boundstones* are mostly highly porous and permeable.

In contrast, porosities and permeabilities are generally lower in the Qatif Arab-D (Fig. 20-14B). Only a few of the reservoir rocks have porosities and permeabilities above 20 percent and 100 millidarcys. The general tendency in the Arab-D reservoir for dolomitized grainstones to be somewhat more porous and substantially more permeable compared with texturally similar but undolomitized grainstones can be seen clearly in Figure 20-14B.

Summary

The Jurassic Arab-C and -D reservoirs in Qatif Field (Saudi Arabia) are carbonate members of the Arab-Hith carbonate-anhydrite cycles. The Arab-D reservoir represents the final shoaling-upward grainstone phase, which filled a restricted shelf basin. It is extensively dolomitized in the field area. The Arab-C reservoir is a mainly transgressive carbonate phase between the Arab-C and -D anhydrite units; it consists, however, of a lower porous oolitic-molluscan grainstone facies separated by a middle, tight regressive lime mudstone facies from an upper porous *Clypeina*-peloidal facies.

The Arab-C and -D nodular anhydrites, which formed from interstitial brines in broad sabkhas, capped the two carbonate reservoirs and thereby protected them from substantial late diagenesis. Much of the diagenesis that affected the reservoir rocks, particularly in the Arab-C, was early and related to depositional environments. High porosities and permeabilities in the reservoirs are due to preserved primary porosity or early diagenetic secondary porosity.

Acknowledgments This paper is published under the auspices of the Saudi Arabian Ministry of Petroleum and Minerals and ARAMCO. My thanks to S. D. Bowers of ARAMCO and Dr. Abdullah Shemlaan of the Ministry for review of the manuscript and to many colleagues in ARAMCO, Exxon Production Research Company, and Chevron Oil Field Research Company for fruitful discussion of Saudi Arabian carbonates.

References

AYRES, M.G., M. BILAL, R.W. JONES, L.W. SLENTZ, M. TARTIR, and A.O. WILSON, 1982, Hydrocarbon habitat in main producing areas—Saudi Arabia: Amer. Assoc. Petroleum Geologists Bull., v. 66, p. 1–9.

DRAVIS, J., 1979, Rapid and widespread generation of Recent oolitic hardgrounds on a high energy Bahamian platform, Eleuthera Bank, Bahamas: Jour. Sedimentary Petrology, v. 49, p. 195–208.

GEBELEIN, C.D., 1978, Predictive models for carbonate-facies distribution (abst.): Amer. Assoc. Petroleum Geologists Bull., v. 62, p. 516.

HALBOUTY, M.T., A.A. MEYERHOFF, R.E. KING, R.H. DOTT, SR., H.D. KLEMME, and T. SHABAD, 1970, World's giant oil and gas fields, geologic factors affecting their formation, and basin classification—Part I, Giant oil and gas fields: *in* Halbouty, M.T., ed., Geology of Giant Petroleum Fields: Amer. Assoc. Petroleum Geologists Mem. 14, p. 502–528.

LUMSDEN, D.J., and J.S. CHIMAHUSKY, 1980, Relationship between dolomite nonstoichimetry and carbonate facies parameters: *In* Zenger, D. H., J. B. Dunham, and R. L. Ethington, eds., Concepts and Models of Dolomitization—a symposium: Soc. Econ. Paleontologists and Mineralogists, Spec. Pub. 28, p. 123–138.

POWERS, R.W., 1962, The Arabian Upper Jurassic carbonate reservoir rocks: *In* Ham, W. E., ed., Classification of Carbonate Rocks—a symposium: Amer. Assoc. Petroleum Geologists Mem. 1, p. 122–192.

SHINN, E.A., 1969, Submarine lithification of Holocene carbonate sediments in the Persian Gulf: Sedimentology, v. 12, p. 109–144.

WILSON, J.L., 1975, Carbonate Facies in Geologic History: Springer-Verlag, Inc., New York, 439 p.

21
Coulommes Field
Bruce H. Purser

RESERVOIR SUMMARY

Location & Geologic Setting Central Paris Basin 20 mi (35 km) east of Paris, France

Tectonics Weak local doming over basement highs, in broad shallow structural basin (late Tertiary)

Regional Paleosetting Shallow-marine shelf (Burgundy Platform)

Nature of Trap Structural with stratigraphic assist; porosity pinchouts on low dome

Reservoir Rocks
  Age Middle Jurassic (Callovian and Bathonian)
  Stratigraphic Units(s) Calcaire de Comblanchien* and Dalle Nacrée**
  Lithology(s) Limestone
  Dep. Environment(s) Grainstone bars and lagoonal muds in shoaling-upward cycles
  Productive Facies Dedolomite (dolomite-mold limestone of diagenetic origin)
  Entrapping Facies Pellet mudstone, packstone and grainstone*; cemented ooid grainstone**
  Diagenesis Porosity Early marine cement and dolomitization; later meteoric-water alteration of dolomite to diagenetic limestone; later burial cementation cementation

  Petrophysics
    Pore Type(s) Rhombic molds, interparticle
    Porosity 5–30%, 15% avg
    Permeability NA
    Fractures Moderate to strong

Source Rocks
  Age Liassic
  Lithology(s) NA
  Migration Time Late Tertiary (?)

Reservoir Dimensions
  Depth 5600 ft (1700 m)
  Thickness 4 lenticular zones, avg 10 ft (3 m)
  Areal Dimensions 3750 acres (15 km^2)
  Productive Area NA

Fluid Data
  Saturations avg S_w = 60%
  Gas-Oil Ratio NA
  API Gravity 32°
  Other —

Production Data
  Oil and/or Gas in Place 59 million BO
  Ultimate Recovery NA
  Cumulative Production 12.94 million BO to July 1, 1983*

Remarks: IP mechanism water drive. Discovered 1958. *Oil and Gas Journal, Dec. 26, 1983, p. 94.

21
Dedolomite Porosity and Reservoir Properties of Middle Jurassic Carbonates in the Paris Basin, France

Bruce H. Purser

Introduction

The studies of dedolomite porosity reported here are based on the Middle Jurassic reservoirs of Coulommes oil field situated near the center of the Paris Basin (Fig. 21-1). However, because insufficient reservoir and production data are available from Coulommes Field, this contribution does not constitute a fully documented reservoir "case study". It is concerned mainly with the petrographic properties and genesis of a particular type of porosity common in Jurassic reservoirs in Coulommes and other Paris Basin fields—porosity resulting from "dedolomitization". Although "dedolomite porosity" is common in the Coulommes Field reservoirs, it is *not* the only pore type present. Its importance with respect to the other pore types present cannot be established exactly from available data, but is believed to be considerable and locally predominant.

Because of limited subsurface data, study of dedolomite fabrics has been extended to outcrops where comparable phenomena are well exposed. The combination of these data should be of help in identifying this pore type in reservoirs elsewhere.

General Geologic Characteristics of Coulommes and Other Middle Jurassic Fields of the Paris Basin

Location and Discovery

Three small oil fields, Coulommes, Chailly-en-Brie and Saint Martin de Bossenay, located in the central part of the Paris Basin (Fig. 21-1), were discovered during the late 1950s. Initial total in-place reserves at Coulommes, the largest field, were approximately 59 million barrels, of which about 12 million barrels had been produced through 1982. Coulommes is located on a weakly defined north-trending dome (Fig. 21-2) with a closure of about 83 feet (25 m). The reservoirs in Coulommes are undersaturated and fractured, resulting in a rapid decline during the early years of production, which is currently only about 100 barrels per day. Pressures remain constant, and the reservoir units are connected via fractures. Average depth of the producing horizons is about 5600 feet (1700 m).

Geology

Coulommes Field, in common with the other fields of the Paris Basin, produces from a series of four intervals located within two distinct Middle Jurassic formations:

Zones A and B, averaging 16 and 6 feet (5 and 2 m) in thickness, occur within an oolitic grainstone unit called the "Dalle nacrée."
Zones C and D, each averaging 10 feet (3 m) in thickness, are located within a muddy pelletal sequence, the "Calcaire de Comblanchien" (Fig. 21-3).

All four reservoir zones vary in thickness across the field, whose average areal extent is about 5.8 square miles (15 km^2). The upper three zones are lenticular and pass laterally into dense limestone on the flanks of the structure. This lenticularity, together with local thinning of the

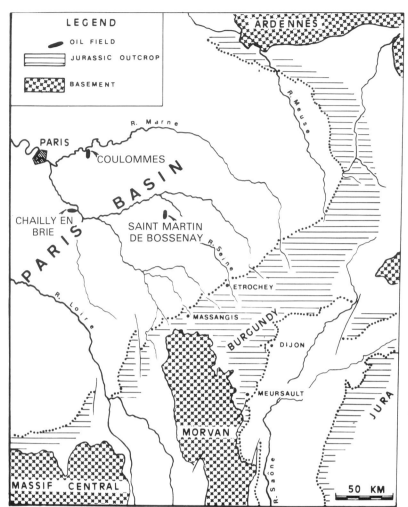

Fig. 21-1. Map of the Paris Basin, showing locations of fields and outcrops discussed.

▼*Fig. 21-2. A.* Structure-contour map on top of Dogger (Middle Jurassic), Coulommes Field. Contours in meters below sea level. *B.* Isopach map of reservoir zone A in Coulommes Field. Contours in meters.

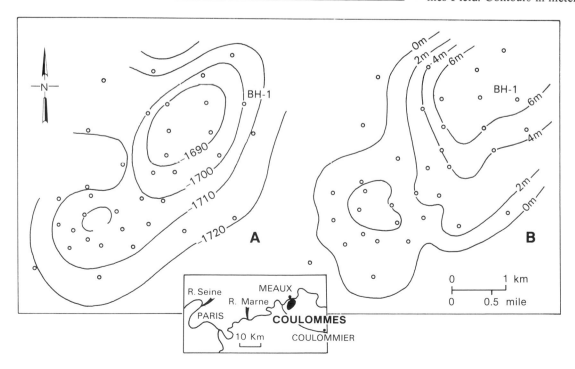

Fig. 21-3. Columnar section showing the distribution of reservoir zones and diagenetic phenomena in well BH-1 (Fig. 21-2), Coulommes Field. Zone C has no porosity and is not well-defined in this well.

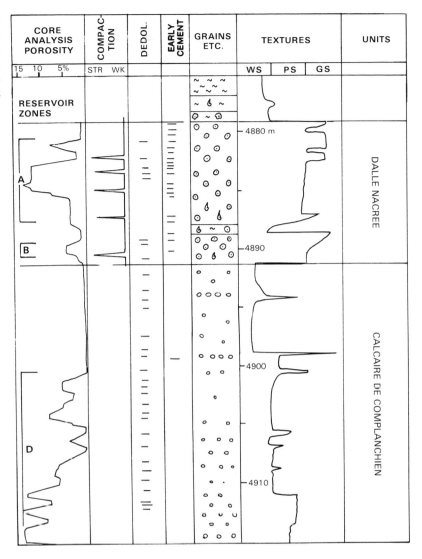

Upper Jurassic cap-rock marls, suggests local paleorelief during the Middle Jurassic. The dedolomite fabrics and other criteria indicate that a paleohigh has emerged repeatedly during the Middle Jurassic.

In addition to these four reservoir zones, porous dedolomites exist elsewhere in the Paris Basin, notably in outcrops along the southeast edge of the basin, where they have been studied in detail in large quarries at Massangis and Etrochey (Fig. 21-1). These subsurface and outcropping porous dedolomites are a minor part of a Middle Jurassic carbonate sequence 660 to 985 feet (200–300 m) thick comprising three major sedimentary cycles, each terminated by a hardground of regional extent. The two upper, Bathonian and Callovian cycles (Fig. 21-4), include

oolitic and bioclastic grainstones containing reservoir zones A and B at Coulommes Field (Fig. 21-3) and porous dedolomite at Massangis quarry. The calcarenites are interpreted to have been barriers protecting lagoonal pelletal limestones, which include the reservoir zones C and D at Coulommes and the dedolomites at Etrochey quarry.

The Middle Jurassic carbonates of the Paris Basin are part of the Burgundy Platform (Purser, 1975a & b) averaging 120 miles (200 km) in width, which extends northwestward for some 300 miles (500 km) from the Jura region through Burgundy and the Paris Basin to Normandy. This elongate platform was one in a series of regional highs that characterized the western European and North Sea region during the Middle Jurassic

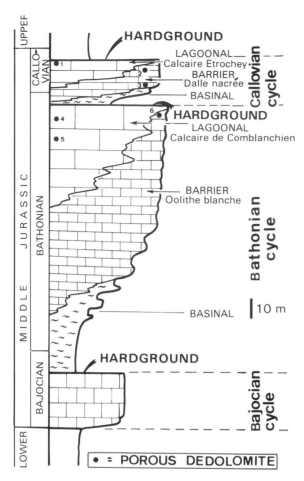

Fig. 21-4. Generalized sequence through the Middle Jurassic of the Paris Basin, showing the distribution of the dedolomite units; 1, Etrochey quarry; 2 and 3, reservoirs A and B at Coulommes field; 4 and 5, reservoirs C and D at Coulommes field; 6, Massangis quarry.

(Purser, 1979). The form and localization of these platforms seem to have been related to pre-existing, late Paleozoic basement blocks created by distensional movements during the opening of the North Atlantic.

The Paris Basin is essentially a late Tertiary structural depression whose weak tectonic deformation has led to exposure of the Middle Jurassic Burgundy Platform along the southeast margins of the basin. Exposures facilitated the study of the subsurface equivalents.

Dedolomite Porosity and Related Diagenetic Fabrics

Certain dolomites, especially those rich in ferrous iron, are relatively unstable and under certain conditions may be replaced by calcite (von Morlot, 1848). This secondary calcite after dolomite is commonly known as "dedolomite", a term that has been criticized (Smit and Swett, 1969). The dedolomitization process does not necessarily modify the porosity of the rock. However, as Evamy (1967) has shown, calcite after dolomite tends to be especially susceptible to dissolution, notably where associated with soil conditions, leading to increased porosity. The process is most active at near-surface temperatures and pressures (De Groot, 1967), and the presence of dedolomite in the subsurface of the Paris Basin is interpreted as proof of Middle Jurassic emergence.

Strictly speaking, only calcite formed by replacement of dolomite should be regarded as "dedolomite"; the subsequent dissolution of this secondary calcite and the development of a series of characteristic petrographic features, including the creation of rhombohedral pore spaces or "dedolomite" molds, belong to other diagenetic processes.

The formation of rhombohedral porosity does not necessarily require that the original dolomite first be replaced by calcite (dedolomite). As Al-Hashimi and Hemingway (1973) have shown, very iron-rich dolomites may be oxidized and dissolved, leaving a network of pores. This also is the case in certain Bathonian carbonates in the Paris Basin.

The easily recognized rhombohedral pore systems described by Shearman *et al.* (1961) and by Evamy (1967) are but the simplest expression of a complex process. An objective of this contribution is to demonstrate that modified dedolomite porosity may have subtle expression and may be more widespread in carbonate reservoirs than is generally recognized.

Petrographic Criteria for the Recognition of Dedolomite

The brownish or pinkish color noted by Shearman *et al.* (1961) in the French Jura also characterizes most Middle Jurassic dedolomites of the Paris

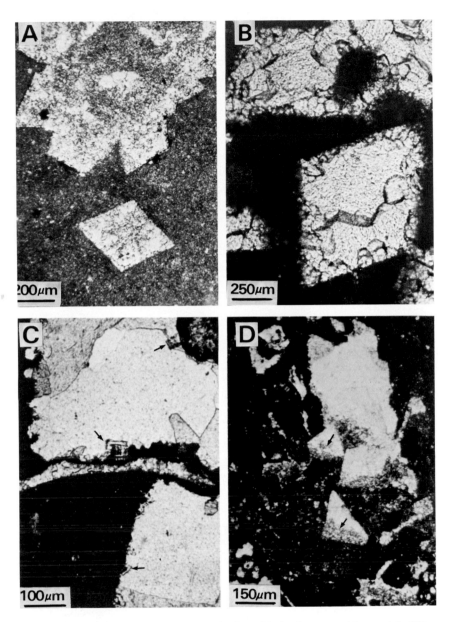

Fig. 21-5. Photomicrographs of typical "dedolomite" fabrics in Middle Jurassic outcrops in Burgundy. *A*. Micritized rhombs. *B*. Sparitized rhombs exhibiting typical drusy mosaic. *C*. Rhombic pseudomorphs (arrows) with iron oxide inclusions. These monocrystals are in optical continuity with the large sparitic crystals filling the cavity. *D*. Rhombohedral pseudomorphs partially filled with Jurassic internal sediment (arrows) and sparitic cement.

Basin; burrows within micritic lagoonal limestones are strikingly visible, their bright pink dedolomite (brown on weathered surfaces) contrasting with the beige-colored host limestones. Microscopic examination of Middle Jurassic dedolomites reveals a variety of distinctive fabrics, discussed in the paragraphs that follow.

Micritized Rhombs

Rhombs "floating" within lagoonal muds locally are composed of micrite (Fig. 21-5A). Interpreted by Evamy (1967) as a recrystallization product, the fine crystallinity of this micrite relative to the pre-existing crystal may have been induced by the numerous impurities in the original dolomite crystal. Clark (1980) has suggested as an alternative that this fabric may be conditioned by rapid nucleation from solutions that have high Ca/Mg ratios. The fabric may also be formed by the filling of rhombohedral pores with fine internal sediment (Fig. 21-5D).

Sparitized Rhombs
Rhombic pseudomorphs may be composed of one or more crystals of sparry calcite (Figs. 21-5B, C). Monocrystalline calcite commonly includes zones of iron oxide inclusions (Fig. 21-5C), suggesting that dedolomitization has occurred in "solid state". Conversion of ferroan dolomite to sparry calcite would not seem to require massive dissolution of the dolomite as suggested by Evamy (1967). Other rhombohedral pseudomorphs may be filled with a sparry calcite mosaic, which of course reduces the rhombohedral porosity.

Rhombohedral Pore Spaces
Perhaps the most characteristic products of "dedolomitization", rhombohedral voids, often within rust-colored limestones, are common on Jurassic outcrops (Fig. 21-6A). More importantly, they result in highly permeable reservoirs in the subsurface of the Paris Basin. Because dedolomitization seems to require near-surface conditions (De Groot, 1967), the rhombohedral voids in the reservoir rocks at Coulommes Field indicate that both dolomitization and dedolomitization are Middle Jurassic in origin. Rhombohedral porosity on modern outcrops along the southeast edge of the Paris Basin may be Jurassic or Recent.

Where dolomitization has been weak and rhombs "float" in sediment, the rhombohedral molds are easily recognized. However, where dolomitization has been more intensive, producing a crystal-supported fabric, later dissolution produces a secondary pore system whose rhombohedral units are not always evident. This is particularly true where the pore system has been modified by dissolution, which enlarged the original rhombohedral porosity (Figs. 21-6C,D) or resulted in vuggy pores whose spaces may attain several centimeters in diameter. Dedolomitized burrows may thus be converted to a cavernous network of voids somewhat resembling *gruyère* cheese, with exceedingly favorable reservoir properties.

The origin of this secondary porosity may be deduced by careful examination of pore boundaries and shapes because parts of the original rhombic pseudomorphs may be preserved locally either as "points" (crystal terminations) or as "flat surfaces" (crystal faces) (Figs. 21-6C,D).

In certain porous oolitic or pelletal "grainstone", both on outcrop and in the Coulommes and Saint Martin de Bossenay fields (Fig. 21-1), the interparticle porosity would appear at first glance to be primary. However, close study reveals irregular grain surfaces with small pointed depressions and local flat surfaces (Figs. 21-7A,B). That these local modifications of an otherwise convex grain surface are the subtle remnants of dissolved rhombs is suggested by the presence of well-preserved rhombohedral pores within adjacent grains. Such porous "grainstone" reservoirs would seem to be the product of a complex diagenetic history involving the following processes (Fig. 21-7C):

1. Partial dolomitization of an oolitic or pelletal *packstone*, the dolomite replacing preferentially the matrix, and the points of certain rhombs replacing the peripheral parts of some adjacent grains; the sediment still lacks porosity other than that possibly present within the matrix.
2. Preferential dissolution of the ferroan dolomite (or dedolomite) but preservation of the calcitic grains, creating a porous "grainstone" whose interparticle voids have minute irregularities in the form of abnormal "points" and "flat surfaces."
3. Finally, the porous *diagenetic grainstone* may be cemented and the rhombic molds filled by blocky calcite, the presence of which tends to further mask the already subtle effects of "dedolomitization".

Petrophysical Properties of Rhombohedral Pore Systems

Total porosity theoretically is the sum of residual sedimentary porosity plus the secondary porosity created, in this case, by dedolomitization and/or dissolution. In the Jurassic limestones of the Paris Basin, primary porosity tends to be minor, and most effective porosity is secondary (Fig. 21-6). Much of this dissolution porosity in the Middle Jurassic reservoirs at Coulommes Field exhibits

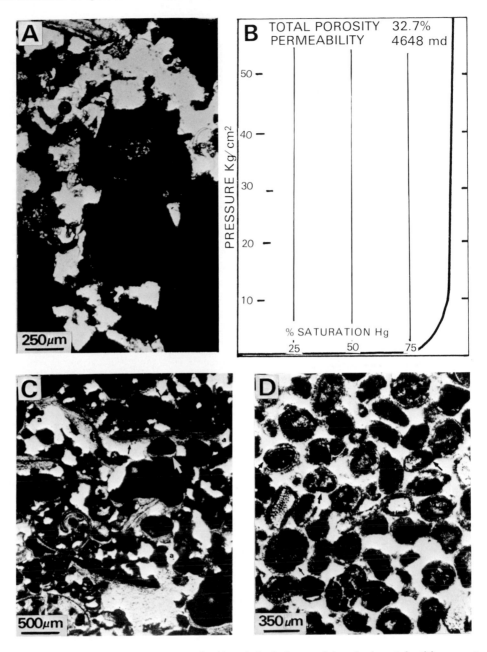

Fig. 21-6. Petrographic and petrophysical properties of rhombohedral voids. *A*. Photomicrograph showing rhombohedral pores (white), Bathonian, outcrop at Massangis. Plane light. *B*. Mercury-air capillary pressure curve of sample shown in *A*. *C*. Photomicrograph of bathonian lagoonal reservoir, Coulommes Field, show-ing rhombohedral pores (a) and microstalactitic cement (arrows). Plane light. *D*. Photomicrograph of callovian oolite barrier reservoir, St. Martin de Bossenay Field. Interparticle porosity is slightly enlarged by local points and flat surfaces (small arrows) interpreted as traces of rhombohedral crystal molds. Plane light.

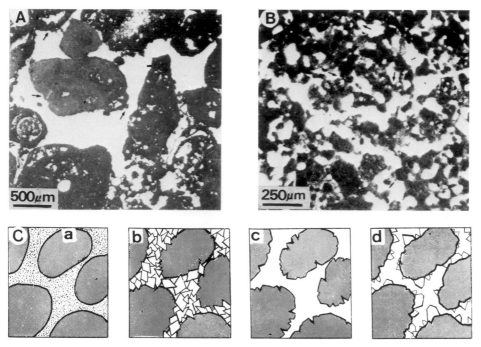

Fig. 21-7. Grainstone reservoirs, Bathonian lagoon, Coulommes Field. *A* and *B*. Plane light. Photomicrographs showing traces of rhombohedral crystal molds (arrows). *C*. Schematic origins of these diagenetic "grainstones."

one or another of the fabrics described above. As such, reservoir quality (apart from that due to fractures) is indirectly dependent on the degree of alteration to ferroan dolomite and directly dependent on the subsequent leaching of this unstable dolomite. Measured porosities attain 35 percent in outcrop and average 15 percent in the reservoirs. Permeabilities clearly are dependent on the extent of dolomitization; in two samples, "floating" rhombohedral porosity (14%) gave a measured permeability of 35 millidarcys, contrasting with a sample whose rhomb-supported fabric (porosity 32.7%) had a permeability of 4648 millidarcys. The mercury injection capillary pressure curve of a similar sample confirms the exceedingly favorable reservoir properties of leached dedolomite (Fig. 21-6B).

Paleogeographic Distribution of Porous Middle Jurassic Dedolomites

As already noted, porous dedolomitized limestone occurs both in the subsurface and in out-

crops along the southeastern part of the basin. The subsurface dedolomite is fossil and of immediate interest. Dedolomitized outcrops are also included in this discussion, not only because they are partly Jurassic in origin, but also because identifying porosity of Quaternary origin may ultimately help in understanding these relatively complex reservoir systems. Porous dedolomites occur in lagoonal, tidal-flat, and oolitic-barrier facies.

Lagoonal and Tidal-Flat Dedolomites

The extensive Bathonian lagoon, some 300 miles (500 km) long and 120 miles (190 km) across, comprising the central parts of the Burgundy Platform, emerged repeatedly as beaches and tidal flats (Purser and Lobreau, 1972). The tidal-flat limestones have numerous desiccation features and commonly are partly replaced by ferroan dolomite attaining maximum thickness of about 16 feet (5 m). On outcrop, the brown stratiform dedolomite is in places exceedingly porous; dedolomitization and related porosity development are probably subrecent. However, the presence of Bathonian internal sediment within cer-

Fig. 21-8. Cross-section of reservoir distribution in the Middle Jurassic limestones of Coulommes Field.

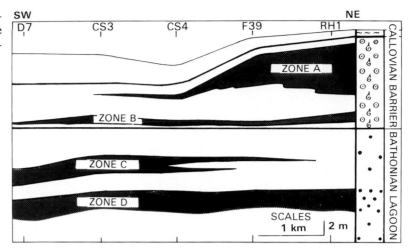

tain rhombohedral pseudomorphs (Fig. 21-5D) clearly indicates that at least part of this dedolomite is Bathonian in origin; dolomitization and dedolomitization within the conformable Bathonian sequence imply a common, related diagenetic system.

Individual porous dedolomite beds in tidal-flat facies may be traced laterally between exposures for distances up to 25 miles (40 km), their three-dimensional geometry being impossible to define because of the monoclinal disposition of outcrops.

Bathonian tidal-flat reservoirs in Coulommes Field (Fig. 21-7) are composed of pelletal packstones and grainstones, comprising zones C and D (Figs. 21-3, 21-8). A secondary porosity commonly exhibits traces of rhombohedral "points." Well-defined rhombohedral crystal molds are also present (Fig. 21-7B), leaving little doubt as to the origins of this secondary pore system. Pelletal "grainstones" have solution-enlarged interparticle pores, the irregularities of which suggest dolomitization and dedolomitization of an original packstone texture (Fig. 21-7). Associated vadose fabrics include microstalactitic cements (Fig. 21-6C), confirming periodic emergence of the Bathonian lagoon.

Average porosities in these two reservoir units are on the order of 15 percent. The uppermost reservoir (C) attains maximum thickness of about 10 feet (3 m), but thins and disappears towards the northeast parts of the field (Fig. 21-7). The underlying zone (D) is continuous throughout the structure, but is generally below the oil/water contact.

A second, lagoonal formation, the Calcaire d'Etrochey, Callovian in age (Fig. 21-4), is topped locally by an erosion surface (Fig. 21-9). The uppermost 2 meters of lime wackestone directly beneath this discontinuity contain numerous leached corals, mollusks, and algal oncoids whose molds are lined with brown iron oxides. Superficial examination suggests recent selective weathering of certain fossils. Close study reveals, however, that similar fossils immediately below the leached zone are selectively replaced by coarse ferroan dolomite; the wackestone matrix sediment was not affected by this grain-selective dolomitization. Similarly, neither the numerous brachiopods nor the scattered ooliths in this rock are dolomitized. Dolomitization only of particular grains that clearly were aragonitic (corals, gastropods, etc.) indicates selective replacement of unstable aragonite. That dissolution of the dolomitized fossils at Etrochey is not solely a recent weathering event is demonstrated by the presence of Callovian internal sediment (including ooliths) within these fossil molds.

In sum, the presence of a porous zone at the top of the Callovian lagoon is the result of Callovian emergence, selective ferroan dolomitization of aragonite grains, and then dedolomitization and dissolution of the unstable ferroan dolomite.

Oolitic Barrier Dedolomites

Both the Bathonian and Callovian lagoonal limestones grade laterally into cross-bedded oolitic and bioclastic limestones that formed marginal-shoal barrier facies and pass laterally into basinal

Fig. 21-9. Photograph of fossil-mold porosity in slab of Callovian lagoonal limestone, Etrochey quarry. Erosional surface at top of lagoon (mottled-looking top surface of sample) is underlain by rust-colored fossil molds (black arrows) with traces of coral structure and partially filled with internal sediment.

marls. These barrier carbonate sands have been dolomitized and dedolomitized locally, notably along the western and northern edges of the platform.

The 65-foot (20-m) thick "Oolithe blanche" has been partially dolomitized at two outcrop localities in Burgundy. Only one unit of these dolomites has been dedolomitized (Fig. 21-10). At Meursault (Fig. 21-1), reputed not only for its dolomite but especially for the exceptional qualities of its white Burgundy wine, the uppermost 40 feet (12 m) of the "Oolithe blanche" has been replaced completely by non-ferroan dolomite. This stable dolomite has *not* been affected by subsequent dedolomitization.

At Massangis, 60 miles (95 km) north-northwest of Meursault (Fig. 21-1), the same barrier oolite has been replaced locally by unstable ferroan dolomite that *has* been dedolomitized and

leached, resulting in unusually porous and permeable limestone (Fig. 21-11). The preferential dedolomitization at Massangis was favored by the ferroan composition of the original dolomite. At this locality, the Bathonian oolite is overlain by 10 feet (3 m) of lagoonal sediment representing the periphery of the Bathonian lagoon (Fig. 21-10). These lagoonal facies also have been replaced by ferroan dolomite and constitute part of the lagoonal dedolomites discussed in the preceding section. Ferroan dolomitization of the oolitic barrier therefore would seem to be the result of downward flow of dolomitizing fluids originating in the lagoon or tidal-flat above; lagoon and barrier seem to have been part of a common dolomitizing system.

High porosity and permeability within the Bathonian oolite at Massangis are due to well-developed rhombohedral porosity (Fig. 21-6A). Although Jurassic dedolomitization cannot be excluded, most porosity seems to have resulted from Quaternary processes. The resulting rhombohedral porosity is several orders of magnitude greater than the interrhombic dolomite porosity at Meursault.

Callovian oolitic/bioclastic barrier sands comprising the "Dalle nacrée" limestone include thin reservoirs in three fields of the Paris Basin (Fig. 21-1). These locally porous grainstones produce from a predominantly secondary interparticle porosity; voids often cut into adjacent grains. Individual pores may have irregular "points" or "flat surfaces" (Fig. 21-6D), the subtle relics of an earlier crystal-mold porosity.

Relationships Between Early Lithification and Dolomite Limits

The dolomitized/dedolomitized lagoonal sediments at Massangis, as well as in many other quarries in Burgundy, are limited at their base by a hardground. The top 8 feet (2-1/2 m) of oolitic grainstone situated immediately below this zone of early cementation have never been dolomitized. An underlying cross-bedded oolite, however, has been dolomitized, the degree of dolomitization tending to increase downwards. Because both dolomite bodies have identical ferroan composition suggesting a common origin, the lack of dolomitization in the uppermost 8 feet of oolite

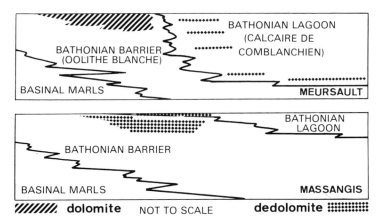

Fig. 21-10. Sections showing general relationships between the principal Bathonian facies and the distribution of Bathonian dolomites at Meursault and Massangis.

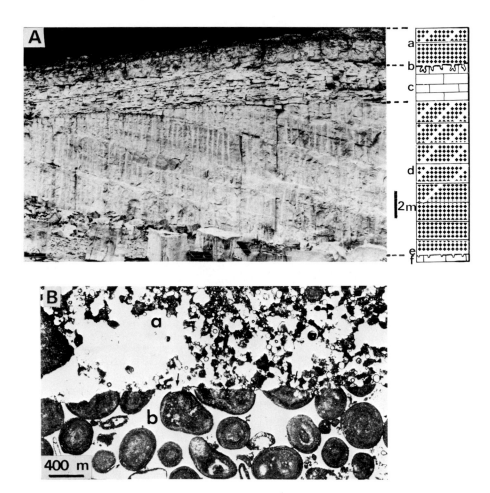

Fig. 21-11. Dedolomitized Bathonian oolite at Massangis Quarry. *A.* Photograph showing general view of quarry: a. dedolomitized lagoon; b. hardground at top of oolite; c. undolomitized oolite; d. dedolomitized oolite; e. hardground; f. undolomitized oolite. *B.* Photomicrograph showing petrographic details of lower hardground: a. porous dedolomite with rhombohedral pores (arrows) above hardground; b. undolomitized oolite below hardground.

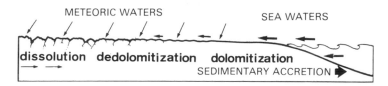

Fig. 21-12. Diagram of possible spatial relationship between synsedimentary dolomitization, dedolomitization, and dissolution as suggested by the Bathonian dedolomites of the Paris Basin.

suggests either that the lithified oolite was not susceptible to dolomitization—possibly because of the presence of protective cement—or that dolomitizing fluids descended obliquely, bypassing the locally lithified oolite to attain the uncemented oolite below. Whatever the explanation, early lithification seems to have prevented dolomitization of the upper part of the Bathonian oolite barrier.

The base of the lower (oolitic) dolomite/dedolomite body also is sharp and coincides with a second hardground below which the oolite is undolomitized (Fig. 21-11B). Again, the limit of the porous dedolomite clearly is determined by early marine lithification.

Discussion

The greater part of the dedolomite porosity within the Middle Jurassic limestones of the Paris Basin probably results from Quaternary weathering of Jurassic ferroan dolomite. However, the presence of porous dedolomite in subsurface reservoirs, plus local filling of rhombohedral crystal molds with Jurassic sediment, also implies a Jurassic age for a series of diagenetic processes involving dolomitization, dedolomitization, and dissolution.

The most logical "conceptual model" (Fig. 21-12) involves tidal-flat or oolite-barrier progradation contemporaneous with dolomitization, a well-known Recent phenomenon. Ferroan dolomitization seems to have been associated with the presence of fresh water, the composition of which evolved following sedimentary progradation. It is not difficult to imagine the landward parts of a coastal plain being affected by increasingly intensive pedogenesis during which the ferroan dolomite was transformed into a network of rhom-

bohedral pores later enlarged by meteoric waters.

The geometry of the porous dedolomite depends on several factors. Dissolution obviously was dependent on the geometry of the original dolomite body. However, where this dolomite varies laterally in composition, the transition into a ferroan facies would appear to have been an important factor influencing subsequent dedolomitization and the development of a secondary leached porosity. Finally, early dedolomitization of unstable dolomite in most cases seems to require exposure to pedological alteration resulting from local emergence. As such, it tends to occur below disconformities affecting dolomitic formations.

The formation of dedolomitic crystal-mold porosity, although complex, does not depend—unlike the intercrystalline porosity of many sucrosic dolomites—on the *degree* of dolomite formation; pore space may be created from any amount of unstable dolomite. However, effective permeability requires that the fabric be rhomb-supported.

Conclusions

Although the existence of dedolomite has been known since the 19th century, the author believes that secondary porosity associated with dedolomitization is more widespread in hydrocarbon reservoirs than is generally believed. Failure to recognize this distinctive type of porosity has been due partly to lack of petrographic criteria other than the fairly well-known rhombohedral crystal-mold pores. Now that such criteria are available, recognition of dedolomite porosity should be facilitated.

Although dedolomite, rhomb-mold reservoir

rocks may be exceptional, carbonates in which leached-fossil porosity is common, as in much of the Cretaceous and Jurassic of the Middle East, should perhaps be reexamined in the light of possible dedolomitization.

Acknowledgments   The author expresses thanks to the director and staff of the French company PETROREP for permission to publish information concerning Coulommes Field.

References

AL-HASHIMI, W.S., and J.E. HEMINGWAY, 1973, Recent dedolomitization and origin of the rusty crusts of Northumberland: Jour. Sedimentary Petrology, v. 43, p. 82–91.

CLARK, D.N., 1980, Replacement of dolomite by calcite (abst.): Int. Assoc. Sedimentologists, 1st Europ. Mtg., Bochum, W. Germ., p. 225–227.

CUSSEY, R., and G.M. FRIEDMAN, 1977, Patterns of porosity and cement in ooid reservoirs in Dogger (Middle Jurassic) of France: Amer. Assoc. Petroleum Geologists Bull., v. 61, p. 511–518.

DE GROOT, K., 1967, Experimental dedolomitization: Jour. Sedimentary Petrology, v. 37, p. 1216–1220.

EVAMY, B.D., 1967, Dedolomitization and the development of rhombohedral pores in limestones: Jour. Sedimentary Petrology, v. 37, p. 1204–1215.

PURSER, B.H., 1975a, Sédimentation et diagenèse précoce des séries carbonatées du Jurassique moyen de Bourgogne. Ph.D. Dissert., Univ. Paris-Sud, Orsay, 383 p.

PURSER, B.H. 1975b, Tidal sediments and their evolution in the Bathonian carbonates of Burgundy, France: *in* Ginsburg, R.N., ed., Tidal Deposits, Springer-Verlag, Inc., New York, p. 335–343.

PURSER, B.H., 1979, Middle Jurassic sedimentation on the Burgundy Platform. Symposium "La sédimentation du Jurassique ouest-Européen": Assoc. Sédimentologistes Français, Spec. Publ. 1, p. 75–84.

PURSER, B.H., and J.-P. LOBREAU, 1972, Structures sédimentaires et diagenétiques précoces dans les calcaires bathoniens de la Bourgogne: Bur. Recherches Géologiques et Minières Bull., v. IV, p. 19–47.

SHEARMAN, D.J., J. KHOURI, and S. TAHA, 1961, On the replacement of dolomite by calcite in some Mesozoic limestones from the French Jura: Geol. Assoc. London, Proc., v. 72, p. 1–12.

SMIT, D.E., and K. SWETT, 1969, Devaluation of dedolomitization: Jour. Sedimentary Petrology, v. 39, p. 379–380.

VON MORLOT, A., 1848, Sur l'origine de la dolomie. Comptes Rendus Acad. Sci. Paris, v. 26, p. 313.

22
Chatom Field
Charles T. Feazel

RESERVOIR SUMMARY

Location & Geologic Setting	Washington Co., T6N-R4W, SE Arkansas, Gulf of Mexico coastal plain, USA
Tectonics	Local closure on north-trending salt anticline; penecontemporaneous growth
Regional Paleosetting	Marine shelf/ramp, south dipping
Nature of Trap	Structural/diagenetic
Reservoir Rocks	
Age	Late Jurassic (Oxfordian)
Stratigraphic Units(s)	Smackover Formation (upper)
Lithology(s)	Dolomite
Dep. Environment(s)	Shallow nearshore(?) shelf
Productive Facies	Pellet-mold dolomite
Entrapping Facies	Buckner Anhydrite (evaporitic limestone and shale)
Diagenesis Porosity	Vadose and phreatic leaching; dolomitization, both early and late
Petrophysics	
Pore Type(s)	Moldic, intercrystal
Porosity	20–35% (average 25%) estimated from thin sections
Permeability	40–200 md (average 63 md) from core-plug analyses
Fractures	Minor, generally healed by calcite
Source Rocks	
Age	Jurassic (Oxfordian)
Lithology(s)	Basal laminated carbonates, Smackover Formation
Migration Time	NA
Reservoir Dimensions	
Depth	16,000 ft (4900 m)
Thickness	0–30 ft (0–9 m) lenses
Areal Dimensions	2.5 × 2 mi (4 × 3.2 km)
Productive Area	3200 acres (12.8 km^2)
Fluid Data	
Saturations	NA
API Gravity	NA
Gas-Oil Ratio	NA
Other	16% H_2S in gas stream
Production Data	
Oil and/or Gas in Place	110 billion CFG (3.11 billion m^3); 18.3 million equivalent BO
Ultimate Recovery	21 billion CFG (0.59 billion m^3), 3.5 million equivalent BO; 0.38 million BGC; 0.16 million BLG; URE = 19.1%
Cumulative Production	NA

Remarks: IP 4.02 million CFGPD (114,000 m^3). Discovered 1970.

22
Diagenesis of Jurassic Grainstone Reservoirs in the Smackover Formation, Chatom Field, Alabama

Charles T. Feazel

Location and Discovery

Chatom Field, which produces gas and gas condensate from the Upper Jurassic Smackover Formation, is located in T6N-R4W, Washington County, southwestern Alabama (Fig. 22-1). The field is situated on a salt ridge that extends northward to encompass several other productive fields, notably South State Line and Copeland.

Seven dry holes, defining closure in Upper Cretaceous strata, were drilled in the immediate area prior to discovery of a Jurassic closure during seismic surveys by Phillips Petroleum Company. The Phillips Williams AA-1 was completed in September 1970, with a potential production of 4.02 million cubic feet (114,000 m^3) of gas per day and 988 barrels of condensate per day. The field was subsequently developed with three additional producing wells and two dry holes.

Reservoir Characteristics

The reservoir consists of three dolomitic lenses, each one 0 to 30 feet (0–10 m) thick, in the uppermost Smackover Formation. The dolomitic lenses are developed on the crestal part of the structure (Fig. 22-2). Off-structure, the Smackover consists of pelleted limestone containing little or no dolomite. Depth to the top of the unit is approximately 16,000 feet (4877 m), and the reservoir occupies an area of approximately 3200 acres (12.8 km^2) defined by structural closure of about 200 feet (60 m). Reservoir porosity is mainly of two types: intercrystalline (dolomitic) and moldic. Overlying evaporitic limestones and shales of the Buckner Formation provide the reservoir seal. On the basis of the composition and pressure of production fluid samples, the hydrocarbons are believed to be in gaseous state in the reservoir.

Production

Daily production per well in Chatom Field (1974 figures) averaged 5.54 million cubic feet (157 thousand m^3) of gas, 1131 barrels of condensate, and 41 long tons (42 $\times$ 10 kg) of sulfur. The gas stream flows 16 percent hydrogen sulfide. Total reserves in place are estimated at 110 billion cubic feet (3.11 billion m^3).

Reservoir Geology

Depositional Environment

Ten Smackover microfacies are recognized in cores from Chatom and nearby fields in Mississippi, Alabama, and northwestern Florida. The upward change in the middle of the Smackover section from laminated lime

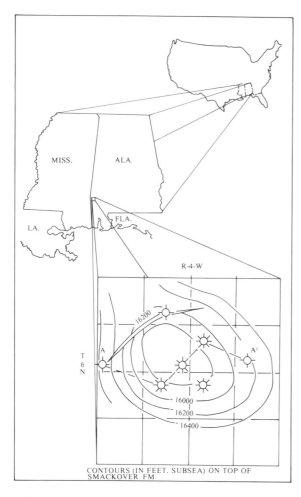

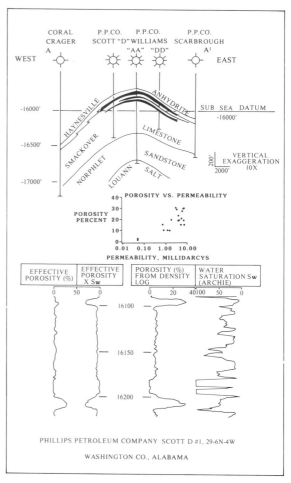

Fig. 22-1. Location map, Chatom Field, Alabama. Line A–A' marks location of the cross-section shown in Figure 22-2. Contours in feet subsea, on top of Smackover Formation, contour interval 200 feet.

Fig. 22-2. Cross-section (top) of Chatom Field, with porous dolomitic lenses at crest of structure shown in black. Crossplot (middle) of porosity and permeability within the dolomite. Typical log interpretation (bottom) of uppermost Smackover Formation in the Phillips Scott D#1, showing two porous intervals.

mudstones to well-sorted grainstones records part of a classic regressive sequence. Deposition of the overlying evaporitic lime mudstones of the Buckner Formation in a sabkha setting (Sigsby, 1976), followed by subaerial detrital sedimentation of the Haynesville Formation, completed the regressive sedimentation format. Smackover depositional environments have been described in greater detail by Bishop (1969, 1971), Dickinson (1969), Ottman *et al.* (1976), Sigsby (1976), and Wakelyn (1977).

Diagenesis

Extensive diagenetic alteration has modified the depositional characteristics of these Jurassic rocks. A spectrum of diagenetic events is inferred from their alteration products and varies from early, syndepositional alteration to late modification by deep brines. A determination of the timing of these events relative to salt tectonics and oil migration is the key to understanding the petroleum potential of the Smackover Formation, since Chatom Field and similar reservoirs occur in combined stratigraphic-diagenetic as well as structural traps.

Cementation

Calcite, dolomite, anhydrite, celestite, and chert cements occluded original pore space, filled secondary voids, and healed post-lithification fractures in these rocks. Calcite cement is present in a range of crystal sizes from 4-micron microspar to void-filling sparry crystals up to 4 millimeters across. Much of this cement has been recrystallized so that its original fabric is no longer discernible, but radiaxial calcite overgrowths on peloids have been preserved in a few intervals. Most calcite cements resemble the products of syndepositional processes in the Holocene marine-phreatic zone, except for the last void-filling crystals, interpreted as late-stage cements derived from subsurface brines through solution transfer (Durney, 1972).

Anhydrite and celestite cements are especially common in dolomitic intervals, where they fill intercrystalline and moldic pores created during dolomitization of lime mud. In some non-dolomitized intervals, anhydrite cement was preceded by a thin rind of calcite crystals on calcareous grains.

Compaction

Compaction of the various Smackover lithotypes occurred in diagenetic realms ranging from very shallow, in which syndepositional plastic flow of soft sediment was the primary means of stress accommodation, to very deep [>20,000 ft (>6 km)] in southern Mississippi, where brittle fracture and pressure solution of lithified carbonates predominated. Shallow burial compaction is recorded by collapsed burrows, stacked dolomitic rims of collapsed moldic pores, and flattened peloids. Some compressed peloids are preserved only as wispy organic films that grade laterally into microstylolites

Plastic deformation of ooids resulted in the formation of slightly interpenetrating grain contacts. Ooids that were more resistant to compaction developed microstylolitic contacts and spalled outer laminae (Fig. 22-3). Cracking and breaking

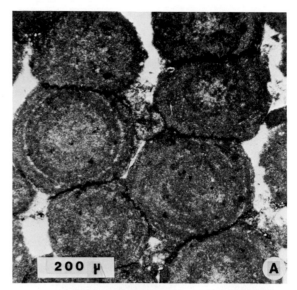

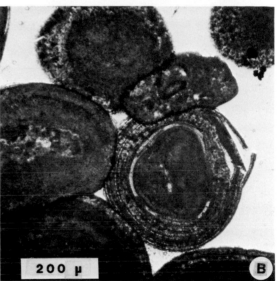

Fig. 22-3. Photomicrographs showing evidence of compaction in upper Smackover grainstone facies. Plane light. *A.* Microstylolitic contacts between ooids, 19,405 feet (5915 m). *B.* Slightly interpenetrating contacts between ooids; spalled outer laminae, 19,178 feet (5876 m). Both examples from Phillips Josephine A#1, Perry Co., Mississippi.

of elongate pisoids may also have occurred in near-surface diagenetic environments.

Deeper burial produced undulatory extinction in detrital quartz grains and in their syntaxial overgrowths. Rarely, this also occurs in crystals of calcite, dolomite, and anhydrite. Fractures in the Smackover carbonates, produced by regional tectonics or compactional stress over salt structures, extend through grains and cements but terminate at stylolites, indicating that stylolitization by pressure dissolution was a very late diagenetic process. Stylolites are generally low-amplitude, undulose surfaces defined by opaque residues containing quartz, micas, clays, bitumen, and other insoluble components. Rarely, calcitic slickensides and botryoidal to rhombic "baroque" dolomite crystals (Folk and Assereto, 1974) are enclosed in stylolitic residues. The calcium carbonate dissolved during the formation of stylolites is a possible source of the late-stage calcite cement in the grainstone microfacies. This late cement contains rare bituminous inclusions that indicate near-synchronous calcite precipitation and oil migration.

Recrystallization

The lime mudstones of the Smackover Formation were deposited as carbonate muds. Recrystallization to finely crystalline calcite obliterated the original crystal form and produced an interlocked fabric of equant crystals. The few bioclasts present in these rocks recrystallized to calcite spar or microspar. In some intervals, this transition involved leaching of aragonite to leave a void later filled by calcite (recognized by centripetal fabric in rim cements); in other intervals, reorientation of the crystal lattice without gross dissolution preserved fine detail. Ooids underwent both micritization, probably through the syndepositional boring of endolithic algae and fungi (Bathurst, 1966), and recrystallization to sparry calcite. As in the case of the bioclastic grains, some of this calcite fills voids produced by the leaching of ooid centers, but much appears to be the product of recrystallization. Calcite cement in some specimens invades the margins of micritic peloids, suggesting that recrystallization was progressive from the centers of pores outward.

Authigenesis

Dolomite

Replacement of Smackover limestones by dolomite, localized over positive structural features, created potential petroleum reservoirs. Some finely crystalline dolomite was early diagenetic or possibly syndepositional in origin (Sigsby, 1976; Bishop, 1968), and clearly formed before the leaching of peloids and other grains. Brecciated dolomitic crusts in the Phillips Williams DD #1 well core resemble those described from subaerial to subsoil Holocene settings (e.g., Shinn et al., 1965).

Structureless dolomitic mud and isolated dolomite crystals in lime mudstone resemble the products of shallow diagenesis beneath Holocene sabkhas (Butler, 1969). The inferred site for all of these dolomitization events is near-surface, and the inferred timing is very early in the rocks' diagenetic history. One such site invoked on both theoretical and geologic criteria is the mixing zone between meteoric waters and brines derived from sea water (Badiozamani, 1973; Land, 1973; Land et al., 1975).

As burial depth increased, dolomite replaced anhydrite nodules, rarely preserving internal textures, and partially replaced calcareous ooids and detrital quartz grains. "Baroque" or "saddle-shaped" dolomite crystals, found in stylolitic residues, have distorted lattices and curved faces. According to Shearman (1977, and personal communication, 1978), such dolomite is the product of deep, hot magnesium-rich brines commonly associated with lead-zinc deposits. In samples containing significant amounts of chert (i.e., from southern Mississippi), this late dolomitization post-dated silicification. Large (300–700 μ), clear dolomite crystals and isopachous vug-lining dolomitic cement (Fig. 22-4) probably represent diagenetic environments of intermediate depth.

Anhydrite and Celestite

Except for rare bedded anhydrite, or nodules with felted-lath internal texture, all sulfate minerals in the Smackover Formation are of diagenetic origin. Some secondary crystals are pseudomorphous after gypsum; others replace calcareous grains and cement. Even "primary" anhydrite may have gone through a gypsum phase, but was altered very early to anhydrite

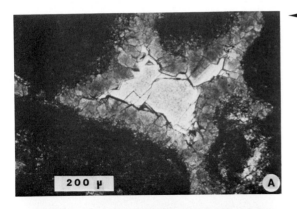

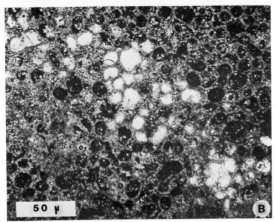

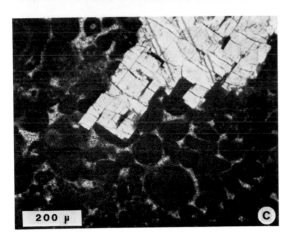

◄ *Fig. 22-4.* Photomicrographs of pore-filling cements in the uppermost Smackover Formation. *A.* Isopachous rim of dolomite crystals surrounding primary (?) vug in peloidal mudstone. Each dark lump comprises several clotted peloids. Center of pore filled with sparry calcite. 15,858 feet (4834 m), Phillips State Regis B#1, Santa Rosa Co., Florida. Plane light. *B.* Void-filling poikilotopic anhydrite in a dolomitic zone with pelmoldic porosity. Light pellet molds contain anhydrite; dark pellet molds are empty. 16,177 feet (4931 m), Phillips Williams DD#1, Chatom Field, Alabama. Cross-polarized light. *C.* Replacive anhydrite in grainstone of peloids and ooids. Anhydrite crystal cuts equally across grains and cement. Cross-polarized light. 19,196 feet (5851 m), Phillips Josephine A#1, Perry Co., Mississippi.

tals engulfed smaller crystals of dolomite or calcite, and pushed aside during crystal growth those minerals that resisted solution-replacement. The outer margins of these crystals commonly contain many more carbonate inclusions than do their centers, reflecting incomplete replacement or displacement of the carbonate matrix.

Sulfur isotopic compositions of anhydrite from the Smackover and Buckner formations (Price and Feazel, 1980) indicate that the ultimate source of the sulfate was Jurassic sea water. Primary anhydrite nodules ($\delta^{34}S_{CDT} = +15.7\% \pm 1.3$) and pore-filling anhydrite ($\delta^{34}S_{CDT} = +16.8\% \pm 1.8$) were precipitated from isotopically similar waters (Jurassic sea water $+15$ to $+19\%$).

Strontium does not form a hydrated sulfate analogous to gypsum. Therefore a syn-depositional origin for any of the celestite in the Smackover Formation is impossible to infer, although minor amounts of celestite have been reported in modern sabkha sediments as a consequence of the release of strontium, which accompanies replacement of aragonite by dolomite (Kinsman, 1969). Celestite crystals intimately intergrown with void-filling anhydrite, or poikilotopically enclosing other minerals, are certainly of secondary origin.

Silica

Silica is volumetrically a minor component of the Smackover, but has contributed to the occlusion of porosity in some reservoir intervals. Syntaxial overgrowths on detrital quartz silt grains are

(Shearman, 1978). Gypsum remaining in the sediment was dehydrated by burial to depths exceeding 3000 feet (~1 km), or to a thermal regime above 42°C (108°F), beyond which anhydrite is the stable sulfate phase (Mossop and Shearman, 1973).

Secondary anhydrite is both void-filling and replacive (Fig. 22-4). Poikilotopic anhydrite crys-

abundant. They penetrate the margins of calcareous grains, but are penetrated by dolomite crystals. In some wells, particularly the Phillips Josephine A #1, located 65 miles (105 km) southwest of Chatom Field, chert has replaced (1) anhydrite nodules, (2) entire beds of peloids or ooids, (3) dolomite rhombs, and (4) individual ooids or bioclasts. Spherulitic chalcedony is common in completely silicified intervals, and in the nuclei of partially replaced pisoids. The last void-filling in many silicified intervals was sparry calcite, indicating that solution transfer of carbonate minerals post-dated silicification. Rare chert cement between grains embays baroque dolomite and is therefore interpreted as a product of late diagenesis.

Pyrite

Pyrite is a very minor constituent of the Smackover Formation, and is most common within microstylolitic grain contacts (of meteoric origin; Becher and Moore, 1976), due to the interaction of sulfate-reducing bacteria, organic matter, and iron-bearing clays in the stylolitic residues. The presence of pyrite is of economic significance in some fields, particularly Chunchula, 48 miles (77 km) south of Chatom, as iron has served as a sink for hydrogen sulfide that would otherwise be produced with petroleum, considerably reducing its value.

Generation of Porosity

Porosity in the Smackover Formation in the study area varies from zero to 35 percent (visual estimates from thin sections). Pores include primary spaces between grains, molds of leached carbonate grains, intercrystal voids between dolomite rhombs, irregular fenestral vugs in dolomitic mud, and very rare open fractures. Primary porosity, common in updip Smackover grainstones of the Arkansas-Louisiana region (Becher and Moore, 1976), is rare in the southeastern region, with the exception of Jay Field, Florida, southeast of Chatom, where Sigsby (1976) reported rare primary porosity as high as 25 percent. Moldic porosity is common in many Smackover intervals, and resulted from leaching of peloids, bioclasts, and ooids (Fig. 22-5). Moldic pores are characteristics of rocks that have been exposed to meteoric waters (Matthews, 1968).

Fig. 22-5. Scanning-electron micrographs of characteristic porosity in dolomites from the Smackover Formation. *A.* Moldic pores where peloids have been leached. Rims of pores consist of dolomite crystals in what was once interparticle space. 16,128 feet (4916 m), Phillips Scott C#1, Chatom Field. *B.* Intercrystalline pore in sucrosic dolomite. 19,772 feet (6027 m), Phillips Josephine A#1, Perry Co., Mississippi.

The restriction of moldic porosity to the crests of structures invites speculation about the paleotopography that permitted entry of meteoric waters. In Chatom Field, for example, the Phillips Scarborough well (on the flank of the domal structure), which penetrated nonporous peloidal lime mudstone, is located less than 2 miles (3 km) from the Phillips Williams AA #1 and Phillips Williams DD #1 (closer to the structural crest), both of which penetrated crystalline dolomite containing anhydrite, with moldic and intercrystalline porosity (Fig. 22-5). The localization of

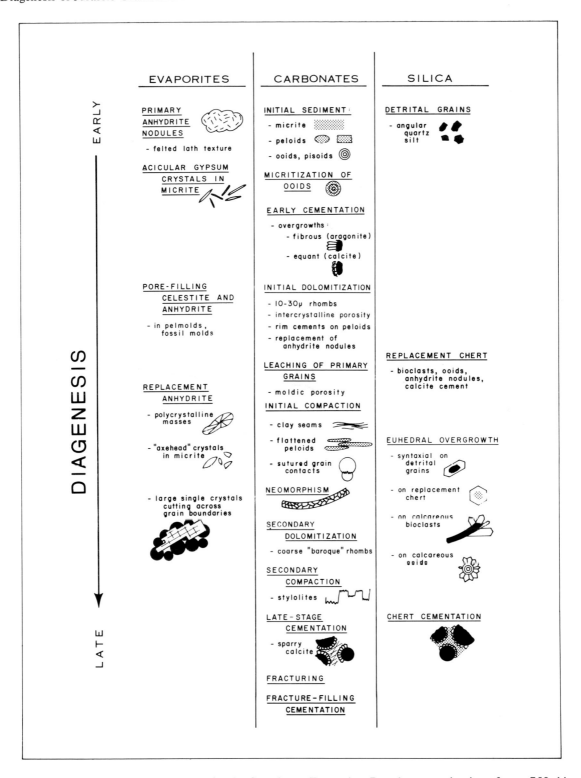

Fig. 22-6. Diagenetic sequence inferred for the Smackover Formation. Based on examination of over 750 thin sections from cores recovered from Chatom Field, Alabama, and surrounding parts of Mississippi, Alabama, and Florida.

these rocks above a salt-cored structure is best explained by inferring that salt movement of the underlying Louann Salt created topographic relief on the Smackover sea floor (Badon, 1974). A modern analog is found in the Persian Gulf, where salt-cored structures emerge as islands (Purser, 1973). Elevated areas of the Smackover carbonates were exposed subaerially, either episodically or by continued regression, and lenses of sabkha brines or meteoric waters were established atop structural highs, leaching metastable components from the crestal sediments.

Cements surrounding moldic pores (Fig. 22-5) are composed of small ($<10\ \mu$) dolomite crystals resembling those ascribed by Folk and Land (1975) to hypersaline environments. Because (1) moldic porosity is common in dolomitized intervals, (2) dolomite is similarly restricted in some fields to structurally high positions, and (3) the mixing zone between fresh and saline waters has been inferred by other workers as a site of dolomitization, the salt-movement model described above explains the origin and areally restricted distribution of both moldic and intercrystalline porosity. The preservation of evaporite minerals in such a setting remains an enigma. It is possible that much of the anhydrite postdates the leaching of metastable carbonates. Fenestral porosity is present in dolomitized peritidal sediments in the Phillips St. Regis B$^{\#}$1, Santa Rosa County, Florida, close to the updip limit of the Smackover carbonates. By analogy with Holocene carbonate settings (Shinn, 1968) the fenestral vugs (1–10 mm) resulted from entrapment of gas bubbles within supratidal sediments.

Petroleum Source

The source of petroleum produced from Smackover reservoirs has long been a mystery. Because the evaporite sequences above (Buckner) and below (Louann) are considered impermeable, and the underlying Norphlet Formation contains no likely source beds, the basal laminated carbonates of the Smackover Formation have been inferred as an oil source by default (Bishop, 1968, 1969). Although no source rock data are available from Chatom Field, a sample of the basal Smackover from the Phillips Godfrey well in Nachitoches Parish, Louisiana, qualifies geochemically as a source rock (1.33 wt% organic carbon, soluble/

total organic carbon = 0.006; average odd-even carbon number predominance = 0.97). The grainstone facies of the upper Smackover has also been suggested as a source (Malek-Aslani, 1968).

Conclusions

Reservoir porosity in the Smackover Formation at Chatom Field and surrounding areas is the result of very early diagenesis in syndepositional to shallow burial environments. The early existence of structures, porous reservoirs, and seals (overlying evaporitic mudstones) implies that these combined structural-stratigraphic-diagenetic traps were available for petroleum entry soon after burial (Fig. 22-6). The presence of oil inclusions in void-filling calcite indicates that petroleum migration into the Smackover reservoirs predated the latest phases of inorganic diagenesis, which undoubtedly continues today in the deep subsurface.

References

BADIOZAMANI, K., 1973, The *dorag* dolomitization model—application to the Middle Ordovician of Wisconsin: Jour. Sedimentary Petrology, v. 43, p. 965–984.

BADON, C.L., 1974, Petrology and reservoir potential of the upper member of the Smackover Formation, Clarke County, Mississippi: Gulf Coast Assoc. Geol. Soc. Trans., v. 24, p. 163–174.

BATHURST, R.G.C., 1966, Boring algae, micrite envelopes and lithification of molluscan biosparites: Geol. Jour., v. 5, p. 15–32.

BECHER, J.W., and C.H. MOORE, 1976, The Walker Creek Field—a Smackover diagenetic trap: Gulf Coast Assoc. Geol. Soc. Trans., v. 26, p. 34–56.

BISHOP, W.F., 1968, Petrology of upper Smackover limestones in North Haynesille Field, Claiborne Parish, Louisiana: Amer. Assoc. Petroleum Geologists Bull., v. 52, p. 92–128.

BISHOP, W.F., 1969, Environmental control of porosity in the upper Smackover limestone, North Haynesville Field, Claiborne Parish, Louisiana: Gulf Coast Assoc. Geol. Soc. Trans., v. 19, p. 155–169.

BISHOP, W.F., 1971, Geology of a Smackover stratigraphic trap: Amer. Assoc. Petroleum Geologists Bull., v. 55, p. 51–63.

BUTLER, G.P., 1969, Modern evaporite deposition and geochemistry of coexisting brines, the sabkha, Trucial Coast, Arabian Gulf: Jour. Sedimentary Petrology, v. 39, p. 70–89.

DICKINSON, K.A., 1969, Upper Jurassic carbonate rocks in northeastern Texas and adjoining parts of Arkansas and Louisiana: Gulf Coast Assoc. Geol. Soc. Trans., v. 19, p. 175–187.

DURNEY, D.W., 1972, Solution-transfer, an important geological deformation mechanism: Nature, v. 235, p. 315–317.

FOLK, R.L., and R. ASSERETO, 1974, Giant aragonite rays and baroque white dolomite in tepee-fillings, Triassic of Lombardy, Italy (abst.): Amer. Assoc. Petroleum Geologists and Soc. Econ. Paleontologists and Mineralogists Ann. Mtg. Abst., v. 1, p. 34–35.

FOLK, R.L., and L.S. LAND, 1975, Mg/Ca ratio and salinity: two controls over crystallization of dolomite: Amer. Assoc. Petroleum Geologists Bull., v. 59, no. 1, p. 60–68.

KINSMAN, D.J.J., 1969, Modes of formation, sedimentary associations, and diagnostic features of shallow-water and supratidal evaporites: Amer. Assoc. Petroleum Geologists Bull., v. 53, p. 830–840.

LAND, L.S., 1973, Contemporaneous dolomitization of Middle Pleistocene reefs by meteoric water, North Jamaica: Bull. Marine Science, v. 23, p. 64–92.

LAND, L.S., M.R.I. SALEM, and D.W. MORROW, 1975, Paleohydrology of ancient dolomites: geochemical evidence: Amer. Assoc. Petroleum Geologists Bull., v. 59, p. 1602–1625.

MALEK-ASLANI, M., 1968, Habitat of oil in carbonate rocks: Gulf Coast Assoc. Geol. Soc. Trans., v. 18, p. 12–25.

MATTHEWS, R.K., 1968, Carbonate diagenesis: equilibration of sedimentary mineralogy to the subaerial environment: coral cap of Barbados, West Indies: Jour. Sedimentary Petrology, v. 38, p. 1110–1119.

MOSSOP, G.D., and D.J. SHEARMAN, 1973, Origin of secondary gypsum rocks: Inst. Mining and Metallurgy Trans. (B), p. 14–154.

OTTMAN, R.D., P.L. KEYES, and M.A. ZIEGLER, 1976, Jay Field, Florida—a Jurassic stratigraphic trap: Amer. Assoc. Petroleum Geologists Mem. 24, p. 276–286.

PRICE, F.T., and C.T. FEAZEL, 1980, A sulfur isotopic study of anhydrites from the eastern Gulf Coast Buckner and Smackover Formations (abst.): Amer. Chem. Soc. 179th Nat'l. Meeting Abst., Geochem. Division, paper 9.

PURSER, B.H., 1973, Sedimentation around bathymetric highs in the southern Persian Gulf: in Purser, B.H., ed., The Persian Gulf: Springer-Verlag, Inc., Heidelberg, p. 151–177.

SHEARMAN, D.J., 1977, Diagenetic processes in oil and ore host rocks: in Garrard, P., ed., Forum on Oil and Ore in Sediments: London, Imperial College, p. 201–205.

SHEARMAN, D.J., 1978, Evaporites of coastal sabkhas: in Dean, W.E. and B.C. Schreiber, eds., Marine Evaporites: Soc. Econ. Paleontologists and Mineralogists Short Course Notes, p. 6–42.

SHINN, E.A., 1968, Practical significance of birdseye structures in carbonate rocks: Jour. Sedimentary Petrology, v. 38, p. 215–223.

SHINN, E.A., R.N. GINSBURG, and R.M. LLOYD, 1965, Recent supratidal dolomite from Andros Island, Bahamas: in Pray, L.C. and R.C. Murray, eds., Dolomitization and Limestone Diagenesis—a symposium: Soc. Econ. Paleontologists and Mineralogists, Spec. Publ. 13, p. 12–123.

SIGSBY, R.J., 1976, Paleoenvironmental analysis of the Big Escambia Creek-Jay-Blackjack Creek Field areas: Gulf Coast Assoc. Geol. Soc. Trans., v. 26, p. 258–278.

WAKELYN, B.D., 1977, Petrology of Smackover Formation (Jurassic): Perry and Stone Counties, Mississippi (abst.): Amer. Assoc. Petroleum Geologists Bull., v. 61, p. 1548–1549.

23
Mt. Vernon Field
Yehezkeel Druckman and Clyde H. Moore, Jr.

RESERVOIR SUMMARY

Location & Geologic Setting	NE Columbia Co., T15S-R21–23W, south Arkansas; Gulf of Mexico coastal plain, USA
Tectonics	Subsiding ramp cut by down-to-basin regional faults and by local monocline
Regional Paleosetting	South-dipping shelf/ramp affected locally by salt tectonics
Nature of Trap	Stratigraphic; updip facies change and cementation trap in homoclinally south-dipping sequence

Reservoir Rocks

Age	Late Jurassic
Stratigraphic Units(s)	Smackover Formation
Lithology(s)	Limestone
Dep. Environment(s)	Carbonate-sand bars and associated muddy shoals
Productive Facies	Oolitic, oncolitic and skeletal grainstones
Entrapping Facies	Anhydrite above, cemented grainstone updip
Diagenesis Porosity	Early and late (burial) dissolution; partial cementation by calcite; some local dolomitization

Petrophysics

Pore Type(s)	Interparticle and moldic; solution-enhanced
Porosity	13–18%, 13% avg
Permeability	0.01–250 md, 108 md avg
Fractures	NA

Source Rocks

Age	Jurassic
Lithology(s)	NA
Migration Time	NA

Reservoir Dimensions

Depth	7900 ft (2410 m)
Thickness	10–20 ft (3–6.1 m)
Areal Dimensions	6.5 × 0.3 mi (10.4 × 0.5 km)
Productive Area	1165 acres (4.66 km^2)

Fluid Data

Saturations	NA
API Gravity	38°
Gas-Oil Ratio	650:1
Other	—

Production Data

Oil and/or Gas in Place	23 million BO and 15 billion CFG (0.42 billion m^3); 25.5 million equivalent BO total
Ultimate Recovery	8 million BO and 5.84 billion CFG (165 million m^3); 8.9 million equivalent BO total; URE = 34.9%
Cumulative Production	5 million BO and 3.65 billion CFG (103 million m^3) through 1983

Remarks: Discovered 1976. Production mechanism active water-drive. Production data courtesy J. Telapek Inc.

23

Late Subsurface Secondary Porosity in a Jurassic Grainstone Reservoir, Smackover Formation, Mt. Vernon Field, Southern Arkansas

Yehezkeel Druckman and Clyde H. Moore, Jr.

Introduction and Location

The Smackover Formation is an Upper Jurassic (Oxfordian) carbonate sequence found entirely in the subsurface throughout the northern Gulf of Mexico region. The Upper Smackover consists of oolitic and associated grainstones that have become known as prolific hydrocarbon reservoirs during the last four decades.

In the southern Arkansas part of the Smack-over trend, exploration has been oriented toward structures, and indeed many of the earlier discoveries were structural traps (Spencer and Peters, 1948; Goebel, 1950). However, many fields discovered in the 1960s were thought to be mainly stratigraphic traps (Bishop 1971a, 1971b; Chimene, 1976). Other investigators · studying many of the same fields have emphasized the significance of diagenetic controls on the distribution of porosity and permeability (*e.g.*, Becher and Moore, 1976; Moore and Druckman, 1981).

The Mt. Vernon Field is located in Columbia County, southern Arkansas, and produces upper Smackover oil and gas from 16 wells (Fig. 23-1). The field was discovered in 1976, and through 1983 had produced 5 million barrels of oil and 3.65 billion cubic feet (103 million m³) of gas. The field is essentially one well wide (approximately 0.3 mi, 0.5 km) and 6.5 miles (10.5 km) long.

Methods of Study

Detailed petrographic study and analyses of trace element and stable-isotope compositions were used to reconstruct the diagenetic history of carbonate rocks in the upper Smackover, considered to be the main controlling factor for trapping hydrocarbons at Mt. Vernon.

Cores from 16 wells served as the basic data source for the study. These cores were taken from three dry holes north of the field (wells 8,10,17), one dry hole south of the field (well 13), and the rest from producing wells. Of the numbered wells in Figure 23 1, cores were unavailable only from wells 6, 7, 14, 20, and 21. Two hundred thin sections of core samples were studied under the petrographic microscope, and in addition, petrophysical logs were used for correlation and definition of gross lithologies. Porosity and permeability data by standard core-analysis methods were available for almost all cores.

Regional Setting

The Smackover Formation is a regional transgressive-regressive carbonate cycle of regional extent that was deposited on a broad, gently Gulfward-sloping ramp (Ahr, 1973). The Smackover overlies the Jurassic Norphlet Formation, a few tens of feet thick, which in turn overlies the

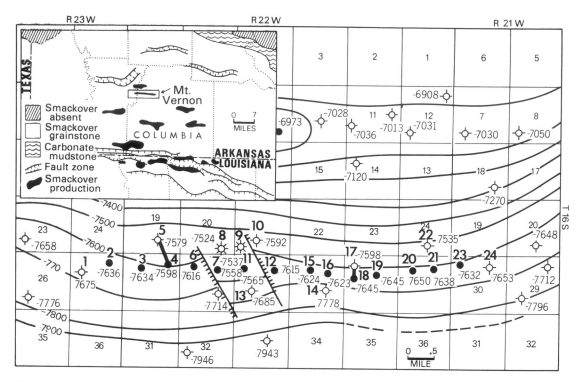

Fig. 23-1. Structural map of Mt. Vernon Field, drawn on top of the Smackover Formation, contour interval 100 feet (33 m). Numbered wells were studied. Grid is 1 × 1 mile (1.6 × 1.6 km), and numbers in centers of squares indicate section numbers. Inset map indicates regional facies distribution of Smackover Formation, production from the Smackover, and the south Arkansas (northern) and Buckner-Haynesville (southern) graben systems.

extensive Louann evaporites, also of Jurassic age (Fig. 23-2).

The Smackover Formation was divided by Dickinson (1968) into three members: the Lower, consisting predominantly of dark-brown, silty to argillaceous limestone, commonly laminated; the Middle, consisting of dense limestone; and the Upper, consisting mainly of oolitic limestone. According to Dickinson (1968), the middle and upper units interfinger laterally in the downdip direction with dark-gray marine shale, which he referred to as the Bossier Formation.

The upper member of the Smackover Formation, also known as the Reynolds Oolite, is basically an oolitic lime grainstone and accounts for most of the production. Akin and Graves (1969) recognized three laterally interfingering facies of this upper member occurring in the area of southern Arkansas and northern Louisiana: (1) a mudstone and pelletal calcarenite facies of low-energy, inner-shelf origin; (2) a restricted-shelf facies consisting of grapestone, which was found later to be a diagenetic belt of altered, oomoldic oolite (Moore and Druckman, 1981); and (3) an open-shelf oolitic facies. According to Dickinson (1968), the shelf lime grainstone facies and basinal shaly facies interfinger just south of the Arkansas-Louisiana state line.

The Smackover Formation is overlain by the Upper Jurassic Buckner Formation, which consists of a sabkha-type sequence mainly of nodular and enterolithic anhydrite, dolomitized lime mudstone, and carbonate algal-mats (Badon, 1973). The Buckner forms the overlying hydrocarbon seal at Mt. Vernon.

Structurally, the Smackover in the southern Arkansas area dips homoclinally basinward (Akin and Graves, 1969) at about 2 degrees on the average. The southward tilting is a cumulative result of differential subsidence basinward throughout the Mesozoic and Cenozoic history of the Gulf Coast region. In the Mt. Vernon area,

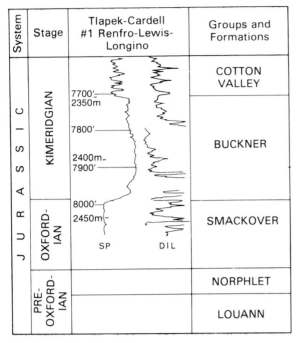

Fig. 23-2. Stratigraphic nomenclature at Mt. Vernon and typical SP (left) and induction (right) log response. Section below logs not to scale.

much of this tilting took place during the Late Jurassic and Early Cretaceous. Locally, there are subtle anticlinal structures generally related to diapirism of Louann Salt (Murray, 1961: Hughes, 1968). Two narrow regional graben systems dissect the area. the South Arkansas system north of Mt. Vernon and the Buckner-Haynesville system along the Arkansas-Louisiana state line, generally parallel to the depositional strike (Fig. 23-1, inset map). These systems are thought to have been active since the Late Jurassic (Murray, 1961; Bishop, 1973).

Mt. Vernon Field

A structural map on top of the Smackover Formation (Fig. 23-1) reveals that Mt. Vernon Field is located on the south-dipping homoclinal regional slope. There is no structural closure of the field to the north and only the slightest closure at the east and west ends of the field. Two minor faults striking northwest-southeast cut across the field, forming a small horst with vertical displacement of about 75 feet (23 m).

The Reservoir Rocks

The Upper Smackover in Mt. Vernon Field consists entirely of high-energy lime grainstones. No mudstones or wackestones were found in any of the cores from producing or dry wells. However, the grainstones are rather varied and can be grouped into four major facies: (1) ooids; (2) mixed oncolites and ooids; (3) mixed ooids, skeletal fragments, and intraclasts; and (4) mixed ooids and pellets. These facies are arranged in cyclically repeated vertical sequences. Up to five consecutive stacked cycles have been observed, with an average thickness of 100 to 130 feet (33–40 m) for each cycle.

The lower part of each individual sequence is commonly composed of even layers 7 to 16 feet (2–5 m) thick of skeletal and oncolitic grains mixed with various amounts of ooids, which grade upward into layers of well-sorted ooids. The oolitic portion of the sequence is generally 0.3 to 3 feet (0.1–1 m) thick, but reaches 16 feet (5 m) in well 13. The oolitic sequence is generally ripple cross-laminated, but in a few cases, high-angle cross-stratification occurs in the upper portions.

Pellets, probably indicating lower-energy conditions, were observed only in well 5, a dry hole just northwest of the field. Considering the thickest and best-sorted oolitic section in well 13 (dry well) south of the field and the pellet-containing sequence to the north, as well as the grain types and their sedimentary structures, one may postulate an east–west-trending oolitic shoal for the Mt. Vernon field, with its steep side facing south (basinward). However, one cannot call upon a sedimentologic porosity-permeability barrier to the north of the field, since wells 10 and 17 contain the same lithofacies as the producing wells.

Porosity and Permeability Distribution

Porosity and permeability within the reservoir sequence show wide variability both vertically and horizontally. Porosities range from 3 to 30 percent and permeabilities from 0.01 to hundreds of millidarcys. Porosity-permeability data are from core analyses.

The most striking data are those of well 13: the best-developed oolitic section (Fig. 23-3) attains a porosity of 25 to 30 percent, but permeabilities average only 0.01 millidarcy. Two south-north

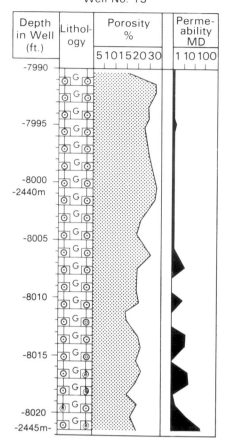

TLAPEK-CARDWELL
#1 RENFRO LEWIS LONGINO
Well No. 13

Fig. 23-3. Porosity and permeability distribution of uppermost Smackover, based on core analyses from well 13. Upper 15 feet (4.5 m) have high porosities and very low permeabilities corresponding to interval of tightly cemented oomoldic grainstone. Increase in permeability in lower part results from secondary late dissolution. Legend for this figure is given in Figure 23-4.

Diagenesis

Descriptions of the products of diagenetic processes observed at Mt. Vernon will generally follow the diagenetic history as interpreted from petrographic relations. They are summarized in Figure 23-6.

Ooid-Selective Dissolution

The earliest observable diagenetic event seems to have been selective dissolution of ooids, leaving behind the outer part of the cortex or, in many cases, only a rounded void within the surrounding cement (Fig. 23-7A). In some transitional cases, only parts of the ooids have been dissolved away. Oomoldic porosity was found in the uppermost 16.4 ft. (5 m) of well 13, and to a minor extent in well 5.

Early selective dissolution of aragonitic Recent and Pleistocene ooids from the Bahamas and Florida was reported by Robinson (1967), and was interpreted as the result of freshwater dissolution. The moldic porosity in Mt. Vernon Field is interpreted in a similar manner. The open structure of the ooids and the fact that their molds are not compacted suggest that dissolution took place early, before compaction.

Circumgranular Bladed-Crust Cement

Circumgranular bladed-crust cement is the earliest observed in the Smackover grainstones. It consists of elongate, well-formed calcite crystals encrusting the grains (Fig. 23-7B). The individual crystals are 20 to 40 microns long and 5 to 10 microns wide, with a general bladed morphology. This cement type is common in all Upper Smackover lithofacies in the Mt. Vernon Field and is not restricted to any particular geographic area or stratigraphic level. However, where it occurs, it makes up less than 5 percent of the rock volume and therefore does not seem to have significantly affected the porosity of the rock. Its effects on permeability may be significant but are hard to assess.

Numerous workers have noted that phreatic-marine cements have a bladed (high-Mg calcite) or a fibrous (aragonite) crystal habit. On the other hand, Halley and Harris (1979) have demonstrated that crusts of circumgranular bladed cements may precipitate in a phreatic freshwater

sections show porosity and permeability distribution across the field (Figs. 23-4, 23-5). Porosity and permeability both decrease sharply from the producing wells in the field to the non-producing wells to the north. As indicated earlier, a lower-energy facies does exist to the north, but the rocks in the non-producing wells are still lime grainstones and therefore the changes in porosity and permeability must be of post-depositional, diagenetic origin. The various products of these changes will be discussed in the sections to follow.

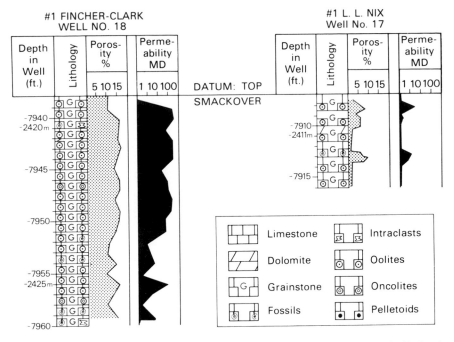

Fig. 23-4. Porosity and permeability distribution based on core analyses across the Mt. Vernon Field, from producing well 18 to non-producing well 17. For location see Figure 23-1. Note the lower porosity and permeability in the dry hole, despite the similarity in depositional characteristics. It is suggested that both porosity and permeability were enhanced by secondary late burial dissolution.

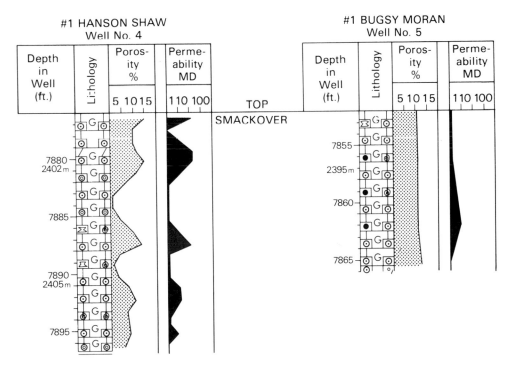

Fig. 23-5. Porosity and permeability distribution based on core analyses across Mt. Vernon Field, from producing well 4 to non-producing well 5. For location refer to Figure 23-1. Porous and permeable zones in producing well are due to secondary late burial dissolution.

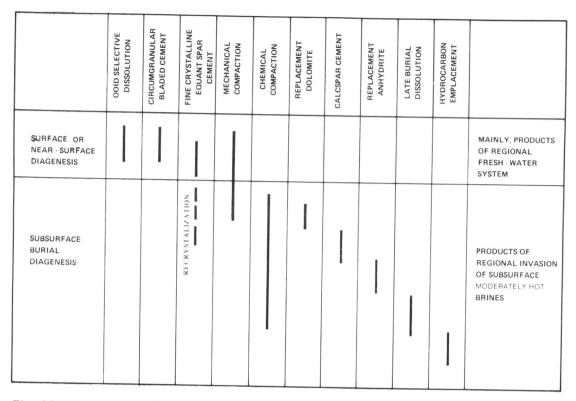

Fig. 23-6. Summary of diagenetic products, their relative time relationships, and suggested diagenetic environments.

environment. Petrographic observations of the circumgranular bladed cements in Mt. Vernon Field did not lead to any solid conclusion concerning the diagenetic environment in which they may have precipitated.

Fine Crystalline Equant Spar

Fine crystalline equant-spar cement is entirely restricted to the oomoldic section, where it has totally occluded interparticle pores (Fig. 23-7A,C). It consists of equant calcite crystals 15 to 30 microns in size. It postdates the bladed circumgranular cement, but also is interpreted as an early diagenetic product because the oomolds it surrounds are uncompacted. The exclusive association of equant-spar cement with oomoldic porosity suggests a genetic relationship between these two diagenetic features.

It seems reasonable to assume that calcium carbonate in the form of aragonite was dissolved from the ooids by fresh water and was precipitated in nearby interparticle voids as calcite, after supersaturation was reached. However,

under these circumstances, one would expect to find some equant-spar crystals filling the molds as well, but such mold fillings are rarely found in these rocks. The absence of calcite cement within the molds is difficult to explain, but may have resulted from a lack of suitable nucleation sites. One possibility is that organic films lining the molds, left behind by dissolved aragonite, prevented nucleation of calcite crystals.

Compaction

Two phases of compaction have been observed at Mt. Vernon: (1) mechanical crushing of grains resulting in spalling-off of oolitic rims, and breakage or deformation of the ooid (Fig. 23-7D); and (2) chemical compaction, both as pressure solution at grain contacts (microstylolites), and as wholesale dissolution along distinct surfaces, forming stylolites. The most destructive phase in terms of porosity loss is pressure solution at grain contacts, which in many cases has completely occluded primary interparticle pore space.

Compacted grainstones appear throughout the

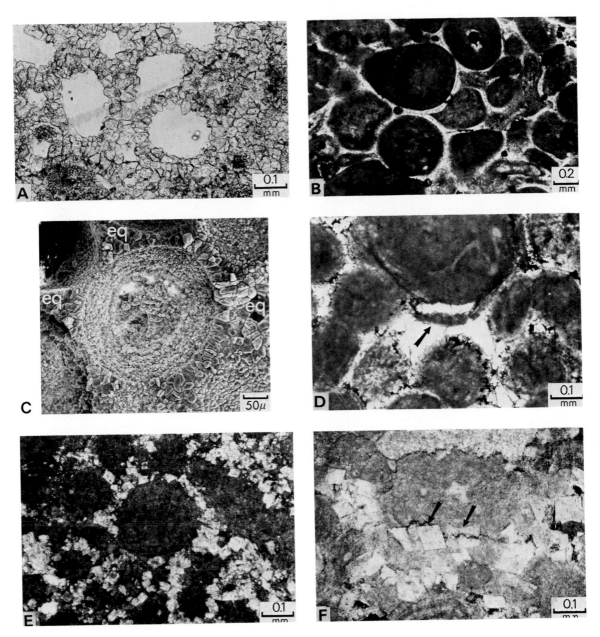

Fig. 23-7. Photomicrographs (in plane light) of thin sections (*A, B, D, E, F*) and SEM photograph (*C*) of textural features in Smackover rocks from Mt. Vernon Field. *A*. Oomoldic pores tightly cemented by equant spar. No. 1 Renfro-Longino-Lewis (well 13), 7998 feet (2437.8 m). *B*. Circumgranular bladed cement. Note significant compaction of grains despite early cement. No. 1 Renfro-Longino-Lewis (well 13), 8018 feet (2443.9 m). *C*. Partially leached ooids, cemented by circumgranular bladed cement and equant (eq) spar cement in the intergranular pore spaces. No. 1 Renfro-Lewis-Longino (well 13), 7996 feet (2437.2 m). *D*. Highly compacted oolitic grainstone. Oolitic rim spalled off (arrow) due to compaction and later cemented by a blocky cement. No. 1 Browning-Danielson (well 19), 7943 feet (2421.0 m). *E*. Partial replacement of oolitic grains by euhedral dolomite. It is not clear whether there was any pre-existing matrix or cement. No. 1 L.L. Nix (well 17), 7909 feet (2410.7 m). *F*. Highly compacted oncolites and ooids with microstylolitic contacts. Note the trace of the microcrystolite in dolomite crystal (arrow) and replacement of microstylolite by another dolomite crystal (arrow). Dolomitization appears to have happened after compaction. No. 1 Wilkerson Strange (well 1), 7946 feet (2421 m).

Upper Smackover. Thin compacted intervals appear even within the producing zones, but compaction is most common beneath the producing interval and in the dry wells north of the field.

It has been stated that early cementation prevents compaction (*e.g.*, Becher and Moore, 1976; Purser, 1978), and indeed most of the highly compacted Smackover grainstones lack early cements, whereas the early-cemented oomolds did not undergo significant compaction. However, in other cases (Fig. 23-7B), grains have undergone severe contact pressure solution, despite having been coated with early circumgranular cement. Moreover, the pressure solution affected mainly the grains rather than the circumgranular cement (see also Bathurst, 1975, p. 465; Cussey and Friedman, 1977). This differential pressure solution of the grains may have been caused by mineralogical differences between grains and cement. For example, if the cement were phreatic freshwater calcite and the grains still retained their aragonitic mineralogy, differential dissolution might have resulted under sufficient overburden pressure. Another possibility is that differences in crystal size between ooids and cement caused different rates of dissolution.

Much, if not most, of the compaction—both mechanical and chemical—appears to have taken place within the first 3300 feet (1000 m) of burial. This is evident from pressure-solution contacts between grains replaced by anhydrite pseudomorphous after gypsum (see following section on replacement anhydrite). Stylolites were relatively late and postdated all other observed diagenetic phenomena (Figure 23-6).

Stylolites are common in the Upper Smackover at Mt. Vernon, averaging 0.4 per foot (1.3 m) of section. The stylolites are horizontal, paralleling the bedding planes; their amplitudes vary from a few millimeters to 2 to 5 centimeters, the latter being most common. Using the assumption that the amplitude of a stylolite represents the minimum amount of dissolution along that surface (Bathurst, 1975), the minimum cumulative dissolution along stylolites was estimated to be 1.6 to 4.2 percent of the total rock volume, averaging 3 percent. The stylolites show no apparent pattern of occurrence, and occur in essentially the same amounts in dry holes and in producing wells. It therefore appears that stylolite development did not play a significant role as a permeability barrier in Mt. Vernon Field.

Replacement Dolomite

Euhedral dolomite crystals partially replace calcite grains. Crystal size is usually 50 to 80 microns, but crystals up to 150 microns have been observed. Replacement by dolomite usually starts at the periphery of a grain and extends inward until most of the grain is replaced (Fig. 23-7E). The original interparticle pore spaces are also partially or entirely filled with dolomite crystals that are commonly somewhat larger, and it is not clear whether these crystals replaced pre-existing cement or a micritic matrix.

The replacement dolomite occurs in thin lenses and layers a few centimeters thick, or locally, only a millimeter thick. Dolomitization is more common in the northern and western wells, but nowhere exceeds 10 percent of the rock volume. However, those intervals that are extensively dolomitized have lost most of their interparticle porosity.

Replacement by dolomite seems generally to have taken place after compaction, as pressure-solution contacts are commonly replaced by dolomite crystals (Fig. 23-7F).

Coarse Sparry Calcite (Calcspar)

Calcspar is the most common cement by far in the Upper Smackover at Mt. Vernon. It comprises up to 15 percent of rock volume and is thus the most significant diagenetic product reducing porosity and permeability. Calcspar cement is most common in the northern dry holes. It is also present in the producing wells, but there it has undergone partial dissolution, which has enhanced primary interparticle porosity and generated late secondary vug porosity.

The distribution of calcspar cement bears no relationship to grain types or depositional environments. Calcspar nucleated directly on grains or over pre-existing early circumgranular cements.

The calcspar consists of two basic types of crystals: large, clear poikilitic calcite crystals up to 1.5 millimeters in size, cementing several grains; and blocky crystals 0.05 to 0.5 millimeter in size, filling partially or entirely one single interparticle pore space.

Calcspar cement postdates mechanical and chemical grain-to-grain compaction, but pre-dates stylolites. On one hand, it has been observed to cement spalled-off oolitic rims (Fig. 23-7D) and crushed ooids; on the other, microstylolites among ooids have never been seen to penetrate into any adjacent calcspar crystals. Calcspar has cemented grainstones with grain-packing densities up to 80 to 90 percent that must have compacted from the original densities of 60 to 65 percent postulated for ooid sands by Coogan (1970) before cementation took place (Fig. 23-8A).

Replacement Anhydrite

Anhydrite is a minor constituent in the Upper Smackover, comprising not more than 1 to 2 percent of rock volume. It occurs either as scattered single crystals or as concentrations of crystals in intervals several centimeters thick. The crystals themselves vary in size from a few hundred microns to a few millimeters, and rarely exceed 1 to 2 centimeters. Most crystals are prismatic in shape, but some are lenticular and probably are pseudomorphous after gypsum (Fig. 23-8B).

The anhydrite crystals replace the pre-existing rock. The replacement rarely has gone to completion, but generally left behind "ghosts" of the replaced grains, their texture, and some of the earlier cements (Fig. 23-8C). Commonly, one sees ghosts of compacted grains as in nearby, unreplaced rock.

Early circumgranular-crust cement, replacement dolomite, and calcspar cements also have been partially or entirely replaced by the anhydrite crystals. These replacement phenomena occur in both prismatic and lenticular anhydrite crystals.

The inversion of gypsum to anhydrite due to dehydration with burial is well documented. Anhydrite pseudomorphous after gypsum has been reported by numerous investigators (*e.g.*, Kerr and Thomson, 1963; West, 1964). The stability fields of the two phases at different temperature, pressure, and fluid salinities were calculated by MacDonald (1953), Hardie (1967) and Berner (1971). Using these data and applying the present-day geothermal gradient in the Mt. Vernon area implies that gypsum will invert to anhydrite in the presence of brines saturated for gypsum at a depth of about 3600 feet (1100 m). In the presence of saltier brines or higher geothermal gradients, this inversion will occur at shallower depth.

The significance of these data is their implica-

tion regarding the timing of some diagenetic processes that were completed while gypsum was still the stable phase—namely, prior to burial below about 3300 feet (1000 m). However, it should be emphasized that the prismatic replacement anhydrite crystals also were precipitated below that depth and that there is no indication of the depth limit at which replacement by anhydrite ceased. The only diagenetic events that postdated anhydrite replacement were stylolitization and late dissolution.

Late Dissolution-Enhanced Porosity

The dominant porosity type in the producing zone at Mt. Vernon, as described more generally by Moore and Druckman (1981), is late dissolution-enhanced porosity (Fig. 23-8D). This porosity postdates all precipitated cements, replacement minerals, and compaction features. No precipitation of any mineral was observed to take place after this dissolution stage, but the emplacement of hydrocarbons did clearly postdate it (Fig. 23-8E).

The resulting voids are equidimensional or lobate pores, hundreds of microns in size. Criteria used for recognition of secondary late dissolution porosity areas are the following:

1. Truncated and corroded grains and cement phases (Fig. 23-8F). This criterion is easily applied in oolitic grainstones, where concentric rims serve as excellent markers for truncation. Truncation of early circumgranular cement also is easily recognized, but truncation of calcspar is more difficult to determine since truncation or incomplete filling may look similar; however, where calcspar crystals are irregularly corroded, the age relationships are clear. It should be noted that corrosion and truncation have not been observed in the case of dolomite or anhydrite crystals; however, anhydrite crystals have undergone slight dissolution at intersections of cleavage planes, leaving micropores.
2. Elongate and lobate shapes of voids. These are excellent indicators of secondary dissolution, particularly when compared with the typical rhomboidal shapes of primary interparticle pores between ooids.
3. Voids among grains that have flat surfaces. Compaction along contacts between grains

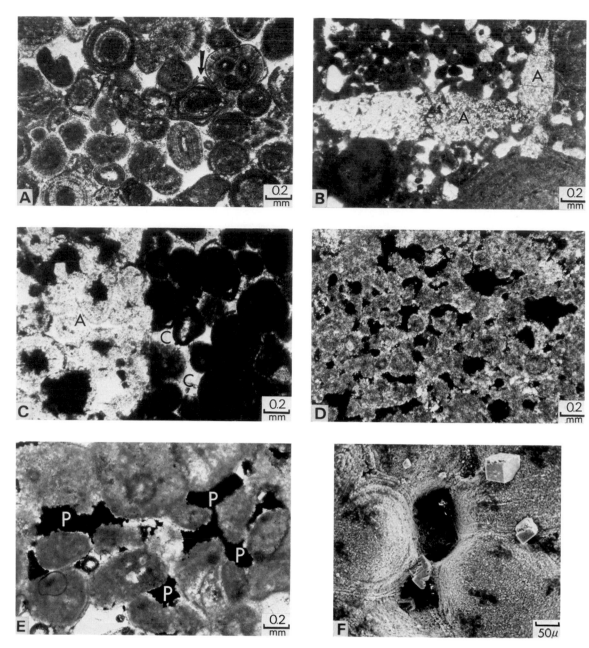

Fig. 23-8. Photomicrographs of thin sections (*A–E*) and SEM photograph (*F*) of textural features in Smack-over rocks from Mt. Vernon Field. *A.* Compacted oolitic grainstone, cemented with poikilitic calcspar. Cementation took place after spalling off of oolitic rims (arrow) and compaction of the grains to their present packing density (90%). No. 1 L.L. Nix (well 17). 7913 feet (2412.0 m). Plane light. *B.* Replacement anhyd-rite (A) that appears to be pseudomorphous after gyp-sum. Ghosts and relics of compacted ooids and dolomite crystals found in anhydrite may indicate that compaction and dolomitization took place prior to in-version of gypsum to anhydrite. No. 1 Lewis-Longino (well 11). 7870 feet (2398.8 m). Plane light. *C.* Replacement anhydrite (A) replacing ooids in compac-ted fabrics, adjacent to non-replaced compacted ooids cemented with calcspar (C). No. 1 L.L. Nix (well 17). 7013.7 feet (2137.8 m). Plane light. *D.* Secondary late porosity (black) resulting from dissolution of both com-pacted grains and cements. No. 1 Browning-Danielson (well 19). 7936 feet (2418.9 m). Cross-polarized light. *E.* Enlarged voids by late dissolution, filled with "pyrobitumen" (P). Hydrocarbon emplacement was the only event that postdated the late dissolution. Note the eroded contacts of both grains and cements with the "pyrobitumen." No. 1 W.E. Dickson (well 23). 7957 feet (2425.3 m). Plane light. *F.* Secondary enhanced porosity in highly compacted grainstone. Note corro-sion of oolitic rims by dissolution. Dolomite crystals are resistant to dissolution. No. 1 Charles Nix (well 15). 7951 feet (2423.5 m).

commonly changes grain shapes from spherical to polyhedral, which are easily recognized in thin sections due to their polygonal intersections. Where the straight boundaries face a void, we infer that the adjacent solid has been removed.

4. Voids as large as or larger than the grains, especially in well-sorted ooids. Even voids smaller than the average grain size but larger than the average primary interparticle pore should be considered as possible enlarged primary pores, but this interpretation can be made only if some corrosion of ooid rims can be observed.

The root cause for localization of late dissolution in a narrow zone parallel to depositional strike is probably linked to a pattern in which preservation of primary porosity was influenced by the original depositional environment. Brock (1980) has shown that certain subtle differences in sediment texture and fabric and grain composition may control porosity preservation during burial; competent grains such as ooids resist compaction and thus retain primary interparticle porosity better than pellets and oncolites, so that preservation of primary porosity depends on the proper proportions of these grains. Such zones of preserved primary porosity may well have been migration conduits for late corrosive fluids, and hence facilitated the ultimate development of economic porosity at Mt. Vernon.

Discussion and Interpretation

Early diagenesis appears to have had minor effects on the reservoir properties of Smackover carbonates at Mt. Vernon. An exception is the total occlusion of permeability in the oomoldic section of well 13 by fine equant spar (Fig. 23-3). However, even this cement may have undergone late recrystallization, as inferred from the common occurrence of two-phase fluid inclusions.

Compaction, late cementation, and replacement appear to have been the main destructive diagenetic phases in Mt. Vernon. Fluid-inclusion homogenization temperatures of 80 to 100°C for both the replacement dolomites and calcspar cements (temperature data from M. J. Klosterman, personal communication) indicate the precipitation of these phases at elevated temperatures; these temperatures were confirmed by oxygen stable-isotope data (Knauth and Moore, 1979). Moreover, depressed freezing points in the same fluid inclusions suggest that the fluids trapped in the inclusions were saline.

Based on the petrographic relations mentioned earlier and the fluid-inclusion and oxygen-isotope data, we suggest that the dolomitization and calcspar cementation as well as recrystallization of some of the early cements took place under relatively deep burial conditions, in moderately hot brines. The occurrence regionally in the Smackover Formation of concentrated brines similar in composition to the brines believed to have been responsible for the late-diagenetic products at Mt. Vernon has been well documented by Collins (1974). Detailed correlation of the regional diagenetic processes and products with these Smackover fluids has been outlined in Moore and Druckman (1981).

Conclusions

Secondary porosity of late subsurface origin has received considerable attention in sandstone reservoirs (Pittman, 1979; Schmidt and McDonald, 1979), but in recent years has been ignored in carbonate reservoirs. Mt. Vernon is the first well-documented case of late subsurface secondary porosity production in a carbonate reservoir.

Porosity enhancement was accomplished late in the diagenetic history of the Mt. Vernon area, and was followed closely by emplacement of hydrocarbons. Dissolution may have been caused by subsurface fluids charged with CO_2 or H_2S, or both, gases commonly generated together with hydrocarbons (Tissot and Welte, 1978). If this mechanism is valid, the development of subsurface solution porosity in carbonate reservoirs may be directly related to hydrocarbon generation, and should be a common phenomenon in the deeper parts of carbonate hydrocarbon provinces.

Acknowledgments This study was made possible by the support of the Applied Carbonate Research Program industrial associates and the Department of Geology, Louisiana State University. We thank George Herman for comments on an earlier version of the manuscript, Marshall Vinet for assistance in acquiring material and input relative to the geologic setting, Mary Jo Klosterman for two-phase fluid inclusion determinations, L. Paul Knauth for stable-isotope analyses, and Marlene Moore for manuscript preparation.

References

AHR, W.M., 1973, The carbonate ramp—an alternative to the shelf model: Gulf Coast Assoc. Geol. Soc. Trans., v. 23, p. 221–225.

AKIN, R.K., and R.W. GRAVES, 1969, Reynolds oolite of southern Arkansas: Amer. Assoc. Petroleum Geologists Bull., v. 53, p. 1909–1922.

BADON, C.L., 1973, Petrology of the Norphlet and Smackover formations (Jurassic), Clarke County, Mississippi: Ph.D. dissert., Louisiana State Univ., Baton Rouge, 197 p.

BATHURST, R.G.C., 1975, Carbonate Sediments and Their Diagenesis: Developments in Sedimentology 12: Elsevier Scientific Publ. Co., 658 p.

BECHER, J.W., and C.H. MOORE, 1976, The Walker Creek Field: a Smackover diagenetic trap: Gulf Coast Assoc. Geol. Soc. Trans., v. 26, p. 34–56.

BERNER, R.A., 1971, Principles of Chemical Sedimentology: McGraw-Hill, Internat. Ser. Earth & Plan. Sci., 240 p.

BISHOP, W.F., 1971a, Geology of a Smackover stratigraphic trap: Amer. Assoc. Petroleum Geologists Bull., v. 55, p. 51–63.

BISHOP, W.F., 1971b, Stratigraphic control of production from Jurassic calcarenites, Red Rock Field, Webster Parish, Louisiana: Gulf Coast Assoc. Geol. Soc. Trans., v. 21, p. 125–137.

BISHOP, W.F., 1973, Late Jurassic contemporaneous faults in north Louisiana and south Arkansas: Amer. Assoc. Petroleum Geologists Bull., v. 79, p. 858–877.

BROCK, F., 1980, Walker Creek revisited: a reinterpretation of the diagenesis of the Smackover Formation of Walker Creek Field, Arkansas: M.S. Thesis, Louisiana State Univ., Baton Rouge, 82 p.

CHIMENE, C.A., 1976, Upper Smackover reservoirs, Walker Creek Field area, Lafayette and Columbia counties, Arkansas: in Braunstein, J., ed., North American oil and gas fields: Amer. Assoc. Petroleum Geologists Mem. 24, p. 177–204.

COLLINS, G.A., 1974, Geochemistry of liquids, gases and rocks from the Smackover Formation: U.S. Bureau Mines, Report of Investigation 7897, 84 p.

COOGAN, A.H., 1970, Measurements of compaction in oolitic grainstones: Jour. Sedimentary Petrology, v. 40, no. 3, p. 921–929.

CUSSEY, R., and G.M. FRIEDMAN, 1977, Patterns of porosity and cement in ooid reservoirs in Dogger (Middle Jurassic) of France: Amer. Assoc. Petroleum Geologists Bull., v. 61, p. 511–518.

DICKINSON, K.A., 1968, Upper Jurassic stratigraphy of some adjacent parts of Texas, Louisiana and Arkansas: U.S. Geol. Survey, Prof. Paper 594E, 25 p.

GOEBEL, L.A., 1950, Cairo Field, Union county, Arkansas: Amer. Assoc. Petroleum Geologists Bull., v. 34, p. 1954–1980.

HALLEY, R.B., and P.M. HARRIS, 1979, Freshwater cementation of a 1,000-year-old oolite: Jour. Sedimentary Petrology, v. 49, no. 3, p. 969–988.

HARDIE, L.A., 1967, The gypsum anhydrite equilibrium at one (atmosphere) pressure: Amer. Mineralogist, v. 52, p. 171–200.

HUGHES, D.J., 1968, Salt tectonics as related to several Smackover fields along the northeast rim of the Gulf of Mexico Basin: Gulf Coast Assoc. Geol. Soc. Trans., v. 18, p. 320–330.

KERR, S.D., and A. THOMSON, 1963, Origin of nodular and bedded anhydrite in Permian shelf sediments, Texas and New Mexico: Amer. Assoc. Petroleum Geologists Bull., v. 47, p. 1726–1732.

KNAUTH, L.P., and C.H. MOORE, 1979, Isotope geochemistry of subsurface Jurassic Smackover carbonates. Geol. Soc. America Abst. with programs, 459 p.

MACDONALD, G.J.F., 1953, Anhydrite-gypsum equilibrium relations. Amer. Jour. Science, v. 251, p. 884–898.

MOORE, C.H., and Y. DRUCKMAN, 1981, Burial diagenesis and porosity evolution, Upper Jurassic Smackover, Arkansas and Louisiana: Amer. Assoc. Petroleum Geologists Bull., v. 65, p. 597–628.

MURRAY, G.E., 1961, Geology of the Atlantic and Gulf Coastal Province of North America: Harper and Bros., New York, 692 p.

PITTMAN, E.D., 1979, Porosity diagenesis and production capability of sandstone reservoirs: in Scholle, P.A. and P.R. Schluger, eds., Aspects of Diagenesis—a symposium: Soc. Econ. Paleontologists and Mineralogists, Spec. Publ. 26..

PURSER, B.H., 1978, Early diagenesis and the preservation of porosity in Jurassic limestones: Jour. Petroleum Geology, v. 1, p. 83–94.

ROBINSON, R.B., 1967, Diagenesis and porosity development in Recent and Pleistocene oolites from southern Florida and the Bahamas: Jour. Sedimentary Petrology, v. 37, p. 355–364.

SCHMIDT, V., and D.A. MCDONALD, 1979, The

role of secondary porosity in the course of sandstone diagenesis: *in* Scholle, P.A. and P.R. Schluger, eds., Aspects of Diagenesis—a symposium: Soc. Econ. Paleontologists and Mineralogists, Spec. Pub. 26, p. 175–208.

SPENCER, L.S., and J.W. PETERS, 1948, Geophysical case history of the Magnolia Field, Columbia County, Arkansas: *in* Nettleton, L.L., ed., Geophysical Case Histories, v. 1, Soc. Exploration Geophysicists, p. 443–460.

TISSOT, B.P., and D.H. WELTE, 1978, Petroleum Formation and Occurrence: Springer-Verlag, Inc., Berlin, 527 p.

WEST, I.M., 1964, Evaporite diagenesis in the Lower Purbeck Beds of Dorset: Yorkshire, Geol. Soc. Proc., v. 34, p. 315–330.

Hico Knowles Field

Paul D. Crevello, Paul M. Harris, David L. Stoudt, and Lawrence R. Baria

RESERVOIR SUMMARY

Location & Geologic Setting	Lincoln Parish, north Louisiana Gulf of Mexico coastal plain, USA
Tectonics	Syndepositional faulting and uplift of horst due to salt diapirism
Regional Paleosetting	Ramp-type Jurassic paleomargin, north of Gulf of Mexico
Nature of Trap	Structural/stratigraphic; updip fault, closure, and lime-mud-rich limestones (mudstone-wackestone) plus terrigenous clastics
Reservoir Rocks	
Age	Late Jurassic (Oxfordian)
Stratigraphic Units(s)	Smackover Formation
Lithology(s)	Limestone
Dep. Environment(s)	Reef and reef-rubble margins
Productive Facies	Coral-stromatoporoid framestone and skeletal rudstone to packstone
Entrapping Facies	Shales, tight sandstones, lime wackestone and mudstone, and updip fault
Diagenesis Porosity	Meteoric-water dissolution and some cementation; deep-burial cementation by ferroan calcite and dolomite; and replacement and cementation by anhydrite
Petrophysics	
Pore Type(s)	Interparticle & skeletal moldic
Porosity	2–21% (14 avg), 10–18% (15% avg) (reef rubble packstone-grainstone)
Permeability	0.1–70 md (2 md), 1–25 md (3 md)
Fractures	Minor, & effects may be negligible
Source Rocks	
Age	NA
Lithology(s)	NA
Migration Time	NA
Reservoir Dimensions	
Depth	10,700 ft (3260 m)
Thickness	20–40 ft (6.1–12.2 km)
Areal Dimensions	1 × 0.5 mi (1.6 × 0.8 km^2)
Productive Area	320 acres (1.3 km^2)
Fluid Data	
Saturations	S_w = 30% avg
API Gravity	55° (condensate)
Gas-Oil Ratio	NA
Other	—
Production Data	
Oil and/or Gas in Place	1 billion CFG and equivalent (159 million m^3); 1.67 million BO equivalent
Ultimate Recovery	~0.5 billion CFG and equivalents (79 million m^3); ~0.84 million BO equivalent; URE ≈ 50%.
Cumulative Production	NA

Remarks: IP 3.9 million CFGPD (0.62 million m^3) and 450 BCPD. IP mechanism water-drive. Discovered 1978.

24
Porosity Evolution and Burial Diagenesis in a Jurassic Reef-Debris Reservoir, Smackover Formation, Hico Knowles Field, Louisiana

Paul D. Crevello, Paul M. Harris, David L. Stoudt, and Lawrence R. Baria

Location and Regional Setting

Hico Knowles Field, located in Lincoln Parish, northern Louisiana, provides the only example to date of Smackover production from a skeletal carbonate buildup (Fig. 24-1; Baria *et al.*, 1982). Reservoirs in the Upper Smackover occur mainly in oolite grainstone shoal facies. Most carbonate buildups drilled to date in the Smackover trend have been either of nonreservoir quality or in water-wet legs of productive intervals (Baria *et al.*, 1982).

Regionally, the Smackover Formation (Fig. 24-2) accumulated during the Upper Jurassic (Oxfordian) on a gently sloping carbonate platform or ramp (Ahr, 1973) that rims the United States Gulf Coast in the subsurface. The pre-Smackover Mesozoic sequence comprises red beds of the Eagle Mills, the Louann Salt, anhydrites and nonmarine sandstones of the Werner Formation, and nonmarine sandstones and shales of the Norphlet Formation. The Smackover itself consists mainly of platform carbonate nonskeletal grainstones to wackestones. Studies of these sedimentary units and of the evolution of the Gulf of Mexico have led to the interpretation that the northern Gulf was a spreading ridge system and passive margin throughout most of the Jurassic until the mid-Cretaceous (Buffler *et al.*, 1980; Dickinson and Coney, 1980). The Late Jurassic also was a time of rising worldwide sea level (Vail *et al.*, 1977). Onlap relationships of these sedi-

mentary units on seismic records indicate that a eustatic transgression of the ancestral Gulf took place during deposition of the Lower Smackover Formation (Todd and Mitchum, 1977). A regressive cycle began in the Upper Smackover and continued with deposition of Buckner evaporites and redbeds. Another transgressive-regressive pulse during deposition of the Cotton Valley Group culminated Jurassic sedimentation in the region.

The Smackover Formation can be subdivided into a lower member of dark fine-grained carbonates, calcareous shales and sandstones, and an upper member of predominantly nonskeletal carbonate sands and muds (Bishop, 1968; Akin and Graves, 1969). Locally, carbonate facies are interbedded with sandstones and shales that complicate well-log correlations of the Upper Smackover.

The regional configuration of depositional environments and facies for the Upper Smackover is well known (Bishop, 1968; Badon, 1974; Collins, 1980) (Fig. 24-1). The upper member is interpreted as a belt of moderate- to high-energy grainstones and packstones that rims much of the Gulf Coast. The belt commonly consists of several offlapping lenses or "shingles" (Chimene, 1976). The shingles thicken basinward (generally southward) and represent deposition as tidal bars and channels, beaches, and islands. The high-energy facies change updip into tidal flat or lagoonal clastics, carbonate muds or evaporites of

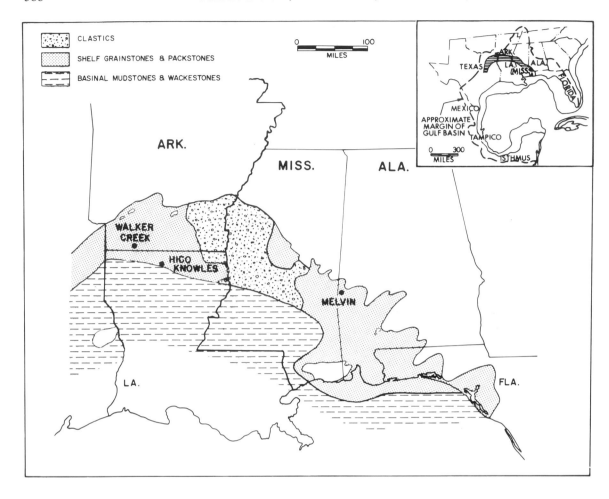

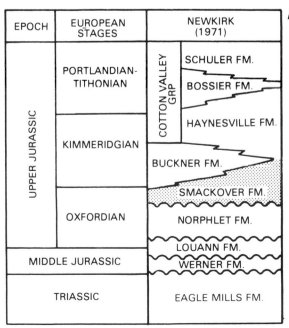

Fig. 24-2. Stratigraphic column for Triassic and Jurassic of United States Gulf Coast.

▲ *Fig. 24-1.* Generalized map of Smackover lithofacies along part of United States Gulf Coast, showing the location of the three Smackover fields discussed in the text. Northernmost facies boundary corresponds to updip limit of Smackover Formation. Inset regional map modified after Ottmann *et al.* (1973).

the Buckner Formation. Reef development occurred seaward of the main belt of oolite shoals. Downdip from the shelf grainstones and reefs, basinal carbonate muds occur along with quartz sandstones, siltstones, and shales.

Discovery

The Smackover discovery well and subsequent wells in Hico Knowles Field (Fig. 24-3) were drilled in 1978 by Bass Enterprises in an attempt to locate a structural trap with fault-bounded closure. Search for these types of traps guided much of the early exploration strategy for Smack-

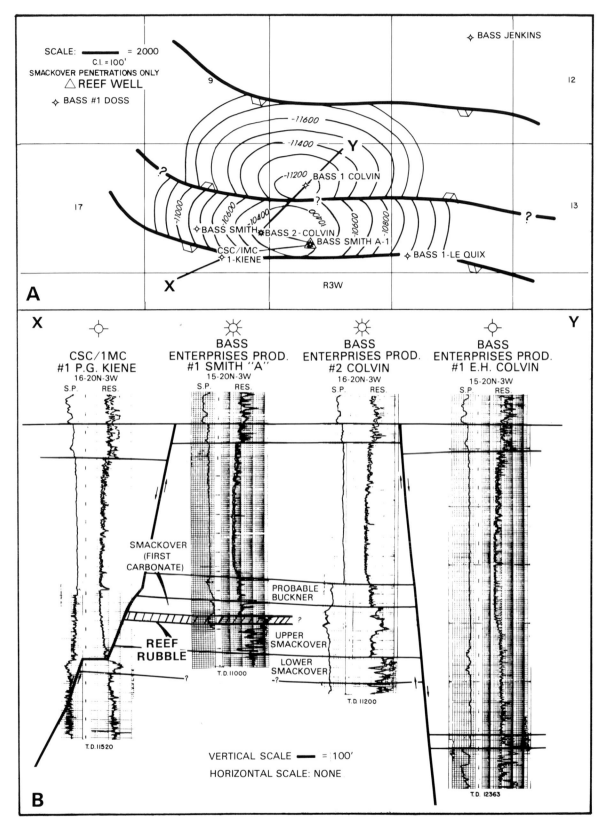

Fig. 24-3. A. Smackover structure map of Hico Knowles Field in northern Louisiana. *B.* Northeast–southwest cross-section through Hico Knowles Field showing position of reef rubble within Upper Smackover on a horst block. Rubble interval is described in cores from Bass #1 Smith "A" well (Fig. 24-5). *A* and *B* modified from Baria *et al.* (1982). Published by permission, American Association of Petroleum Geologists.

over reservoirs because many of the highs, which were produced by salt diapirism or deeper-seated faulting, developed across the Smackover shelf contemporaneously with deposition and localized high-energy, shallow-water grainstone facies (Chimene, 1976; Becher and Moore, 1976; Baria *et al.*, 1982).

Six wells defined the limits of the field on a narrow east-west horst with fault closure to the north and south, and structural dip closure to both east and west (Fig. 24-3A). Well-log correlations show several shale-sandstone-carbonate cycles in the Upper and Lower Smackover (Figs. 24-3B, 24-4). Updip, but off structure, log correlations with the Bass #1 E. H. Colvin show a thinner and shalier Upper Smackover, about 200 feet (61 m) thick compared to 270 and 250 feet (82 m and 76 m), respectively, in the Bass #1 Smith "A" and #2 Colvin (Fig. 24-3B). In addition, thin Buckner sections in these latter two on-structure wells, 130 and 120 feet (40 m and 36 m), respectively, and dramatic thickening of the Buckner off structure to 235 feet (72 m) in the Bass #1 E. H. Colvin, indicate structural movement (salt diapirism?) contemporaneous with deposition of the Smackover and Buckner. Gentle westward tilting of the structure postdated Buckner deposition.

Production was limited to the two wells located on the crest of the structure. The Bass #2 Colvin was drilled to a total depth of 11,200 feet (3414 m) subsea (Figs. 24-3B, 24-4). Upper Smackover lithologies in this well consist of alternating cycles of shale, sandstone, and oncolite-oolite grainstone and packstone. Initial production by water drive from perforations at 10,822 to 10,830 feet (3298–3301 m) tested 3.946 million cubic feet (0.112 million m^3) of gas per day and 450 barrels of condensate per day through a 16/64-inch (0.64-cm) choke from moderately quartz-rich oolite and oncolite grainstones. A second well drilled on the structure, the Bass #1 Smith "A," reached a total depth of 11,000 feet (3353 m) subsea. This well encountered 45 feet (14 m) of skeletal rubble, grainstones, and packstones between 10,725 and 10,770 feet (3269–3283 m) and cored 35 feet (11 m) of the unit. Perforations between 10,731 and 10,771 feet (3271–3283 m) tested 4,711 million cubic feet (133.3 million m^3) of gas per day and 576 barrels of condensate per day through a 14/64-inch (0.56-cm) choke. Drillstem test data were not available to deter-

mine whether a single, continuous reservoir exists between the two wells. However, similar production test rates suggest that the two productive but lithologically different reservoirs were in communication. Both wells were shut in because of limited reservoir pressures.

General Reservoir and Trap Characteristics

The reservoir at Hico Knowles Field occurs in oolite grainstones (Bass #2 Colvin) and, more unique to the Upper Smackover, in skeletal reef rubble and sand (Bass #1 Smith "A"). Discovery of skeletal debris in only one well does not allow the size of the carbonate buildup to be determined, and seismic data was not available to resolve buildup size or reservoir geometry. However, the overall reservoir geometry was estimated to be considerably less than 1-1/3 miles (2 km) long east to west and 2,000 feet (600 m) wide. This estimate was based on the absence of productive reservoir lithologies in the Bass Smith well drilled on the western flank of the structure, and on the width between the updip and downdip faults (*i.e.*, north and south bounding faults) of the structure (Fig. 24-3). Potential porous reservoir thickness of the reef rubble approaches 45 feet (14 m) (Fig. 24-4). By contrast, the oolite intervals in the Bass #2 Colvin are individually less than 20 feet (6 m) thick but combine for a cumulative porous interval approaching 40 feet (12 m). However, log data suggest that a gas column only 15 to 20 feet (5–6 m) thick occurs in the uppermost porous intervals of both wells.

Geology of the Bioclastic Rubble

Hico Knowles Field produced from oncolite-oolite grainstones and skeletal reef rubble and lime sand which accumulated on a topographic high of the Smackover shelf. The skeletal unit is the only lithofacies discussed here because it is an excellent example of a biotically diverse and complex bioclastic lithofacies representing debris that accumulated in an area where patch reefs were probably developed (Crevello and Harris, 1982). During diagenesis, the skeletal limestone unit retained primary interparticle porosity of reser-

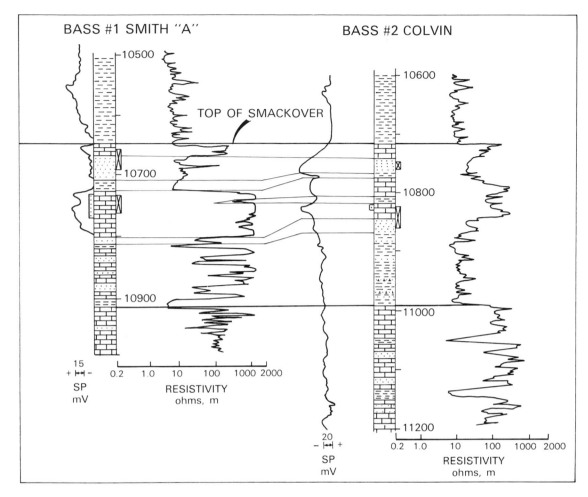

Fig. 24-4. Log correlation of lithologies in the Bass #1 Smith "A" and Bass #2 Colvin wells. Lithologies, represented by limestone, sandstone, and shale symbols, are based on core descriptions and log characteristics. Perforated and cored intervals designated on the left and right, respectively, of the graphic lithology column.

voir quality after deposition, developed extensive moldic porosity through leaching of skeletal grains in the meteoric-phreatic environment, and retained porosity of reservoir quality through several episodes of cementation in the deep subsurface.

The skeletal rubble was encountered near the seaward margin of the horst which, during deposition of the Upper Smackover, was the open-ocean-facing side of a sea floor paleohigh. Borehole log correlations of Smackover thicks and Buckner thins over the structure indicate that carbonate deposition was localized on this high. Borehole logs of the Bass #1 Smith "A" and #2 Colvin (Fig. 24-4) show that the skeletal rubble is capped by a shale-sandstone-carbonate cycle and

overlies terrigenous sandstone-carbonate couplets deposited in subtidal marine conditions. The skeletal rubble correlates to oncolite-oolite grainstones and packstones in the Bass #2 Colvin that overlie 140 feet (43 m) of shale and sandstone. The log signature of the thick terrigenous interval in the Upper Smackover of the Bass #2 Colvin well identifies events of silt or calcareous sand lithologies that correlate with shale-sand-carbonate cycles in the Bass #1 Smith "A."

Ahr and Palko (1981) incorporate the Bass #1 E. H. Colvin well (Fig. 24-3) into a borehole-log cross section that extends northward nearly 10 miles (16 km) to Shiloh Field. Their core data and log correlations show that shales and sandstones accumulated in topographic depressions,

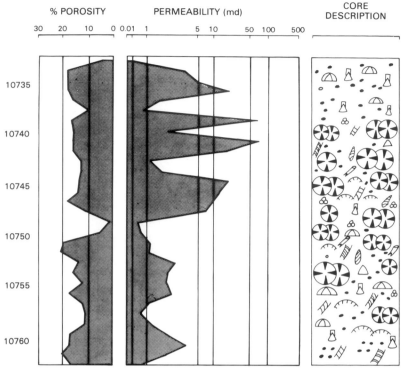

Fig. 24-5. Core description of the reef rubble in Bass #1 Smith "A" well, Hico Knowles Field (modified from Baria *et al.*, 1982. Published by permission, American Association of Petroleum Geologists.).

Symbols in the graphic log are after Swanson (1981; see Fig. 24-12). See Figure 24-6 for detailed description of the Bass #1 Smith "A" reef rubble core.

most likely salt-withdrawal troughs, whereas carbonate-sandstone cycles accumulated in positions intermediate to the crests and flanks of paleotopographic highs on which carbonate-shoal lithofacies were being deposited.

The cyclic nature of the major lithologies carries over to smaller depositional packages where, in the cored interval of the Bass #1 Smith "A" well (Figs. 24-5, 24-6), skeletal rudstones, grainstones, packstones, and sandy siltstones alternate in beds several feet thick. Though dominated by carbonate lithologies, the makeup of the cored interval is variable. Depositional cycles in the carbonates are recognized by variations in grain size and texture or by the occurrence of thin beds of terrigenous siltstones and sandstones. Roughly seven cycles can be recognized in the Bass #1 Smith "A" core (Fig. 24-6).

Bioclastic rudstones are the coarsest sediments present and form the basal beds of depositional units. Rudstones consist primarily of gravel to pebble-sized whole and fragmented colonial scleractinian corals, calcareous algae (*Parachaetetes*

sp., *Cayeuxia* sp., and coralline red algae), and mollusks. The matrix of the rudstones consists of sand-sized, comminuted skeletal fragments of corals, mollusks, skeletal algae, peloids, and echinoid spines deposited as grainstones and packstones. The grainstone-packstone matrix has patchy distribution and often occurs as sediment that filtered in between the gravel-sized skeletal components filling sediment-floored shelter cavities. The fine- to medium-size bioclastic grainstones and packstones also overlie the rudstones in several places, capping each depositional sequence. Terrigenous siltstones and very fine-grained sandstones often cap packstones, but also occur at random through the rudstones.

The skeletal interval in the Bass Smith "A" core is interpreted as reworked detritus from a nearby patch reef, based on several lines of evidence. These include: (1) the combination of fragmented, gravel-supported skeletal debris with bioclastic grainstones and packstones filtered in between clasts, and variable depositional sequence of bioclastic rudstones, grainstones, packstones, and terrigenous silt/sandstones, (2) the absence of reef-building organisms in growth position, and (3) extensive marine cementation. Crevello and Harris (1982 and 1984) discuss criteria used to recognize various types of Smackover buildups and related lithofacies in cores.

Porosity Evolution and Progressive Burial Diagenesis

The paragenetic sequences and relative importance of diagenetic events affecting the porosity of the skeletal rubble in Hico Knowles Field are outlined in Figure 24-7, along with interpreted diagenetic environments. The sequential timing of events outlines a diagenetic history that reflects progressive burial. Marine cements and internal sediment were important in reducing primary interparticle porosity. Exposure to meteoric water caused leaching of corals and other aragonite skeletal grains and resulted in precipitation of early-stage cements. Neomorphism of the remaining metastable carbonate grains to low-magnesium calcite must have been temporally related to flushing by meteoric waters. Collapse of skeletal molds, grain fracturing, and minor fracturing of rock matrix along with several late-stage cements, affected the reef-rubble porosity prior to emplacement of hydrocarbons.

Stage 1: Marine Phreatic

Interparticle pores and shelter cavities in skeletal reef rubble rudstones and grainstones are commonly lined with isopachous calcite druses of fibrous crystals up to 0.3 millimeter thick (generally 0.1–0.14 mm) that exhibit radial-fibrous or sweeping extinction (Fig. 24-8A–C). Circumgranular bladed calcite spar crystals up to 80 microns in length also exhibit sweeping extinction. Generally, the cement types occur in close association within the same pore. Fibrous cements quite commonly totally occlude individual pores. Where fibrous and bladed cements occur together, they are present only as isopachous druses. These cement fabrics, which we interpret as early marine, are similar to those reported from many Holocene reefs, where their marine origin is well documented (Ginsburg *et al.*, 1971; Schroeder, 1973; James *et al.*, 1976).

Stage 2: Mixed Phreatic and Shallow Meteoric Phreatic

Molds of corals, gastropods, serpulids, and layers within pelecypod shells suggest that aragonite was preferentially dissolved (Figs. 24-8C–F and 24-9A,B,C,E). By contrast, marine cements and skeletal red algae, echinoids, and certain layers within mollusk shells all were preserved, implying that these were magnesian calcite. Selective removal of aragonite portions of skeletal grains probably occurred in meteoric waters before the metastable mineralogies of the skeletal grains had time to stabilize to calcite.

Temporally related to leaching of skeletal aragonite grains was precipitation of clear, bladed to equant nonferroan calcite crystals 0.015 to 0.2 millimeter in size (Figs. 24-7 and 24-8C,E,F). These crystals occur as druses lining intraskeletal pores, as isopachous coatings on grains, and as isolated or patchy crystals within molds and replacing skeletal grains. These cements display unit extinction and rarely coarsen toward the centers of skeletal molds and interparticle pores. The circumgranular and pore-filling bladed to equant calcite cements are interpreted as freshwater cements due to their similarity to cements forming from fresh waters in recent carbonate environments (Halley and Harris, 1979), and as phreatic rather than vadose because of their commonly equant forms. The absence of ferrous iron in these cements (unstained by potassium ferricyanide), and their relationship suggesting that the cements predated compaction-fracturing, point toward precipitation in a shallow (probably oxygenated) phreatic diagenetic environment rather than in a deeper phreatic environment.

Stage 3: Deep Meteoric Phreatic to Deep Subsurface Brines

Short vertical fractures, commonly connecting skeletal molds and minor collapse of coral and gastropod molds, developed from compaction (Fig. 24-9A–C). Nearly all fractures cross-cut equant nonferroan calcite cements that partly replaced and filled corals (Fig. 24-9C). The fractures and collapse molds themselves are filled most commonly with late-stage ferroan calcite, baroque dolomite, or anhydrite cements.

Coarse calcspar cements, similar to those

Fig. 24-6. Whole-core photograph and graphic description of the cored interval in the Bass #1 Smith "A," Hico Knowles Field, from 10,732 to 10,767 feet (3271–3282 m). (1) rudstone; (2) grainstone; (3) grainstone/packstone; (4) packstone; (5) siltstone to very fine sandstone. (a) corals; (b) calcareous red algae. Bases of probable depositional cycles labeled I through VII.

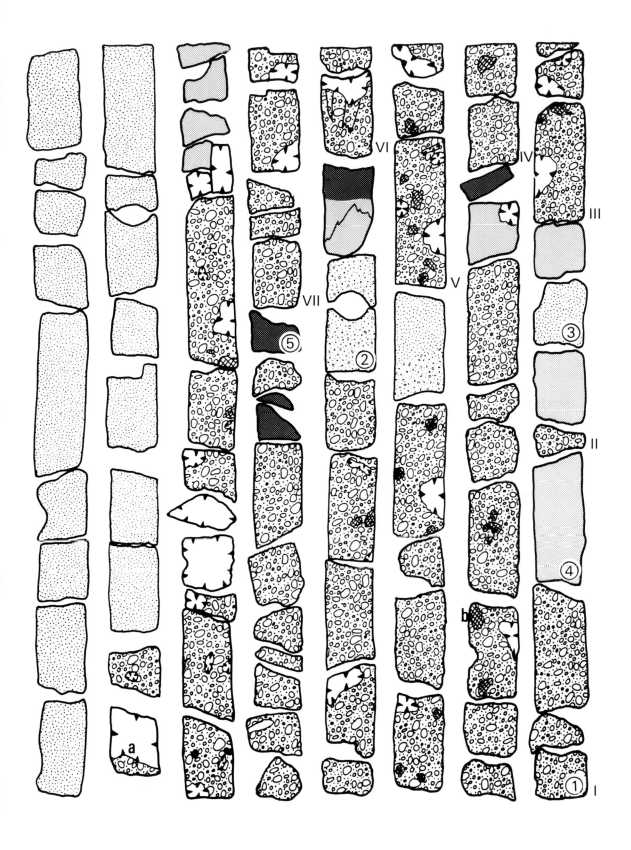

DIAGENETIC STAGE / EVENT	MARINE	MIXED PHREATIC	SHALLOW METEORIC PHREATIC	DEEP METEORIC PHREATIC TO DEEP SUBSURFACE MIXED BRINES
INTERNAL SEDIMENTATION	◇—			
CEMENT. BY BLAD. & FIB. ARAG. & MG CAL.	◇—			
LEACHING OF SKELETAL ARAGONITE		- - -	▱ - - -	
CEMENTATION BY EQUANT CALCITE			▱	
NEOMORPHISM			- - ▱ - - -	
CEMENTATION BY SYNTAXIAL CALCITE			- - ▱ - -	
MECHANICAL COMPACTION			- - ▱ - - -	
CEMENTATION BY FERROAN CALCITE			- — ▱ -	
CEMENTATION BY FERROAN DOLOMITE				▱
ANHYDRITE				- ▱ - - -
CHEMICAL DISSOLUTION (STYLOLITIZATION)			- - - - - - - - -	- - - - -
CORROSION OF FE- CALCITE DOLOMITE				- - - - - —— - - -

Fig. 24-7. Diagenetic events and interpreted diagenetic environments (stage #1) for the reef rubble in Hico Knowles Field.

Fig. 24-8. Photographs of thin sections and a core slab showing diagenetic features. A. Fibrous cement lining a primary shelter pore (a) and cementing internal sediment (b). Later generation cements are non-ferroan equant calcite (c) and coarser ferroan calcite (d). Bass #1 Smith "A" well in Hico Knowles Field, Louisiana, 10,755 feet (3278 m). Photomicrograph in plane light. B. Isopachous rim of fibrous crystals (a) lining an intrareef void. Crystals exhibit radial-fibrous extinction. Note the scalloped indentations of both the cement rim (b) and matrix suggestive of marine borings. Late-stage baroque dolomite (c) fills remaining pore space. ARCO #1 McFadden well in Walker Creek Field, Arkansas, 10,985 feet (3348 m). Photomicrograph in cross-polarized light. C. Cement-reduced interparticle (a) and moldic porosity (b) of gastropod and coral in coarse rudstone. First generation of interparticle pore fill is a druse of fibrous marine cement (c). Freshwater, bladed to equant, non-ferroan calcite crystals (d) overlay the marine cements between c and d in the interparticle pores and also partially fill moldic pores. Late-stage ferroan calcite (d) occurs as isolated crystals through the pore. Baroque dolomite (e) occurs in moldic pore of a bivalve, which is outlined by a micrite envelope. Remaining porosity is oil coated. Bass #1 Smith "A" in Hico Knowles Field, Louisiana, 10,755 feet (3278 m). Photomicrograph in plane light. D. Moldic (a) and preserved primary shelter porosity (b) in skeletal-coral-algal rudstone. Massive coral *Actinastrea* sp. (c) has been partially leached and its mold filled with calcite spar. Moldic porosity is well-developed in reef rubble of Hico Knowles Field, Louisiana. Core-slab photo from Bass #1 Smith "A" well, 10,752 feet (3277 m). E. Cement-filled interparticle porosity in coarse rudstone. Pores were initially lined by rims of equant, freshwater-phreatic, non-ferroan calcite (a) and then completely filled with coarse calcspar (b). Note sediment-filled gastropod mold (c) that was coated with equant calcite after the shell was leached. Bass No. 1 Smith "A" in Hico Knowles Field, Louisiana, 10,748 feet (3276 m). Plane light. F. Large coral head that has undergone various degrees of leaching. Secondary porosity was filled with equant non-ferroan calcite (a) in upper part and ferroan calcite (c) in lower half. Note cement-reduced intraskeletal porosity (b) within the coral. ARCO #1 McFadden in Walker Creek Field, Arkansas, 10,970 feet (3344 m). Photomicrograph in plane light.

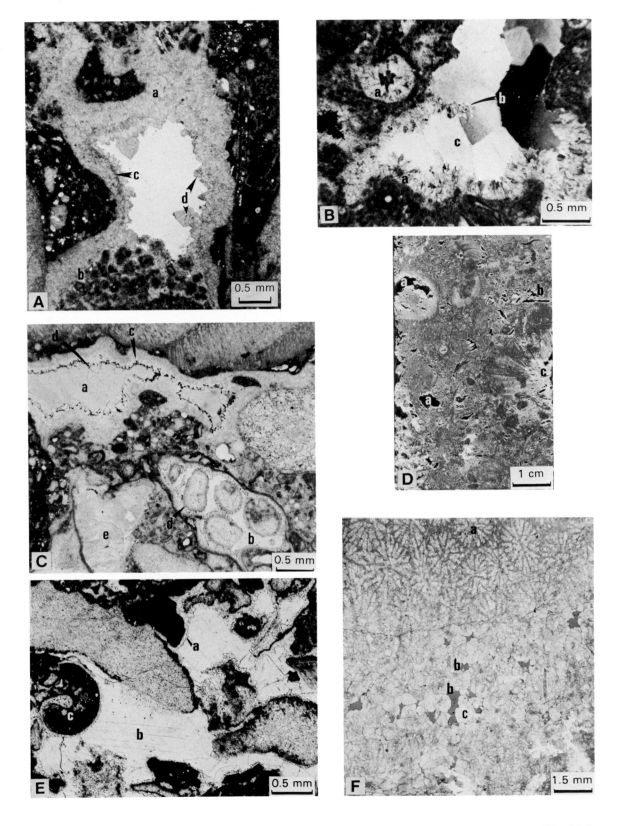

Fig. 24-8.

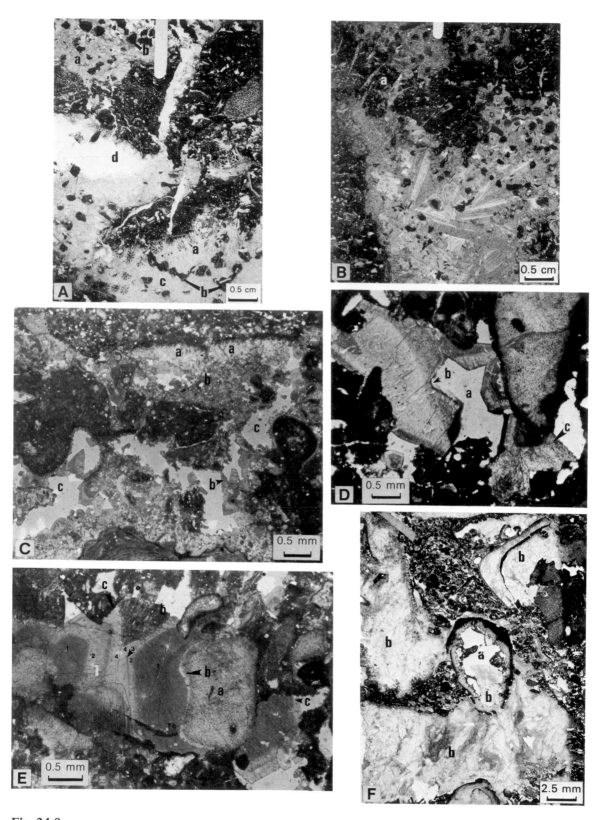

Fig. 24-9.

reported by Moore and Druckman (1981), occur as isolated rhombic crystals in skelmoldic and intraskeletal pores (Fig. 24-8A,C); as subhedral to euhedral crystals occluding pore space (Figs. 24-8E,F, and 24-9C,D); as coarse poikilitic cements in grainstones; and as syntaxial overgrowths on crinoid grains (Fig. 24-9E). The calcspar cements

---

Fig. 24-9. Photomicrographs of thin sections showing intermediate and late-stage diagenetic features. *A.* Reef fabric of algal-*Tubiphytes* boundstone and multiple generations of skeletal-peloidal packstones and grainstones as internal sediment in shelter cavities and growth voids. Large *Actinastrea* sp. coral (a) is extensively bored (b) by sponges and bivalves. Borings are filled with dark, fine-peloid packstones and wackestones. Coral and fractures between corals infilled partly by ferron calcite (c) and partly replaced by lath-shaped anhydrite (d). ARCO #1 McFadden well in Walker Creek Field, 10,974 feet (3345 m). *B.* Anhydrite laths replacing non-ferroan equant calcite in coral molds. Microfractures between coral molds in stromatolite-*Tubiphytes* framework (a) are healed with baroque dolomite and calcspar. ARCO #1 McFadden in Walker Creek Field, Arkansas, 10,978 feet (3346 m). Plane light. *C.* Fractured skeletal mold partially replaced by non-ferroan calcite (a—lighter, clear color and encased in zoned ferroan calcite (b) which also partially occludes moldic porosity (c). Bass #1 Smith "A" in Hico Knowles Field, Louisiana, 10,751 feet (3277 m). Plane light. *D.* Cement-reduced, interparticle porosity in coarse skeletal rudstone. The coarse ferroan calcspar cements are partly coated with oil. The jagged crystal face (a) and rounded crystal face (b) suggest a post-calcspar dissolution event. Note the truncated outer ferron-poor calcspar zone (c) in contact with ferroan baroque dolomite due to either leaching or pressure solution. Bass #1 Smith "A" in Hico Knowles Field, Louisiana, 10,759 feet (3279 m). *E.* Syntaxial overgrowth on echinoderm (a) showing first-stage of non-ferroan calcite and 4 stages (1–4) of varying ferric iron content in the calcspar cement. Ferroan baroque dolomite (b) partially fills the intergranular porosity, and calcspars show effects of microstylolitization (c). Bass #1 Smith "A" in Hico Knowles Field, Louisiana, 10,752 feet (3277 m). Plane light. *F.* Extensive development of shelter and secondary moldic porosity (a) in rudstone and grainstone. Most of the porosity is infilled with coarse baroque dolomite cement (b), but some porosity remains unfilled. Bass #1 Smith "A" in Hico Knowles Field, Louisiana, 10,757 feet (3279 m). Plane light.

vary in crystal size from place to place, from 0.05 to 2.4 millimeters. Emplacement of calcspar cements followed compaction because these cements envelop many broken grains in the grainstones and fill some of the early compaction features within the reef rubble. Staining with potassium ferricyanide reveals at least four crystal-growth zones (Fig. 24-9E). The presence of ferrous iron indicates that these cements were precipitated from phreatic waters under reducing conditions, probably below the vadose or aerated phreatic zone.

Pore-filling dolospar or coarse baroque dolomite postdates the calcspar cements. The dolomite cements range in crystal size from 0.1 to 2.0 millimeters and are distributed like the calcspar cements (Figs. 24-8B,C, 24-9D,F, and 24-10). Baroque dolomite crystals exhibit undulose, sweeping extinction, ferroan compositions, and occasional zonation with respect to Fe^{+2} contents.

Pore space also is occluded by laths of anhydrite that occur as cement or by replacment of carbonate grains and early- (Fig. 24-9A,B) to late-stage cements (Fig. 24-10). Anhydrite postdates the calcspar and baroque dolomite cements and consequently has a similar distribution. It is common for anhydrite to fill skeletal grain molds and pores between grains.

Intercrystal boundaries between baroque dolomite and anhydrite are jagged and irregular (Fig. 24-10A–C), whereas boundaries between crystals of baroque dolomites (Figs. 24-8B,C, 24-9F, and 24-10B,C), calcspars (Figs. 24-8E,F, and 24-9D,E), and dolomite and calcspar (Fig. 24-9D,E) all are generally planar with sharp edge offsets. The irregular intercrystal boundaries suggest either corrosion of the dolomite prior to anhydrite precipitation or, more likely, corrosion during replacement of the baroque dolomite by anhydrite. Occurrences of anhydrite laths penetrating baroque dolomite crystals (Fig. 24-10A–C) appear to be unequivocal evidence for replacement origin. Corrosive events are recorded, however, by the presence of jagged and rounded crystal faces of ferroan calcspar cements (Fig. 24-9D). A corroded and truncated outer iron-poor zone of a calcspar cement appears to predate emplacement of baroque dolomite (Fig. 24-9D). Corrosion of the ferroan calcspar is volumetrically insignificant.

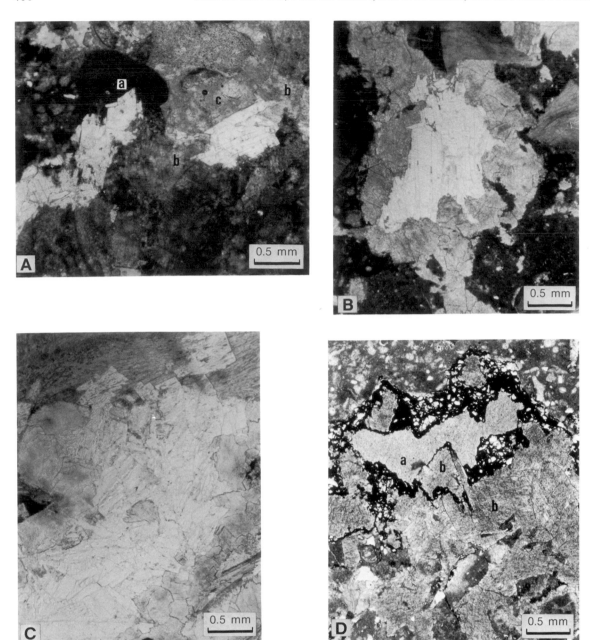

Fig. 24-10. Photomicrographs of thin sections showing late stage diagenetic features. *A.* Anhydrite replacing nonskeletal grain (a) and baroque dolomite (b). Calcspar overgrowth (c). Bass #1 Smith "A" Hico Knowles Field, Louisiana, 10,752 feet (3277 m). Plane light. *B,C.* Anhydrite replacement of ferroan baroque dolomite. Note the irregular and jagged outline of intercrystalline boundaries of dolomite and anhydrite as well as the anhydrite laths penetrating the dolomite. ARCO #1 McFadden in Walker Creek Field, Louisiana: *B.* 10,980 feet (3347 m), *C.* 10,985 feet (3348 m). Plane light. *D.* Stylolite cross-cutting coarse ferroan calcspar (a) and ferroan baroque dolomite (b). Bass #1 Smith "A" in Hico Knowles Field, Arkansas, 10,751 feet (3277 m). Plane light.

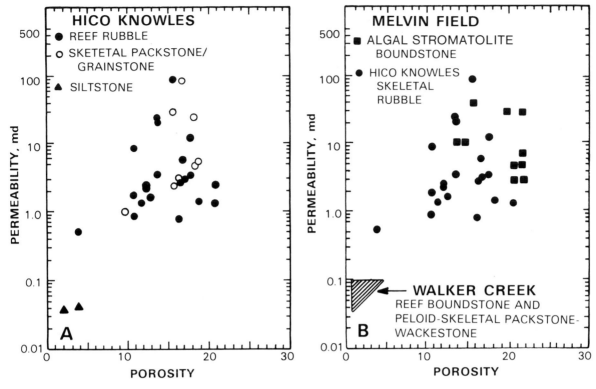

Fig. 24-11 A. Porosity and permeability in skeletal rubble from Bass #1 Smith "A" cored interval (10,732–10,765 ft; 3271–3281 m) in Hico Knowles Field. *B.* Porosity and permeability variations in Smackover buildups and related lithologies, based on standard core analysis. Hico Knowles rubble from the Bass #1 Smith "A," cored interval, included for comparison with algal stromatolite boundstone from the cored interval in the Tesoro #1 C. M. Land well in Melvin Field (Fig. 24-12). All reef lithologies from the ARCO #1 McFadden, Walker Creek Field, Arkansas, plot within the shaded area.

Stylolites cross-cut all cement types, including baroque dolomites (Fig. 10D), indicating that pressure-dissolution continues into the deep subsurface and may locally provide a carbonate source for cements.

Porosity and Permeability

Petrophysical data from 1-inch-diameter core plugs taken from the Bass #1 Smith "A" core show porosities ranging between 1.8 and 21 percent and permeabilities between 0.04 and 84 millidarcys (Fig. 24-11A). High porosities and permeabilities of reef rubble and grainstones/packstones reflect the excellent reservoir characteristics produced by combinations of preserved primary interparticle pores and secondary leached skeletal molds. High porosities combined with low permeabilities reflect the presence of only skeletal moldic porosity and the absence of well-developed matrix porosity in grainstones (*i.e.*, preserved intergranular) or packstones (*i.e.*, leached matrix). The sandstones and siltstones have the poorest reservoir characteristics, with porosities less than 4 percent and permeabilities below 0.1 millidarcy. If laterally extensive, they could serve as vertical permeability barriers.

Comparison of Diagenesis and Porosity Evolution With Other Smackover Reefs

The paragenetic sequences for the Hico Knowles reef rubble, a sponge-coral-algal reef 90 miles (144 km) to the northwest at Walker Creek Field (Figs. 24-1, 24-12A), and an algal-stromatolite reef in the eastern part of the Smackover trend

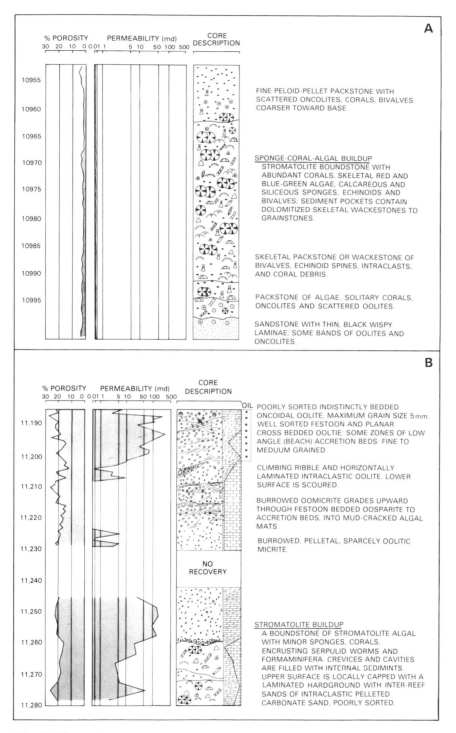

Fig. 24-12 A. Petrophysical log and core description of Upper Smackover buildup in ARCO #1 McFadden well from Walker Creek Field. Entire sequence is limestone. Reef interval comprises a well-developed algal boundstone framework with numerous coral heads and sticks. Symbols used in graphic log are from Shell Standard Legend (Swanson, 1981) except for symbol for *Tubiphytes* (T). *B.* Upper Smackover buildup in C. M. Land Brothers Tesoro #1 well from Melvin Field, showing a reef interval with good boundstone framework. Note the high permeabilities of the dolomitized intervals. Reef interval is of reservoir quality but contains no hydrocarbons. Both figures modified from Baria *et al.* (1982). Published by permission, American Association of Petroleum Geologists.

Table 24-1. Paragenetic Sequences for Three Smackover Builups and Interpreted Diagenetic Environments

Hico Knowles Reef Rubble	Walker Creek Sponge-Coral-Algal Reef	Melvin Stromatolite Reef
Marine sedimentation	Marine cementation and synsedimentary fracturing	Marine cementation and synsedimentary fracturing
	Minor dolomitization	Extensive dolomitization
Leaching of skeletal aragonite	Leaching of skeletal aragonite	Leaching of skeletal, peloidal and ooidal grains
Cementation by fine-medium calcite	Cementation by fine-medium calcite	
	Silicification	
Compaction of grains and skelmolds	Compaction of grains and skelmolds, fracturing	
Coarse syntaxial calcite (non-ferroan-ferroan zoned)	Coarse syntaxial calcite (non-ferroan-ferroan zoned)	Coarse syntaxial calcite
		Stylolitization
Cementation by medium-coarse ferroan-zoned calcite (calcspar)	Cementation by medium-coarse ferroan-zoned calcite (calcspar)	Fracturing
Cementation by coarse ferroan baroque dolomite (dolospar)	Cementation by coarse ferroan baroque dolomite (dolospar)	Dolomite cements along stylolites
Pressure dissolution (stylolitization and corrosion of calcspar and dolospar)	Pressure dissolution (stylolitization and corrosion of calcspar and dolospar)	Dissolution enlargement along fractures
Cementation and replacement by anhydrite	Cementation and replacement by anhydrite	Cementation and replacement by anhydrite
Pyrite		

from Melvin Field (Figs. 24-1, 24-12B) are all compared in Table 24-1 and their petrophysical properties plotted in Figure 24-11B. The depositional textures of the three reef or reef-related lithologies and several stages in their diagenetic history affected the porosity and permeability of the reef facies differently. The original interframework and interparticle porosities of Walker Creek and Melvin fields were reduced to 5 to 20 percent by micrite, fibrous and bladed marine cements, and by binding stromatolites and internal sediment, whereas the primary interparticle porosity of skeletal rudstones and grainstones in Hico Knowles Field was reduced only to 15 to 25 percent by marine cementation. Selective dissolution of skeletal grains also occurred in all three reservoirs, but was most significant in Hico Knowles Field, where formerly aragonitic skeletal debris makes up a large percentage of the rubble. In Hico Knowles, primary interparticle and skelmoldic porosities combine to produce the good reservoir quality of the near-reef sediments.

Compaction, stylolitization, and, most important, several later stages of cementation all contributed to the poor reservoir quality of the reef in Walker Creek Field (Fig. 24-11B) (Harris and Crevello, 1983). Compaction of leached corals that were partially filled or replaced by nonferroan equant calcite, and fracturing of the reef matrix, were common in the buildup cored in the Arco McFadden well. The oolitic grainstones immediately overlying the buildup, however, do not appear to contain the early-stage nonferroan, equant calcite. Consequently, the ooids show extensive mechanical compaction and chemical dissolution.

Brock and Moore (1981), Moore and Brock

(1982), and Moore and Druckman (1981) state that the only diagenetic event predating compaction in Smackover limestones of Walker Creek Field was sea-floor cementation by fibrous crusts. Yet, Becher and Moore (1976) reported fine equant calcite cement (their Figs. 6A,C) in Walker Creek that has the same *petrographic* character as equant calcite cements in the ARCO Bodcaw Unit #1 well from that field, which Wagner and Matthews (1982) interpret as early meteoric. The Bodcaw well is located about one mile (1.6 km) west of the ARCO McFadden well that contains the buildup discussed here. Wagner and Matthews (1982) show that these cements fill leached grain molds in nuclei of ooids, and also occur as isopachous cements, presumably predating compaction. Our study illustrates that equant nonferroan calcite cements occur throughout the Smackover buildups at Hico Knowles and Walker Creek, and also in North Haynesville Field (Crevello, personal observation). These equant cements predate compaction. Perhaps the mineralogically more complex (or varied) and metastable reef interval provided a local internal source for the equant calcite through leaching of its abundant skeletal aragonite components. This more varied mineralogical makeup of skeletal grains may have provided more sensitive indications of the presence of meteoric pore waters than the overlying nonskeletal lithologies.

In Walker Creek and Hico Knowles, calcspar and baroque dolomite were the cements most responsible for occluding porosity. Moore and Druckman (1981) suggested that these cements in Walker Creek were precipitated in the deep subsurface from brines migrating out of the North Louisiana salt basin, at temperatures of 85°C to 112°C, based on fluid inclusions. The brines predated but may have been closely associated with hydrocarbon migration, because some inclusions contain both saline brines and hydrocarbons (Moore and Druckman, 1981). Had there been no late-stage cements, the reef interval at Walker Creek would have made an attractive reservoir rock, with porosities of 5 to 15 percent rather than 1 to 5 percent, as at present.

The reef boundstones in Melvin Field display the best reservoir characteristics (Fig. 24-11B) of any Smackover reef studied to date, because they were partially dolomitized early in their diagenetic history and then were not subjected to as varied and extensive a late-stage history of pore-filling cementation as at Walker Creek. As shown in Figure 24-12B, good porosities occur throughout the cored interval, whereas the best permeabilities are found in ooidal-peloidal grainstones where those are most dolomitic.

Summary

Bioclastic rubble derived from a patch reef that accumulated on the seaward margin of a Smackover paleohigh has produced hydrocarbons from preserved primary interparticle and shelter porosity and secondary skeletal-moldic porosity. The rubble was altered in three main diagenetic stages: (1) Deposition of internal sediment and marine cementation reduced primary depositional porosities of nearly 35 percent to porosities of 15 to 25 percent. (2) Leaching of aragonite skeletal grains, probably by mixed marine and meteoric or meteoric-phreatic pore fluids increased porosities back to 25 to 30 percent, with minor cementation. (3) Finally, late-stage cementation by iron-rich calcite and dolomite and by anhydrite, and replacement of late-stage cements, probably in deep meteoric-phreatic and deep subsurface mixed brines, reduced porosity in the reef rubble to its present values of 15 to 20 percent. Similar diagenetic trends in the Smackover reef at Walker Creek and the reef debris at Hico Knowles suggest similar pore fluid histories, both in early and late diagenetic stages. This implies the development of a regional pore fluid system, as Moore and Druckman (1981) have suggested. The paragenetic sequence inferred for the Hico Knowles and Walker Creek buildups is similar to the sequence seen in the oolitic grainstones of this region by Moore and Druckman (1981), except that the reefs saw an early stage of meteoric flushing. Exposure of the Walker Creek buildup prior to deposition of the lower grainstone unit is not indicated from core studies, so does not appear to be an alternate explanation for early meteoric diagenetic events. Studies by Becher and Moore (1976) and Wagner and Matthews (1982) suggest that the pre-compaction equant calcite occurs in the oolitic grainstones overlying the reef lithologies, implying that both lithologies saw early meteoric flushing.

Acknowledgments We thank Marathon Oil Company, Gulf Oil Exploration and Production Company, and Mosbacher Production Company for permission to publish. The senior author expresses thanks to Robert Halley for thoughtful discussions on burial diagenesis.

References

AHR, W.M., 1973, The carbonate ramp—an alternative to the shelf model: Gulf Coast Assoc. Geol. Soc. Trans., v. 23, p. 221–225.

AHR, W.M., and G.J. PALKO, 1981, Depositional and diagenetic cycles in Smackover limestone-sandstone sequences, Lincoln Parish, Louisiana: Gulf Coast Assoc. Geol. Soc. Trans., v. 31, p. 7–17.

AKIN, R.H., and R.W. GRAVES, 1969, Reynolds oolite of southern Arkansas: Amer. Assoc. Petroleum Geologists Bull., v. 53, p. 1909–1922.

BADON, C.L., 1974, Petrology and reservoir potential of the upper member of the Smackover Formation, Clarke County, Mississippi: Gulf Coast Assoc. Geol. Soc. Trans., v. 24, p. 163–174.

BARIA, L.R., D.L. STOUDT, P.M. HARRIS, and P.D. CREVELLO, 1982, Upper Jurassic reefs of the Smackover Formation, United States Gulf Coast: Amer. Assoc. Petroleum Geologists Bull., v. 66, p. 1449–1482.

BECHER, J.W., and C.H. MOORE, 1976, The Walker Creek Field—a Smackover diagenetic trap: Gulf Coast Assoc. Geol. Soc. Trans., v. 26, p. 34–56.

BISHOP, W.F., 1968, Petrology of upper Smackover limestones in North Haynesville Field, Claiborne Parish, Louisiana: Amer. Assoc. Petroleum Geologists Bull., v. 52, p. 92–128.

BROCK, F.C., JR., and C.H. MOORE, 1981, Walker Creek revisited: a reinterpretation of the diagenesis of the Smackover Formation of Walker Creek Field, Arkansas: Gulf Coast Assoc. Geol. Soc. Trans., v. 31, p. 49–58.

BUFFLER, R.T., J.S. WATKINS, F.K. SCHAUB, and J.L. WORZEL, 1980, Structure and early geologic history of the deep central Gulf of Mexico basin: in Pilger, R.H., Jr., ed., The origin of the Gulf of Mexico and the early opening of the central north Atlantic Ocean: Proc. and Symp. at Louisiana State Univ., Baton Rouge, p. 3–16.

CHIMENE, C.A., 1976, Upper Jurassic reservoirs, Walker Creek Field area, Lafayette and Columbia Counties, Arkansas: in Braunstein, J., ed., North American oil and gas fields: Amer. Assoc. Petroleum Geologists Mem. 24, p. 177–204.

COLLINS, S.E., 1980, Jurassic Cotton Valley and Smackover reservoir trends, East Texas, North Louisiana, and South Arkansas: Amer. Assoc. Petroleum Geologists Bull., v. 64, p. 1004–1013.

CREVELLO, P.D., and P.M. HARRIS, 1982, Depositional models and reef-building organisms, Upper Jurassic reefs of the Smackover Formation: Third Annual Research Conf., Gulf Coast Section, Soc. Econ. Paleontologists and Mineralogists: Jurassic of the Gulf Rim, Program and Abst., p. 25–28.

CREVELLO, P.D., and P.M. HARRIS, 1984, Depositional models for Jurassic reefal buildups: in Proc. of the Third Annual Research Conf., Gulf Coast Section, Soc. Econ. Paleontologists and Mineralogists: Jurassic of the Gulf Rim, p. 57–102.

DICKINSON, W.R., and P.J. CONEY, 1980, Plate tectonic constraints on the origin of the Gulf of Mexico: in Pilger, R.H., Jr., ed., The origin of the Gulf of Mexico and the early opening of the central north Atlantic Ocean: Proc. and Symp. at Louisiana State Univ., Baton Rouge, p. 27–36.

GINSBURG, R.N., D.S. MARSZALEK, and N. SCHNEIDERMANN, 1971, Ultrastructure of carbonate cements in a Holocene algal reef of Bermuda: Jour. Sedimentary Petrology., v. 41, p. 472–482.

HALLEY, R.B., and P.M. HARRIS, 1979, Freshwater cementation of a 1,000-year-old oolite: Jour. Sedimentary Petrology, v. 49, p. 969–988.

HARRIS, P.M., and P.D. CREVELLO, 1983, Upper Jurassic Smackover reefs—an example from Walker Creek Field, Arkansas: in Harris, P.M., ed., Carbonate buildups—a core workshop: Soc. Econ. Paleontologists and Mineralogists, SEPM Core Workshop No. 4, p. 366–380.

JAMES, N.P., R.N. GINSBURG, D.S. MARSZALEK, and P.W. CHOQUETTE, 1976, Facies and fabric specificity of early subsea cements in shallow Belize (British Honduras) reefs: Amer. Assoc. Petroleum Geologists Bull., v. 46, p. 523–544.

MOORE, C.H., and F.C. BROCK, JR., 1982, Porosity preservation in the Upper Smackover (Jurassic) carbonate grainstone, Walker Creek Field, Arkansas: Response of paleophreatic lenses to burial processes—discussion: Jour. Sedimentary Petrology, v. 52, p. 19–23.

MOORE, C.H., and Y. DRUCKMAN, 1981, Burial diagenesis and porosity evolution, Upper Jurassic Smackover, Arkansas and Louisiana: Amer. Assoc. Petroleum Geologists Bull., v. 65, p. 597–628.

OTTMANN, R.D., P.L. KEYES, and M.A. ZIEGLER, 1973, Jay Field—a Jurassic stratigraphic trap: Gulf Coast Assoc. Geol. Soc. Trans., v. 23, p. 146–157.

TODD, R.G., and R.M. MITCHUM, JR., 1977, Seismic stratigraphy and global changes of sea level—Part 8, Identification of upper Triassic, Jurassic, and lower Cretaceous seismic sequences in Gulf of Mexico and offshore West Africa: *in* Seismic stratigraphy—applications to hydrocarbon exploration: Amer. Assoc. Petroleum Geologists Mem. 26, p. 145–164.

SCHROEDER, J.H., 1973, Submarine and vadose cements in Pleistocene Bermuda reef rock: Sedimentary Geol., 10, p. 179–204.

SWANSON, R.G., 1981, Sample Examination Manual: Methods in Exploration Series, Amer. Assoc. Petroleum Geologists, Tulsa, OK.

VAIL, P.R., R.M. MITCHUM, and S. THOMPSON III, 1977, Seismic stratigraphy and global changes of sea level—Part 4, Global cycles of relative changes of sea level: *in* Amer. Assoc. Petroleum Geologists Mem. 26, p. 83–97.

WAGNER, P.D., and R.K. MATTHEWS, 1982, Porosity preservation in the Upper Smackover (Jurassic) Carbonate Grainstone, Walker Creek Field, Arkansas: Response of paleophreatic lenses to burial processes: Jour. Sedimentary Petrology, v. 52, p. 3–18.

25
La Paz Field

T. J. A. Reijers and P. Bartok

RESERVOIR SUMMARY

Location & Geologic Setting	NW Venezuela, onshore Maracaibo Basin
Tectonics	Wrench-fault tectonics, NNE-trending fault-block highs; middle Eocene orogeny, later epeirogeny
Regional Paleosetting	Shallow-marine paleoshelf
Nature of Trap	Structural-stratigraphic; lenticular, shale-capped reservoirs on fault-block high
Reservoir Rocks	
Age	Early Cretaceous (Aptian-Cenomanian)
Stratigraphic Units(s)	Cogollo Group
Lithology(s)	Limestone
Dep. Environment(s)	Grainstone bars and pelecypod banks on paleoshelf
Productive Facies	Coated-grain and skeletal grainstone, and biostromal pelecypod wackestone-packstone
Entrapping Facies	Shales, lime mudstone-wackestone (laterally and vertically)
Diagenesis/Porosity	Dissolution and local cementation near surface (meteoric) and during burial diagenesis; strongly fractured
Petrophysics	
Pore Type(s)	Moldic, solution-enhanced interparticle, intraparticle
Porosity	2–12%
Permeability	0–80 md, avg 2 md
Fractures	Abundant near faults, many open
Source Rocks	
Age	Cenomanian to Coniacian
Lithology(s)	Marine shales and marls
Migration Time	Eocene
Reservoir Dimensions	
Depth	<10,000 ft (<3000 m)
Thickness	1320 ft (400 m) gross; net unknown, fractured basement rocks also productive
Areal Dimensions	NA
Productive Area	25,000 acres (100 km^2)
Fluid Data	
Saturations	NA
API Gravity	30–34°
Gas-Oil Ratio	600–960:1
Other	—
Production Data	
Oil and/or Gas in Place	NA
Ultimate Recovery	~1.0 billion BO (Halbouty *et al.*, 1970)
Cumulative Production	834 million BO to July 1, 1983*

Remarks: IP mechanism natural flow, reservoirs overpressured. Discovered 1944. Includes other reservoir lithologies (sandstone, schist, granite), but principal reservoir rocks are carbonates. *Oil and Gas Journal, Dec. 26, 1983. p. 109.

25
Porosity Characteristics and Evolution in Fractured Cretaceous Carbonate Reservoirs, La Paz Field Area, Maracaibo Basin, Venezuela

T. J. A. Reijers and P. Bartok

General Characteristics and Discovery

The shallow-marine Cretaceous Cogollo Group in the Maracaibo Basin (Figs. 25-1, 25-2) is oil-productive from fractured carbonates of low porosity and permeability. These carbonates have been moderately folded and faulted, as recorded by seismic record sections (Figs. 25-3–25-5). Tectonic deformation, including fracturing, is particularly intense along major faults. Oil accumulations in the carbonates are localized in a large number of structural traps (Fig. 25-6).

The first Cretaceous discovery occurred on land in the La Paz structure, which was found in 1945 by surface mapping. From there, the search for Cretaceous oil moved to Lake Maracaibo, where a number of other fields were discovered and delineated. Initial production from these fields, including La Paz Field, is generally by natural flow since the Cretaceous reservoirs are highly overpressured. Oil gravities vary from well to well within and between all fields. The inter-field variation is 27° to 43° API. Gas-oil ratios are highly variable, ranging from 600:1 to 3200:1, and individual reservoir fluids are generally described as volatile oil or as gas condensate.

This Study

It was evident early in the development of these fields that production came from fractures (Dikkers, 1964). However, it also became apparent that productivities could not be due solely to natural and induced fracture systems, and the presence of porous and permeable layers was suspected. Consequently, a study was undertaken that involved cores up to 900 feet long (275 m) from several wells (Fig. 25-1), and four outcrop sections, to identify the presence of texture-related porosity and determine its origins. Two facies of the Cogollo Group exhibit such porosity: grainstone bars and pelecypod biostromes.

The Cogollo Group

The Cogollo Group ranges in age from Aptian to Cenomanian and varies in thickness from 800 to 2000 feet (250–610 m) (Reijers and Bartok, 1977; Bartok et al., 1981). It consists mainly of carbonates deposited on a broad and stable shelf (Fig. 25-7). Grainstone and packstone bars are the principal shelf deposits and reflect shallow-marine, high-energy environments. Associated with these bars are slightly argillaceous carbonate inter-bar deposits that reflect more restricted conditions, and biostromal units of pelecypod packstone along the leeside of the bars. Back-bar conditions were restricted-marine, with low wave-energy conditions. The lagoonal carbonate sediments are locally dolomitized and widely sapropelic, and in places include siliciclastic intervals. The Cogollo Group is overlain regionally by the bituminous calcareous shales of the La Luna

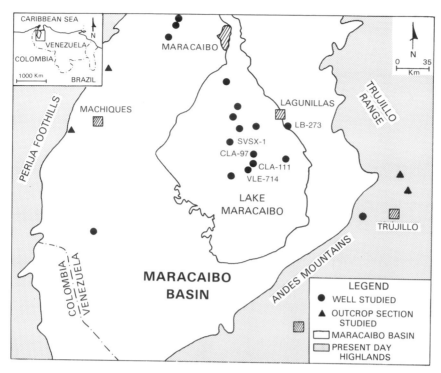

Fig. 25-1. Map of the study area for the Cretaceous Cogollo Group and outline of the present-day Maracaibo Basin. From Bartok *et al.*, 1981. Published by permission, American Association of Petroleum Geologists.

Formation, which is the source for both the Cretaceous and the Tertiary oil accumulations in the Maracaibo province (Figs. 25-2, 25-6).

The core study included measuring the porosity and permeability values of plugs taken from cores at approximately 2-foot (0.6-m) intervals, and petrographic study of representative samples. This study revealed that only the grainstone/packstone bar and pelecypod biostrome facies contain significant amounts of porosity and permeability (Fig. 25-8).

To indicate genetic types of carbonate porosity, a numerical system has been used to which have been added the porosity terms of Choquette and Pray (1970). Pore types 1, 2, and 3 are texture-related; 4 and 6 are texture-related in specific cases.

Diagenetic Models

Texture-related porosity in these carbonates is the result of the cumulative effect of diagenetic processes on sediments of particular constituent/mineralogical compositions. Specific texture-related porosity in the Cogollo Group has therefore depended on the determination of areal distribution of these deposits and an understanding of their porosity history. Areal distribution of sedimentary facies has been modeled and related to porosity history, and a succession of diagenetic processes has been defined that affected these porous facies through time and space (Figs. 25-9, 25-10). These processes are separated into those that create, destroy, or do not affect porosity. Time is expressed in steps or sequences of pro-

Genetic Pore Types Used in Diagenetic-Model Diagrams				
Primary	Interparticle	1a	Intraparticle	1b
Secondary	Intraparticle	2		
	Intraskeletal	3a	Intrapellet,	3b
	Inter- and/or		intraooid*	
	intraparticle	4		
	Fracture	5		
	Intercrystal	6		

*Pellet-moldic and oomoldic by Choquette and Pray (1970) classification

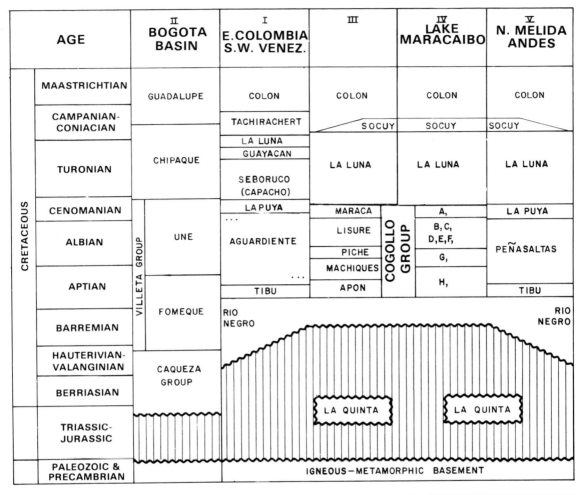

	AGE	II BOGOTA BASIN	I E.COLOMBIA S.W. VENEZ.	III	IV LAKE MARACAIBO	V N. MELIDA ANDES
CRETACEOUS	MAASTRICHTIAN	GUADALUPE	COLON	COLON	COLON	COLON
	CAMPANIAN- CONIACIAN		TACHIRACHERT	SOCUY	SOCUY	SOCUY
	TURONIAN	CHIPAQUE	LA LUNA	LA LUNA	LA LUNA	LA LUNA
			GUAYACAN			
			SEBORUCO (CAPACHO)			
	CENOMANIAN	UNE	LA PUYA	MARACA	A,	LA PUYA
	ALBIAN		AGUARDIENTE	LISURE	B, C, D, E, F,	PEÑASALTAS
				PICHE	G,	
				MACHIQUES		
	APTIAN	FOMEQUE	TIBU	APON	H,	TIBU
	BARREMIAN		RIO NEGRO			RIO NEGRO
	HAUTERIVIAN- VALANGINIAN	CAQUEZA GROUP		LA QUINTA	LA QUINTA	
	BERRIASIAN					
	TRIASSIC- JURASSIC					
	PALEOZOIC & PRECAMBRIAN		IGNEOUS—METAMORPHIC BASEMENT			

(Villeta Group and Cogollo Group are labeled vertically; Cogollo Group spans columns III and IV)

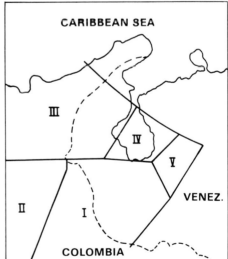

Fig. 25-2. Stratigraphic chart showing the Cretaceous litho-stratigraphic units in outcrop and subcrop in western Venezuela. From Bartok *et al.*, 1981. Published by permission, American Association of Petroleum Geologists.

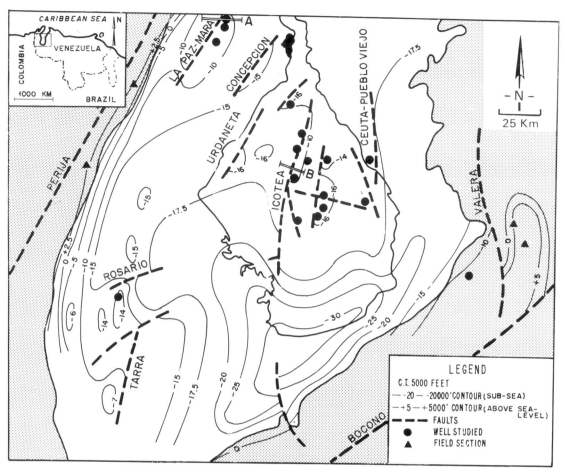

Fig. 25-3. Structure contour map of top Cretaceous limestone in Lake Maracaibo. Horizon depicted is the Socuy Member of the Colon Formation (Fig. 25-2). Map is based on well and seismic data. From Bartok *et al.*, 1981. Published by permission, American Association of Petroleum Geologists.

cesses, as revealed by petrographic analysis; space is described in terms of diagenetic environments such as near-surface, shallow-burial, or deep-burial.

Figures 25-9 and 25-10 show the inferred diagenetic pathways for two facies from an initial deposit to an end product, here termed *carbonate fabric unit* (*cf.* Fig. 25-11). The overall transformation of the porosity is the result of the impact of each individual process on the sediment.

The degree and intensity of alteration of Cogollo carbonate sediments depended on a large number of factors, the most important of which were: (1) original mineralogical composition; (2) residence time in a diagenetic environment; (3) number and sequence of diagenetic environments through which the porous carbonate units have passed; and (4) susceptibility of these units to each diagenetic process.

Carbonate Fabric Unit

In order to relate porosity history to the sedimentation model, the *carbonate fabric unit* is employed. This unit is defined here as a carbonate interval characterized by a uniform depositional facies and a specific post-depositional history. The subdivision of Cogollo carbonate rocks into units with specific characteristic carbonate fabrics illustrates a mosaic of potential reservoir intervals that result from a specific diagenetic overprint on a specific deposit (Figs. 25-9–25-11). Some of these units are porous; others are tight. Cogollo carbonate-fabric units are named after their major diagenetic and depositional feature. For non-porous units, the term "tight" is used. The Cogollo units are basically rock-stratigraphic in character and hence mappable. Grainstone bars and pelecypod biostromes have been recognized

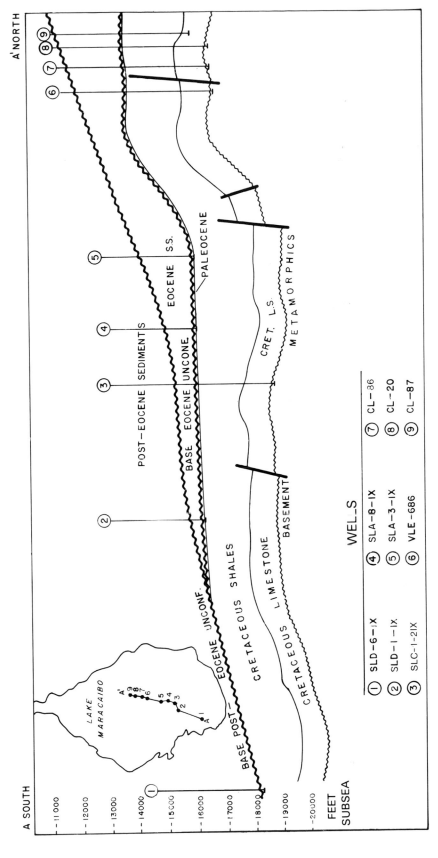

Fig. 25-4. Structural cross-section based on seismic data and tied in to wells. Based on L. Rodriguez, personal communication (1976).

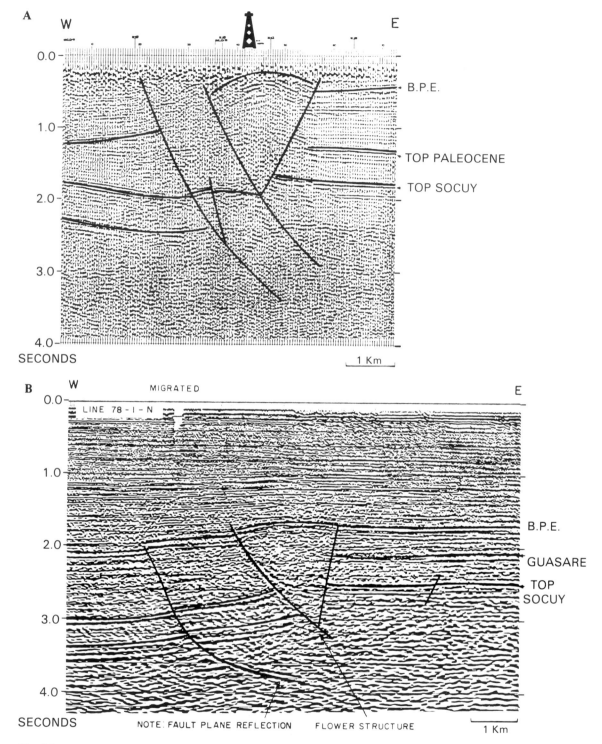

Fig. 25-5 A. Seismic section (unmigrated) over the La Paz-Mara Field (section A in Fig. 25-3). *B.* Seismic section migrated across the land fault in Lake Maracaibo (section B in Fig. 25-3).

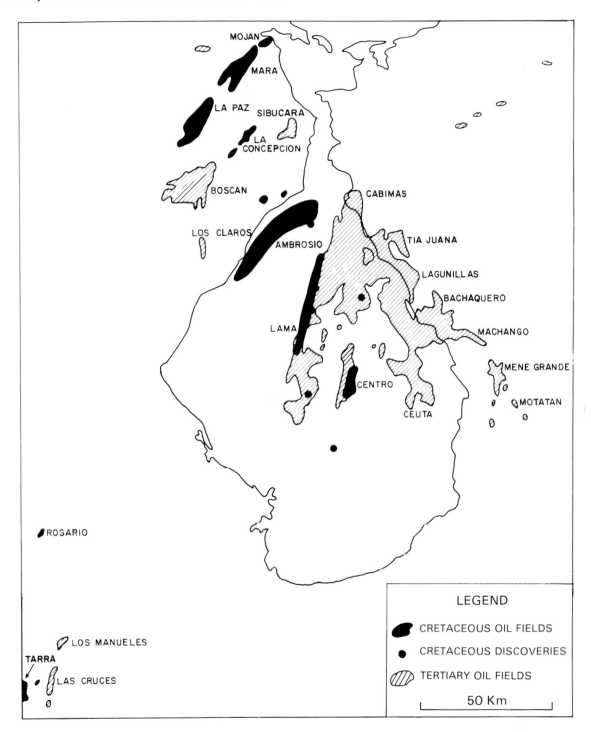

Fig. 25-6. Cretaceous and Tertiary oil accumulations in the Maracaibo oil province.

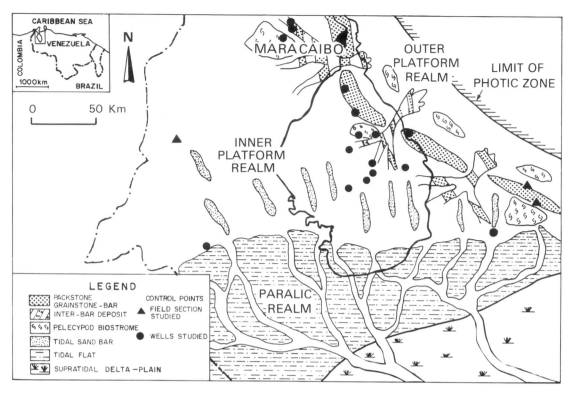

Fig. 25-7. Schematic paleographic setting for the uppermost Cogollo Group, indicating the main depositional environments and sediment associations of the upper Albian. From Bartok *et al.*, 1981. Published by permission, American Association of Petroleum Geologists.

as significant porous carbonate fabric units in the Cogollo; others exist but are less significant.

Diagenetic Overprints on the Porous Deposits

Grainstone/Packstone Bars

The bars initially had high original interparticle porosity (1a), which has been destroyed in virtually all cases. Many sedimentary particles acquired some intraparticle porosity during diagenesis, but except for a few cases this too has been destroyed. The diagenetic or "lithogenetic" pathway of grainstone/packstone deposits is shown in Figure 25-9. The resulting fabric unit has porosity values up to 9 percent and permeability values reaching 4 millidarcys (Fig. 25-8A).

Near-Surface Environment

Step 1   Selective micritization of skeletal particles during deposition or within the near-surface diagenetic environment.

Micritization has been observed mainly in sediments that also have undergone vadose diagenesis.

Step 2   Dissolution of initially unstable carbonate particles by pore waters undersaturated with respect to $CaCO_3$ (*e.g.*, aragonitic pelecypod fragments and ooids; Fig. 25-12A).

The grainstone bars suffered primarily from diagenetic processes in fully water-saturated pores. However, they occasionally display signs of having gone through one or more vadose diagenetic environments in which the pores contained both air and water (uneven druse; Fig. 25-12A).

Step 3a   In the vadose zone, preferred cement growth was confined to grain contacts and the lower parts of grains, locations favored by interstitial fluids under the influence of gravity. Dripstone or microstalactitic cements were formed (Fig. 25-12A).

Step 3b   Equant drusy cements commonly line

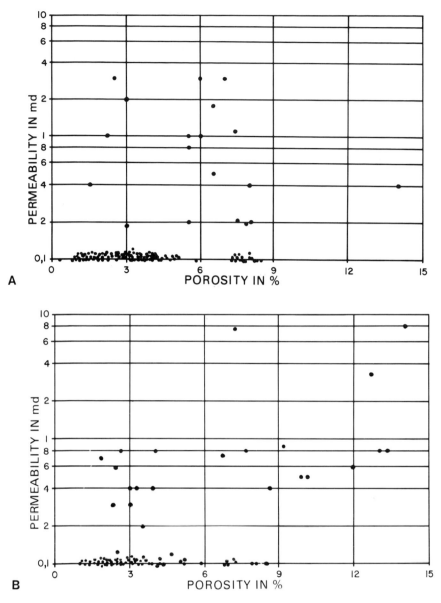

Fig. 25-8. A. Porosity/permeability plot for 174 plug measurements in grainstone/packstone bar carbonate fabric units in wells VLB-704, SVSX-1, and CLA-97.

B. Porosity/permeability plot for 135 plug measurements in pelecypod biostrome carbonate fabric units in wells LB-273, VLB-704, and CLA-111.

Step 3c   the pores with an even layer of isopachous cement. Around monocrystalline fragments, such as echinoderm segments, additional cement growth in optical continuity with the enclosed fragment is commonly observed, and syntaxial overgrowths fill the entire pore spaces, enclosing other grains. In the Cogollo sediments, however, overgrowth cement occasionally postdates a poorly developed, uneven drusy meniscus cement, which may reflect an earlier vadose diagenetic environment. Rapid addition of overgrowth cement on echinoderm fragments arrested formation of the drusy crystals.

Shallow-Burial Environment

Step 4   The processes in steps 3a–c were followed by precipitation of blocky cement crystals. After first growing out

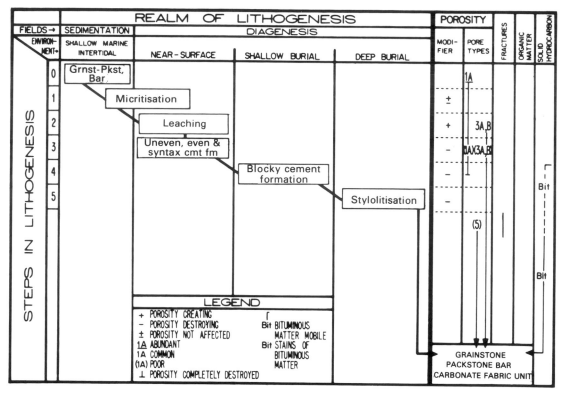

Fig. 25-9. Diagenetic model of grainstone/packstone bar carbonate fabric unit. See Figure 25-8A and B, for porosity-permeability distribution. From Bartok *et al.*, 1981. Published by permission, American Association of Petroleum Geologists.

freely from the drusy cement lining, these crystals became increasingly restricted by mutual interference, resulting in a blocky appearance. The well-developed natural crystal faces indicate growth in fully water-filled pores.

In the grainstone bars, only a limited number of features reflecting deep-burial diagenetic processes have been observed. The clearest features are the stylolites, for which the first occurrence is tentatively set at around 1900 feet (600 m).

Deep-Burial Environment

Step 5   It is clear from interpenetrating grains with fitted or sutured interparticle contacts and from fractured grains, that overpacking and mechanical breakage occurred. The fractures appear to be related to the major fractures and faults in the Maracaibo Basin, which apparently formed in the period from late Cretaceous to Eocene. Thus, the

fractures observed in cores are assumed to have been formed during periods of tectonic activity. Stylolites and fractures apparently served as pathways along which bitumen was mobilized.

Hydrocarbons were generated in the overlying La Luna Formation from Eocene time onward (*cf.*, Bartok *et al.*, 1981). The capping Colon shales prevented vertical escape. Consequently, the hydrocarbons moved laterally through fractures in the La Luna and through texture-related porosity and fractures in the grainstone bars of the Cogollo, to be trapped in structural closures. Bituminous matter is now present as solid bitumen stains in many of the wells studied (*cf.* Fig. 25-13).

Pelecypod Biostromes

Pelecypod biostrome carbonates occasionally contain rudistid remains with body cavities retaining high initial intraparticle porosity, but primary

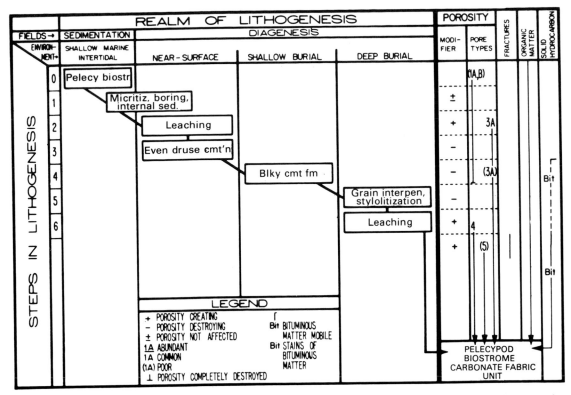

Fig. 25-10. Diagenetic model of pelecypod biostrome carbonate fabric unit. See Figure 25-8B for porosity-permeability distribution. From Bartok *et al.*, 1981. Published by permission, American Association of Petroleum Geologists.

intraparticle porosity was virtually absent in pelecypod biostromes composed of oyster shells. The lithogenetic pathway of the deposit is shown in Figure 25-10. The resulting porous pelecypod biostrome unit (Fig. 25-8B) has porosity values up to 15 percent and permeability values as high as 8 millidarcys.

Near-Surface Environment

Step 1   Micritization and boring by large scavenging organisms occurred during deposition or in the near-surface (diagenetic) environment. Body cavities and borings were sometimes open, but in most cases were partly or wholly filled with internal sediment.

Step 2   Skeletal particles were leached, and this created a fair amount of porosity and permeability in some cases (Fig. 25-12B).

Near-Surface and Shallow-Burial Environment

Steps 3, 4   Interparticle porosity was destroyed by blocky cement (Fig. 25-12B); of-

ten, solution porosity was destroyed by this cement as well. In such cases a thin, uneven drusy cement precedes the blocky cement, suggesting that leaching was followed by surface-related cementation. Thus "early" leaching occurred and at least two periods of near-surface diagenesis are implied.

Deep-Burial Environment

Step 5   Up to 5 percent of kaolinite and other clay minerals are frequently found. We speculate that such clay admixtures might have triggered grain interpenetration and microstylolite formation in these sediments (Fig. 25-14B).

Step 6   The microstylolites might have acted as paths along which leaching fluids entered the rock system a second time, creating porosity as a result of "late leaching." Existing pores were enlarged and microstylolites widened.

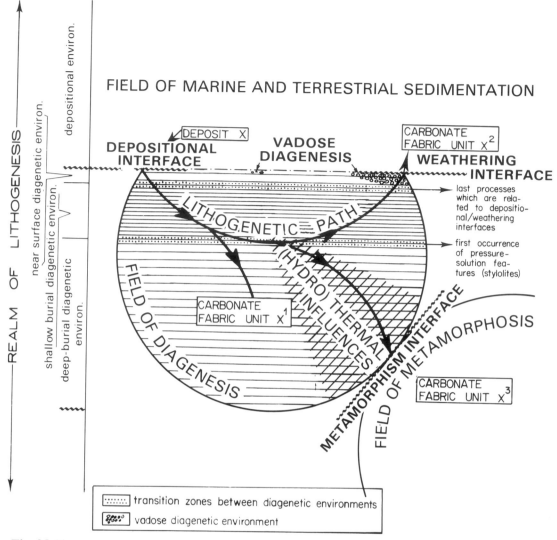

FIELD OF MARINE AND TERRESTRIAL SEDIMENTATION

DEPOSIT X

DEPOSITIONAL INTERFACE

VADOSE DIAGENESIS

CARBONATE FABRIC UNIT x^2

WEATHERING INTERFACE

REALM OF LITHOGENESIS

depositional environ.

near surface diagenetic environ.

shallow burial diagenetic environ.

deep-burial diagenetic environ.

LITHOGENETIC PATH

FIELD OF DIAGENESIS

(HYDRO)THERMAL INFLUENCES

CARBONATE FABRIC UNIT x^1

last processes which are related to depositional/weathering interfaces

first occurrence of pressure-solution features (stylolites)

METAMORPHISM INTERFACE

FIELD OF METAMORPHOSIS

CARBONATE FABRIC UNIT x^3

:::::: transition zones between diagenetic environments

vadose diagenetic environment

Fig. 25-11

Fig. 25-11. Schematic diagram suggesting lithogenetic pathways of a deposit through diagenetic environments and resulting carbonate fabric units.

Fig. 25-12 A. Photomicrograph of a grainstone containing ooids and skeletal fragments. Some ooids display voids (P porosity) as a result of partial leaching by undersaturated pore waters. Arrows point to meniscus cement of vadose origin. Well SVSX-1, 13,207.2 to 13,209.2 feet (4038.9–4039.5 m). Thin section, plane light. *B.* Photomicrograph of a limestone with predominantly large pelecypod skeletons. The intergranular porosity has been largely destroyed by blocky cement. Aragonitic skeletal fragments and some ooids have been leached, leaving some intraskeletal and intraooidal porosity (P porosity). 12% porosity and 3.2 md permeability were measured. Well SVSX-1, 13,201.0 to 13,203.3 feet (4037.0–4037.7 m). Thin section, plane light.

Fig. 25-13. Photomicrograph of a packstone showing a large number of interpenetrating grains indicating that intense pressure solution has taken place. Where grains do not touch, some blocky cement is visible. Note the bituminous stains along the stylolites. Well LB-273, 14,460 feet (4422 m). Thin section, plane light.

Fig. 25-14 A. Photomicrographs. Packstone/wackestone with preferentially leached skeletal fragments (P porosity). In places grains interpenetrate. Note the bituminous stains around the pores. Well SVSX-1, 13,088.5 to 13,091.4 feet (4002.6–4003.5 m). Thin section, plane light. *B.* Grain-interpenetrated limestone. Along grain interpenetrations (arrows), some solution seams and cavities (P porosity) are present. 10% porosity and 0.5 to 0.6 md permeability were measured. Well SVSX-1, same depth as in *A.* Thin section, plane light.

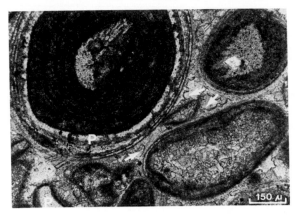

Fig. 25-12A

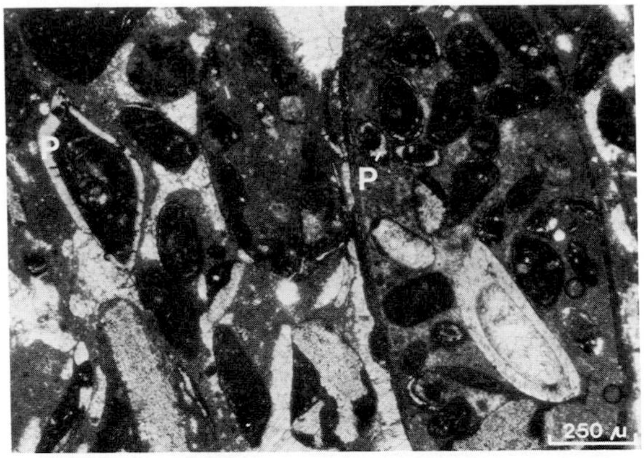

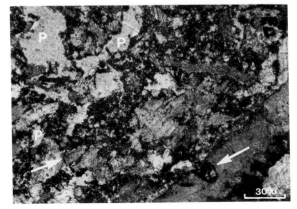

Fig. 25-12B

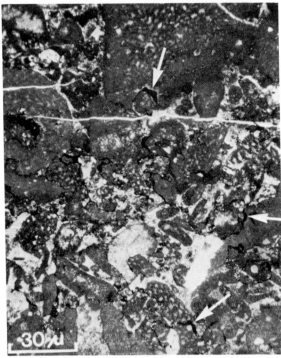

Fig. 25-13

Fig. 25-14A

Fig. 25-14B

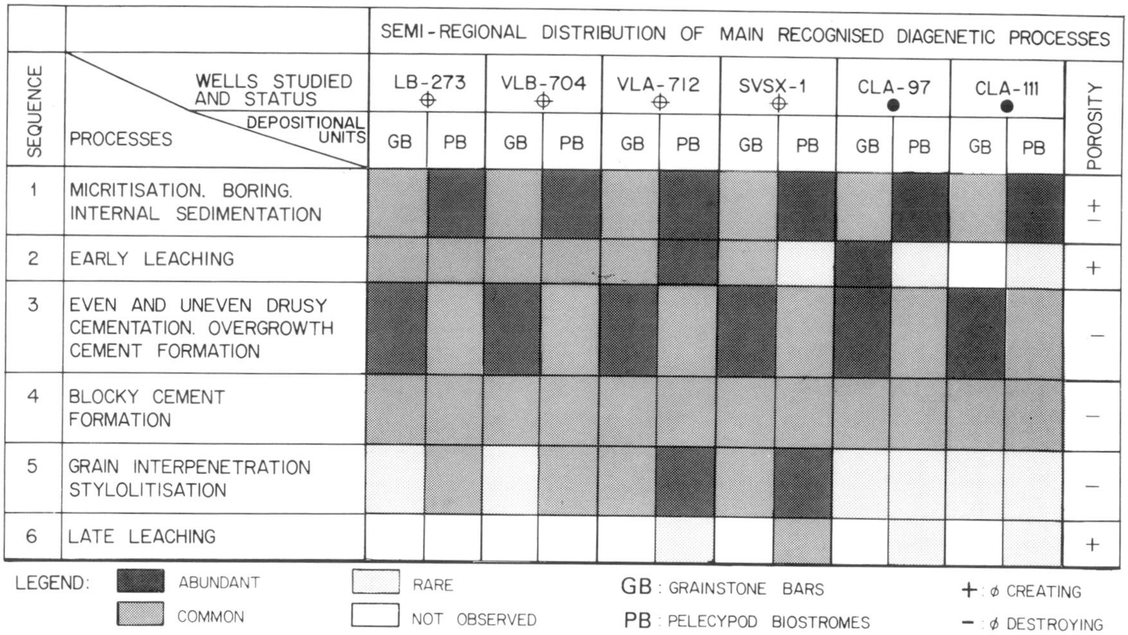

SEQUENCE	PROCESSES / DEPOSITIONAL UNITS	SEMI-REGIONAL DISTRIBUTION OF MAIN RECOGNISED DIAGENETIC PROCESSES												POROSITY
	WELLS STUDIED AND STATUS	LB-273 ⊕		VLB-704 ⊕		VLA-712 ⊕		SVSX-1 ⊕		CLA-97 ●		CLA-111 ●		
		GB	PB	GB	PB	GB	PB	GB	PB	GB	PB	GB	PB	
1	MICRITISATION. BORING. INTERNAL SEDIMENTATION													+
2	EARLY LEACHING													+
3	EVEN AND UNEVEN DRUSY CEMENTATION. OVERGROWTH CEMENT FORMATION													−
4	BLOCKY CEMENT FORMATION													−
5	GRAIN INTERPENETRATION STYLOLITISATION													−
6	LATE LEACHING													+

LEGEND: ■ ABUNDANT   ▨ COMMON   ▧ RARE   □ NOT OBSERVED   GB : GRAINSTONE BARS   PB : PELECYPOD BIOSTROMES   + : ø CREATING   − : ø DESTROYING

Fig. 25-15. Semi-quantitative plot suggesting the impact of diagenetic processes on Lower Cretaceous grainstone bars and pelecypod biostromes in selected wells, Lake Maracaibo, Venezuela.

As a result, the rocks occasionally have some reservoir potential (Fig. 25-14A,B). Such late leaching is more prevalent in the pelecypod biostromes than in the grainstone/packstone bars.

In the pelecypod biostromes, as in the grainstone bars, there are tectonic fractures, probably of Late Cretaceous-Eocene age. It is probable that bitumen, closely related in time to open fractures, was introduced both from the La Luna source rocks and from organic-rich intervals within the Cogollo. Movements of bitumen are indicated by stains and oil occurrences in the wells studied.

Reservoir Characteristics of the Cogollo Group Carbonates

The main diagenetic processes in the grainstone bars and pelecypod biostromes of the central Maracaibo Platform area (Fig. 25-15) are documented by rock samples from six wells. The sedimentary units have been altered into porous carbonate fabric units (*cf.*, Figs. 25-9, 25-10). The porosity-destroying processes of cementation and grain interpenetration were clearly intense, and eventually occluded most depositional and diagenetic texture-related porosity, as confirmed by very low porosity and permeability (Fig. 25-8). Only in a few cases has late leaching created texture-related porosity that is still preserved today.

The diagnetic model of the grainstone bars (Fig. 25-9) and the semiquantification of the impact of the diagenetic processes on these deposits (Fig. 25-15) show that early cementation, particularly overgrowth cementation, destroyed the high depositional porosity in bars at an early stage. Intense cementation also destroyed porosity in the pelecypod biostromes. Shells in the biostromes are almost invariably oysters with hardly any primary intraparticle porosity.

Significance of Texture-Related Reservoir Space

Oil with low sulfur content has been produced for more than 30 years from the Cretaceous carbonates in the Maracaibo Basin, but until recently, the opinion that such production comes mainly from fractures remained unchallenged (Dikkers,

1964). This opinion is supported by the common occurrence of bitumen-stained fractures in cores of the Cogollo. Some facts, however, are not easily explained by assuming production only from fractures. The La Paz Field, for example (Fig. 25-6), has produced over 800 million barrels of oil, partly from the Cogollo carbonates and partly from fractured basement. This oil volume considerably surpasses the combined storage potential of the Cogollo carbonates and the basement within the La Paz structure, if only fractures are assumed to have produced oil. In fact, fractures rarely account for more than 0.5 percent porosity. Similar reasoning is valid for several other fields producing from the Cogollo Group in the Maracaibo Basin (Fig. 25-6). We conclude that the matrix stores the oil and that the fractures enhance the production potential.

The porous carbonate fabric units recognizable within the Cogollo Group provide the storage space required. Modest texture-related porosities and permeabilities permit relatively high-gravity oil to gradually drain through the highly permeable fracture system into producing wells. This study is dependent on the location of well cores analyzed, and consequently, only a limited number of producing wells have been included. This might have a negative influence on the modest texture-related porosity and permeability values recorded: maximum porosity less than 12 percent, average porosity less than 8 percent, maximum permeability less than 3.3 millidarcys,

average permeability less than 1 millidarcy (Fig. 25-8). However, even these low porosity/permeability values go a long way toward explaining the apparent excessive productions of some Cogollo fields.

References

BARTOK, P., T.J.A. REIJERS, and I. JUHASZ, 1981, Lower Cretaceous Cogollo Group, Maracaibo basin, Venezuela—sedimentology, diagenesis, and petrophysics: Amer. Assoc. Petroleum Geologists Bull. v. 65, no. 6, p. 1110–1134.

CHOQUETTE, P.W., and L.C. PRAY, 1970, Geologic nomenclature and classification of porosity in sedimentary carbonates: Amer. Assoc. Petroleum Geologists Bull. v. 54, p. 207–250.

DIKKERS, A.J., 1964, Development history of the La Paz Field, Venezuela: Inst. Petroleum, v. 50, p. 330–333.

HALBOUTY, M.T., A.A. MEYERHOFF, R.E. KING, R.H. DOTT, SR., H.D. KLEMME, and T. SHABAD, 1970, World's giant oil and gas fields, geologic factors affecting their formation, and basin classification—Part I, Giant oil and gas fields: in Halbouty, M.T., ed., Geology of Giant Petroleum Fields: Amer. Assoc. Petroleum Geologists Mem. 14, p. 502–528.

REIJERS, T.J.A., and P. BARTOK, 1977, Depositional patterns and diagenetic sequences in Cretaceous Cogollo Group, Maracaibo Platform, Venezuela: Amer. Assoc. Petroleum Geologists Bull. v. 61, no. 5, p. 821–822.

26
Fateh Field

Clifton F. Jordan, Jr., Thomas C. Connally, Jr., and Harry A. Vest

RESERVOIR SUMMARY

Location & Geologic Setting	Arabian Gulf offshore Dubai, UAE., flank of Arabian Shield
Tectonics	Salt dome on stable shelf
Regional Paleosetting	Khatiyah Embayment of Mesozoic Rub Al Khali Basin
Nature of Trap	Structural
Reservoir Rocks	
Age	Mid-Cretaceous
Stratigraphic Units(s)	Mishrif Formation
Lithology(s)	Limestone
Dep. Environment(s)	Reef, forereef
Productive Facies	Reef, forereef
Entrapping Facies	Laffan shale overlying post-Mishrif unconformity
Diagenesis/Porosity	Primary reefal and secondary vadose-phreatic
Petrophysics	
Pore Type(s)	Inter- & intraparticle, biomoldic, vug
Porosity	1–25%, 19% avg
Permeability	1–102 md, 30 md avg
Fractures	Vertical, mainly healed
Source Rocks	
Age	Mid-Cretaceous
Lithology(s)	Khatiyah Shale
Migration Time	Eocene
Reservoir Dimensions	
Depth	8000–8500 ft (2440–2600 m)
Thickness	900 ft closure (274 m)
Areal Dimensions	6 × 9 mi (9.6 × 14.4 km)
Productive Area	18,000 acres (71.6 km^2)
Fluid Data	
Saturations	$S_o = 88$, $S_w = 12\%$
API Gravity	32°
Gas-Oil Ratio	350:1
Other	—
Production Data	
Oil and/or Gas in Place	2.36 billion BO
Ultimate Recovery	1.02 billion BO; URE = 43.2%
Cumulative Production	555.9 million BO through May 1980 for Ilam, Mishrif & Thammama reservoirs of which 397.7 million BO was from Mishrif Formation

Remarks: IP mechanism weak water-drive; IP 20,000–30,000 BOPD. Average production 120,000 BOPD, 43 producing wells. Discovered 1966.

26
Middle Cretaceous Carbonates of the Mishrif Formation, Fateh Field, Offshore Dubai, U.A.E.

Clifton F. Jordan, Jr., Thomas C. Connally, Jr., and Harry A. Vest

Location

Fateh is a giant oil field in offshore Dubai producing primarily from Middle Cretaceous carbonates (Fig. 26-1), with 12 platforms, 43 producing wells, 3 producing horizons, and estimated ultimate reserves of 1.02 billion barrels of oil. It is located on the north side of the Rub Al Khali Basin (Fig. 26-2A), the sediments of which onlap the Arabian Shield to the west.

Discovery

In 1966, a Conoco-operated group including Dubai Petroleum Company discovered Fateh Field based on the analysis of marine seismic data. Productive intervals in the A-1 discovery well were established in the Upper Cretaceous Ilam Formation and in the Lower Cretaceous Thamama Formation. The Mishrif Formation, later to become the principal reservoir, was absent by erosion in the discovery well (Fig. 26-2B).

The first stepout well, the B-1, was drilled downdip and discovered oil in the Mishrif Formation in early 1967. A second stepout well, the C-1, tested oil in the Ilam and Thamama formations but again encountered no Mishrif. Development drilling in late 1969 and early 1971 of the D-1 and F-1 wells confirmed Mishrif oil on the west and east flanks of Fateh Field. Later, it was demonstrated that the Mishrif reservoir had an elliptical shape due to erosion of the Mishrif from the crest of the structure.

Reservoir Characteristics

The Mishrif Formation is the main reservoir at Fateh Field and lies at about 8000 to 8500 feet (2440–2600 m). The field is elliptical in map view, measuring about 9 miles (15 km) by 6 miles (10 km), and has vertical closure of about 900 feet (275 m). The productive area of Fateh Field is about 18,000 acres (71.6 km^2), and production is obtained from three reservoirs, with 71 percent coming from porous Middle Cretaceous carbonates of the Mishrif Formation (Table 26-1). From February 1970 to June 1980, the Mishrif here produced nearly 400 million barrels of oil, and all three reservoirs produced 556 million barrels of oil. The field has a weak water-drive, and water injection began in June 1974. The cumulative total of water injected into the Mishrif by the end of 1979 was 454,366,142 barrels. Total original oil in place in the Mishrif Formation is estimated to be 2.36 billion barrels. The oil is 32°API, and typical Mishrif saturations are 88 percent oil and 12 percent water.

Tectonic History

The Arabian Peninsula is located on the eastern flank of the Arabian Shield (Fig. 26-2A). During the latest Precambrian, an evaporative basin

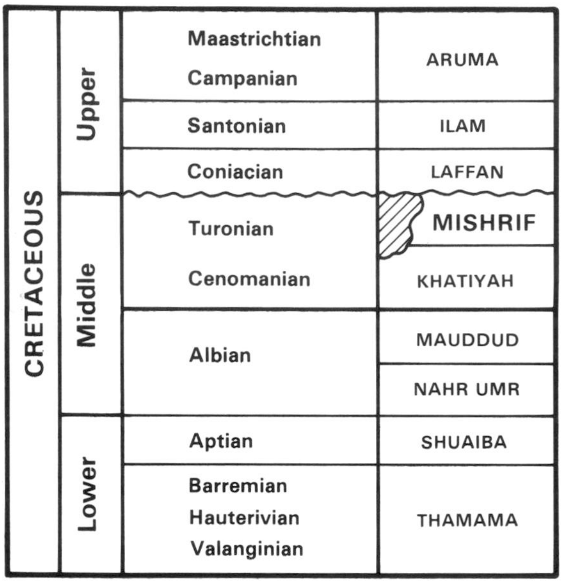

Fig. 26-1. Stratigraphic section for the Cretaceous of the southern Persian Gulf.

formed in the Dubai area, resulting in the deposition of the Hormuz Salt, which later was mobilized to form large salt domes and plugs throughout the southeastern Persian Gulf.

Tectonic stability persisted throughout the entire Paleozoic. Broad, stable shelf areas developed, and numerous transgressive/regressive cycles are recognized. By the end of the Permian, the Tethys Sea had formed northeast of the Arabian Peninsula (Fig. 26-2A). The Devonian through the Early Permian was a period of nondeposition.

The Mesozoic was a period of prolific carbonate deposition whose reservoirs are shoal facies, commonly reef-related. The widespread Paleozoic platform differentiated into the Zagros Trough and Uplift. In addition, the Qatar Arch (a second-order tectonic feature) rose to divide the Zagros Trough into two large sub-basins (Fig. 26-2A): (1) the Basrah Basin in the northern Gulf and (2) the Rub Al Khali Basin in the southern part. Both originated as large-scale tectonic depressions, most likely caused by faulted continental basement. The somewhat oval-shaped Rub Al Khali Basin is an intracratonic feature over 600 miles (1000 km) wide; the reef-rimmed Khatiyah Embayment (Fig. 26-2A) is interpreted as a large re-entrant on the north side of this basin.

Early in the Cretaceous, diapiric movement of Hormuz salt began the formation of large, salt-

Table 26-1. Fateh and Related Offshore Dubai Fields: Statistics for Estimated Recoverable Reserves and Cumulative Production

	Reservoirs	Fateh	Southwest Fateh	Falah	Rashid
Estimated Ultimate Reserves (MMBO)	Ilam	62.6	—	—	gas cond.
	Mishrif	950.6	900.9	40.1	—
	Thamama	170.0	159.4	—	21.5
	Total	1,183.2	1,060.3	40.1	21.5
Production through June 1980 (MMBO)	Ilam	43.7		—	—
	Mishrif	397.7	331.9	4.9	—
	Thamama	114.5	49.8	—	0.9
	Total	555.9	381.7	4.9	0.9
Percent of Original Oil in Place	Ilam	39	—	—	
	Mishrif	43	44	7.7	?
	Thamama	37	41	—	
	Average	40	44	7.7	
Platforms		12	8	1	1
Wells		43	29	6	6

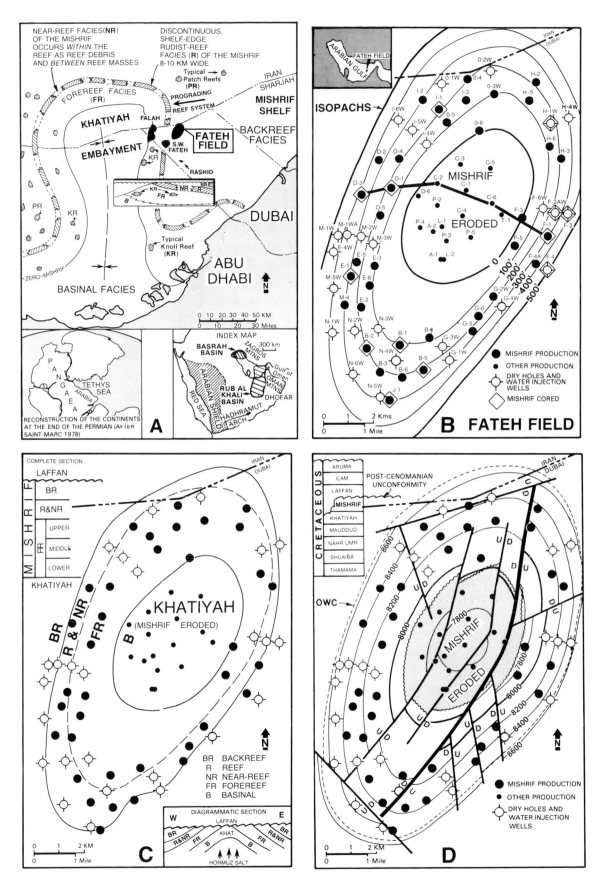

Fig. 26-2A. Regional development and depositional facies setting. *B.* Isopach map of Mishrif Formation. Contours in feet. *C.* Mishrif Formation facies distribution. *D.* Structure contour map. Paleogeographic reconstruction in *A* after Saint-Marc, 1978, published by permission, Elsevier Scientific Publishing Company.

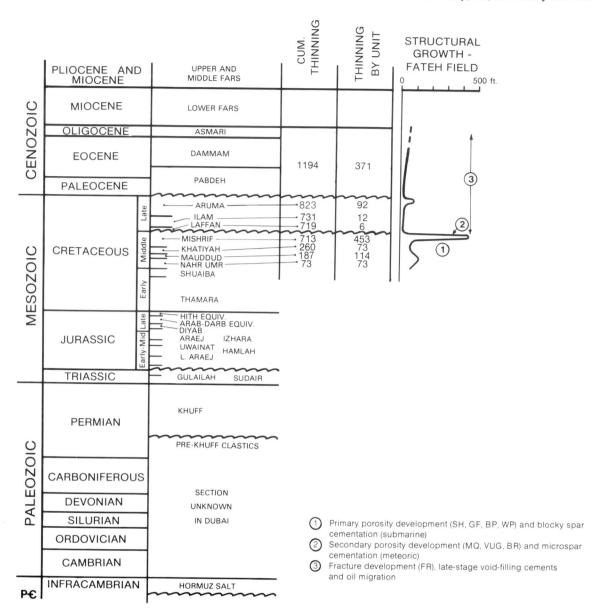

Fig. 26-3. Summary of the influence of salt tectonics on sedimentation over the Fateh structure.

dome structures in the southern Gulf, one of which is Fateh Field. Episodic movement of the Hormuz salt has periodically created unconformities across the top of the Fateh structure. Here the earliest record of salt movement is recorded by 73 feet (22 m) of thinning in the Middle Cretaceous Nahr Umr Formation over the crest of the structure (Fig. 26-3). At the end of Turonian time, strong movement of the salt struc-

ture caused the stripping of the entire Mishrif Formation, a total thinning of over 453 feet (138 m) in this pulse alone. The end of the Cretaceous is marked by an unconformity at the top of the Aruma Formation, where a total of 823 feet (251 m) of thinning at Fateh is attributed to structural growth caused by local salt tectonics. Closure increases at depth in these structures because of their sporadic growth.

Cretaceous Stratigraphy

Lower Cretaceous shelf carbonates of the Thamama Group were deposited as cycles of porous grainstones and tight wackestones and packstones. The uppermost member of the Thamama, the Shuaiba, has well-developed rudist-reef facies similar to those of the Mishrif. These sediments were then transgressed by the Middle Cretaceous Nahr Umr Shale, which grades upward into the Mauddud Limestone. Next, the basinal limestones of the Khatiyah Formation were deposited; this was followed by Mishrif deposition in the Cenomanian and a major unconformity of post-Cenomanian age. These relationships plus facies analysis of the Mishrif are shown in an east–west cross-section across Fateh Field (Fig. 26-4). The Upper Cretaceous Laffan Shale formed as a transgressive deposit overlying the Mishrif. It grades upward into the regressive Ilam Limestone, which, in turn, is overlain by another transgressive unit, the Aruma Shale.

A major unconformity separates Cenozoic and Mesozoic strata of the Arabian Peninsula. This was followed by a transgression that continued through the Eocene, depositing the Pabdeh Shale and Dammam Limestone. The overlying Oligocene Asmari Limestone is a regressive shelf sequence. The Asmari is overlain by the Fars evaporites, deposited as the closing of the Tethyan Sea formed a restricted basin. The degree of restriction gradually diminished, and normal-marine conditions are in the process of being re-established in the modern-day Gulf.

Middle Cretaceous Facies

Deposition of the Khatiyah and Mishrif occurred as a single transgressive/regressive sequence, some 750 feet (230 m) in thickness. The Khatiyah represents the transgressive phase, having been deposited first in the axis of the Khatiyah Embayment and later having spread across the shelf areas surrounding the basin. The Mishrif is basically a rudist-reef complex that prograded basinward with several well-developed stages of reef advancement. The facies discussed below are illustrated in Figure 26-5 by polished core slabs.

Stratigraphic relationships are shown in Figure 26-4, and petrographic details in Figure 26-7.

Khatiyah Formation—Basinal Facies

The Khatiyah consists of calcareous shales grading upward into argillaceous lime mudstones that contain abundant calcispheres (*Oligostegina* sp.) and planktonic foraminifers (Fig. 26-8). Thin-walled pelecypods and ostracods also occur, but are generally scarce. The shales are laminated and locally burrowed; stylolites are common. Geochemical data indicate that the Khatiyah is a mature, oil-prone source rock and the source for the hydrocarbons trapped in the Mishrif.

Mishrif Formation—Reef Related and Shelf Facies

Mishrif facies are interpreted as backreef, reef/near-reef, and forereef (Figure 26-2c and Table 26-2). Paleo-environmental reconstructions imply a wide, shallow-marine carbonate shelf that experienced moderate wave energies. The outer shelf was rimmed by a series of radiolitid rudist reefs (Fig. 26-2A). The reefs are discontinuous and may be more a series of near-reef shoals than ecologic reefs. The dominance of grainstones over boundstones suggests a low-relief feature similar to the "coppice" form of Kauffman and Sohl (1974), which is a very low-relief form of rudist reef. Depositionally, the Mishrif displays a ramp profile, as described by Ahr (1973) and Wilson (1975), with gentle dips of 1 to 2 degrees at the shelf edge. Forereef sediments consist almost entirely of fragmented rudist material. Small satellite reefs, termed "knoll reefs" (Fig. 26-6), occur on forereef slopes as low-relief features (coppices) that grew in deeper waters away from the main reef mass. In some cases, these smaller features may have been large blocks of slumped reef material. They are interpreted from several boundstone intervals, each no more than 2 to 4 meters thick. The hollow, delicately partitioned walls of rudists apparently broke down readily by mechanical and biological processes to provide sand-size material for extensive slope deposits.

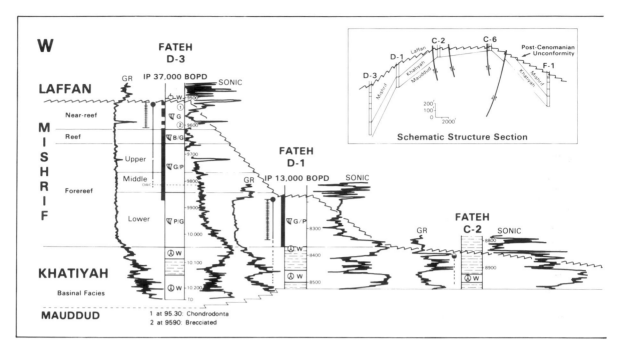

Fig. 26-4. Stratigraphic and structural cross-section of the Mishrif across Fateh Field. Line of section is shown in Figure 26-2B.

Forereef Facies

Mollusk-fragment grainstones and packstones comprise the forereef facies. Identification of grain types in these sediments is commonly difficult because of very fine grain sizes and freshwater diagenetic alterations. Thin-section and SEM (scanning-electronic microscope) analyses indicate that over 95 percent of the sand- and gravel-size material in the forereef is derived from breakdown of rudists, predominantly, and minor amounts of associated pelecypods and gastropods. The remaining 5 percent consists of echinoids (indicative of normal marine salinity), ostracods, and foraminifers derived from the shelf.

The forereef facies is divided into three somewhat gradational microfacies—upper, middle, and lower forereef—based on a combination of the following criteria:

1. Sediment textures become muddier in the downslope direction, with grainstones dominating in the upper forereef and packstones in the lower forereef.

2. The average grain size of forereef sediments decreases downslope. This trend results from two factors, both of which relate to the natural structural breakdown of rudists into sand- and gravel-size fractions. First, the average size of sand shed from the reefs and transported downslope ranges from fine to very fine. Second, a coarse gravel fraction (composed of large rudist fragments and whole rudists) occurs in the upper forereef and extends downslope into the middle forereef. This coarse, bioclastic admixture results in a poorly sorted, bimodal sediment and is an obvious proximity indicator for reef or near-reef facies.

3. Churned burrow textures are common in the middle forereef, and somewhat less common in the lower forereef environments. The textures are similar to those commonly created in modern carbonate sediments by *Callianassa*, the mud shrimp.

4. Occurrences of thick, dark wisps or streaks are

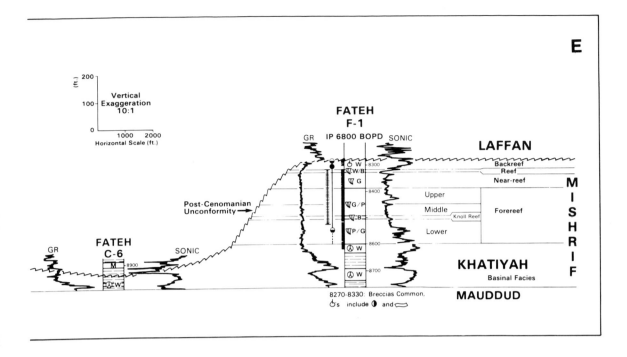

greatest in lower forereef sediments, suggesting gradation into basinal facies.

Reservoir characteristics of the forereef facies are good to excellent. Average porosity is 20 to 25 percent for the upper and middle forereefs and 10 to 15 percent for the lower forereef. Since thick sections of the reef facies were eroded at Fateh, the forereef facies is the most extensive and productive.

Reef and Near-Reef Facies

The reef facies of the Mishrif (Fig. 26-5) is poorly preserved at Fateh Field, owing to post-Cenomanian erosion. The facies consists of rudist boundstones and associated rudist grainstones. The close spacing of rudists and their apparent growth position in cores indicate true boundstone textures. However, it is doubtful if these reefs developed much vertical relief. Paleoecological reconstructions indicate a series of low-lying rudist reefs that grade into grainstone shoals of shelly reef debris (Fig. 26-8).

The near-reef facies consists of medium- to coarse-grained rudist grainstones. Volumetrically, near-reef facies account for as much as five times the amount of section as the reef facies. This suggests that through time, coarse shell beds in outer shelf environments may be more common than true reefs.

Radiolitids dominate the rudist population in reefs and shoals at the outer shelf. Caprinids and monopleurids are also common (see Fig. 26-8). *Chondrodonta*, a large, thick-shelled, deeply ribbed Cretaceous pelecypod, is considered by Lozo *et al.* (1959) to be an indicator of the crestal portion of rudist reefs; our observations at Fateh support this interpretation. *Tylostoma*, a large gastropod, also occurs (but rarely) in near-reef facies. *Cladophyllia*, a small, branching coral, is found in the reef core itself. The result of tremendous sediment productivity by the rudist reef/shoal complex belt 5 to 6 miles (8–10 km) wide is a gently dipping apron of forereef sediment, ex-

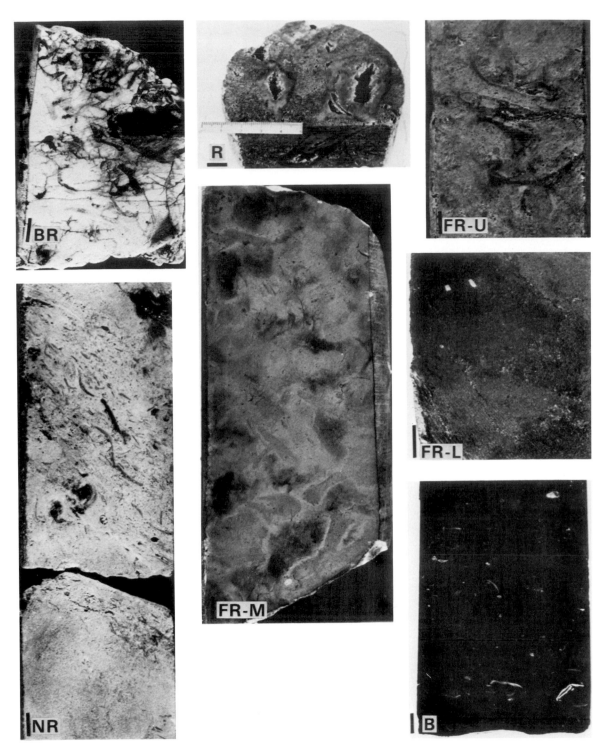

Fig. 26-5. Mishrif and Khatiyah facies core sample photographs of scale bar equals 1 cm. BR-Backreef facies: brecciated bioclastic wackestones, Fateh E-4W, 9543 ft. R-Reef facies rudist boundstone, Fateh D-3, 9539 ft. NR Near-reef facies: rudist-fragment grainstone (with whole rudists), Fateh E-1, 8446 ft. FR-U-Upper forereef facies: rudist-fragment grainstone with large rudist fragments, Fateh F-2, 9652 ft. FR-M-Middle forereef facies: burrowed rudist-fragment grainstone/packstone, Fateh B-2, 9730 ft. FR-L-Lower forereef facies: rudist-fragment packstone with argillaceous wisps, Fateh B-1, 8302 ft. B-Basinal facies: argillacaeous mudstone, from the Khatiyah Formation, Fateh F-1; all other cores shown are from the Mishrif. Well locations are shown in Figure 26-2B.

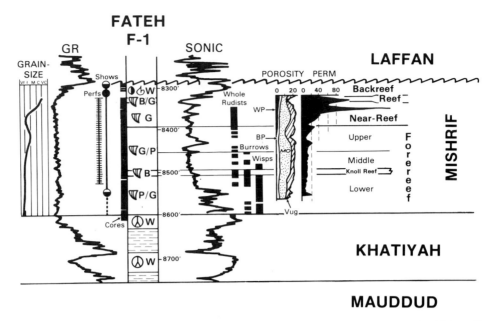

Fig. 26-6. Typical Mishrif stratigraphic section and wireline log as shown in the F-1 well.

tending some 12 to 15 miles (20–25 km) out into the Khatiyah Embayment. Reef and near-reef facies have the best porosities (25–30%) and excellent matrix permeabilities that range between 30 and 200 millidarcys, averaging 50 millidarcys.

Backreef Facies

Muddy, bioclastic limestones containing miliolids are the dominant backreef facies in the broad shelf lagoon behind the outer shelf complex of rudist reefs and shoals. The occurrence of the *Halimeda*-like codiacean algae, *Bacinella* (also called *Lithocodium*; Johnson, 1964), is typical of warm, shallow, well-lit waters of the middle shelf. In addition, the presence of dasyclad algae is interpreted to indicate water depths generally less than 15 feet (4.6 m). The abundance of miliolids is characteristic of shelf environments and compares well with the Holocene distribution of miliolids across the modern, carbonate shelves of south Florida and the Great Bahama Bank (Rose and Lidz, 1977). Minor echinoid fragments indicate normal marine salinity.

Two associated lithofacies occur in the back-reef: (1) pelletal packstones and wackestones, which are typical of shelf facies, and (2) rudist boundstones/packstones, interpreted to have been small, basically circular, patch reefs growing in shallow waters of the middle shelf. Rudist patch reefs across the Lower Cretaceous shelf in Texas are similar to these and show a zonation of species distribution controlled by water circulation and salinity. In the Glen Rose of central Texas, patch reefs on the inner shelf (i.e., those with the least exposure to open oceanic circulation) were dominated by oysters; middle shelf reefs by monopleurids; and outer shelf reefs by caprinids and radiolitids (Perkins, 1974). The requienids, such as *Toucasia* with its characteristically dark-brown shell wall, occurred across the entire shelf with no apparent zonation.

Porosity and permeability in backreef facies are erratic, from near zero to more than 25 percent for porosity (average less than 10%), and from near zero to 100 millidarcys for permeability (average less than 10 md). The inconsistent reservoir quality is due to the original muddy textures and also to the effects of freshwater diagenesis during subaerial exposure of the Mishrif. Solution breccia is common immediately below the post-

Table 26-2. Mishrif Facies (diagnostic properties listed in **bold** type)

| | MISHRIF FORMATION | | | | | KHATIYAH FORMATION |
| | | | FOREREEF FACIES | | | |
	Backreef Facies	**Reef and Near-Reef Facies**	Upper	Middle	Lower	**Basinal Facies**
LITHOFACIES	**algal foram wackestones to mudstones (algae = platy green algae and forams = miliolids**	**rudist boundstones to packstones and rudist grainstones to packstones with whole rudists (C)**	**rudist grainstones to packstones**	**rudist grainstones to packstones**	**rudist packstones to grainstones**	**mudstones and gray calcareous shale**
ASSOCIATED LITHOLOGIES	rudist boundstones to packstones (R-C) and pelletal packstones to wackestones (C) dolomite (R)	pelletal grainstones to packstones	pelletal grainstones to packstones	pelletal grainstones to packstones (R)	dolomite (R)	dolomite (R)
COLOR	very light gray (some dark gray)	very light brown	very light brown	very light brown	very light brown	medium brown and gray
GRAIN-SIZE	mud-size matrix with bioclasts of all sizes	**bimodal: fine & crs to v crs**	**bimodal: fine & crs to v crs**	**fine** (with minor amts of crs)	**very fine to fine**	mud-size
SORTING	very poor	poor-good	**med-good**	**excellent**	**good**	—
ROUNDING	subrounded to rounded	rounded	**well rounded**	**well rounded**	**well rounded**	—
ARGILLACEOUS CONTENT	low to medium — argillaceous wisps (R-C)	low	low	low	medium — argillaceous wisps (C-A)	**high**
FOSSILS MOLLUSCS	caprinids and monopleurids occur in patch reefs	*Chondrodonta* (R-C) radiolitids† caprinids† monopleurids Tylostoma R	radiolitids (C)	Rudist and pelecypod fragments are very abundant; many are too fine to identify.		thin-walled pelecypods (R)

FORAMS	**miliolids** (C) *Praealveolina* (R) small benthonics (R) agglutinated forams (R)	miliolids (VR)	miliolids (VR)			**planktonics**
ALGAE	***Bacinella*** (C—R) dasyclads (R)	—	—	—	—	**Oligostegina** (C—A)
Others	echinoids	echinoids *Cladophyllia* (R—C) (coral)	echinoids	echinoids	echinoids	ostracods (R) pelecypods R
SEDIMENTARY STRUCTURES	burrowing (R/C) brecciation (C) vert fractures (R/C) geopedals (R) stylolites (R—C) internal fracture-fill (R—C) pyrite (C)	burrowing (R) fractures (R—C) geopedals (R)	burrowing (R) fractures (R) geopedals (R)	burrowing (A) argill. wisps (R) fractures (R)	**argill. wisps (C)** burrowing (A) fractures (R)	burrowing (C) fractures (R—C) stylolites (C)
POROSITY TYPES	BP, MO	BP, WP, MO, VUG, SH	BP, WP, SH, VUG, MO	BP, MO, VUG	BP, VUG	—
AMOUNT (%)	0-25	20-25	20-25	20-25	10-15	—
PERMEABILITY (md)	5-10	50	5-30	5-20	5-10	—
AVERAGE THICKNESS (in feet)	30-50	100	75	75	150	200
UNCONFORMITY AT THE TOP OF THE MISHRIF	erosion and deep weathering	most of the reef facies was eroded	may be completely removed by erosion	very little effect	very little effect	very little effect
RESERVOIR QUALITY	poor to zero	**good to excellent**	**excellent (relatively homogeneous units)**		poor (argill. wisps & stylolites are C)	zero (this is a source rock)
PROFILE POSITION	middle shelf	outer shelf	slope	slope	slope	basin

Very Rare (VR)   Rare (R)   Common (C)   Abundant (A)

* Dark brown if oil-stained

† Some rudists are in growth position.

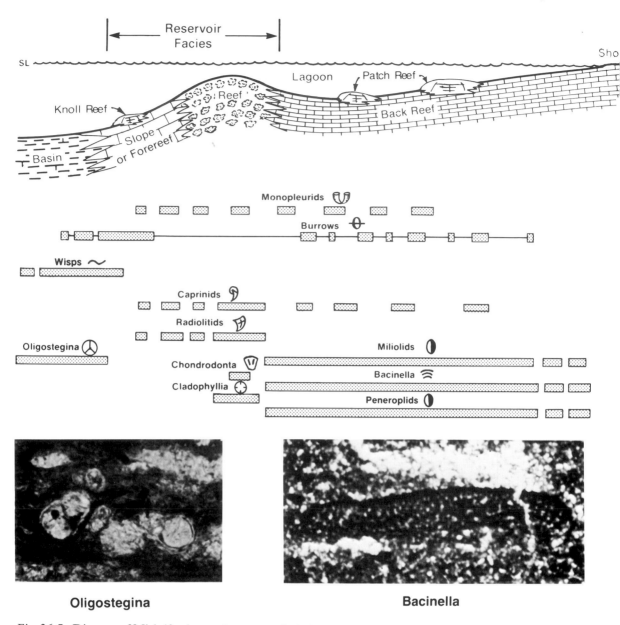

Fig. 26-7. Diagram of Mishrif paleo-environments. A shelf-to-basin profile showing the distribution of fossils and sedimentary structures together with photomicrographs of microfauna.

Cenomian unconformity and commonly reaches depths 20 to 30 feet (6–9 m) below it. Intertidal and supratidal environments must have occurred locally in the backreef facies, as indicated by the presence of root traces, oxidized iron nodules, and coarse, unsorted, intraclast-bearing storm beds.

Reservoir Facies

From base to top, reservoir facies in the Mishrif can be divided into three fundamental units:

1. The lower forereef facies consists of about 30 feet (9 m) of very fine-grained rudist pack-

Fig. 26-8. Diagenetic features of the Mishrif. Photo-micrographs and SEM photos of Formation. *A.* Moldic porosity developed in rudist-fragment grain-stone, Fateh D-3, 9539 ft, ×30. *B.* Interparticle and moldic porosity; note partially filled molds; Fateh D-3, 9686 ft, ×50. *C.* Equant spar, rim cement, and coarse mold-filling spar, Fateh-4A, 9450 ft, ×125. *D.* Vug-filling cements—thin, vug-lining cement and coarse vug-filling spar; Fateh F-1, 8445 ft, ×30. RC rim cement; ES equant spar; CS coarse spar; and MR micritized rim.

stone and grainstone, which overlie the Khatiyah. Porosities range from 1 to 15 percent and permeabilities from 1 to 20 millidarcys.

2. Middle forereef sediments form a fine-grained rudist grainstone unit approximately 75 feet (23 m) thick. This interval has consistent porosities of 23 percent and permeabilities ranging from 14 to 20 millidarcys, with an average of 16 millidarcys.

3. The reef/near-reef and upper forereef facies are fine- to very coarse-grained rudist grainstones and boundstones ranging up to 90 feet (27 m) thick, with porosities of 25 percent. Permeabilities in this zone range from 21 to 102 millidarcys, with an average of 49 millidarcys.

The reservoir behaves as a homogeneous unit. Core plug analyses show that vertical permeability in the Mishrif is as good as or better than horizontal permeability. Pressure data indicate that faults over the crest of the Fateh structure (Fig. 26-2D) are non-sealing and have no effect on continuity in the Mishrif reservoir. Had the bulk of the reef facies not been eroded across Fateh, the size of the reservoir might have been doubled.

Diagenesis

The combined effects of salt tectonics, a major unconformity at the top of the Mishrif, and meteoric diagenesis have produced excellent quality reservoir facies at Fateh Field.

Porosity

Several types of porosity are developed in the Mishrif (Fig. 26-6). Secondary pore types (i.e., moldic and vuggy) account for the majority. Moldic porosity (MO) is actually biomoldic and formed by the dissolution of aragonite fossils, mostly mollusks. Primary porosity types, mainly interparticle (BP) and intraparticle (WP) porosity, also occur but are much less important than secondary porosity development. In addition, minor porosities are associated with three rudist-related porosity types: (1) intraparticle porosity (WP) occurring within the hollow microstructure

of rudist wall material and in large body cavities and accessory cavities within rudists, (2) growth-framework porosity (GF) occurring between rudists as they encrust one another in boundstone fabrics, and (3) shelter porosity (SH) resulting from the sheltering effect of whole rudists and large rudist fragments.

Open and partially cemented fractures (FR) add to the overall porosity, together with a very localized breccia porosity (BR) occurring in the back-reef that developed between breccia fragments beneath the unconformity at the top of the Mishrif.

Cements

Calcite-cemented rudist-fragment grainstones are common in these rocks (Fig. 26-7). The cements are of three main types: (1) bladed, isopachous cement (calcite); (2) equant spar, with subhedral to anhedral calcite crystals averaging 30 to 40 microns; and (3) coarse, euhedral calcite crystals, filling (or partially filling) vugs, molds, large pores and cavities within rudists, and fractures.

The most common is equant spar, a weak cement that imparts a powdery, somewhat friable texture to the rock. Subhedral to anhedral crystals are formed in mainly primary interparticle pores, but also in molds, vugs, and intraparticle pores. This spar is believed to have grown in an active freshwater phreatic environment, a phase of cementation that was incomplete, leaving considerable porosity.

Bladed isopachous cement pre-dates equant spar cement. The bladed crystals are about 80 microns long and are interpreted as an earlier freshwater phreatic cement.

During the course of burial diagenesis, certain molds, vugs, and fractures were cemented with coarse, blocky calcite spar with well-formed crystals as large as 1.5 millimeters. This is interpreted to be a late-stage cement and is supported by fluid-inclusion analysis.

Diagenetic Sequence

Mishrif sediments were deposited in shallow, warm, tropical seas. The earliest indication of diagenesis is in the form of micrite rims (Bathurst, 1966) around many mollusk fragments. Subsidence and burial followed, as approximately 550 to 600 feet (168–183 m) of Mishrif was deposited, this thickness being the regional, off-structure average. Strong upward diapiric movements at the end of the Cenomanian established the Fateh dome and initiated the erosion of the entire Mishrif from the top of the structure. This probably caused the fracturing seen in the older rocks and placed the flanking Mishrif beds in the freshwater phreatic zone, where bladed isopachous cements and extensive equant spar cements occurred in sequence.

Increased uplift raised the Mishrif into the upper part of the freshwater phreatic or possibly into the vadose zone, where extensive dissolution of aragonite fossil material took place, creating considerable biomoldic and vuggy porosity. Some etching of equant spar can be seen in SEM photographs (Fig. 26-7) and accounts for the powdery texture of the rock. This was followed by continued subsidence throughout the Cenozoic, whereby the Mishrif was lowered to its present depth of about 8500 feet (2600 m). Coarse blocky calcite void-filling was emplaced at present depths of burial.

Seal

The Laffan Shale averages about 90 feet (27 m) thick in the Fateh Field area and is the seal for the Mishrif reservoir. It was deposited on the erosional surface that marks the post-Cenomanian unconformity. At depths of about 7800 feet (2380 m), these shales are hard, brittle, and splintery, with several thin stringers of bioclastic wackestone. In cores, the Laffan/Mishrif contact shows a deep weathering profile of siliceous soil zones at the top of the limestone.

Source and Migration

The underlying Khatiyah Shale is the likely source of Mishrif oil. The Khatiyah is an oil-prone source rock lying within the oil window with respect to kerogen maturity (vitrinite reflectance, R_o, of about 0.9 %), rich in organics (total organic carbon of about 1.5 %), and 800+ feet (240 m) thick in the basin. The facies contact between lower forereef facies of the Mishrif and the Khatiyah forms an adequate interface for the updip passage of hydrocarbons and lateral migra-

tion into the more porous upper and middle forereef facies. Migration into porous Mishrif facies progressed out of the basin from west to east and into reef-related and shelf facies sometime during the Eocene. A subsequent eastward dip then concentrated hydrocarbons in reef and forereef facies to the west along the outer edge of the Mishrif shelf.

Summary

Fateh Field, a giant oil field in offshore Dubai, has 12 platforms, 43 producing wells, and original oil in place of 2.3 billion barrels of oil, with estimated ultimate recoverable reserves of 1.02 billion barrels of oil. Production is from three formations, with 71 percent coming from porous Cretaceous limestones of the Mishrif Formation. Since production began in 1969, 398 million barrels of oil have been produced from the Mishrif, averaging 120,000 barrels per day. Reservoir facies in the Mishrif are fine-grained, mollusk-fragment grainstones and packstones deposited mainly in forereef environments. Porosities average 20 to 25 percent, and permeabilities average 15 to 50 millidarcys. These facies are part of a large rudist reef complex that rimmed the Khatiyah Embayment west of Fateh Field. The overlying 90-feet-thick (27-m) Laffan Shale provides the seal for the Mishrif reservoir. The Khatiyah Formation, which is the basinal equivalent of the Mishrif, is the likely source of hydrocarbons. The trap at Fateh is stratigraphic and structural, formed by truncation of reef, near-reef, and forereef carbonate facies beneath a post-Cenomanian unconformity on a salt structure.

References

AHR, W.M., 1973, The carbonate ramp—an alternative to the shelf model: Gulf Coast Assoc. Geol. Soc. Trans., v. 23, p. 221–225.

BATHURST, R.G.C., 1966, Boring algae, micrite envelopes and lithification of molluscan biosparites: Geol. Jour., v. 5, p. 15–32.

JOHNSON, J.H., 1964, The Jurassic algae: Quarterly, Colorado School of Mines, v. 59, no. 2, p. 129.

KAUFFMAN, E.G., and N.F. SOHL, 1974, Structure and evolution of Antillean Cretaceous rudist frameworks: *in* Contributions to the geology and paleobiology of the Caribbean and adjacent areas: Naturforsch. Ges. Basel, Verh., v. 84, no. 1, p. 399–467.

LOZO, F.E., H.F. NELSON, K. YOUNG, O.B. SHELBURNE, and J.R. SANDIDGE, 1959, Symposium on the Edwards Limestone in Central Texas: Bureau Econ. Geology, Publ. 5905, p. 62–104.

PERKINS, B.F., 1974, Paleoecology of a rudist reef complex in the Comanche Cretaceous, Glen Rose Limestone, central Texas: *in* Perkins, B.F., ed., Aspects of Trinity Division Geology—a symposium: Louisiana State Univ., Geoscience and Man, v. 8, p. 131–173.

ROSE, P.R., and B. LIDZ, 1977, Diagnostic foraminiferal assemblages of shallow-water modern environments: South Florida and the Bahamas: Sedimenta VI, the Univ. of Miami, Comp. Sed. Lab., p. 55.

SAINT-MARC, P., 1978, The Mesozoic: *in* Moullade, M. and A.E. Nairn, eds., Phanerozoic Geology of the World, v. II, Part A: Elsevier Scientific Publ. Co., New York, p. 435–462.

WILSON, J.L., 1975, Carbonate Facies in Geologic History: Springer-Verlag, Inc., New York, 439 p.

27
Sunniland Field
Robert B. Halley

RESERVOIR SUMMARY

Location & Geologic Setting	Collier Co., south Florida, on Florida Platform, USA
Tectonics	Stable subsiding continental-margin carbonate platform
Regional Paleosetting	South Florida Basin, landward from Lower Cretaceous reef trend
Nature of Trap	Structural/stratigraphic; gentle anticline, anhydrite and lime mudstone seals

Reservoir Rocks
Age	Early Cretaceous (Early and Middle Albian)
Stratigraphic Units(s)	Sunniland Limestone
Lithology(s)	Limestone, dolomite
Dep. Environment(s)	Bank-margin sand body fringing shallow platform-interior basin
Productive Facies	Grainstones, rudites, sucrosic dolomites
Entrapping Facies	Lime mudstone, anhydrite
Diagenesis/Porosity	Dolomitization, submarine cementation, early dissolution of aragonitic bioclasts; early to late burial cementation

Petrophysics
Pore Type(s)	Interparticle, intercrystal, moldic
Porosity	0–30%, avg 18%
Permeability	1–1000 md, avg 65 md
Fractures	Negligible

Source Rocks
Age	Early Cretaceous Sunniland Limestone
Lithology(s)	Lime mudstone(?)
Migration Time	Late Tertiary to (?)present

Reservoir Dimensions
Depth	11,615 ft
Thickness	NA
Areal Dimensions	3.5 × 1 mi (5.6 × 1.6 km)
Productive Area	2250 acres (9 km^2)

Fluid Data
Saturations	NA
API Gravity	26°
Gas-Oil Ratio	100:1
Other	3.95% sulfur

Production Data
Oil and/or Gas in Place	37.69 million BO
Ultimate Recovery	18.84 million BO; URE = 50.0%.
Cumulative Production	NA

Remarks: Approximate reservoir temperature 200°F (93°C). Discovered 1945.

27
Setting and Geologic Summary of the Lower Cretaceous, Sunniland Field, Southern Florida

Robert B. Halley

Location and Discovery

The Sunniland Field, located 80 miles (130 km) west-northwest of Miami in Collier County, Florida, was discovered in 1943. Subsequently, nine other fields, some of them single-well fields, have been discovered in the same limestone unit along a northwest–southeast trend some 60 miles (100 km) in length (Fig. 27-1). The productive unit is the Sunniland Limestone, a sequence named and defined by Pressler (1947) and Applin (1960) and known only in the subsurface.

Reserves and Production

The Sunniland Field contains an estimated 19 million barrels of recoverable oil. The largest field found to date in the Sunniland Limestone is West Felda with estimated recoverable oil reserves of 50 million barrels (Pontigo *et al.*, 1979). Average production from wells in the Sunniland Limestone is about 200 barrels of oil per day, and some wells produce as much as 400 barrels of oil per day.

All commercial oil production in South Florida has been from the Sunniland Limestone, a unit 200 to 300 feet (60–100 m) thick that lies between 11,000 and 12,000 feet (3600–4000 m) below sea level in southern Florida. Most oil is produced from reservoirs in porous grainstones and in dolomite from the upper 150 feet (50 m) of the formation. One well produces oil from fractured limestone in the lower half of the Sunniland Limestone.

Geologic Setting

Over 10,000 feet (~3100 m) of shallow-water carbonate and sulfate rocks accumulated in the South Florida Basin during the late Jurassic and early Cretaceous. The basin, which lies within the Florida Platform, is a structural basin filled with sediments deposited in relatively shallow water, at most a few hundred feet deep ($\gtrsim 100$ m). From time to time, the platform contained salinity-stratified interior lagoons in which carbonate muds rich in organic matter accumulated. In general, however, sedimentation kept pace with subsidence of the basin, and shallow-marine deposition prevailed.

As illustrated in Figure 27-1, the South Florida Basin lay shelfward (landward) from the main Lower Cretaceous reef trend that bordered the west side of the basin, fronting on the ancestral Gulf of Mexico (Meyerhoff and Hatten, 1974). The reef trend occupies the approximate position of the West Florida escarpment. This paleogeographic setting is similar to that of the Maverick and North Texas-Tyler basins of Texas, which also developed landward of an Early Cretaceous (Stuart City) barrier reef system (Rose, 1972). The strata in the basin are generally flat-lying, although Winston (1971) has identified a number of subtle structural features in the area (Fig. 27-2).

Reservoir limestones are mostly grainstone with varying amounts of skeletal pebbles and cobbles. Reservoir dolomites are mostly sucrosic in

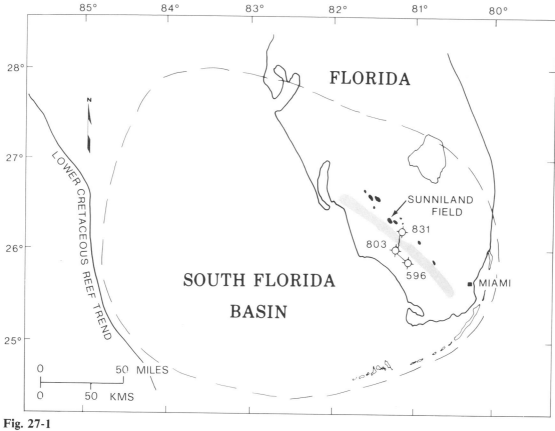

Fig. 27-1

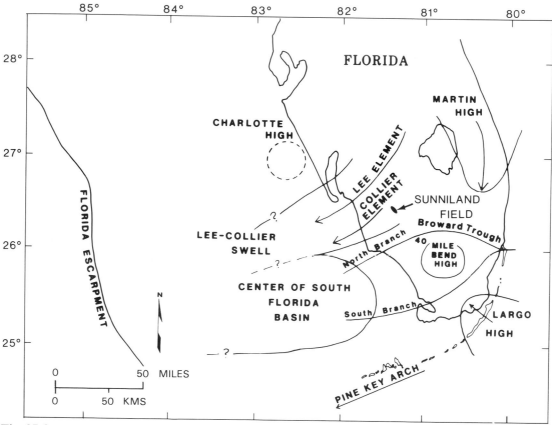

Fig. 27-2

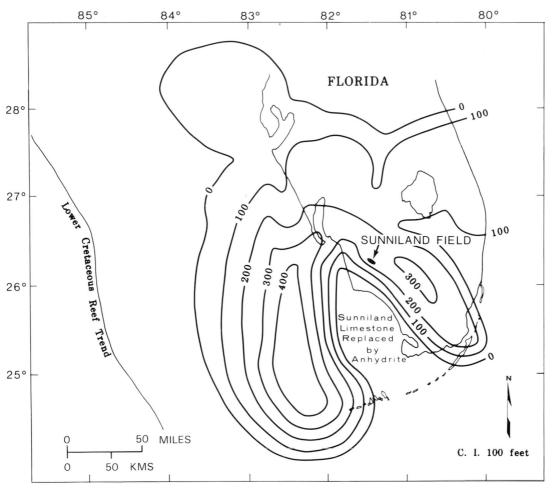

Fig. 27-3. Thickness isopach and distribution map of the Sunniland Limestone in the South Florida Basin, after Oglesby (1965).

texture and were originally skeletal sand grainstone, packstone, and wackestone. Common constituents of the grainstones are pellets, foraminifers, mollusk fragments (mostly gastropods and rudists), ooids, and peloids. Larger skeletal fragments are almost exclusively whole or frag-

---

◀ Fig. 27-1. Map showing paleogeographic setting of the Sunniland Field. The field is one of several (in black) in carbonate-sand bodies rimming the northeastern side of the South Florida Basin. A Lower Cretaceous reef trend occupies the approximate position of the West Florida Escarpment and defines the western edge of the South Florida carbonate platform (after Meyerhoff and Hatten, 1974). Wells 813, 803, and 596 (state lease numbers) indicate position of cross section shown in Figure 27-4.

◀ Fig. 27-2. Map showing structural features of the South Florida Basin, after Winston (1971).

mented rudists. Reef facies as such (boundstones) are absent. These carbonate sands and gravels accumulated as sand banks along the northeast margin of the South Florida basin.

Stratigraphic relationships (Figs. 27-3, 27-4) suggest that the Sunniland grainstones formed along the northeast edge of a platform-interior lagoon less than 150 feet (<50 m) deep. The lagoon was filled with evaporites at the end of Sunniland deposition. Figure 27-3 shows the thickness and distribution of the Sunniland Limestone as well as that portion supplanted by anhydrite (after Oglesby, 1965).

Stratigraphy

The Sunniland Limestone is early and middle Albian in age. It is underlain by the Punta Gorda

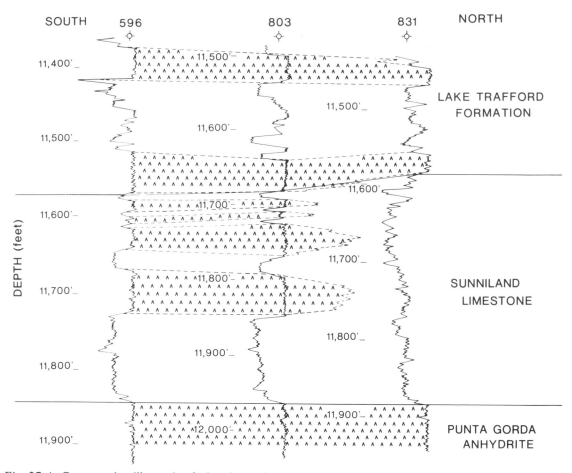

Fig. 27-4. Cross-section illustrating facies change in upper Sunniland Limestone from predominantly carbonate in the area of oil fields to predominantly anhydrite to the southwest. Correlation is based on sonic logs from wells located in Figure 27-1.

Anhydrite and overlain by a sequence of anhydrite, dolomite, and limestone named the Lake Trafford Formation (Oglesby, 1965). The contacts between the Sunniland Limestone and formations above and below are most frequently determined from characteristics on wire-line logs. This practice works well for the lower Sunniland, which has a sharp and regionally widespread contact with the underlying Punta Gorda Anhydrite. The top of the Sunniland Formation is less distinct and grades into the overlying Lake Trafford Formation. This gradation has resulted in the inclusion by some authors of several nodular anhydrite beds (Fig. 27-5) into the top of the Sunniland (Feitz, 1976; Tyler and Erwin, 1976; Means, 1977). These anhydrites serve as convenient marker beds for the top of the Sunniland, but may be grouped with the other anhydrites of the Lake Trafford, leaving the Sunniland an entirely carbonate unit in the type area.

A rapid facies transition to anhydrite, illustrated in Figure 27-4, occurs in the upper half of the Sunniland Limestone about 15 miles (24 km) southwest of the trend of fields (Applin, 1960). Much of this anhydrite, illustrated in Figure 27-6, may represent an evaporite filling of a shallow, restricted lagoon in late Sunniland time. Dark, organic-rich limestones comprise the lower half of the Sunniland and underlie the evaporites to the southwest. These limestones are interpreted to be shelf-lagoon deposits that accumulated in a restricted, shallow, platform interior probably less than 100 feet (<30 m) deep.

Source Rocks

Source-rock characteristics of the Sunniland have been investigated by Palacas (1978) and Pontigo et al. (1979). Petroleum source rocks occur

Fig. 27-5. Photograph of nodular anhydrite (core slab) typical of anhydrites in the Lake Trafford and uppermost Sunniland Limestone.

Fig. 27-6. Photograph of banded anhydrite (core slab) typical of anhydrites occupying middle and upper Sunniland Limestone southwest of the Sunniland trend and filling the Sunniland basin. Note the contrasting fabric of this anhydrite with that illustrated in Figure 27-5.

throughout the Lower Cretaceous of the South Florida Basin and within the Sunniland Limestone itself. Palacas (1978) reported organic carbon contents as high as 3.14 weight percent for the lower Sunniland (avg. 0.57%) and 2.95 percent for the upper Sunniland (avg. 0.44%). Pontigo *et al.* (1979) reported that the entire Sunniland section exhibits a strong predominance of algal, oil-prone, amorphous sapropel-type kerogen. Analyses of cores from upper Albian rocks

(Fredericksburg of the Gulf Coast) suggest that these rocks are thermally immature (Pontigo *et al.*, 1979), but the presence of local-source petroleum in Sunniland reservoirs suggests that lower and middle Albian rocks (late Trinity of the Gulf Coast) have generated hydrocarbons (Palacas, 1978). It seems probable that the source of petroleum in Sunniland reservoirs was either the Albian sequence of the South Florida Basin or possibly the Sunniland Limestone itself. It is even possible that oil in upper Sunniland reservoirs had its source in upper Sunniland carbonates (Palacas, 1978).

Sunniland Field

The Sunniland Field lies within a gentle, asymmetrical anticline trending northwest–southeast, with a slope of about 100 feet per mile (19 m/km) on its northeast flank and 50 feet per mile (9 m/km) on its southwest flank (Figs. 27-7 and 27-8). Rainwater (1974) believed this structure to be inherited from underlying Triassic horst blocks and grabens. Puri and Banks (1959) and Applin and Applin (1965) documented a gradual decrease in the structure of successively higher units. Applin and Applin (1965) believed this was evidence of gradual and continued growth of the structure during Comanche (Aptian-Albian) time. They also suggested that the steep northeast side of the structure is bordered by a fault. Puri and Banks (1959) and Fietz (1976), on the other hand, interpreted the structure as the result of depositional and compactional processes. This latter interpretation of the subtle structure in the Sunniland seems reasonable, but the possibility of gentle deformation during and after deposition cannot be ruled out. Limited wire-line log data suggest that the Sunniland Limestone as a whole, and individual units within it, do not vary in thickness across the field anticline. Thinning of some units toward the anticline crest would be expected if that area had been topographically high prior to Sunniland deposition. Thickening would be expected if the structure were due to biologic or hydrologic buildup of carbonate sediments during Sunniland deposition. Rather than thickening or thinning, beds are quite uniform and may be traced in a simple anticlinal structure across the field.

Four units can be recognized immediately overlying and including the productive zone of the

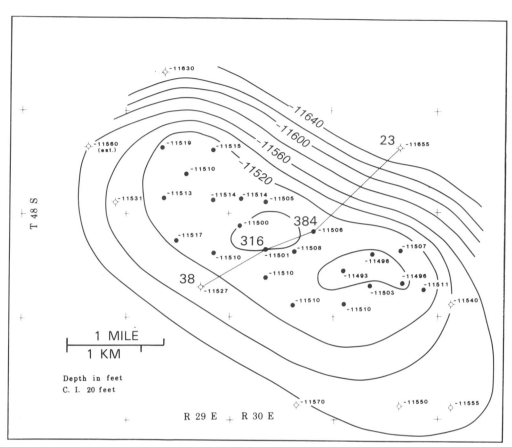

Fig. 27-7. Structure contour map of Sunniland Field drawn on anhydrite bed approximately 50 feet (15 m) above the Sunniland Limestone. The steep flank in the northeast may be fault-induced. Producing wells lie within the −11,520-foot (3510-m) contour line; datum is sea level.

field and are illustrated together with overlying units in Figure 27-8. These units are: (1) a nonporous dolomite 3 to 6 feet thick (1–2 m); (2) an anhydrite bed 6 to 9 feet thick (2–3 m) generally thought to be the reservoir seal; (3) a porous dolomite 6 to 18 feet thick (2–6 m); and (4) a grainstone unit more than 100 feet thick (>30 m).

The uppermost dolomite, unit A of Figure 27-8, is light gray, finely crystalline (individual dolomite rhombs are 10 to 50 microns), and evenly laminated, and is interpreted to be due to early diagenetic dolomitization in supratidal settings modified by later diagenetic processes.

The anhydrite bed, unit B in Figure 27-8, is composed of nodular anhydrite. Individual nodules are rimmed with dolomite and organic matter, and this unit like the dolomite above is thought to be a supratidal deposit. Oil stains immediately below the anhydrite are rare, suggest-

ing that permeability changes in the carbonate rock near the top of the reservoir may also provide part of the reservoir seal.

Below the anhydrite is a sucrosic, medium-crystalline, porous dolomite of unit C, with dolomite rhombs 40 to 250 microns on an edge, and averaging 100 microns (Figs. 27-8, 27-9). This dolomite contains both intercrystal pores and larger irregular pores several hundred microns in diameter. The larger pores are of sizes and shapes that suggest dissolved carbonate sand grains. This dolomite is probably a dolomitized carbonate grainstone and packstone. In several cores, unit C is only partially dolomitized, and the remaining limestone is packstone or wackestone. These observations suggest that permeability was a major factor controlling dolomitization and that more permeable rocks (grainstones) were preferentially dolomitized relative to less permeable rocks (mudstones).

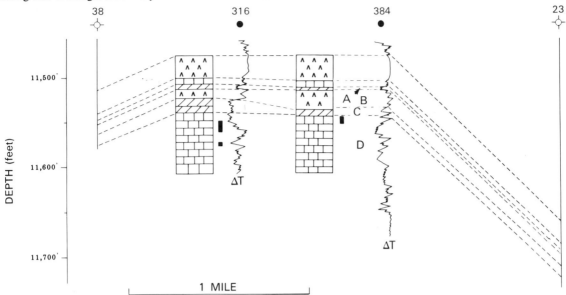

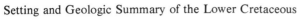

Fig. 27-8. Diagrammatic cross-section through Sunniland Field. Location of wells indicated on Figure 27-7. Perforated intervals of wells 316 and 384 are indicated by black rectangles. Correlation to wells 38 and 23 is based on electric log characteristics (not shown).

Lithologic interpretation is from cuttings, sonic logs, and cores from these and nearby wells. Well numbers are state lease numbers. Letter designated units are described in text.

0.5 mm

Fig. 27-9. Photomicrograph of porous dolomite exhibiting intercrystal (1) and larger pores (2), the latter probably at positions of dissolved skeletal grains. Plane light.

Underlying the porous dolomite is a thick grainstone unit (unit D of Fig. 27-8) that contains the main producing horizons of the reservoir. These rocks are mollusk (mostly rudist and other pelecypod and gastropod fragments), foraminiferal, pellet grainstones with larger rudist fragments and some whole rudists. Cross-bedding is

not apparent in cores. The grainstone contains patches of dolomite, packstone, and wackestone in some cores. Large secondary pores after leached rudists are illustrated by Banks (1960), in cores with pores large enough that pencils can be placed through the holes. Although casual examination of cores leaves the impression that secondary porosity is the significant type in this unit, detailed study of thin sections has demonstrated that primary porosity is more abundant in approximately a 60:40 ratio (Halley, 1979). Although less obvious, primary porosity enhances permeability by providing connections between relatively isolated secondary pores.

Brief Diagenetic History of Sunniland Grainstones

Petrographic study of these rocks reveals a complex diagenetic history for the main producing intervals of grainstones in the Sunniland reservoir. This history is divisible into: (1) an early phase (pre-compaction) of aragonite dissolution and calcite cementation; (2) a later phase of compaction followed by calcite and dolomite cementation; and (3) a latest phase of calcite replacement by anhydrite.

These diagenetic phases are largely distinguished from the recognition of processes that take place before, during, and after burial compaction and late-stage cementation. None of these processes is considered to be a distinct single event in the rock history or the reservoir diagenesis. Rather, they operated for unknown but apparently varying (possibly even overlapping) periods of time. "Early" and "late" processes refer only to their timing relative to features associated with burial compaction. These processes did not act equally on all portions of the reservoir, but varied in effectiveness from foot to foot, indeed in some instances from grain to grain, within the rock. In part, the processes followed the natural inhomogeneities of porosity, permeability, grain packing, grain shape and mineralogy inherent in the original sand body; and, in part, the diagenesis left its own heterogeneous imprint on the reservoir.

Early Phase Dissolution and Cementation

The earliest diagenetic process recognized in Sunniland grainstones is marine cementation. Marine cement is rare and occurs in the interior cavities of skeletal fragments, especially in gastropods and rudists (Fig. 27-10). The cement can be recognized as bladed to fibrous calcite, which is often cloudy due to the presence of submicroscopic inclusions. It has not been found to occur between grains. This marine cement is believed to be essentially syndepositional in origin.

The first post-depositional cement is an isopachous fringe of "dog-tooth" calcite (Fig. 27-11). It tends to occur in rocks that contain evidence of aragonite dissolution, but not all rocks with secondary porosity after aragonite contain this early cement (Fig. 27 8). Evidence of dissolved aragonite includes moldic porosity from "leached" gastropods, rudists, and other mollusks, often enclosed by micritic envelopes surrounding voids. It is probable that this early fringing calcite cement formed simultaneously with aragonite dissolution in some portions of the Sunniland grainstone. In other areas, aragonite dissolved without concurrent cementation, probably because of variations in the chemical composition of pore fluids.

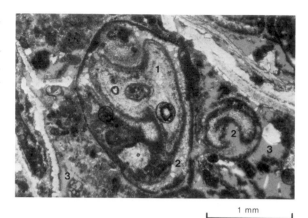

Fig. 27-10. Photomicrograph of Sunniland reservoir grainstone illustrating cloudy, bladed calcite cement partially filling gastropod cavity (1). Gastropod shell has been dissolved and resulting secondary porosity partially filled with clear calcite cement (2). Primary pore space is largely free of cement (3). Plane light.

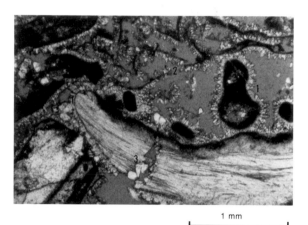

Fig. 27-11. Photomicrograph of Sunniland reservoir grainstone illustrating early, fringing, calcite cement (1), fractured micrite envelopes (2) and late calcite cement in fractured grains (3). Plane light.

Compaction and Late Cementation

Evidence of both physical and chemical compaction is widespread in Sunniland grainstones. Compaction started essentially during sedimentation and continued with increased burial. In reservoir rocks, compaction is evinced by fracturing and disruption of grains and early, fringing calcite cement (Fig. 27-11). Where early fringing cement is absent and aragonite grains are dissolved, com-

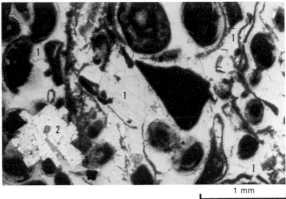

Fig. 27-12. Photograph of core slab showing internal molds of rudists (light) surrounded by fine-grained, oil-stained, peloidal, skeletal sand (dark). Rudist shells have been dissolved and much secondary porosity lost due to compactional displacement of matrix. Most matrix movement has occurred during displacement and breakage of grains, but some has occurred along short fractures evident in the upper half of the photo.

Fig. 27-13. Photomicrograph of coarse, poikilotopic, late calcite cement (1) surrounding fractured grains, and anhydrite laths (2) replacing grains, early fringing cement, and late calcite cement. Plane light.

Fig. 27-14. Photomicrograph of coarse, clear, late dolomite cement precipitated on broken fringe of early calcite cement. Plane light.

paction resulted in unusual fabrics such as that illustrated in Figure 27-12. In this example, structures that appear to be burrows are in reality the mud infillings (probably slightly marine-cemented) of rudist cavities. The skeletons themselves were removed by dissolution and the uncemented matrix moved by compression to partially fill secondary porosity.

There is abundant evidence for pressure solution of calcite in the non-reservoir limestones of Sunniland. Evidence includes stylolites, solution seams, and interpenetrating grain contacts. In reservoir rocks with substantial porosity, evidence of pressure solution is rare. Grains are, however, commonly fractured, rotated, and translated to new positions by overburden pressure.

Two late-stage cements, precipitated during and after significant compaction, are distinguished in reservoir rocks. One is a coarsely crystalline calcite cement, occurring in fractures, that may reach 500 microns in diameter. Single crystals may surround several grains (Fig. 27-13). The other is a dolomite cement that occurs as relatively large rhombs up to 1 millimeter on an edge, filling primary and secondary pore space. The dolomite is white and often exhibits curved crystal faces typical of "saddle" or "baroque" dolomite. It too fills fractures caused by burial

compaction (Fig. 27-14). The source for these late carbonate cements seems to have been material released during pressure solution in surrounding non-reservoir rocks.

Sulfate Replacement

The latest, petrographically distinct process in Sunniland carbonates is replacement of calcite by anhydrite. Early cements, grains, and late cements are replaced by coarse crystal laths of anhydrite during this process (Fig. 27-13). Sulfate replacement occurs without the development of microscopically visible pore space. There appears to be no selectivity according to carbonate grain size. Since the process is replacive, it is relatively insignificant in porous reservoir rocks and accounts for only about 5 percent of mineral substitution.

Summary

The Sunniland reservoir originated as one of many carbonate sand bodies bordering a small, restricted, platform-interior basin. The oil reservoir lies on a structural high partly formed by depositional and compactional processes and perhaps partly by faulting. The major reservoir seal is a thin anhydrite above the reservoir. Impermeable carbonate rocks below the anhydrite may act locally as seals. Source rocks probably are organic-rich carbonates within the Sunniland Limestone. Reservoir rocks are predominantly grainstones with subordinate sucrosic dolomites. Although skeletal moldic porosity is obvious in cores, primary interparticle porosity accounts for most of the storage capacity of the reservoir. Porosity and permeability were reduced in reservoir grainstones by calcite and dolomite cements of both burial-compaction and post-compaction origins.

References

APPLIN, P.L., 1960, Significance of changes in thickness and lithofacies of the Sunniland Limestone, Collier County, Fla.: in Short papers in the Geological Sciences: U.S. Geol. Surv. Prof. Paper 400-B, Art. 91, p. B209–B211.

APPLIN, P.L., and E.R. APPLIN, 1965, The Comanche series and associated rocks in the subsurface in central and southern Florida: U.S. Geol. Surv. Prof. Paper 447, 84 p.

BANKS, J.E., 1960, Petroleum in Comanche (Cretaceous) section, Bend Area, Florida: Amer. Assoc. Petroleum Geologists Bull., v. 44, p. 1737–1748.

FEITZ, R.P., 1976, Recent developments in Sunniland exploration of South Florida: Gulf Coast Assoc. Geol. Soc. Trans., v. XXVI, p. 74–78.

HALLEY, R.B., 1979, Pore types in Sunniland Limestone (lower Cretaceous): Amer. Assoc. Petroleum Geologists Bull., v. 63, p. 460.

MEANS, J.A., 1977, Southern Florida needs another look: Oil and Gas Jour., Jan. 31, 1977, p. 212–225.

MEYERHOFF, A.A., and C.W. HATTEN, 1974, Bahamas salient of North America: in Burk, C.A. and C.L. Drake, eds., The Geology of Continental Margins: Springer-Verlag, Inc., New York, p. 429–446.

OGLESBY, W.R., 1965, Folio of South Florida basin, a preliminary study: Florida Geol. Surv. Map Series, No. 19, 3 p., 10 maps.

PALACAS, J.G., 1978, Preliminary assessment of organic carbon content and petroleum source rock potential of Cretaceous and lower Tertiary carbonates, South Florida Basin: Gulf Coast Assoc. Geol. Soc. Trans., v. XXVIII, p. 357–381.

PONTIGO, F.A., JR., A.V. APPLEGATE, J.H. ROOKE, and S.W. BROWN, 1979, South Florida's Sunniland oil potential: Oil and Gas Jour., July 30, p. 226–232.

PRESSLER, E.D., 1947, Geology and occurrence of oil in Florida: Amer. Assoc. Petroleum Geologists Bull., v. 31, p. 1841–1862.

PURI, H.S., and J.E. BANKS, 1959, Structural features of the Sunniland oil field, Collier County, Florida: Gulf Coast Assoc. Geol. Soc. Trans., v. IX, p. 121–130.

RAINWATER, E.H., 1974, Possible future petroleum potential of peninsular Florida and adjacent continental shelves: in Cram, I.H., ed., Future Petroleum Provinces of the United States, v. 2: Amer. Assoc. of Petroleum Geologists Mem. 15, p. 1311–1341.

ROSE, P.R., 1972, Edwards Group, surface and subsurface, central Texas: Bur. Econ. Geology, Univ. Texas, Austin, Rept. of Invest. No. 74, 198 p.

TYLER, A.N., and W.L. ERWIN, 1976, Sonoco-Felda Field, Hendry and Collier counties, Florida: in Braunstein, J., ed., North American Oil and Gas Fields: Amer. Assoc. Petroleum Geologists Mem. 24, p. 287–299.

WINSTON, G.O., 1971, Regional structure, stratigraphy and oil possibilities of the South Florida basin: Gulf Coast Assoc. Geol. Soc. Trans., v. 21, p. 15–29.

Poza Rica Field

Paul Enos

RESERVOIR SUMMARY

Location & Geologic Setting 135 mi (215 km) NE of Mexico City in Veracruz state, Gulf coastal plain, central eastern Mexico

Tectonics SE-plunging anticline: reservoirs confined to NE limb by pinchout; regional post-Laramide eastward tilting

Regional Paleosetting Distal forereef slope, 9 miles (15 km) SW of Golden Lane Atoll

Nature of Trap Stratigraphic/structural; basinward pinchout and facies change of debris-flow carbonates into basin facies in conjunction with eastward tilting

Reservoir Rocks
  Age Mid-Cretaceous (Albian-Cenomanian)
  Stratigraphic Units(s) Tamabra Limestone
  Lithology(s) Limestone, some dolomite
  Dep. Environment(s) Deep-basinal marine
  Productive Facies Skeletal-fragment grainstone, intraclast breccia, gravity-flow limestones (likely includes turbidites or grain-flow deposits) locally dolomitized
  Entrapping Facies Basinal pelagic lime mudstone
  Diagenesis/Porosity Early deep phreatic(?) cementation by calcite; dissolution; local pervasive dolomitization

Petrophysics
  Pore Type(s) Skeletal moldic, vuggy, some interparticle
  Porosity 8% (3.7–9.7% weighted avg)*
  Permeability 0.01–700 md (0.3–0.6 md avg)*
  Fractures Rare in cores. Probable minor permeability enhancement

Source Rocks
  Age Upper Jurassic (Oxfordian)?
  Lithology(s) Santiago Formation, dark shales
  Migration Time Oligocene?

Reservoir Dimensions
  Depth 6400–8850 ft (1950–2700 m), SE plunge
  Thickness 575 ft (175 m) avg oil column at axis of field (max 740 ft, 225 m); avg gross reservoir thickness 400 ft (120 m)
  Areal Dimensions 30,000 acres (120 km^2)
  Productive Area 19 × 3 mi (30 × 5 km)

Fluid Data
  Saturations S_w = 1–56%, avg 19%; S_w[†] = 2.8–27%, avg 10%
  API Gravity 35°
  Gas-Oil Ratio 180:1 (1949)
  Other —

Production Data
  Oil and/or Gas in Place 2.84 billion BO Poza Rica
  Ultimate Recovery 1.20 billion BO Poza Rica *sensu stricto***; URE = 42.3%
  Cumulative Production 1.98 billion BO for 12 fields, Greater Poza Rica area, and 865 million BO for Poza Rica ss, to July 1, 1983p[††]

Remarks: Discovered 1930. *Range of average values for 5 main reservoir lithologies **Estimates for entire field range from 2.04 to 2.76 billion BO recoverable [†] Possibly a percentage of bulk volume. [††] Oil and Gas Journal, Dec. 16, 1983, p. 100.

28
Cretaceous Debris Reservoirs, Poza Rica Field, Veracruz, Mexico

Paul Enos

Location and Discovery

The giant Poza Rica Field was the backbone of Mexican oil production from shortly after its discovery in 1930 until eclipsed by spectacular new discoveries in the Isthmus of Tehuantepec and Campeche Shelf in the 1970s. Poza Rica (Fig. 28-1) is located in northeastern Veracruz state, Mexico, 20 miles (30 km) southwest of the Gulf of Mexico, 120 miles (190 km) south of Tampico, and 135 miles (215 km) northeast of Mexico City. The field was discovered in May 1930, with the drilling of the Poza Rica #2 well, which produced from the gas cap at 6714 feet (2047 m) subsea for about 3 years (Salas, 1949). Initial drilling was based on a torsion balance survey in 1923 that showed a gravity high along the Río Cazones about 2.5 miles (4 km) west of the village of Poza Rica. Mapping of the outcropping Oligocene and Miocene rocks gave a hint of structure at depth. An additional gravity survey in 1930, followed by a seismic survey, led to recognition of a broad domal structure and additional drilling that penetrated the oil reservoir in November 1932 (Salas, 1949).

Reservoir and Reserves

The reservoir is a wedge of skeletal debris and breccia of mid-Cretaceous age that pinches out across the nose of a broad southeast-plunging anticline (Figs. 28-2, 28-3, and 28-4; Barnetche and Illing, 1956). Time-equivalent and younger dense basinal limestones provide seals up the structural dip and overlying the reservoir. The greater Poza Rica Field is about 19 miles (30 km) northwest–southeast by 3 miles (5 km) northeast–southwest, encompassing approximately 30,000 acres (120 km^2). In developing the field, 444 wells have been drilled; 336 are currently producing (Oil and Gas Jour., 1980). The oil column, including gas cap, is as much as 740 feet (225 m) thick; 575 feet (175 m) is a typical value along the axis of the field. The top of the reservoir is 6400 feet (1950 m) deep at the northwest end of the field and 8850 feet (2700 m) deep at the southeast end.

Estimated ultimate production is from 2.04 billion barrels (Petróleos Mexicanos, 1969, *in* Enos, 1977a, p. 276) to 2.76 billion barrels (Halbouty *et al*, 1970). Production through July 1980 was 1.17 billion barrels (Oil and Gas Jour., 1980). Production peaked in 1951 at 150,000 barrels of oil per day (Barnetche and Illing, 1956), which accounted for more than half of Mexico's total production that year. Production rate was reduced to 100,000 barrels per day in the 1950s as a conservation measure (Acuña, 1957). Although the field has been enlarged, current production has declined to 55,000 barrels (Oil and Gas Jour., 1980). Reservoir energy is derived from dissolved gas (evaluated at 56% of the drive; A. Romero Júarez and J. Hefferan, *in* Barnetche and Illing, 1956), free gas (16%), and

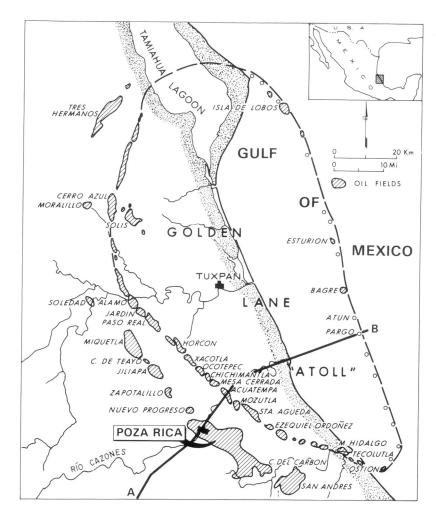

Fig. 28-1. Location map of Poza Rica and Golden Lane trends, showing fields and off-shore wells. Cross-section A–B is Figure 28-2. After Guzman (1967)*.

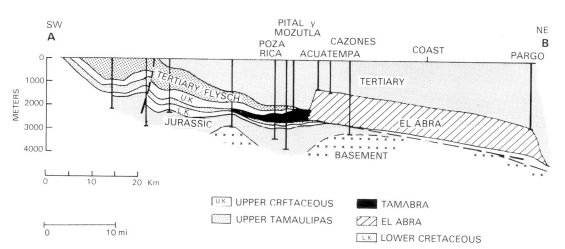

Fig. 28-2. Regional cross-section of the Tampico Embayment, showing mid-Cretaceous formations. For location of section, see Figure 28-1. From files of Shell Oil Company; original data courtesy of Petróleos Mexicanos.

Fig. 28-3. Structural contour map on top of mid-Cretaceous formations (Upper Tamaulipas, Tamabra, El Abra). ➤ Golden Lane escarpment in northeast corner. Contours in meters below sea level. After Enos (1977a).

Fig. 28-4. Isopach map of Tamabra Limestone, northern part of Poza Rica Field. After Barnetche and Illing ➤ (1956). Contour interval 25 m.

*Figures 28-1, 28-2, 28-3, 28-5, 28-6A, 28-7, 28-8, and 28-11 from Enos, 1977a. Published by permission, Society of Economic Paleontologists and Mineralogists.

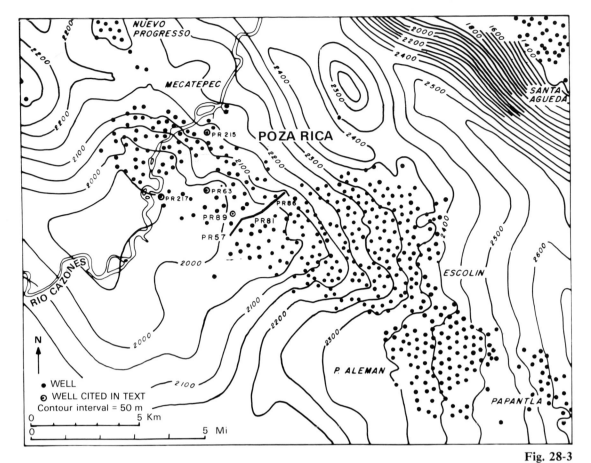

Fig. 28-3

Legend within Fig. 28-3:

- • WELL
- ◉ WELL CITED IN TEXT

Contour interval = 50 m

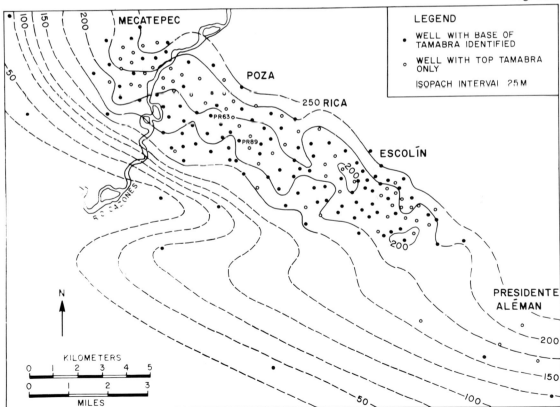

LEGEND

- • WELL WITH BASE OF TAMABRA IDENTIFIED
- ○ WELL WITH TOP TAMABRA ONLY

ISOPACH INTERVAL 25 M

Fig. 28-4

natural water drive (28%). Currently, only 41 wells are flowing, and artificial lift is used in the remaining 295 (Oil and Gas Jour., 1980). A combined program of water and gas injection was begun in 1951 to stabilize reservoir pressures (Barnetche and Illing, 1956).

General Geology

Geologically, Poza Rica is located in the Tampico Embayment, an intracratonic downwarp adjacent on the west to the Gulf of Mexico. The Tampico Embayment is bounded on the west by the Sierra Madre Oriental fold belt, on the south by the Jalapa uplift north of Veracruz, and on the north the remnant Tamaulipas Arch separates it from the Burgos Basin and the Rio Grande Embayment in south Texas (Murray and Krutak, 1963). Basinal pelagic limestones of the Lower Tamaulipas Formation and La Peña/Otates Formation (Fig. 28-5) were deposited over vast areas of eastern Mexico during the Early Cretaceous. Shallow-water platforms rimmed by rudistid molluscan reefs and surrounded by pelagic deposits developed during the mid-Cretaceous (Bonet, 1952; Enos, 1974). Notable among these are the Golden Lane "atoll" immediately east of the Poza Rica Field and the large Valles-San Luis Potosí platform exposed in the Sierra de El Abra and the Sierra Madre Oriental to the west (Griffith *et al.*, 1969; Carrillo-Bravo, 1971; Aguayo, 1978). These platforms range from a few tens of miles to more than a hundred miles across and attained relief of up to 3300 feet (1000 m) with slopes locally as steep as 45 degrees by the end of the Cenomanian (Enos, 1977a; 1980). The platforms were intermittently exposed during and after deposition before being generally submerged to pelagic depths during a major Turonian transgression.

Development and Petrophysics of Reservoir Rocks

Spectacular oil fields were developed in cavernous reef or platform-margin limestones at the edge of the Golden Lane "atoll" (Guzmán, 1967; Viniegra and Castillo-Tejero, 1970). Larger fields with much lower production rates are developed along the basin margin west of the Golden Lane "atoll." The Poza Rica Field (including the coalesced Mecatepec, Poza Rica (s.s.), Escolín, Presidente Alemán, Papantla, Tolaxco, and Cerro del Carbon or Cazuelas fields) is by far the largest of these, but geologically similar major fields (Fig. 28-1) include Tres Hermanos (140 million bbls recoverable reserves), Moralillo (19 million bbls), Miquetla (63 million bbls), and Jiliapa (38 million bbls.).

Characteristics of Reservoir Rocks

Reservoir rocks throughout this trend occur in the Tamabra Limestone (Fig. 28-5) and consist of thick-bedded skeletal-fragment (predominantly rudist) lime grainstones and packstones (Fig. 28-6); thick, discontinuous intervals of polymictic lime breccia (Fig. 28-7); and thin beds of lime wackestone with pelagic foraminifers and calcispheres (probably from pelagic algae). Pay thickness is as much as 820 feet (250 m), and the average thickness is 400 feet (120 m) (Barnetche, 1949). Barnetche and Illing (1956) showed that mechanical log markers, which generally correspond to the pelagic wackestones and nonporous portions of the skeletal-fragment limestones, can be widely correlated within the Poza Rica Field (Fig. 28-8). The most extensive horizon (*f* in Fig. 28-8) apparently divides the Tamabra Limestone into an upper and lower reservoir with separate oil-water contacts, although it is unlikely that any single bed extends throughout the field.

Locally, parts of the Tamabra interval are partially to completely dolomitized. Any of the above lithologies may be dolomitized, but the general order of decreasing susceptibility appears to be breccia matrix, wackestone intraclasts, pelagic wackestone beds, skeletal lithoclasts, skeletal-fragment packstones, and grainstones.

These lithologies are interpreted to have been deposited by sediment-gravity flows (turbidity currents and debris flows) transporting shallow-water skeletal fragments and lithoclasts from the adjacent Golden Lane escarpment and mixing them with intraclasts of basinal ooze formed during intermittent periods of "background" deposition. An alternative interpretation, based largely on paleontologic arguments, is that the skeletal fragments represent *in situ* accumulations from rudist reefs mixed with locally eroded lithoclasts

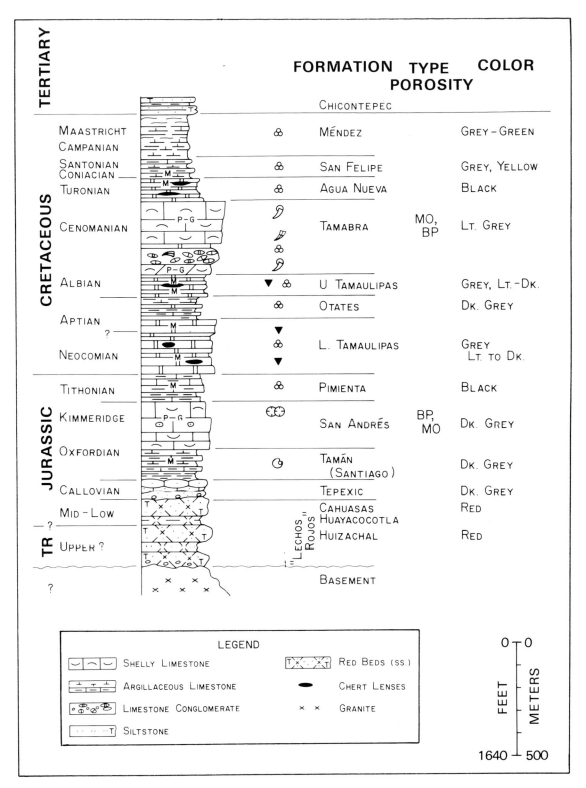

Fig. 28-5. Generalized Mesozoic stratigraphic column, Posa Rica area. Thickness data sparse below Tamabra Limestone. Abbreviations for lithologies are: P-G, packstone and grainstone; M, mudstone; limestone with a double vertical brick, wackestone; inverted black triangles, chert. The fossil symbols follow the Shell Oil Company Standard Legend. Data from Carrillo-Bravo (1971), Enos (1977a), Miramontes (1972). From Enos, 1983. Published by permission, Rocky Mountain Section, Society of Economic Paleontologists and Mineralogists.

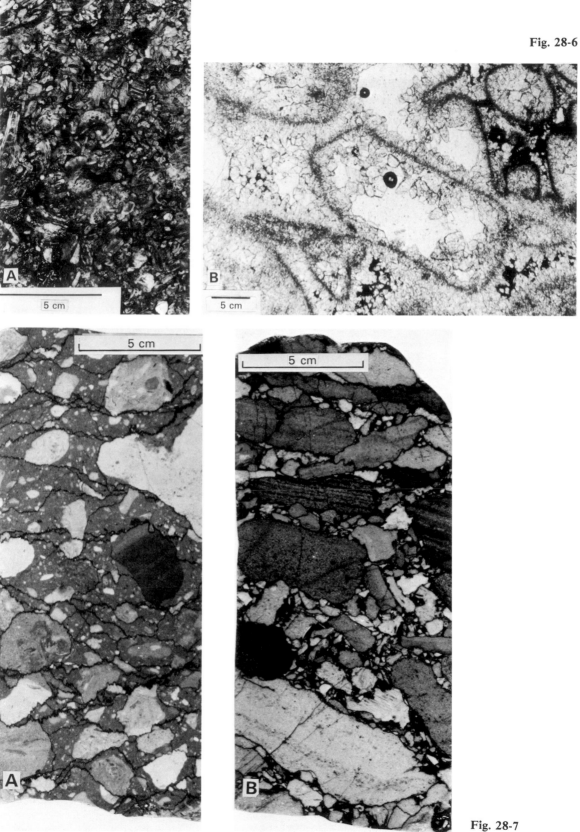

Fig. 28-6

Fig. 28-7

Table 28-1. Diagenetic Sequence and Porosity Evolution of the Tamabra Limestone, Poza Rica Field (after Enos, 1977b). Porosity (%) at the close of each diagenetic stage is the mean value estimated from thin section point counts. Porosity types are indicated using the symbols of Choquette and Pray (1970).

Diagenetic Stage and Porosity Types	Depositional Texture				
	Grainstone	Packstone	Wackestone	Breccia	Dolomite (Mudstones)
Sedimentation (primary) BP, WP, SH, BC (matrix)	37.3	36.6	50.5	50.3	69.1
Matrix lithification BP, WP, SH, BC	35.2	16.6	1.3	7.1	0.02
Early cementation cr(BP, WP, SH) BC	4.0	1.7	0.2	0.4	0.02
Fracture add. FR_1	4.02	1.72	0.27	1.7	3.6
Solution add. MO, VUG, sxBP	24.4	16.2	6.3	11.1	20.3
Fracture add. FR_2		16.3	6.32		
Cementation-present ϕ cr(MO, VUG, FR, BP)	9.6	8.3	3.7	5.8	6.1
Number of samples	17	56	20	6	6

Fig. 28-6. Grain-supported lithologies, Poza Rica area. *A*. Rudist fragment lime grainstone; good primary intergranular porosity (Pe cr-BP; Choquette and Pray, 1970), locally enhanced by leaching. Oil impregnated. Core slab, Pital y Mozutla #102 well, 7144 ft (2178 m). *B*. Rudist fragment lime grainstone with primary interparticle (Pe cr-BP 5%) and intraparticle porosity (Pe cf WP) and extensive moldic porosity (Sm sx-cr-Mo 6%). Both primary and secondary porosity show residual oil stain. Photomicrograph of Pital y Mozutla #102 well, 7127 ft (2173 m). From Enos and Moore, 1983. Published by permission, American Association of Petroleum Geologists.

Fig. 28-7. Photographs of polymict breccias from Tamabra Limestone, Poza Rica Field. *A*. Clasts are rudist-fragment lime grainstone and pelagic-microfossil lime wackestone; matrix is micritic with pelagic microfossils. Note abundant stylolites. Core slab, Poza Rica #217 (see Fig. 28-3), 6988 feet (2173 m). *B*. Sutured clasts of laminated pelagic-microfossil lime wackestone and rudist-fragment lime grainstone. Sparse matrix. Note that suturing occurred at slightly shallower depth than breccia in *A*. Inclined fabric of clasts suggests imbrication; horizontal orientation unknown. Core slab; same scale as *A*. Poza Rica #217, 7065 ft (2154 m).

and occasional products of pelagic deposition; that is, that the Poza Rica trend represents the true shelf margin of the Golden Lane Platform (Coogan *et al.*, 1972; Barnetche and Illing, 1956; Salas, 1949). For a one-sided discussion of the stratigraphic and sedimentologic evidence favoring deep-water origin of the Tamabra Limestone, see Enos (1977a).

The skeletal-fragment limestones were deposited with about 35 percent primary inter- and intraparticle porosity (Table 28-1) (Enos, 1977b). Only about 2 percent primary porosity on the average remains open, but it locally contributes to reservoir quality. The breccias and pelagic wackestones were deposited with muddy matrix, so had little preservable porosity and poor interconnection of pores. Stylolitization (Fig. 28-7A) further reduced reservoir quality. Extensive leaching to produce skeletal moldic and, locally, vuggy porosity is responsible for the reservoir quality of the Tamabra Limestone (Fig. 28-6B) (Enos, 1977b). Grain-supported skeletal rocks remain the best reservoirs, with average porosities of 8 to 11 percent (Figs. 28-9, 28-10). Extensive development of the skeletal-fragment limestones within the Poza

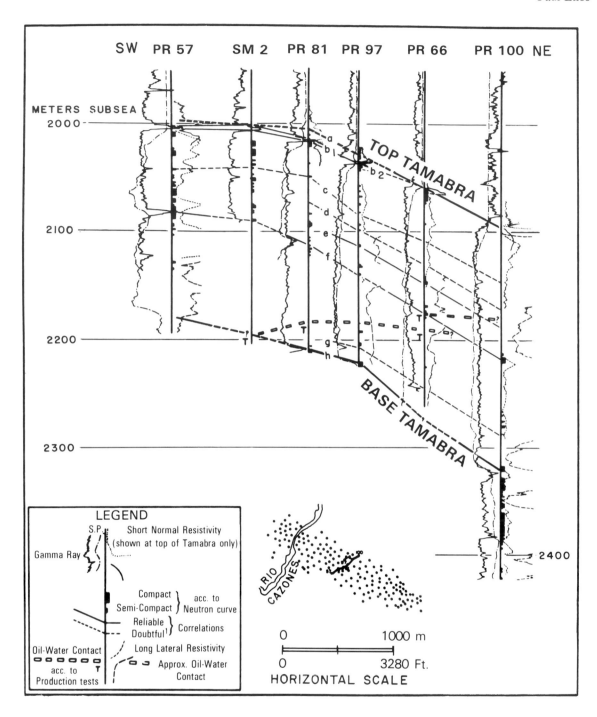

Fig. 28-8. Representative cross-section of Poza Rica Field showing typical mechanical logs; correlations of "compact" horizons, some of which are pelagic- microfossil wackestones; and the oil-water contact. Location of section in Figure 28-3 (Barnetche and Illing, 1956).

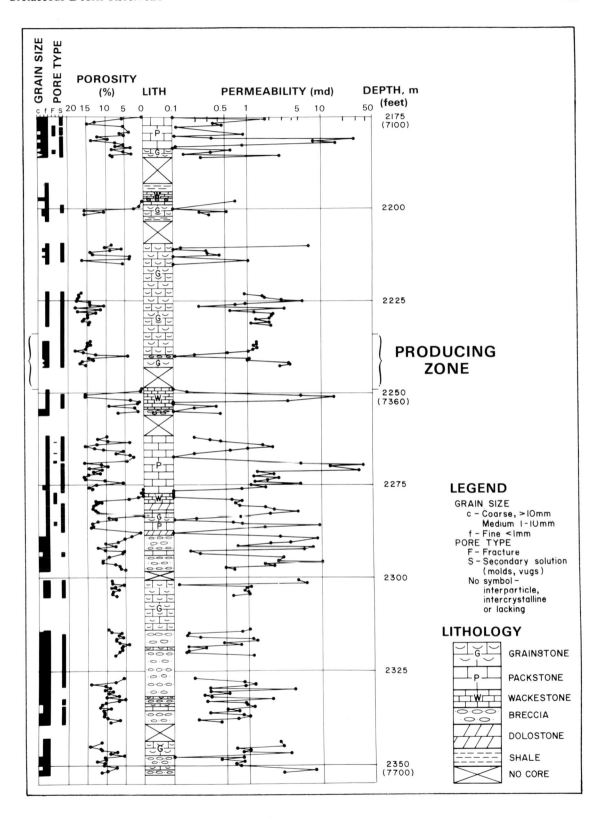

Fig. 28-9. Lithologic log of cored interval from Poza Rica #215 well showing grain size, pore types, and perm-plug porosity and permeability (location in Fig. 28-3).

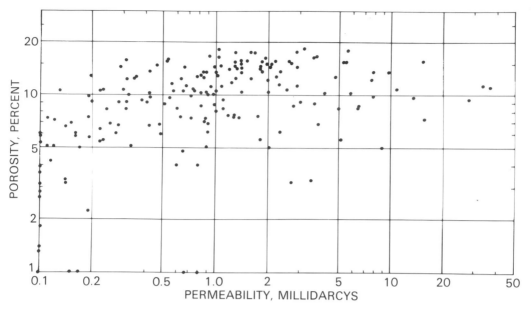

Fig. 28-10. Logarithmic plot of porosity *vs.* permeability from Poza Rica #215 well (Fig. 28-9).

Rica Field may be a major factor in its huge re-serves compared to smaller fields along the trend where breccias apparently form proportionally larger volumes of the reservoir.

Dolomitization has been overprinted on pre-existing moldic and vuggy porosity and has not significantly improved or reduced reservoir quality (Table 28-1) (Barnetche and Illing, 1956; Enos, 1977b). Fractures are visible in many core samples of Tamabra Limestone, but they are most prominent in thoroughly cemented rocks with low porosity. Their possible contribution to permeability and large-scale reservoir intercon-nection is hard to evaluate, but they are relatively rare in highly porous rocks or even in dolo-stones.

Average porosity of the pay zone is estimated at 10.6 percent (Barnetche, 1949), and values range as high as 25 percent. The correlation be-tween porosity and permeability shown in Figure 28-10 is typical of producing wells from the Tamabra trend. Porosity and permeability are poorly, but positively, correlated at low values, but porosities lie between 10 and 15 percent for all permeabilities above 1 to 10 millidarcys. This continued increase in permeability without ap-preciable increase in porosity reflects better inter-connection of pore space, which could result from fractures. Such interconnection has not been ob-served visually; moreover, the data are derived

from 1-inch perm plugs that are unlikely to detect large-scale permeability. Permeability calculated from flowing wells ranges as high as 114 milli-darcys (Barnetche, 1949). Permeabilities in ex-cess of 100 millidarcys have been measured in a few perm plugs in which fractures were noted—for example, 651 millidarcys in Poza Rica #89 and 228 millidarcys in Escolín #3. It is not re-corded whether these are *in situ* fractures or may have been developed in laboratory preparation. Exceptional permeabilities of 2000 millidarcys and greater than 5000 millidarcys have been recorded for single samples from Jiliapa #61 and Mecatepec #80, respectively, in which fractures were not noted. Jiliapa is another Tamabra field north of the greater Poza Rica Field (Fig. 28-1).

Reservoir Trap

The trapping mechanism in the Poza Rica Field is a basinward facies change from clastic limestone to pelagic wackestone (Fig. 28-11), which is most pronounced across the crest of a broad plunging anticline (Fig. 28-2). The facies change appears to form an effective lateral seal. Thus the reversal of original depositional dip, presumably westward from the Golden Lane escarpment, to the present eastward regional dip (Fig. 28-2), is a more im-portant result of the structure than is the actual

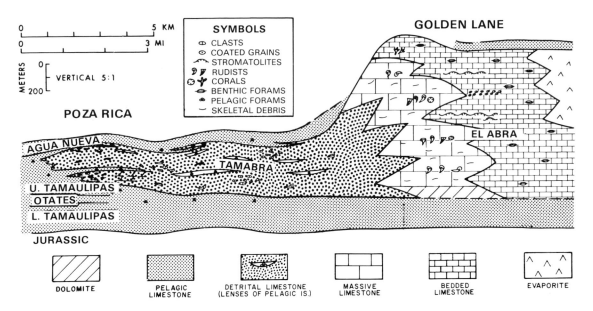

Fig. 28-11. Interpretive cross section, Poza Rica to the Golden Lane.

closure. Both the structure and the dip reversal are probably related to the formation of the Sierra Madre Oriental during the Laramide orogeny. The dip reversal can be dated as post-Eocene by stratigraphic relations along the Golden Lane escarpment (Sotomayor, 1954; Coogan et al., 1972). An important practical result of the structure is the formation of a positive gravity anomaly in the vicinity of, but not exactly coincident with, the crest of the anticline, which led to the discovery of the Poza Rica Field.

The Tamabra Limestone is overlain by the Agua Nueva Limestone (Turonian) or locally by the San Felipe Formation (Coniacian-Santonian), both fine-grained pelagic limestones capable of forming very effective seals (Fig. 28-5).

Petroleum Source

The source for this giant field, as well as those for the adjacent Golden Lane fields and other giant fields in the Tampico Embayment, remains uncertain. The basinal rocks laterally equivalent to the Tamabra Limestone have been cited as possible source rocks, but analyses from outcrops in the Sierra Madre Oriental show them to be quite low in organic matter, generally less than 0.5 percent (Scholle and Arthur, 1980). The dark color of these rocks apparently derives from dis-

seminated pyrite. Basinal limestones interbedded with the Tamabra Limestone are much lighter in color and presumably even lower in organic content. The Taraises Formation (Neocomian) is rich in organic matter in outcrops at the east edge of the Sierra Madre Oriental, but is not known to extend into the subsurface as far as Poza Rica. The Santiago Formation (Upper Jurassic, Oxfordian) is favored as the source rock by geologists of Petróleos Mexicanos because it is of source-rock quality in outcrops 35 miles (55 km) southwest of the Poza Rica Field and extends beneath the field (René Cabrera, personal communication, 1978).

The Golden Lane and Poza Rica fields probably have common source beds. The apparent spill-point filling along the southern Golden Lane trend suggests filling from the south (Acuña, 1957, Fig. 28-11). The lower API gravity (greater density) of the Golden Lane fields (avg. 22.4° vs. 35° for Poza Rica) (Enos, 1977a) probably results from the less effective seals of Tertiary terrigenous clastics overlying much of the Golden Lane in contrast to the Upper Cretaceous basinal limestones that overlie Poza Rica. If the source was indeed common, migration may have postdated late Oligocene, the age of the youngest seals in the Golden Lane (Viniegra and Castillo-Tejero, 1970). Migration at Poza Rica was almost certainly later than Eocene, the period when stratigraphic closure was formed by eastward tilting to reverse regional depositional dip.

Conclusion

Poza Rica and related fields are of particular interest in that they provide a model for entrapment of hydrocarbons beyond the shelf margin, which is commonly the seaward limit of exploration in carbonate provinces. Perhaps the largest reserves in the western hemisphere have been discovered in predominantly basin-margin-debris reservoirs of mid-Cretaceous to Paleocene age in the Isthmus of Tehuantepec and Campeche shelf (Franco, 1976; Metz, 1978; Meyerhoff, 1980). Little information is currently available on these reservoirs, but highly dolomitized depositional analogs of the Tamabra Limestone are reported (Flores, 1978; Meyerhoff and Morris, 1977), adding new stimulus to understanding the Poza Rica trend.

References

ACUÑA G., A., 1957, El distrito petrolero de Poza Rica: Bol. Asoc. Mexicana Geólogos Petroleros, v. 9, p. 505–553.

AGUAYO C., E., 1978, Sedimentary environments and diagenesis of a Cretaceous reef complex, eastern Mexico: An. Centro Cienc. del Mar y. Limnol., Univer. Nacional Auton, Mexico: v. 5, no. 1, p. 83–140.

BARNETCHE, A., 1949, Reservoir studies in Poza Rica Field: Oil and Gas Jour., v. 47, p. 161–169.

BARNETCHE, A., and L.V. ILLING, 1956, The Tamabra Limestone of the Poza Rica oil field, Veracruz, Mexico: 20th Internat. Geol. Cong. Proc., Mexico City, 38 p.

BONET, F., 1952, La facies Urgoniana del Cretácio medio en la región de Tampico: Bol. Assoc. Mexicana Geólogos Petroleros, v. 4, no. 5–6, p. 153–262.

CARRILLO BRAVO, J., 1971, La plataforma Valles-San Luis Potosi: Bol. Asoc. Mexicana Geólogos Petroleros, v. 23, no. 1–6, p. 1–113.

CHOQUETTE, P.W., and L.C. PRAY, 1970, Geologic nomenclature and classification of porosity in sedimentary carbonates: Amer. Assoc. Petroleum Geologists Bull., v. 54, p. 207–250.

COOGAN, A.H., D.G. BEBOUT, and C. MAGGIO, 1972, Depositional environments and geological history of Golden Lane and Poza Rica trends, Mexico, an alternative view: Amer. Assoc. Petroleum Geologists Bull., v. 56, p. 1419–1447.

ENOS, Paul, 1974, Reefs, platforms, and basins of middle Cretaceous in northeast Mexico: Amer. Assoc. Petroleum Geologists Bull., v. 58, p. 800–809.

ENOS, Paul, 1977a, Tamabra limestone of the Poza Rica trend, Cretaceous, Mexico, in Cook, H.E. and Paul Enos, eds., Deep-Water Carbonate Environments: Soc. Econ. Paleontologists and Mineralogists, Spec. Publ. 25, p. 273–314.

ENOS, Paul, 1977b, Diagenesis of a giant: Poza Rica trend, Mexico (abst.): Gulf Coast Assoc. Geol. Soc. Trans., v. 27, p. 438.

ENOS, Paul, 1983, Late Mesozoic paleogeography of Mexico: in Reynolds, M.W. and E.D. Dolly, eds., Mesozoic Paleogeography of West-Central United States: Rocky Mtn. Sect., Soc. Econ. Paleontologists and Mineralogists, p. 133–157.

ENOS, Paul, and C.H. MOORE, 1983, Fore-reef slope: in Scholle, P.A., D.G. Bebout, and C.H. Moore, eds., Carbonate Depositional Environments: Amer. Assoc. Petroleum Geologists Mem. 33, p. 508–537.

FLORES, V.O., 1978, Paleosedimentologia en la zona de Sitio Grande—Sabancuy: Petroleo Internacional, v. 36, p. 44–48.

FRANCO, A., 1976, Reforma finds push Mexico to new oil heights: Oil and Gas Jour., v. 74, p. 71–74.

GRIFFITH, L.S., M.G. PITCHER, and G.W. RICE, 1969, Quantitative environmental analysis of a Lower Cretaceous reef complex, in Friedman, G.M., ed., Depositional Environments in Carbonate Rocks—a Symposium: Soc. Econ. Paleontologists and Mineralogists, Spec. Publ. 14, p. 120–138.

GUZMÁN, E.J., 1967, Reef type stratigraphic traps in Mexico: 7th World Petrol. Cong., Proc., v. 2, p. 461–470.

HALBOUTY, M.T., A.A. MEYERHOFF, R.E. KING, R.H. DOTT, SR., H.D. KLEMME, and T. SHABAD, 1970, World's giant oil and gas fields, geologic factors affecting their formation, and basin classification: in Halbouty, M.T., ed., Geology of Giant Petroleum Fields: Amer. Assoc. Petroleum Geologists Mem. 14, p. 502–555.

METZ, W.D., 1978, Mexico: the premier oil discovery in the western hemisphere: Science, v. 202, p. 1261–1265.

MEYERHOFF, A.A., 1980, Geology of the Reforma-Campeche shelf: Oil and Gas Jour., v. 78, p. 121–124.

MEYERHOFF, A.A., and A.E.L. MORRIS, 1977, Central America petroleum potential centered mostly in Mexico: Oil and Gas Jour., v. 75, p. 104–109.

MIRAMONTES-E., D.E., 1972, Las formaciones del Jurasico superior en el subsuelo del Atolon de la

Faja de Oro: Bol. Asoc. Mexicana Geólogos Petroleros, v. 24, no. 1–3, p. 29–44.

MURRAY, G.E., and P.R. KRUTAK, 1963, Regional geology of northeastern Mexico: *in* Geology of Peregrina Canyon and Sierra de El Abra, Mexico: Corpus Christi Geol. Soc. Ann. Field Trip Guidebook, p. 1–10.

OIL AND GAS JOURNAL, 1980, v. 78, no. 52, p. 110.

SALAS, G.P., 1949, Geology and development of Poza Rica oil field, Veracruz, Mexico: Amer. Assoc. Petroleum Geologists Bull., v. 33, no. 8, p. 1385–1409.

SCHOLLE, P.A., and M.A. ARTHUR, 1980, Carbon isotope fluctuations in Cretaceous pelagic limestones: Potential stratigraphic and petroleum exploration tool: Amer. Assoc. Petroleum Geologists Bull., v. 64, no. 1, p. 67–87.

SOTOMAYOR-C.A., 1954, Distribucion y causas de la porosidad en las calizas del Cretacico medio en la region de Tampico, Poza Rica: Bol. Assoc. Mexicana Geólogos Petroleros, v. 6, p. 157–206.

VINIEGRA-O.F. and C. CASTILLO-TEJERO, 1970, Golden Lane fields, Veracruz, Mexico: Amer. Assoc. Petroleum Geologists Mem. 14, p. 309–325.

29
Garoupa and Pampo Field
Albert V. Carozzi and F. U. H. Falkenhein

RESERVOIR SUMMARY

Location & Geologic Setting	Offshore SE Brazil, 22°30′S and 40°30′W Campos Basin, rifted continental margin
Tectonics	Growth faulting and salt domes with related faults and collapse structures
Regional Paleosetting	Continental-margin basin
Nature of Trap	Structural/stratigraphic; salt-dome structures and diagenetic and growth-fault traps
Reservoir Rocks	
Age	Mid-Cretaceous (Albian-Cenomanian)
Stratigraphic Units(s)	Macae Formation
Lithology(s)	Limestone
Dep. Environment(s)	Shallow-marine bars on shallow carbonate platform
Productive Facies	Oncolitic lime packstone
Entrapping Facies	Pelagic argillaceous lime mudstone
Diagenesis Porosity	Dissolution in meteoric water associated with subaerial exposure
Petrophysics	
Pore Type(s)	Moldic, vuggy, and solution-enlarged interparticle
Porosity	18–30%, avg 20%
Permeability	50–2450 md, avg 200 md
Fractures	Common but calcite-filled
Source Rocks	
Age	Late Cretaceous to Paleocene
Lithology(s)	Marine shales
Migration Time	NA
Reservoir Dimensions	
Depth	6270–6650 ft (1910–2030 m), Pampo; 10,400 10,800 ft (3170–3290 m), Garoupa
Thickness	295 ft (90 m) avg, Pampo; 330 ft (100 m) avg, Garoupa net pay
Areal Dimensions	1.2 × 2.4 mi (2 × 4 km), Pampo; 1.8 × 3 mi (3 × 5 km), Garoupa
Productive Area	3840 acres (15 km²), both fields
Fluid Data	
Saturations	S_w = 12–25%
API Gravity	20–30°
Gas-Oil Ratio	34–83:1
Other	—
Production Data	
Oil and/or Gas in Place	NA
Ultimate Recovery	630 million BO*
Cumulative Production	NA

Remarks: *For all 8 fields presently in Campos Basin, including Pampo and Garoupa; total production 228500 BOPD (Oil & Gas Jour., May 5, 1982). IP 1800–2500 BOPD. Discovered 1975 & 1977.

29

Depositional and Diagenetic Evolution of Cretaceous Oncolitic Packstone Reservoirs, Macaé Formation, Campos Basin, Offshore Brazil

Albert V. Carozzi and F. U. H. Falkenhein

Location

The Campos Basin is located approximately 120 miles (200 km) northeast of Rio de Janeiro (Fig. 29-1) and covers an area of about 12,000 square miles (30,000 km^2) extending from a small onshore portion to the 660 foot (200-m) isobath, between 21° and 24° south latitude. The Garoupa and Pampo fields (Fig. 29-2) produce from structures in the Macaé Formation, an entirely offshore carbonate sequence which occurs in the central and eastern part of the Campos Basin.

Discovery

The first oil discovery in the Macaé Formation was made by PETROBRAS in 1975 during the drilling of well RJS-9 and led to the development of the Garoupa Field. The field is located on a dome-like structure associated with growth faults, and was defined by conventional seismic reflection mapping (Fig. 29-3). In 1977, following enhancement of seismic reflection techniques (CDP and others), a second accumulation was found by well RJS-40 in the southeastern part of the Campos Basin. This discovery led to the development of the Pampo Field located on a semidome-like structure limited by growth faults (Fig. 29-4).

Reservoir Trap Characteristics

The reservoirs of the Macaé Formation are located in the upper 30 feet (100 m) of oncolitic packstones and sealed by overlying argillaceous pelagic lime mudstones. The traps are of mixed structural and stratigraphic origin resulting from closure provided by domes and growth faults in combination with favorable depositional and diagenetic conditions. On dipmeter logs, packstone beds commonly display increasing dips that increase with depth from 2 to more than 20 degrees, a feature attributed to draping over salt pillows. One well penetrated more than 150 feet (45 m) of halite immediately below the carbonates.

In Garoupa Field, the reservoir is located between 10,400 and 10,800 feet (3150–3280 m) depth, with a total net pay of up to 300 feet (90 m). In the Pampo Field, the reservoir occurs at depths between 6270 and 6650 feet (1910–2020 m), with a total net pay slightly over 330 feet (100 m) thick.

Fig. 29-1. Location map of the Campos Basin.

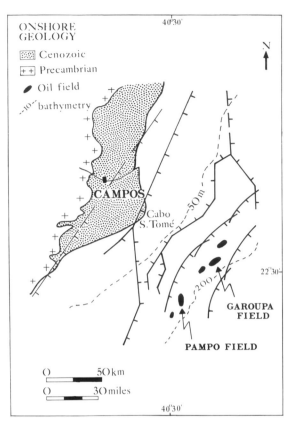

Fig. 29-2. Generalized structural maps of the Campos Basin.

The variations of porosity as observed from logs and cores, and of permeability measured from drill-stem tests and cores, are tabulated as follows:

	Porosity (%)		
Fields	Minimum	Maximum	Average
Garoupa	16	25	19
Pampo	18	30	25
Other areas*	5	10	8

	Permeability (md)		
	Minimum	Maximum	Average
Garoupa	126	200	175
Pampo	110	2400	800
Other Areas	5	10	7

*Porosity partially occluded by phreatic cementation.

Maximum bottom-hole temperatures observed during drill-stem tests are around 170°F (76.6°C) at 6560 feet (2000 m) depth and 260°F (126.6°C) at 13,120 feet (4000 m) depth, with a geothermal gradient of 4.59°F/100 m or 25.9°C/km.

The Garoupa Field is approximately 3 miles long by 1.8 miles wide (5 × 3 km), and the Pampo Field is about 2.4 miles long by 1.2 miles wide (4 × 2 km) and displays a semidome-like geometry.

Production Data

Water saturation in the reservoir rocks ranges from 12 percent (Garoupa Field) to 25 percent (Pampo Field). Present production of individual wells varies from 1500 to 2500 barrels of oil per day with a 27 to 30° API oil. Gas-oil ratios range from 34:1 to 83:1.

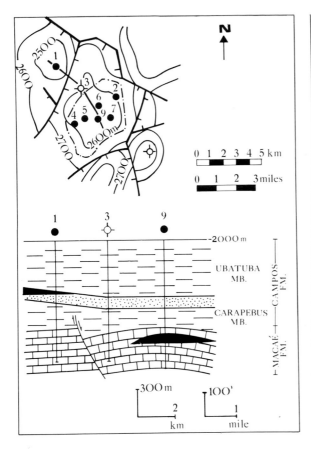

Fig. 29-3. Structural map (top of Macaé Formation) and cross-section of Garoupa Field.

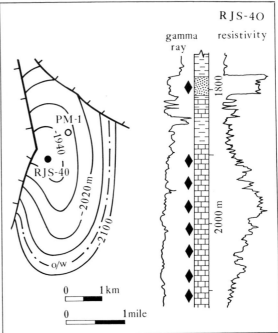

Fig. 29-4. Structural map (top of Macaé Formation) of Pampo Field and simplified column of well RJS-40. Solid diamonds indicate oil-producing intervals.

The production of both fields, which were under development was around 150,000 barrels of oil per day in 1983.

Production Mechanism

Formation pressures in drill-stem tests and in production tests are normal hydrostatic pressures and very close to the normal hydrostatic gradient. They are 2940 psi (200 kg/cm^2) at 6600 feet (~2000 m) and 5260 psi (370 kg/cm^2) at 11,500 feet (3500 m) depth. At present, all wells produce by natural flow without secondary stimulation through a temporary production system. The output of several wells is collected through a central manifold and pumped into tankers. This temporary system was adopted because of production logistics related to the depth of water (up to 400 ft or 120 m) in which the oil fields occur.

Tectonic Setting and Evolution

The Campos Basin displays a stratigraphic sequence (Fig. 29-5) that records the progressive tectonic stages characteristic of rifted continental margins discussed in detail by Ponte et al. (1980) and Ojeda (1982). The principal Mesozoic stages envisioned by these authors for the evolution of the South Atlantic and the Brazilian continental margin are shown diagrammatically in Figure 29-6. By early Cretaceous time (Neocomian) a great rift valley system had evolved, similar to the present-day East African rift. Fluvio-deltaic conglomerates and sandstones of the lower Lagoa Feia Formation (Fig. 29-5), together with shales, filled lakes occupying structural lows within the rift valley. In later Aptian time, the first marine transgression of the proto-South Atlantic began to invade the rift system from the south, forming a narrow and elongated embayment. This proto-Atlantic was restricted and subjected to an arid

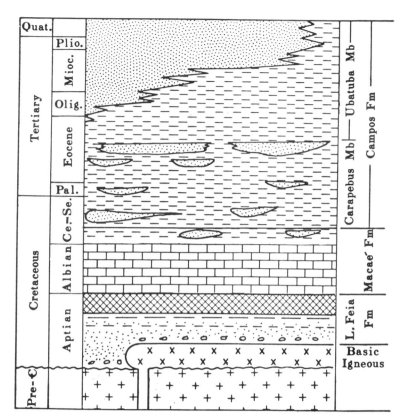

Fig. 29-5. Generalized stratigraphic diagram for the Campos Basin (Ce.-Se. Cenomanian to Senonian).

climate, becoming thereby an ideal site for the deposition of a thick evaporite sequence, the upper Lagoa Feia Formation. This sequence represents, therefore, the transition between a rift valley and a continental shelf. It was a time of tectonic quiescence and of low relief on the continent. In Albian time, the continental shelf underwent further subsidence and seaward tilting and became the site of a carbonate platform whose deposits were predominantly the oncolitic packstones and mudstones of the lower part of the Macaé Formation (Fig. 29-5). Only small amounts of terrigenous sediments were carried to the continental shelf at this time. The sequence of events inferred during and subsequent to deposition of the Macaé Formation carbonates is outlined later.

Progressive widening of the ocean ensued during Cenomanian time (Ponte *et al.*, 1980; Ojeda, 1982) and was accompanied by further seaward tilting of the continental shelf, in which the platform carbonates subsided as a unit and were

transgressed by a sequence of argillaceous pelagic mudstones and intercalated turbidite sandstones of the upper Macaé Formation (Fig. 29-5). In turn, this sequence became buried beneath younger Senonian to Eocene deep water basinal shales and intercalated turbidite sandstones of the Carapebus Member of the Campos Formation. The basinal shales have source-bed characteristics, as indicated by detailed geochemical evaluation (PETROBRAS, unpublished). Deposition of this predominantly shaly sequence preceded renewed seaward tilting of the continental margin and associated topographic rejuvenation leading to an increase of terrigenous clastics represented by deltaic progradation deposits of the upper Campos Formation (Ubatuba Member).

The Macaé Formation

Chronostratigraphic horizons within the Macaé Formation consistently show the sedimentological juxtaposition of offshore bars. These bars are

Fig. 29-6. Diagrammatic sequence of the geologic evolution of the Brazilian continental margin (modified from Ponte *et al.*, 1980). Published by permission, Canadian Society of Petroleum Geologists.

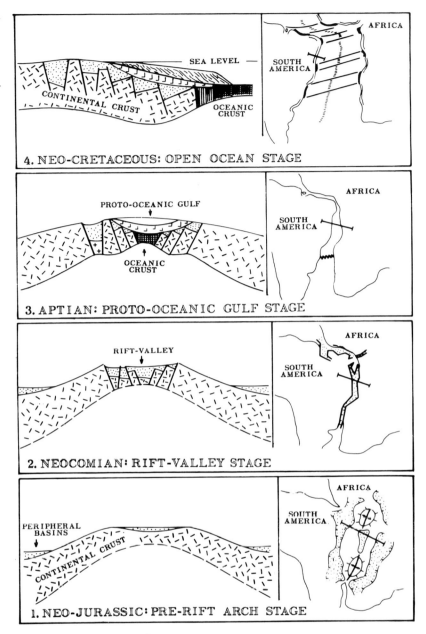

4. NEO-CRETACEOUS: OPEN OCEAN STAGE

3. APTIAN: PROTO-OCEANIC GULF STAGE

2. NEOCOMIAN: RIFT-VALLEY STAGE

1. NEO-JURASSIC: PRE-RIFT ARCH STAGE

composed of infratidal to intertidal, blue-green, oncolitic packstones that parallel depositional strike in association with interbar bands of infratidal mudstones (Fig. 29-7). The bars display little areal variation in composition, except for the landward occurrence of ooids and the seaward occurrence of crinoids and bryozoans.

The mudstone bands are very uniform petrographically and, with the exception of small quantities of pelagic and benthic foraminifers, are totally devoid of any visible carbonate-producing organisms. They represent the final accumulation, in quiet infratidal conditions, of fine carbonate mud perhaps derived from the mechanical disintegration of the blue-green algal material forming the oncolitic bars. Ratios of pelagic to benthic constituents increase in a seaward direction, indicating increasingly open marine conditions. The mudstones eventually grade into argillaceous basinal mudstones.

Shoreward, contemporaneous terrigenous clastics formed coalescent fan-deltas with intervening

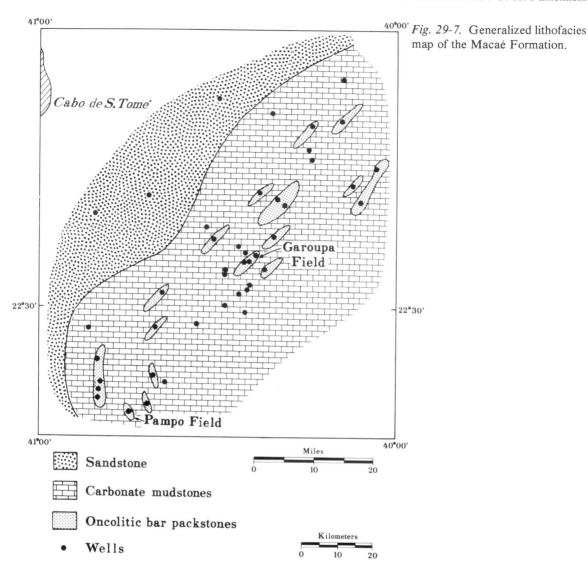

Fig. 29-7. Generalized lithofacies map of the Macaé Formation.

Sandstone

Carbonate mudstones

Oncolitic bar packstones

• Wells

tidal-flat carbonates. (Ponte *et al.*, 1980; Ojeda, 1982).

Depositional-Diagenetic Sequence

The vertical sequence in the Macaé Formation represents a shallowing-upward sequence of events locally reaching emergence (Falkenhein *et al.*, 1981). Partial occlusion of once highly porous oncolitic packstone reservoirs occurred during this gradual change of environmental conditions. Six stages of deposition and/or alteration can be distinguished on the basis of textural evidence.

Stage 1 represents the deposition of the initial unconsolidated sediments, consisting of a frame-

work of abraded large oncolites (1–1.5 cm in diameter) with interstitial sand-size micro-oncolites, bioclasts of crinoids and bryozoans, and mudstone intraclasts. All of these are set in a fine matrix interpreted as "oncolitic flour" derived from the abrasion of the large oncolites (Fig. 29-8). Early lithification of the oncolites is shown in thin section by individuals broken by impacts, with margins variably eroded and also at all stages of mechanical disintegration releasing smaller debris ranging from sand size to matrix size (less than 60 μ). This sediment was probably deposited under medium-energy conditions in a low-intertidal environment.

Stage 2 also was depositional and resulted in a redistribution of interstitial constituents into an

Stage 1- Depositional

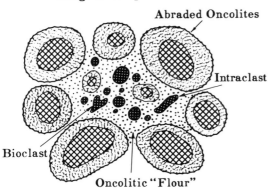

Abraded Oncolites

Intraclast

Bioclast

Oncolitic "Flour"

Stage 2-Depositional

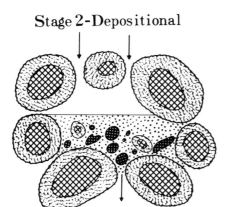

Stage 3- Diagenetic

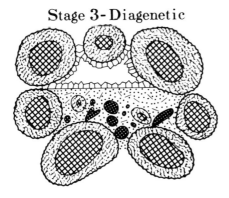

Fig. 29-8. Interpreted depositional-diagenetic sequence of the Macaé Formation. *Stage 1*: Depositional, medium energy, low-intertidal environment. *Stage 2*: Depositional, high-energy, high-intertidal environment, geopetal organization of internal sediment. *Stage 3*: Diagenetic, high-energy, high-intertidal beachrock environment, cementation of internal sediment (1), deposition of isopachous rim cement (2), primary reduced interparticle porosity.

internal sediment with geopetal features and horizontal surfaces (Fig. 29-8). The redistribution within a grain-supported sediment resulted in localized larger open spaces between the larger oncolites that formed the framework of the sediment. The general energy of this stage was high, and the environment is interpreted to have been high-intertidal, probably adjacent to a beachrock. The rearrangement of the internal sediment indicates a predominant downward vertical circulation of sea water related to the action of tides. This particular texture occurs systematically in time and space before the development of the following stage and therefore cannot be attributed to localized gravity filtration and settling.

Stage 3 was a diagenetic modification of stage 2 wherein the internal sediment became lithified with introduction of interparticle calcite-spar cement and an isopachous rim cement of aragonite and/or high-magnesium calcite (now calcite) was deposited along the boundaries of the open spaces generated during the previous stage (Figs. 29-8 and 29-10A). Variably reduced primary interparticle porosity was produced and was preserved upon burial in places, but led to non-commercial reservoirs. The environment of stage 3 remained high intertidal and is interpreted as a beachrock setting. This interpretation is supported by the fact that some of the isopachous rim cement shows pendular texture beneath oncolites, and also by the time sequence, in which stage 3 is positioned between the marine environment and the freshwater-vadose environment.

Stage 4A was also a diagenetic event, the preserved results of which are reservoir rocks. An appreciable proportion of the rim cement is dissolved, leaving a margin of corroded crystals (Figs. 29-9 and 29-10B,C). This dissolution generated a secondary porosity by appreciably enlarging the reduced primary porosity of stage 3. These new conditions fostered the preservation upon burial of a reservoir with 10 to 15 percent porosity and about 100 millidarcys permeability. The environment of stage 4A was high-intertidal beachrock exposed to incipient dissolution.

The best reservoir rocks, however, resulted from *stage 4B*. In this situation, the rim cement was almost entirely dissolved, the margins of the oncolites corroded, and the upper horizontal surfaces of cemented internal sediment deeply etched (Fig. 29-9). Further solution-enlarged interparticle to vuggy porosity, fabric to non-fabric

Stage 4A-Diagenetic [Reservoir]

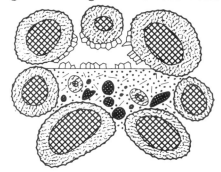

Stage 4B-Diagenetic [Reservoir]

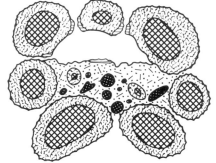

Stage 5-Diagenetic

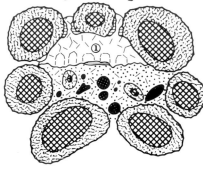

Fig. 29-9. Interpreted depositional-diagenetic sequence of Macaé Formation, second three stages. *Stage 4A*: Diagenetic (reservoir generation), high-intertidal beachrock environment, incipient dissolution, secondary interparticle porosity. *Stage 4B*: Diagenetic (reservoir generation), high-intertidal beachrock environment, intense dissolution, solution-enlarged interparticle to vuggy porosity. *Stage 5*: Diagenetic (reservoir occlusion), meteoric phreatic cementation by clear blocky calcite (1).

selective, was generated and was in places augmented by intraparticle porosity resulting from the differential solution of oncolite nuclei (Figs. 29-9 and 29-10D,E). After preservation by burial, this stage is represented by a reservoir with measured 25 to 30 percent porosity and 200 to 400 millidarcys permeability. The environment of stage 4B was high-intertidal, which included an event of intense dissolution of beachrock.

Stage 4B beachrock retained its higher porosity and reservoir features because rapid subsidence and subsequent burial beneath the transgressive argillaceous pelagic mudstones precluded proximity to pore-occluding waters of meteoric origin.

Stage 5 represents a period of porosity occlusion which affected certain oncolitic bars. The solution-enlarged interparticle to vuggy porosity of the previous stage was completely occluded by clear blocky calcite cement (Figs. 29-9 and 29-10F). The diagenetic environment is interpreted as phreatic, which developed when a freshwater lens was generated during an interlude of subaerial exposure of some bars. Calcrete crusts and associated casts of grass rootlets provide positive evidence. Porous zones originating from the previous beachrock stage, including any porosity enhanced by subsequent vadose conditions were transferred by subsidence into the meteoric, phreatic zone where precipitation of blocky calcite occluded the reservoir prior to burial under the transgressive pelagic mudstones.

Generation of reservoirs in the Macaé Formation occurred only in certain oncolitic bars which underwent shoaling up to the beachrock phase (stages 4A and 4B) and subsequently were buried beneath transgressive pelagic carbonates. Other bars underwent greater shoaling, and were subjected to vadose diagenesis. Consequently, this beachrock porosity (stages 4A and 4B) was occluded by freshwater phreatic cementation during final subsidence (stage 5).

Exploratory wells have demonstrated that reservoirs occur only at the top of the Macaé carbonates and that porous but tight oncolitic bars occur in a random pattern within a regional seismic trend. This randomness, which for the time being prevents the development of any rationale for prediction, may be only apparent. It is influenced by two major factors which are difficult to evaluate: (1) the intensity of uplifting of the oncolitic bars by salt tectonics (to be discussed

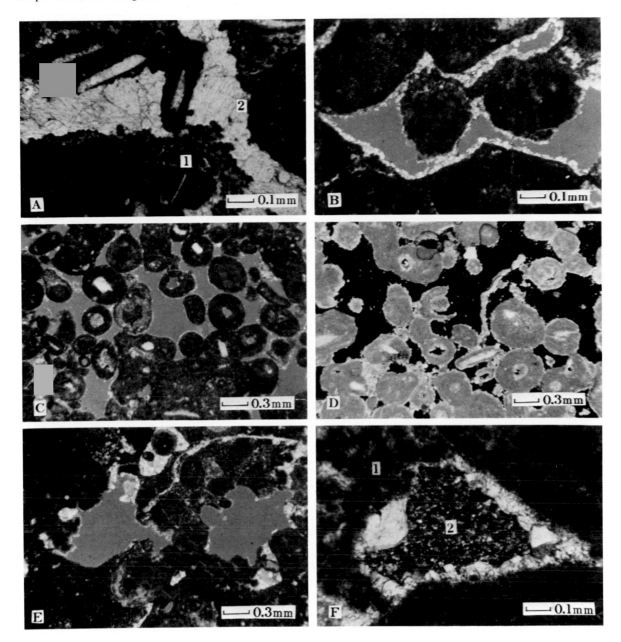

Fig. 29-10. Photomicrographs of depositional-diagenetic features of the Macaé Formation. *A. Stage 3:* Geopetal arrangement of cemented internal sediment (1) consisting of micro-oncolites, bioclasts, mudstone intraclasts and "oncolitic flour", and overlain by blocky sparite cement. Notice abraded margin of oncolite (2). Plane light. *B. Stage 4A:* Framework of oncolites surrounded by a partially dissolved rim cement. Cross-polarized light, porosity in gray. *C. Stage 4A:* General view showing some oversized pores. Notice intraparticle porosity due to dissolution of oncolite nuclei. Cross-polarized light, porosity in gray. *D. Stage 4B:* Rim cement is largely dissolved, margins of oncolites are deeply corroded, and numerous oncolites are partially to almost completely dissolved. Resulting solution-enlarged interparticle and vuggy porosity combine fabric selectivity and non-fabric selectivity. Additional intraparticle porosity is due to dissolution of some oncolite cores. Cross-polarized light, porosity in black. *E. Stage 4B:* Well-developed examples of megavugs, fabric selective to non-fabric selective. Cross-polarized light, porosity in gray. *F. Stage 5:* Vug of stage 4B cutting across internal sediment (1) with partially dissolved rim cement and filled with a single crystal of blocky sparite at extinction (2). Cross-polarized light.

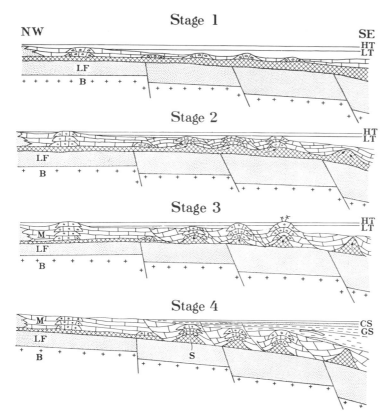

Fig. 29-11. Interpretation of accumulation and burial of Macaé Formation carbonates. Structural evolution of the Brazilian continental margin modified from Ponte *et al.* (1980) and Ojeda (1982). B basement; LF Lagoa Feia fluvio-deltaic clastics; S Lagoa Feia evaporites; M Macaé Formation; GS Source bed shales of Carapebus Member; CS turbidite sandstones of Carapebus Member. *Stage 1*: Predominant infratidal environment with packstone bars separated by lime-mud belts. *Stage 2*: Predominant low-intertidal environment. Some bars have built to low tide (LT) level or are emergent (at left). Incipient salt flowage. *Stage 3*: Predominant high-intertidal beachrock environment with subaerial exposure of some bars. Salt diapirism. *Stage 4*: Transgression over downward tilted carbonate platform.

below); and (2) the morphology of the bars which in turn results from the ecology of the blue-green algae forming the oncolites and the physical conditions of the environment.

Evolution of Macaé Carbonates

Examination of seismic sections in the Campos Basin reveals that the sedimentological evolution of the Macaé carbonates has been strongly influenced by salt tectonics involving diapiric flow of the underlying evaporite sequence in the upper Lagoa Feia Formation. No data are available on possible participation of the underlying faulted basement in the proposed tectonic evolution. Interpretation of the Macaé carbonate sedimentation can be summarized in four major schematic stages (Fig. 29-11).

The *first stage* corresponds to the establishment for the first time, in early Albian time, of a carbonate system which consisted from northwest to southeast of the following belts: tidal-flat carbonates, shoreface oncolitic-oolitic banks, platform oncolitic bars with interbar mudstone bands, slope mudstones, and basinal argillaceous mudstones.

The initial oncolitic bars are interpreted as having been established over highs on an evaporite surface, the subjacent top of the Lagoa Feia Formation. Carbonate muds perhaps derived, as mentioned above, from the mechanical disintegration of the blue-green algal material forming the oncolitic bars concentrated in the lows of the evaporite surface and led to overthickening of the mudstones compared with the oncolitic packstones. This situation was enhanced by the high degree of compaction of the mudstones by comparison with the almost non-compacting oncolitic packstones. Overthickening and overcompaction initiated a gravitational disequilibrium which triggered salt tectonics (Fig. 29-11, stage 1).

The *second stage* illustrates the vertical effect of salt tectonics and related rotational movements along initial growth faults which affected the carbonate system, while an additional horizontal component tended to move some of the salt basinward, where large salt domes were produced. The differential behavior between thickening and compacting carbonate muds in the interbar areas and thinner, almost non-compacting, oncolitic packstones continued during this stage and accentuated the rising of salt beneath the bars. Consequently, the mudstone areas remained subtidal and became further restricted, while the oncolitic bars rose into shallower intertidal conditions and increased their barrier effect (Fig. 29-11, stage 2). The growth faults have well-defined positions seaward of the major oncolitic bars, separating the bars from the thickest parts of the mudstone troughs. This preferred position, clearly displayed in seismic and structural sections (Fig. 29-3), can be used for predicting of the general location of major oncolitic bars.

In the *third stage*, salt tectonics and growth faulting are believed to have been essentially completed, with the available salt concentrated as pillows or incipient domes beneath the oncolitic bars. The interbar mudstone bands remained infratidal and attained maximum restriction. Some oncolitic bars reached the critical high-intertidal environment as beachrock and became reservoirs when subjected to dissolution. Other bars underwent additional uplifting into a zone of vadose condition and subaerial exposure and consequent destruction of their reservoirs by blocky calcite upon subsidence into the phreatic zone (Fig. 29-11, stage 3).

The fourth and *last stage* included oceanward tilting and rapid subsidence of the carbonate system with related transgression of pelagic argillaceous mudstones that formed seals. The mudstones are, in turn, overlain by deep-water basinal shales (GS stage 4, Fig. 29-11) which are the assumed source beds associated with intercalated turbidite sandstones (CS). Both lithologies belong to the Carapebus Member of the Campos Formation. Some growth faults were apparently reactivated during deposition of the Campos Formation, and their upward extension established the necessary physical connection for a lateral migration of the hydrocarbons between shale source beds and the top of underlying oncolitic bar reservoirs (Fig. 29-11, stage 4).

Summary

Oil accumulations in oncolitic calcarenites of the Macaé Formation (Lower Cretaceous) of the Campos Basin were discovered in 1975 (Garoupa Field) and 1977 (Pampo Field) by seismic methods. Reservoir trap characteristics are due to a combination of structural closure and favorable diagenetic conditions. Reservoirs occur in offshore sites at depths ranging from around 6200 to 10,800 feet (1900–3300 m) and each accumulation involves an average area of 6 square miles (15 km^2), with a net pay of about 300 feet (90 m). Porosity ranges from 18 to 30 percent and permeability from 5 millidarcys to 2.4 darcys. The production mechanism is by natural flow through temporary submarine completion systems. Oil production of the Garoupa and Pampo fields is estimated at around 150,000 barrels per day in 1983.

The Campos Basin displays a stratigraphic evolution which expresses the tectonic stages of a rifted continental margin. The last stage of this evolution was an open-marine continental shelf, which began with the deposition of the Macaé platform carbonates. These consist of juxtaposed oncolitic packstone bars and intervening mudstone bands extending parallel to depositional strike. All oncolitic bars underwent a shoaling-upward evolution. When the oncolitic bars developed into intertidal beachrock and were exposed to dissolution by meteoric water, reservoirs of economic interest were generated in them and preserved after subsidence and burial beneath transgressive argillaceous mudstones. Whenever further emergence of oncolitic bars led to the development of a freshwater lens, their beachrock porosity was occluded upon subsidence by phreatic cementation and the reservoirs destroyed.

The shoaling-upward evolution of the Macaé carbonates and the structural features of the Campos Basin (domes and growth faults) were strongly influenced by salt tectonics involving the underlying evaporites of the Lagoa Feia Formation. Reactivation of growth faults during deposition of the Campos Formation allowed lateral migration of the hydrocarbons from the basinal shale source beds into the underlying oncolitic bar reservoirs.

Acknowledgments   This paper presents parts of the results of a special project undertaken for PETROBRAS by the authors in collaboration with C. F. Lucchesi, R. de O. Mercio, B. Ansaloni, and N. Uesugui. The authors are grateful to the Board of Directors of Petróleo Brasileiro S. A. PETROBRAS for permission to publish.

References

FALKENHEIN, F.U.H., M.R. FRANKE, and A.V. CAROZZI, 1981, Petroleum geology of the Macaé Formation (Albian-Cenomanian), Campos Basin, Brazil—carbonate microfacies—depositional and diagenetic models—natural and experimental porosity: Ciencia Tecnica Petroleo, PETROBRAS-CENPES, v. 11, 140 p.

OJEDA, H.A.O., 1982, Structural framework, stratigraphy and evolution of Brazilian marginal basins: Amer. Assoc. Petroleum Geologists Bull., v. 66, p. 732–749.

PONTE, C.F., J. DOS REIS FONSECA, and A.V. CAROZZI, 1980, Petroleum habitats in the Mesozoic-Cenozoic of the continental margin of Brazil: *in* Miall, A.D., ed., Facts and Principles of World Oil Occurrence: Canadian Soc. Petroleum Geologists Mem. 6, p. 857–886.

30
Fairway Field
C. W. Achauer

RESERVOIR SUMMARY

Location & Geologic Setting	Henderson and Anderson Co.'s, east Texas, USA
Tectonics	Interdomal high, fault-bounded, between 3 salt domes
Regional Paleosetting	Paleoshelf and reef trend
Nature of Trap	Stratigraphic/structural; atoll reef complex crossing structural high between salt domes

Reservoir Rocks

Age	Early Cretaceous
Stratigraphic Units(s)	James Limestone, upper member
Lithology(s)	Limestone
Dep. Environment(s)	Atoll reef and lagoon
Productive Facies	Rudistid boundstone and wackestone, skeletal grainstone
Entrapping Facies	Calcareous shale and argillaceous lime mudstone
Diagenesis/Porosity	Dissolution in meteoric water associated with post-reef subaerial exposure

Petrophysics

Pore Type(s)	Skeletal moldic, solution-enlarged interparticle
Porosity	11% avg
Permeability	3–27 md
Fractures	sparse

Source Rocks

Age	NA
Lithology(s)	NA
Migration Time	NA

Reservoir Dimensions

Depth	9000–10,000 ft (2750–3050 m)
Thickness	~100 ft (~30 m)
Areal Dimensions	NA
Productive Area	24,000 acres (115 km^2)

Fluid Data

Saturations	S_w = 18% avg
API Gravity	48°
Gas-Oil Ratio	1425:1
Other	—

Production Data

Oil and/or Gas in Place	395 million BO
Ultimate Recovery	230 million BO; URE = 58.2%
Cumulative Production	168.5 million BO thru March 1983

Remarks: Reservoir temperature = 130°C (265°F). Viscosity = 0.133 cp @ 20°C. Discovered 1960.

30
Facies, Morphology, and Major Reservoir Controls in the Lower Cretaceous James Reef, Fairway Field, East Texas

C. W. Achauer

Introduction

Fairway Field, located 85 miles (142 km) southeast of Dallas, Texas, is one of the important carbonate reservoirs in the East Texas Basin (Fig. 30-1). As of April 1983, Fairway had produced 168.5 million barrels of oil (26.8 million m^3) from the Lower Cretaceous James Limestone. Good well control and extensive coring of this significant reservoir, which attains net pay thicknesses ranging to about 100 feet (30 m), have provided an opportunity for close examination of the facies, morphology, and reservoir controls of a Cretaceous reef complex.

In this paper, the depositional facies and morphology of the James reef complex at Fairway Field are described, as well as the major depositional and diagenetic controls on porosity and permeability in the James reservoir. The main data base comes from examination of slabbed cores and 180 thin sections from 23 wells in Fairway Field. In addition, based on a control network of 122 wells, including 68 from which cores and cuttings were examined, the broad picture of James depositional environments and lithofacies trends in a portion of the East Texas Basin is presented.

Relation of Fairway Field to Paleogeography

Figure 30-1, a generalized lithofacies map, reflects James paleogeography in the East Texas Basin. On this map, two belts of shallow-marine grainstone are prominent in the James Limestone. One belt extends in a northeastward direction and is differentiated into large bodies of oolitic grainstone, oolitic bars and patch reefs. The other belt is composed predominantly of reef-derived skeletal grainstone and the James atoll reef complex. This belt projects southeastward for 35 miles (58 km) from the oolitic grainstone belt. Argillaceous lime mudstone and calcareous shale of deeper shelf origin were deposited seaward, to the southeast of these two belts.

Local Stratigraphy

At Fairway Field, the reef complex occurs in the upper member of the James Limestone and exhibits a sharp upper boundary and a transitional lower boundary (Fig. 30-2). The uppermost lagoonal facies of this reef complex is sharply overlain by a black carbonaceous shale member

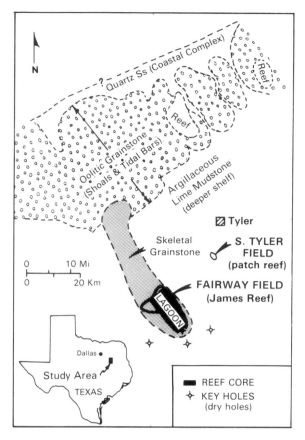

Fig. 30-1. Map showing location of Fairway Field and paleogeographic facies of the upper James Limestone Member, in part of the east Texas Embayment.

of the Rodessa Formation. In contrast, the base of the reef complex is marked by a gradational boundary between reef boundstone or grainstone and an underlying argillaceous lime mudstone facies that comprises the lowermost part of the upper member of the James Limestone.

Local Structure

The hydrocarbon trap at Fairway Field is a large interdomal high between three salt domes (Fig. 30-3). This structural high is bounded by faults on its northwest, northeast, and southern margins and is crossed by the northwest-trending James reef complex. The faults are believed to have developed in response to salt piercement.

Reef Facies and Morphology

Reef-Core Facies

The James reef complex at Fairway can be divided into three principal facies: reef-core, reef-derived skeletal grainstone, and lagoonal. The positions of reef cores and reef-derived skeletal grainstone are shown in Figure 30-4, a facies map and cross-section of Fairway Field. On this map, the reef facies are effectively delineated by wells that penetrate a distinctly zoned, vertical se-

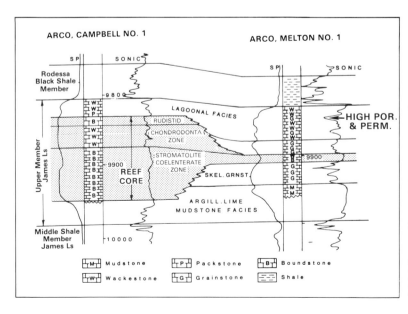

Fig. 30-2. Representative SP and sonic-log profiles through reef-core facies, reef-derived skeletal grainstone facies, and lagoonal facies of James Reef Complex at Fairway Field. Note particularly the sharp sonic-log deflection (arrow) opposite the very porous and permeable miliolid-pellet grainstone unit in the upper part of the lagoonal facies, ARCO Melton No. 1 well. Lithology plot corresponds to cored intervals in each well.

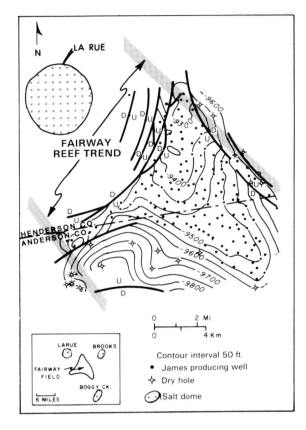

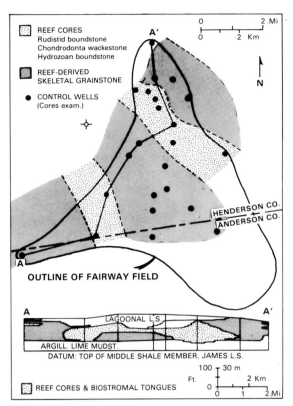

Fig. 30-3. Structure map on top of James Limestone at Fairway Field, Texas. Adapted from Terriere (1976), and published by permission, American Association of Petroleum Association of Petroleum Geologists. Note that a fault-bounded structural high is crossed by northwest-trending James Reef Complex. Inset map shows location of Fairway Field with respect to three major salt domes.

Fig. 30-4. Facies map (excluding lagoonal limestones) and cross-section of the James Reef Complex at Fairway Field, Texas.

quence of boundstones and wackestones comprising over 50 percent of the reef facies in the James upper member. The reef-derived skeletal grainstone facies is also effectively defined by the wells that penetrated a reef facies in the James upper member consisting of over 50 percent grainstone.

Reef facies of the James Limestone can be further subdivided stratigraphically into three biofacies zones based on content and, to some degree, SP-sonic log response (Fig. 30-2). These are, in ascending order:

Stromatolite-Coelenterate Boundstone

This zone is the lowest of the reef-core facies. The coelenterates and stromatolites show an intimate growth association (Fig. 30-5A) (Achauer and Johnson, 1969). Formal paleontologic studies of the coelenterates have not been made; however, their internal structure resembles that of the hydrozoan *Spongiomorpha*.

Chondrodonta Limestone

The middle zone of the reef-core facies has common *Chondrodonta* (pelecypod) specimens oriented with their long axes at high angles to stratification (Fig. 30-5B). It is not apparent from this orientation whether *Chondrodonta* specimens assumed a burrowing habit on the reefs or whether they attached themselves along their ventrals to the reef surface with their beaks pointing upward.

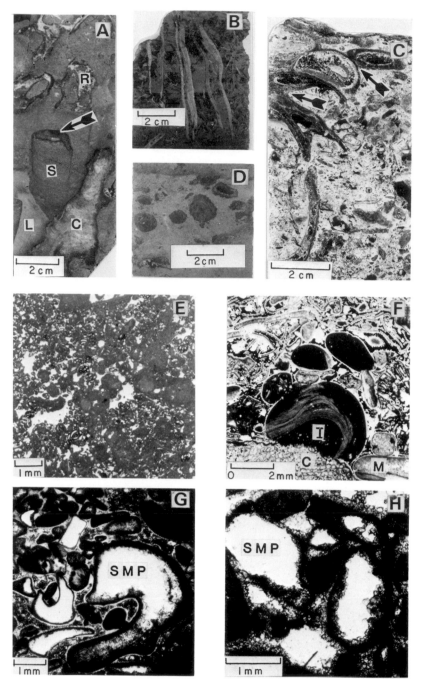

Fig. 30-5. A. Photograph of a slabbed core of the stromatolite-coelenterate boundstone facies. Note the intimate growth association between a vertically oriented coelenterate structure (C) and a columnar algal stromatolite (S). Note also that a coelenterate structure (arrow) is perched on top of the columnar stromatolite. Lime mud matrix (L) and rudistid shells (R) are also shown. ARCO Truitt No. 1, 10255.5 feet (3108 m). *B*. Photograph of a slabbed core of the *Chondrodonta* wackestone facies. Valves of *Chondrodonta* specimens are at high angle to stratification, near vertical position.

Cities Service Truitt No. 1, 9870 feet (3010 m). *C*. Photograph of slabbed core of rudistid boundstone facies. Portion of organic framework of rudistid shells (arrows) is preserved. Cities Ser vice Truitt No. 1, 9836 feet (2998 m). *D*. Photograph of slabbed core of algal nodule wackestone, a principal lagoonal facies. The algal nodules display a rude concentric layering. Cities Service Emerson No. 1, 9823 feet (2994 m). *E*. Photomicrograph of miliolid-pellet grainstone, a principal lagoonal facies. White areas are primary in-

Rudistid Boundstone

The upper zone is comprised of rudists. Figure 30-5C reveals that the individual rudist shells are intergrown, thereby forming a rigid skeletal framework. However, the rudistid framework has been bored extensively by bivalves, sponges, and other organisms, and as a result preservation of a continuous organic framework is rarely complete; only vestiges of the original framework remain.

The reefs were constructed in two morphologic stages closely related to their vertical biofacies zonation. In the earliest stage, the coelenterates and stromatolites constructed at least two discrete, elongate bioherms along the eastern trend of reefs and a linear buildup along the western trend of reefs (Fig. 30-6). During the latest stage, *Chondrodonta* and the rudistids built a prominent ridge-like feature superimposed upon the older bioherms of the eastern trend, and also constructed a low-relief, linear feature along the western trend. These features are defined in Figure 30-7, a thickness map of the combined *Chondrodonta* wackestone and rudistid boundstone facies.

The overall extent of reefs is shown in Figure 30-1. Within the productive limits of Fairway Field, lithologic data from 23 wells permit the detailed mapping of two northwest-trending reefs. To the south of Fairway Field, cores and cuttings of the James from a number of scattered dry holes (shown in Fig. 30-1), are instrumental in demonstrating that the reef facies at Fairway join to the southeast, thereby forming a *reef atoll* that partly encloses a lagoon. Whether the reefs join to the northwest and form a completely enclosed atoll is not known.

Reef-Derived Skeletal Grainstone

Two prominent lenticular bodies of reef-derived skeletal grainstone extend away from the reef facies. Figure 30-4 depicts the lenticularity of these skeletal grainstone bodies in cross-section, and Figure 30-8 shows the distribution and thickness of the uppermost grainstone body as related to the axial positions of reef cores. Derivation of the skeletal debris in the grainstone bodies from the reef cores is indicated by (1) interfingering between the reef facies and the peri-reef skeletal grainstone (Fig. 30-4), and (2) the presence of abraded skeletal fragments of reef-forming organisms and intraclasts of reef boundstone and wackestone (Fig. 30-5F).

Lagoonal Limestone Facies of the Reef Complex

The contemporary emergence of the two parallel reef trends led to the deposition of lenticular bodies of reef-derived skeletal grainstone in an intervening natural topographic low of the reef complex. This low was a lagoon which became the site for the deposition of two distinctive limestone facies: miliolid-pellet grainstone and algal-nodule wackestone. The miliolid-pellet grainstone consists of miliolid foraminifers and lime mud pellets (Fig. 30-5E). In contrast, the algal-nodule wackestone is a mud-supported rock in which algal nodules (algal structures around a skeletal nucleus) are scattered in a lime mud matrix (Fig. 30-5D). These lagoonal facies are regularly interbedded throughout the vertical and lateral extent of the Fairway reef lagoon.

Depositional thickening of the lagoonal facies away from reef cores into the central portion of the lagoon is evident in Fairway Field proper (Fig. 30-9). However, abnormal apparent thickening of the lagoonal facies occurs north of the field, where it is believed to be a result of steep dips associated with faulting in proximity to the La Rue salt dome.

---

terparticle void spaces. This is the interval of high porosity and permeability whose sonic-log response is illustrated in Figure 30-2. Porosity 20.8%, permeability 800 md. ARCO Westbrook No. 1, 9904.5 feet (3018.9 m). Plane light. *F.* Photomicrograph of reef-derived skeletal grainstone facies. Note skeletal fragments of corals (C) and mollusks (M), and intraclasts (I) of reef core rock. Bitumen (black) impregnates much of pore space and reduces permeability. Porosity 6%, permeability 2.6 md. ARCO Westbrook No. 1, 9945 feet (3031.2 m). Plane light. *G.* Photomicrograph of reef-derived skeletal grainstone showing well-developed skeletal-moldic porosity (SMP) in the Fairway Field reef. Porosity 11.9%, permeability 1.6 md. Cities Service Emerson No. 1, 9887 feet (3044.0 m). Plane light. *H.* Photomicrograph of reef-derived grainstone showing well developed skeletal-moldic porosity (SMP) in the James patch reef at South Tyler Field, Smith County, Texas. Phillips Vannie No. 1, 9976 feet (3040.6 m). Plane light.

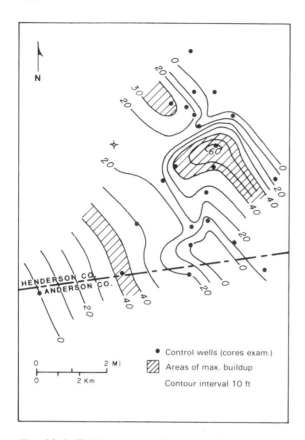

Fig. 30-6. Thickness map of stromatolite-coelenterate (hydrozoan) boundstone facies.

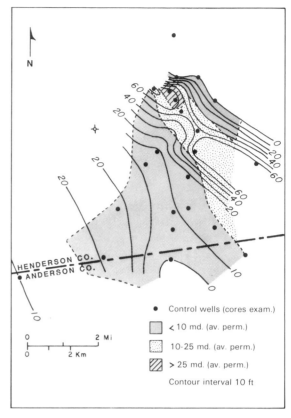

Fig. 30-7. Combined thickness and average permeability map of Chondrodonta and rudistid wackestone and boundstone facies.

Major Reservoir Controls

Porosity

Skeletal-moldic porosity is prevalent in the reef complex. For example, in the reef-derived skeletal grainstone facies, skeletal molds, largely of dissolved rudist fragments (Fig. 30-5G), are particularly well-developed. In addition, the limemud matrix of the *Chondrodonta* wackestone facies and the rudistid boundstone facies is fairly well riddled with the skeletal molds of fine to coarse skeletal debris. The porosity ranges from 8 to 20 percent in the majority of samples analyzed in this study.

Three lines of evidence and reasoning strongly suggest that skeletal-moldic porosity in James reefs was created as the result of freshwater movement through the reefs during early subaerial exposure. To begin with, Allan (1977)

reported a thin subaerial crust at the top of the James oolitic grainstone belt in the Sun Retherford No. 1 well, 40 miles (64 km) north of Fairway Field. This provides direct evidence of a paleoexposure surface at the top of the James Limestone in the East Texas Basin.

Secondly, it is well known from studies of Pleistocene reefs that selective dissolution of skeletal fragments commonly occurs when metastable reef debris is exposed to subaerial weathering and freshwater diagenesis (Matthews, 1967; Steinen, 1974; Friedman, 1975). Since the distinctive skeletal-moldic porosity of Pleistocene reef limestones is practically identical to that found in the Fairway James reef, it follows that the James porosity may also have been created during subaerial exposure (Fig. 30-5G).

Finally, skeletal-moldic porosity is also well-developed in a small James patch reef at South Tyler Field, about 15 miles (25 km) northeast of

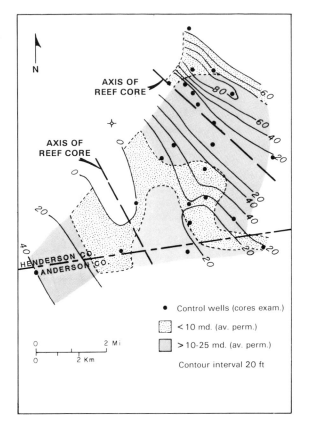

Fig. 30-8. Combined thickness and average permeability map of upper reef-derived skeletal grainstone unit.

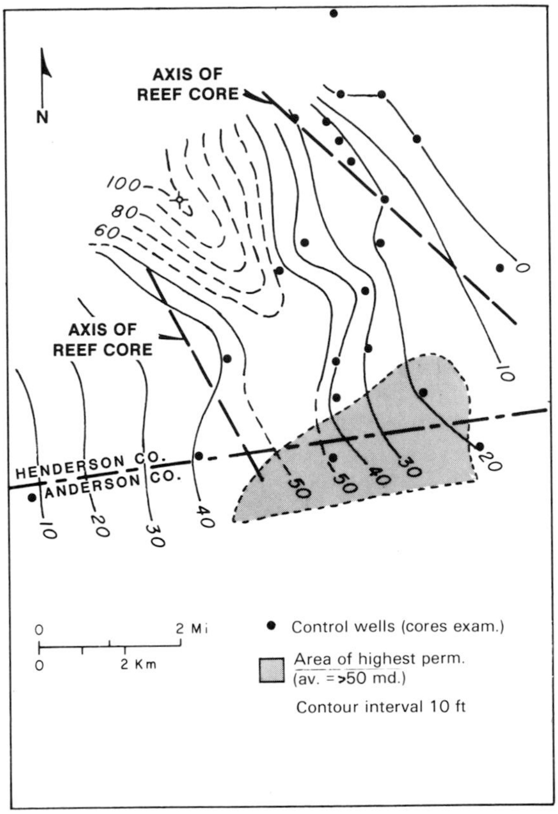

Fig. 30-9. Thickness map of the lagoonal facies.

Fairway Field (Figs. 30-1, 30-5H). The patch reef is encased in argillaceous lime mudstone and is well separated from other major James grainstone belts. This isolation would have retarded the entry of ground water into the patch reef during its burial history. Therefore, the most opportune time for movement of fresh water through the patch reef would have been during early subaerial exposure and prior to burial.

Although skeletal-moldic porosity dominates the reef facies, the highest porosities and permeabilities are found in a miliolid-pellet grainstone unit in the upper part of the lagoonal facies where primary, interparticle void space has been preserved (Figs. 30-2, 30-5E). This particular grainstone unit ranges upward to 10 feet (3 m) in thickness and has porosities as high as 22 percent and permeabilities up to 800 millidarcys. The typical sonic log response of this miliolid-pellet grainstone is shown in Figure 30-2.

Permeability

Local permeability increases in the reef facies are fairly common and result from two conditions: (1) extensive dissolution of locally abundant skeletal parts and shells, and (2) local concentration of organisms with high primary porosity. For instance, local permeability increases in the easternmost reef core of the *Chondrodonta* and rudistid facies are attributable to extensive leaching of abundant skeletal debris and whole shells and to an abundance of corals whose chambered structures still retain their primary porosity. These local high-permeability areas along the easternmost reef core are shown in Figure 30-7.

Permeability is also reduced locally in reef facies as a result of impregnation by black, gilsonite-like bitumen (Fig. 30-5F). As shown in Figure 30-8, substantial impregnation of a reef-derived grainstone body by this bitumen has created low-permeability zones that cut across the depositional strike of the grainstone body. The bitumen, therefore, serves as a permeability barrier that divides the James reservoir into areas of relatively high and low permeability.

Mapping of average permeabilities in the lagoonal facies reveals that the highest values occur in the southern portion of the lagoonal area and lowest in the northern portion (Fig. 30-9). Average permeabilities are highest in the southern portion because of exceptionally high permeabilities of the miliolid-pellet grainstone unit (averaging 50 md) in the upper half of the lagoonal facies (Fig. 30-2). In this particular unit, lime-mud matrix and calcite cement occur only sparsely, allowing excellent communication between pore spaces (Fig. 30-5E). However, in the northern lagoonal area, several factors combine to decrease permeability, especially within the miliolid-pellet grainstone unit. These factors include an increase in lime-mud matrix between pellets and miliolids, overly close packing of the pellets, and impregnation of existing void space by bitumen.

Summary

The James Limestone reservoir at Fairway Field is an atoll reef complex situated near the southern end of a seaward-projecting grainstone shoal. A structural high, probably due to salt tectonics, exerts the strongest control on hydrocarbon accumulation at Fairway. Porosity in the reef is mainly in the form of skeletal molds, created during early subaerial exposure and freshwater diagenesis. Although skeletal-moldic porosity dominates in reef cores and reef-derived grainstone, the highest porosities and permeabilities are associated with a miliolid-pellet grainstone that still retains its primary pore space in the southern reef-lagoonal area. In the northern lagoonal area, porosity and permeability in the miliolid-pellet grainstone are reduced by increases in lime-mud matrix, by compaction of pellets, and by impregnation with bitumen. Bitumen had a major destructive effect on permeability by impregnating pores and clogging pore throats, especially in reef-derived skeletal grainstone, and resulted in belts of low permeability that cut across the depositional strike of the skeletal grainstone facies.

References

ACHAUER, C.W., and J.H. JOHNSON, 1969, Algal stromatolites in the James Reef Complex (Lower Cretaceous), Fairway Field, Texas: Jour. Sedimentary Petrology, v. 39, no. 4, p. 1466–1472.

ALLAN, J.R., 1977, Carbon and oxygen isotopes as diagenetic and stratigraphic tools: five examples from ancient limestone sequences: Ph.D. Dissert., Brown Univ., Providence, RI, 252 p.

FRIEDMAN, G.M., 1975, The making and unmaking of limestones, or the downs and ups of porosity: Jour. Sedimentary Petrology, v. 45, no. 2, p. 379–398.

MATTHEWS, R.K., 1967, Diagenetic fabrics in biosparites from the Pleistocene of Barbados, West Indies: Jour. Sedimentary Petrology, v. 36, no. 4, p. 1147–1153.

STEINEN, R.P., 1974, Phreatic and vadose diagenetic modification of Pleistocene limestone—Petrographic observations from subsurface of Barbados, West Indies: Amer. Assoc. Petroleum Geologists Bull., v. 58, p. 1008–1024.

TERRIERE, R.T., 1976. Geology of Fairway Field, east Texas: in North American Oil and Gas Fields: Amer. Assoc. Petroleum Geologists Mem. 24, p. 157–176.

31
Greater Ekofisk Field
Charles T. Feazel, John Keany, and R. Michael Peterson

RESERVOIR SUMMARY

Location & Geologic Setting	Central North Sea, Norwegian sector; central graben of North Sea Basin
Tectonics	Reservoirs localized on salt-flowage structures
Regional Paleosetting	Early phase of modern setting
Nature of Trap	Structural
Reservoir Rocks	
Age	Late Cretaceous (Maastrichtian) and Paleocene (Danian)
Stratigraphic Units(s)	Ekofisk Formation and Tor Formation
Lithology(s)	Limestone
Dep. Environment(s)	Deep-water (sub-lysocline), shelf to slope
Productive Facies	Chalks
Entrapping Facies	Paleocene shales (overpressured)
Diagenesis/Porosity	Compaction (mechanical and chemical), cementation by calcite, clays, silica
Petrophysics	
Pore Type(s)	Interparticle
Porosity	0–45%, avg 32%
Permeability	0.1–1000 md, avg 1.0 md
Fractures	Abundant in productive intervals
Source Rocks	
Age	Late Jurassic (Kimmeridgian)
Lithology(s)	Kimmeridge Clay (shales)
Migration Time	NA
Reservoir Dimensions	
Depth	*9600–10,000 ft (29,30–3080 m)
Thickness	Chalk several hundred meters thick; upper 0–300 m productive
Areal Dimensions	*8 $\times$ 5.5 mi (11 $\times$ 8 km)
Productive Area	*12,071 acres (49 km^2)
Fluid Data	
Saturations	Variable, min S_w < 5%, avg 24%
API Gravity	*36°
Gas-Oil Ratio	*1547:1
Other	—
Production Data	
Oil and/or Gas in Place	*5.4 billion STBO
Ultimate Recovery	*1.2 billion STBO, 4.41 billion CFG (700 billion m^3), 110 million BNGL; URE = 22.2%
Cumulative Production	715.7 million BO to July 1, 1983**

Remarks: *Ekofisk Field *sensu stricto*. IP of discovery well, 10,000 BOPD. 38 producing wells. Discovered 1969 (Ekofisk). **Oil and Gas Journal, Dec. 16, 1983, p. 104.

31
Cretaceous and Tertiary Chalk of the Ekofisk Field Area, Central North Sea

Charles T. Feazel, John Keany, and R. Michael Peterson

Location and Regional Setting

Large volumes of petroleum are entrapped in Upper Cretaceous and lower Tertiary strata in the Central Graben, Norwegian sector, North Sea (Fig. 31-1). Reservoir intervals up to several hundred meters thick are present in seven major oil fields constituting the Greater Ekofisk area. Hydrocarbon-productive intervals are within the Tor Formation (Maastrichtian) and the Ekofisk Formation (Danian).

Discovery

The Ekofisk structure (Fig. 31-2), a broad dome overlying a diapir of Zechstein salt, was mapped from common reflection point seismic data. The discovery well, the Phillips 2/4A-1X, was drilled in December 1969, at a time when the petroleum industry was rapidly losing interest in North Sea exploration. The well came in with an initial potential of over 10,000 barrels per day of low-sulfur, 36° API gravity oil with a gas-oil ratio of 1547:1. Four wells were placed on production within 18 months, and produced at a combined rate of 40,000 barrels of oil per day. These initial wells produced 28 million barrels of oil before construction of the large network of platforms that now constitutes the Greater Ekofisk area development complex (Van den Bark and Thomas, 1981).

Reservoir and Reserves

The Ekofisk structure, shown in Figure 31-2, is approximately 8 miles long north–south by 5 miles wide (11 × 8 km), with a producing area of 12,071 acres (49 km²). Average depth to the top of the reservoir (top Ekofisk Fm) is about 10,000 feet (~3050 m). A well-defined oil-water contact does not exist in the Ekofisk Field. The base of hydrocarbons appears domed, possibly by post-accumulation salt movement. The seal on the reservoir is provided by highly overpressured Paleocene shales. The gross oil column at Ekofisk is over 1000 feet (305 m) thick; the average net pay is 590 feet (180 m) (Van den Bark and Thomas, 1981). The field presently includes 38 development wells.

Porosity within the Ekofisk and Tor reservoirs is commonly 20 to 35 percent and in places as high as 50 percent. Recoverable reserves from Ekofisk Field have been estimated at 1.2 billion barrels of oil, 4410 billion cubic feet of gas (125 billion m³), and 110 million barrels of natural gas liquids (Van den Bark and Thomas, 1981).

Reservoir Facies

The Upper Cretaceous to Lower Tertiary hydrocarbon-bearing strata in the Ekofisk area of the North Sea are chalks. Chalk is very fine-grained limestone composed largely of minute skeletons

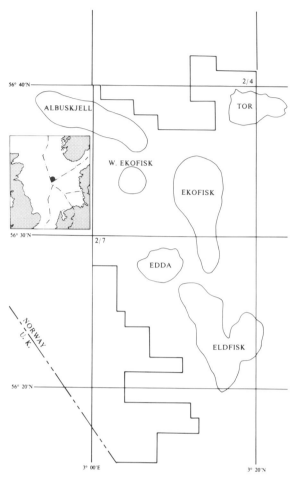

Fig. 31-1. Map of North Sea, with location of the Greater Ekofisk area. From Van den Bark and Thomas, 1981. Published by permission, American Association of Petroleum Geologists.

of coccolithophores. Subordinate amounts of planktonic and benthonic foraminiferal tests, fecal pellets, sponge spicules, bryozoans, echinoderm fragments, and calcispheres are present. Variable amounts of detrital and authigenic clay minerals, framboidal pyrite, glauconite, and quartz silt of probable eolian origin are also found in the chalk.

In cores, the chalk is light-gray to medium-tan, depending in part on the amount of oil-staining, moderately soft to hard, burrowed (*Zoophycos*, crustacean, and annelid burrows have been recognized), with common stylolites and solution seams. Replacement chert nodules are common, usually concentrated in burrowed intervals. Some intervals are highly fractured; only rubble is recovered in core barrels from many productive intervals.

Depositional Setting

The distribution of calcareous nannoplankton in the world ocean is well known (Berger, 1976). Coccolithophores, the primary contributors to chalks, live within the photic zone of the ocean and are subject to climatic, latitudinal, and oceanographic controls. Figure 31-3 is a schematic representation of chalk deposition, showing the two most abundant particle types, the skeletal remains of calcareous nannoplankton and foraminifers. Upon death of the nannoplankton, through either completion of the life cycle or predation, the coccospheres break up, producing coccoliths. These tiny platelets of low-magnesium calcite (average size 1–10 μ) settle slowly toward the sea floor. During settling, the coccoliths are subject to dissolution, which tends to remove selectively those species that are the least solution-resistant. This dissolution, combined with the chemical stability of low-magnesium calcite, results in a carbonate sediment which, in comparison to shallow-water limestones, undergoes less dramatic diagenetic alteration.

Once the particles reach the sea floor, they become incorporated into a sediment that may have porosity as high as 70 percent (Scholle, 1977). This sediment is usually soft and intensely burrowed. It is likely, however, that some sea floor cementation may occur. Sea floor cementation of chalks is intimately related to the sedimentation rate. In areas where sedimentation is rapid, the degree of lithification is lower than in areas of slow sedimentation because the available cement is dispersed through a greater thickness of skeletal grains. Since sedimentation rate is dependent on both skeletal productivity in surface waters and water column chemistry, we emphasize that paleo-oceanographic conditions are responsible for most of the porosity distribution in chalk reservoirs.

Origin of Porosity

Since carbonate dissolution rates increase with depth, calcareous tests are typically better preserved on elevated areas of the sea floor (Berger, 1976). The boundary between carbonate and noncarbonate sediments, discovered by Murray and Renard (1891), is called the calcite compensation depth (CCD), because at this level the rate

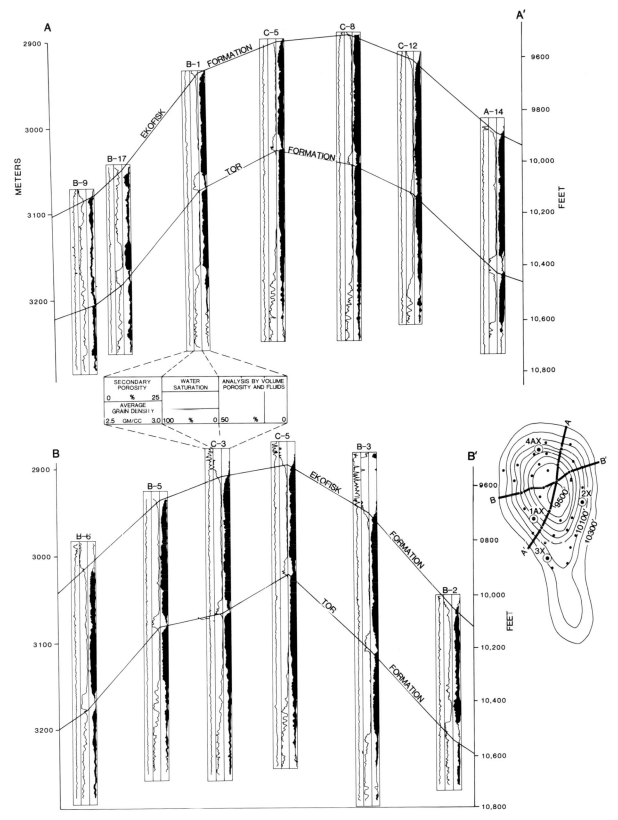

Fig. 31-2. Cross-sections through the Ekofisk structure, compiled from computer-processed electric logs (from Van den Bark and Thomas, 1981). Ekofisk Formation is chalk of Danian age; Tor Formation, chalk of Maastrichtian age. Map shows structure on top of Ekofisk Formation. Black area to right of one set of curves shows amount of oil-filled porosity.

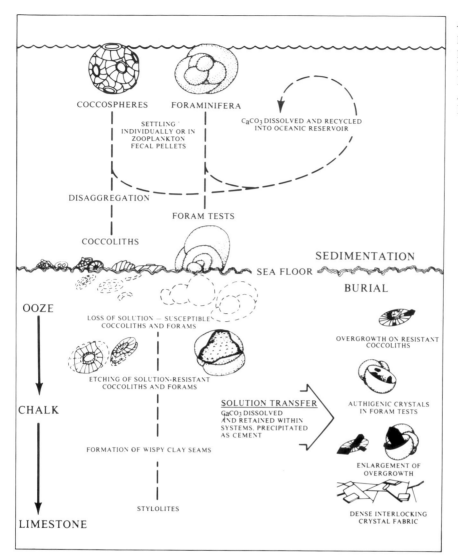

Fig. 31-3. Schematic representation of chalk deposition. After Van den Bark and Thomas, 1981. Published by permission, American Association of Petroleum Geologists.

of supply of calcareous tests is equal to the rate of dissolution. The CCD is a product of various factors and reflects conditions in both the upper waters and the deep sea (Bramlette, 1961; Berger, 1970, 1971). In any one basin, the level of the CCD may change through time in response to changing conditions of productivity, oceanography, geology, and climate.

Ruddiman and Heezen (1976) and Berger (1968, 1970) introduced the concept of the lysocline (the depth at which dissolution of calcium carbonate becomes apparent). Above the lysocline, saturation with respect to calcite prevents significant dissolution. Although the factors controlling the lysocline are not well understood, its level varies in different basins from coincidence with the CCD to several hundred meters above it. The greatest separation of the two horizons is generally found in regions of high planktonic productivity, where the abundance of carbonate depresses the CCD while higher concentrations of dissolved carbon dioxide, resulting from oxidation of organic matter, raise the lysocline. The high skeletal productivity represented by chalk hundreds of meters thick suggests that dissolved organic material was abundant in the Cretaceous-Tertiary sea. We therefore conclude that, during the time of chalk deposition, separation of the lysocline and CCD was at a maximum. For a small region of the sea floor, the major controlling factor for carbonate preservation was water depth (Fig. 31-4). Carbonate sediments

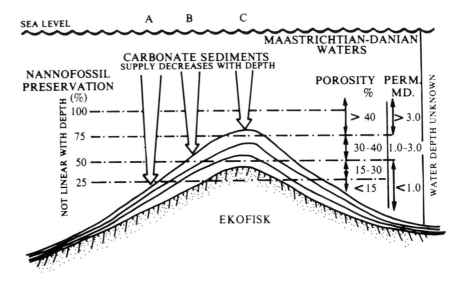

Fig. 31-4. Depositional model for central North Sea chalk reservoirs. Illustrated relationships between chalk thickness, reservoir properties, and nannofossil preservation are based on data from three cores designated A, B, C.

from depths shallower than the lysocline would exhibit little or no evidence of dissolution. Between the lysocline and the CCD, dissolution effects such as fragmentation and etching would increase downward. Below the CCD, dissolution effects are extreme, and few biogenic carbonate particles would be preserved.

Arrhenius (1952), Berger (1970) and Thunell (1976) established the degree of foraminiferal fragmentation as an important criterion in recognizing carbonate dissolution patterns in deep sea sediments. In the North Sea chalk, the degree of carbonate dissolution is also reflected by preservation of calcareous nannofossils, as described for Pacific pelagic sediments by Johnson *et al.,* (1977). Nannofossil preservation is expressed as:

$$\frac{\text{whole tests}}{(\text{whole tests} + \text{fragmented tests})} \times 100$$

for each sample. A positive correlation exists between nannofossil preservation and measured porosity and permeability (Figs. 31-5, 31-6). High-porosity intervals contain well-preserved nannofossils (minimum fragmentation), and the lowest-porosity intervals contain poorly preserved nannofossils (maximum fragmentation). Inter-well correlation reveals that nannofossil fragmentation increases off-structure within synchronous horizons. Although the cause of the cor-

relation between carbonate preservation and rock properties is not clear, the data suggest that the relationship is primary, and reflects depositional processes.

Occlusion of Porosity

The primary porosity imprinted by paleo-oceanographic conditions was modified by diagenetic events. Once the pelagic sediments reached the sea floor, all diagenetic processes acted to occlude the pore spaces between them. The only mechanism for introducing secondary porosity (more strictly, permeability) in this setting is post-lithification fracturing; the chalk deposited within the Central Graben of the North Sea shows no evidence of subaerial exposure or leaching.

Abundant evidence of various pore-filling processes is found in the chalk, although the occlusion mechanisms are poorly understood. Mechanical compaction might reduce the porosity of a calcareous ooze from an initial 70 percent to approximately 50 percent by dewatering. Some intervals in the North Sea chalk have porosity values approaching 50 percent (Fig. 31-5B); apparently little alteration occurred in these sediments other than expulsion of interstitial waters until a grain-supported matrix was achieved.

Low-magnesium calcite cement in chalk is believed to be derived internally through solution-

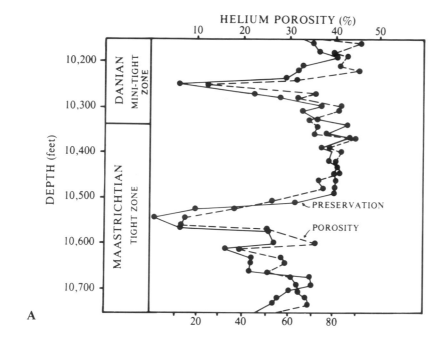

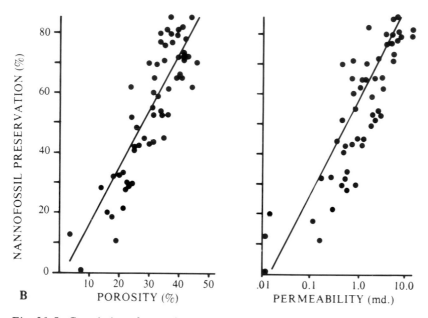

Fig. 31-5. Correlation of nannofossil preservation and rock properties as measured from core samples: (*A*) down-hole relationship in a single core; (*B*) cross-plots demonstrating linear relationships within a single reservoir.

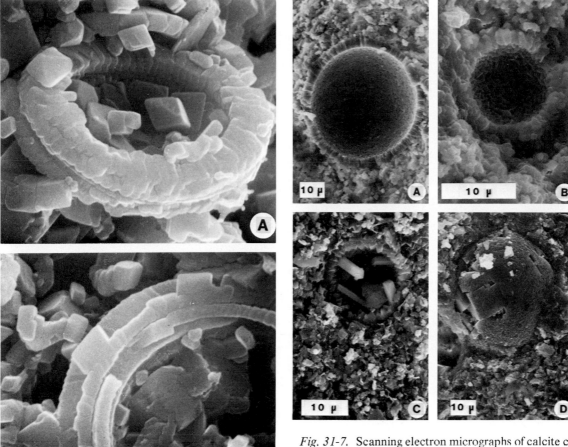

Fig. 31-6. Scanning electron micrographs of authigenic calcite overgrowths that begin as enlarged crystallites in the margins of coccoliths (*A and B*) and as euhedral crystals filling the central coccolith cavity (*A*). Both micrographs to same scale. Figure 31-6A from Van den Bark and Thomas, 1981. Published by permission, American Association of Petroleum Geologists.

Fig. 31-7. Scanning electron micrographs of calcite cement that occludes intraparticle porosity, primarily within the chambers of foraminiferal tests. A progressive sequence is inferred, from smooth, relatively unaltered chamber walls (*A*), through inward growth of secondary calcite crystals (*B,C*), to the complete filling of chambers to produce micro-steinkerns (*D*). *A* from Van den Bark and Thomas, 1981. Published by permission, American Association of Petroleum Geologists.

transfer (Durney, 1972). This process of dissolution at points of greatest stress (grain contacts) and precipitation in regions of lower stress (pore walls) may reduce the porosity of chalk to near zero, given the proper combination of pore fluids and overburden thickness (Scholle, 1977). In some chalks, however, there is geochemical evidence (Sr and $\delta^{18}O$ distribution) that fluid flow has introduced additional carbonate cement from an external source, despite the low permeability of the matrix (L. S. Land, personal communication, 1978). Calcite cement began as overgrowths on coccolith elements (Fig. 31-6) and on the in-

terior walls of foraminiferal tests (Fig. 31-7). Progressive enlargement of these overgrowths formed the single crystals in the skeletal matrix shown in Figure 31-8. Continued crystal growth resulted in the dense mosaic of calcite crystals characteristic of low porosity intervals (Fig. 31-9), in which only a few skeletal fragments remain identifiable.

Another process that occludes porosity in chalk is the growth of authigenic clay minerals. Kaolinite, chlorite, and mixed-layer clays have grown between coccoliths and inside the chambers of foraminiferal tests (Fig. 31-10). The effect of

Fig. 31-8. Secondary calcite crystals in matrix of much smaller calcareous bioclasts. Scanning electron micrographs.

Fig. 31-9. Dense mosaic of calcite crystals, often the only distinguishing feature visible with the SEM in low-porosity intervals. Note that few nannofossils remain identifiable. Both scanning electron micrographs to same scale.

these clays on reservoir quality is not as great as in clastic rocks, due to the relatively large size of the clay platelets ($1–5\ \mu$) compared to the size of the pores ($0.5–2\ \mu$, with pore throats considerably smaller).

Scholle (1975) and Van der Lingen and Packham (1975) described a rounded coating on skeletal components of chalk that resembles "icing sugar." Such coatings cover calcareous nannofossils in a few intervals of the North Sea chalk (Fig. 31-11). Energy-dispersive X-ray analysis revealed these coatings to be siliceous, although their mineralogy (or even crystallinity) is unknown. Authigenic silica is abundant in the chalk, and commonly appears as chert in isolated irregular nodules and as internal molds of thalassinidean shrimp burrows and smaller ichnofossils. A likely source for the secondary silica was biogenic opal; siliceous sponge spicules and radiolarian tests are abundant in some intervals within the chalk. The siliceous coatings shown in Figure 31-11 reduce pore space, but their effect

on reservoir properties is diminished by their restricted distribution.

Preservation of Porosity

Because the diagenetic processes described above would normally destroy all porosity in pelagic sediments after relatively shallow burial, some mechanism must be sought to explain the preservation of porosity in the central North Sea chalk reservoirs, where porosity approaching 50 percent has been retained despite burial to depths exceeding 10,000 feet (3 km). Two types of pore fluids have played a role in the retention of primary porosity: (1) connate fluids, primarily brines derived from sea water, and (2) migrated fluids, primarily petroleum. If connate fluids are prevented by permeability barriers from escaping as overburden stress is increased, pore pressures build up beyond hydrostatic levels. In such overpressured situations, part of the overburden is

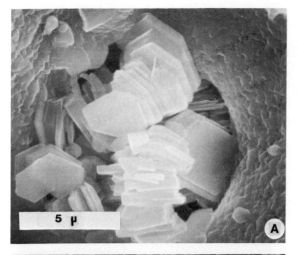

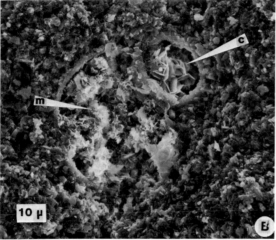

Fig. 31-10. Scanning electron micrographs showing authigenic clay minerals inside chambers of foraminiferal tests. (*A*) Kaolinite. (*B*) Chlorite (c) and mixed-layer clays (m).

Fig. 31-11. Rounded siliceous coatings, of uncertain mineralogy, on calcareous nannofossils.

Scholle (1975, 1977), using data from the Ekofisk area, demonstrated the coexistence of overpressuring and anomalously high porosity at depth. Certainly connate fluid pressures are capable of preserving porosity. Figure 31-12 shows that exclusion of water by petroleum results in similar porosity preservation in the oil-wet chalk, whereas water-wet chalk in the same pressure regime is tightly cemented by secondary calcite.

Petroleum Source

Van den Bark and Thomas (1981) presented evidence from the analysis of biological marker compounds that the source of the Ekofisk area oils was the Upper Jurassic Kimmeridge Clay. Vitrinite reflectance data ($R_o = 0.93 - 1.16$) indicate that the Kimmeridge Clay is presently in the thermal regime of maximum petroleum genesis. Scholle (personal communication, 1977) suggested that the stable isotopic composition of the chalk ($\delta^{18}O_{PDB} = -3$ to $-4‰$) indicates that cementation in the Ekofisk reservoir was arrested after 3000 to 5000 feet ($\sim$1–1.5 km) of burial. This depth may correspond to the onset of overpressure and the timing of oil entry.

Summary

The prolific hydrocarbon production from North Sea chalk reservoirs is due to the preservation of high primary porosity and the enhancement of permeability by post-lithification fracturing. Paleo-oceanographic conditions controlled the

supported by the fluids; the intergranular stress is lower than in normally pressured rocks, and the physical drive toward pressure solution and generation of pore-filling cement is reduced (Scholle, 1975, 1977). Thus, in overpressured reservoirs, both mechanical and chemical compaction are retarded and porosity is preserved.

Migrating fluids displace entrapped pore waters. Petroleum entering a pore forces out the aqueous phase that serves as a pathway for ions during the formation of authigenic minerals. Oil, in some parts of the North Sea chalk that are oil-wet, has been particularly effective at excluding water from the pores; irreducible water saturation is lower than 10 percent in many porous intervals.

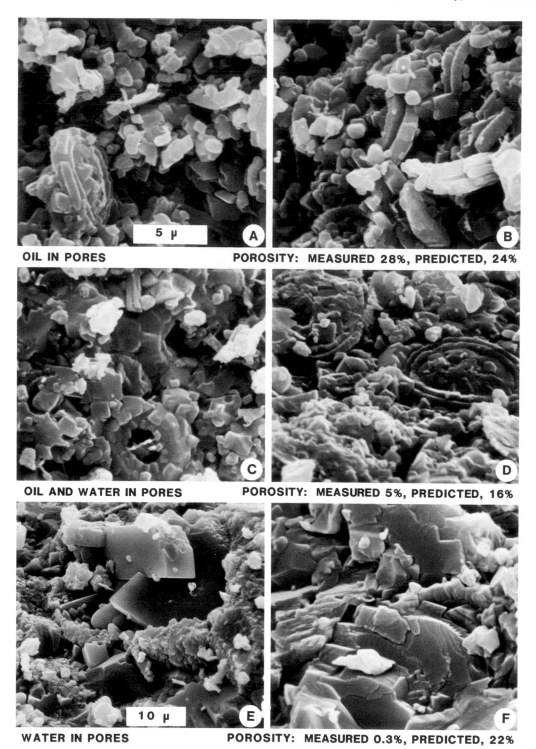

OIL IN PORES   POROSITY: MEASURED 28%, PREDICTED, 24%

OIL AND WATER IN PORES   POROSITY: MEASURED 5%, PREDICTED, 16%

WATER IN PORES   POROSITY: MEASURED 0.3%, PREDICTED, 22%

Fig. 31-12. Scanning electron micrographs of chalk from a well in the Ekofisk area. Sample locations: *A*,*B* high within; *C*,*D* middle of; and *E*,*F* below the base of the hydrocarbon zone. Preservation of calcareous nannofossils suggests that primary porosity was greater than 15% in all three intervals. Note that extensive secondary calcite in water-saturated chalk has all but obliterated any trace of skeletal particles. Pore pressures are thought to be approximately equal in all three intervals. All micrographs except *E* to scale shown in *A*.

distribution of primary porosity in the chalk. All diagenetic processes are acting to occlude the pores; the preservation of porosity approaching 50 percent at a burial depth exceeding 10,000 feet (3 km) is due to a combination of overpressuring of the pore fluids and exclusion of water from the pores by migrating oil.

References

ARRHENIUS, G., 1952, Sediment cores from the east Pacific: Repts. Swedish Deep-Sea Exped., v. 5, p. 1–227.

BERGER, W.H., 1968, Planktonic foraminifera: selective solution and paleoclimatic interpretations: Deep-Sea Res., v. 15, p. 31–43.

BERGER, W.H., 1970, Planktonic foraminifera: selective solution and lysocline: Marine Geology, v. 8, p. 111–138.

BERGER, W.H., 1971, Sedimentation of planktonic foraminifera: Marine Geology, v. 11, p. 325–358.

BERGER, W.H., 1976, Biogenous deep-sea sediments: production, preservation, and interpretation: in Riley, J.P. and R. Chester, eds., Chemical Oceanography, 5, Academic Press, London, p. 265–388.

BRAMLETTE, M.N., 1961, Pelagic sediments: in Sears, M., ed., Oceanography: Amer. Assoc. Advmt. Sci., v. 67, p. 345–366.

DURNEY, D.W., 1972, Solution-transfer, an important geological deformation mechanism: Nature, v. 235, p. 315–317.

JOHNSON, T.C., E.L. HAMILTON, and W.H.

BERGER, 1977, Physical properties of calcareous ooze: control by dissolution at depth: Marine Geology, v. 24, p. 259–277.

MURRAY, J., and A.F. RENARD, 1891, Report on deep-sea deposits based on the specimens collected during the voyage of H.M.S. Challenger in the years 1872 to 1876: in Report on the Scientific Results of the Voyage of the H.M.S. Challenger, 525 p.

RUDDIMAN, W.F., and B.C. HEEZEN, 1976, Differential solution of planktonic foraminifera: Deep-Sea Res., v. 14, p. 801–808.

SCHOLLE, P.A., 1975, Application of chalk diagenetic studies to petroleum exploration problems (abst.): Amer. Assoc. Petroleum Geologists Bull., v. 59, p. 2197–2198.

SCHOLLE, P.A., 1977, Chalk diagenesis and its relation to petroleum exploration: oil from chalks, a modern miracle: Amer. Assoc. Petroleum Geologists Bull., v. 61, p. 982–1009.

THUNELL, R.C., 1976, Optimum indices of calcium carbonate dissolution in deep-sea sediments: Geology, v. 4, p. 525–528.

VAN DEN BARK, E., and O.D. THOMAS, 1981, Ekofisk, first of the giant oil fields of western Europe: in Halbouty, M.T., Giant Oil and Gas Fields of the Decade 1968–1978, Amer. Assoc. Petroleum Geologists Mem. 30, p. 195–224.

VAN DER LINGEN, G.J., and G.H. PACKHAM, 1975, Relationships between diagenesis and physical properties of biogenic sediments of the Ontong-Java Plateau (Sites 288 and 289, Deep Sea Drilling Project): in Curray, J.R., D.G. Moore et al., eds., Initial Repts, Deep Sea Drilling Project, v. 30: U.S. Govt. Printing Office, Washington, D.C., p. 443–381.

Cenozoic Reservoirs

Gachsaran and Bibi Hakimeh Fields
Harry McQuillan

RESERVOIR SUMMARY

Location & Geologic Setting	SE Khuzestan province of SW Iran, in open-folded portion of Zagros Mts.
Tectonics	Large asymmetrical folds of Late Tertiary (Pliocene) Alpine origin
Regional Paleosetting	Carbonate shelf on NE margin of Arabian Plate
Nature of Trap	Structural
Reservoir Rocks	
Age	Oligocene to Early Miocene
Stratigraphic Units(s)	Asmari Formation
Lithology(s)	Limestone and some dolomites
Dep. Environment(s)	Shallow-marine shelf
Productive Facies	Fractured shelf lime wackestone-packstone and sucrosic dolomites
Entrapping Facies	Anhydrite "caprock" of Gachsaran Formation
Diagenesis Porosity	Dolomitization of some intervals in lower Asmari; anhydrite pore and fracture filling in upper Asmari
Petrophysics	
Pore Type(s)	Fracture; intercrystal and moldic in some zones
Porosity	NA
Permeability	NA
Fractures	Major importance for productivity
Source Rocks	
Age	Santonian to Eocene (Pabdeh and Gurpi Fms)
Lithology(s)	Shales
Migration Time	Late Tertiary, syn-Alpine tectonic phase
Reservoir Dimensions	
Depth	3035 ft (925 m) subsea*, 725 ft (220 m) subsea**
Thickness	5000 ft (1525 m)*, 6000 ft (1830 m)**
Areal Dimensions	54 × 3 mi (86 × 5 km)*, 50 × 6 mi (84 × 9.6 km)**
Productive Area	86,200 acres (345 km^2)*
Fluid Data	
Saturations	S_w ~33% avg*, ~29% avg**
API Gravity	30°*, 31°**
Gas-Oil Ratio	NA
Other	—
Production Data	
Oil and/or Gas in Place	31 billion BO*, 32 billion BO**
Ultimate Recovery	4.5 billion BO*, 9.0 billion BO** (Halbouty *et al.*, 1970); URE = 14.5%*, 25.0%**
Cumulative Production	NA

Remarks: Oil and gas columsn (Asmari) 1050 ft (350 m) thick*; oil column (Asmari) 1600 ft (450 m) thick**. (*Bibi Hakimeh, **Gachsaran Field). Discovered 1940 (Gachsaran) and 1961 (Bibi Hakimeh). Ultimate recovery estimates from Halbouty *et al.* (1970); reference in INTRODUCTION.

Fracture-Controlled Production from the Oligo-Miocene Asmari Formation in Gachsaran and Bibi Hakimeh Fields, Southwest Iran

Harry McQuillan

Introduction

Since the 1908 discovery of oil in the Zagros Mountains foothills of southwest Iran, the main production has been from limestones of the Asmari Formation (Fig. 32-1), a sequence 1050 to 1600 feet (320–488 m) thick and Oligo-Miocene in age. In recent decades, oil and gas have also been produced from Mesozoic and Paleozoic reservoirs, but the Asmari Formation still remains the most important single producing reservoir. The mechanism of production is related mainly to fracturing, and variations in productivity across oil fields within which lithologies are relatively constant have been interpreted to be due to variations in fracture density or spacing (Lees, 1948; O'Brien, 1953). The fracturing referred to is at the scale of field observation on surface anticlines and as recorded from borehole cores.

The Asmari Formation is extensively exposed along the Zagros foothills of Iran in huge surface anticlines that are replicas of the associated buried structures forming the producing oil fields (Fig. 32-2). Studies of fracturing over surface anticlines in the foothills and in more tightly folded structures to the northeast have provided results that, instead of revealing a relationship between structural position and fracture density, have shown that their fracture density is constant.

In the search for an explanation of variations in productivity within the giant Gachsaran and Bibi Hakimeh fields, fracture patterns developed in surface Asmari anticlines and in outcrops above the two producing fields were studied using air photos and confirmed by field work (McQuillan, 1973a, 1974). The results reveal that fracture patterns are controlled by basement features.

General Features of the Area

Structural and stratigraphic features of the Zagros Mountain foothills have been well-documented over the past 60 years of oil exploration in Iran. Structural deformation of the northwest–southeast-trending fold belt increases in a northeasterly direction toward the Zagros Suture Zone (Haynes and McQuillan, 1974). Falcon (1967 and 1969) divided the Zagros into three structural belts: a simply folded belt, an imbricated belt, and a thrust belt (Fig. 32-1). The oil fields lie within the simply-folded belt, which is subdivided by the mountain front, a major topographic feature. Northeast of the front, the Asmari and older formations are exposed in high-relief folds, while to the southwest, the low-relief foothill badlands expose disharmonically folded evaporitic and clastic rocks of post-Asmari age (McQuillan, 1973b), which rest on productive structures like those exposed to the northeast. Such a situation makes it possible to examine anticlines involving the Asmari Formation, which simulate the structure of adjacent oil fields. The anticlines are typically asymmetrical, with steeper southwest flanks which in places are thrust-faulted (Fig. 32-3).

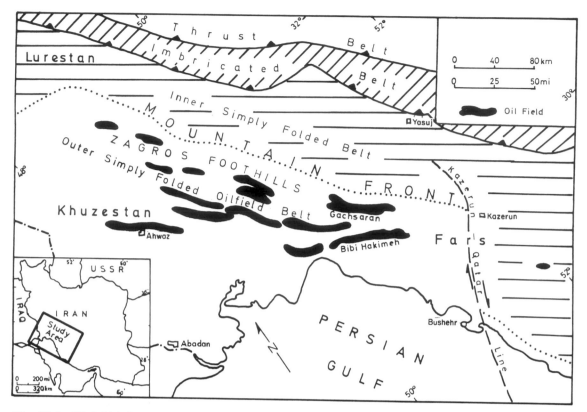

Fig. 32-1. Oil-field belt and general structural features of the Zagros foothills of southwest Iran. Adapted from McQuillan (1973a).

Asmari Formation Stratigraphy

The Asmari Formation is present in the Zagros Mountains as a prominent limestone commonly forming the huge "whaleback" anticlines so characteristic of the simply-folded belt (James and Wynd, 1965). In the main oil field belt of Khuzestan Province, the formation consists of 1050 to 1600 feet (320–488 m) of well-indurated wackestones and packstones. These were deposited in an elongate shallow basin which extended from Iraq through Lurestan and Khuzestan into Fars Province during the Oligocene and early Miocene (Dunnington, 1958). The rich foraminiferal fauna characteristic of the formation (Fig. 32-4) provides the basis of its division into lower, middle, and upper Asmari units coincident with the *Eulepidina-Nephrolepidina-Nummulites, Miogypsinoides-Archaias-Valvulinid,* and *Borelis melo-Meandropsina iranica* assemblage zones. Shelf sediments with larger

foraminifers forming most of the 50 to 75 percent grain content of the lower Asmari give away to muddy wackestones with 25 to 50 percent grains in the middle and upper Asmari. The upper third of the formation is thin- to well-bedded, but passes into more massive, locally dolomitized units in its lower levels. Southwest, toward the head of the Persian Gulf, sandstones derived from the Arabian Shield interfinger with the middle Asmari as the Ahwaz Member, whereas to the northwest in Lurestan, evaporites of the Kalhur Member come in at a similar stratigraphic level.

The Asmari Formation is conformably overlain by evaporites of the Gachsaran Formation (James and Wynd, 1965; Gill and Ala, 1972), the lower member of which is the important and areally extensive 130-foot (40-m) sequence of anhydrite comprising the cap rock for the Asmari reservoir. Above follow thousands of feet of halite and anhydrite interbedded with red and gray marls and thin limestone stringers, all contorted into a chaotic mass, the result of incompetent

Fig. 32-2. Huge surface anticlines exposing the Asmari Formation along the Zagros mountain front. Paler-colored rocks lying above the carapace of Asmari limestones show badland topography typical of the Gachsaran Formation evaporites. Adapted from McQuillan (1973a).

folding during the Zagros orogenic movements late in the Tertiary (Stocklin, 1968), and later gravitational slumping (McQuillan, 1973c). Succeeding the Gachsaran Formation are the shallow-marine gray marls and limestones of the Mishan Formation, which give way upward to more competent red sandstones and marls of the Agha Jari Formation (Fig. 32-5). The Asmari Formation is underlain by upwards of 2000 feet (609 m) of gray marine marlstones with a rich planktonic fauna. These rocks of the Pabdeh and Gurpi formations are considered to be a major source for Asmari oil (Lees, 1948).

Fracturing

Early discoveries of oil in Khuzestan were in the northwest, but by the 1960s, major discoveries extended to the southeast at Gachsaran and Bibi Hakimeh. High initial well productions from low porosity limestones, often in excess of 80,000 barrels per day, were explained by appealing to the presence of fractures (O'Brien, 1953). However, the variations in production between adjacent development wells within any one field were problematical because they showed no relation-

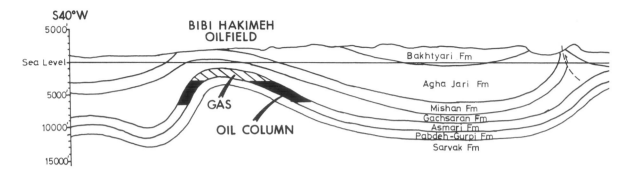

Fig. 32-3. Structural cross-section through the Bibi Hakimeh and Gachsaran oil fields. Thickness variations of the mobile evaporite rocks of the Gachsaran Formation result in disharmony between surface and sub-surface structure. Adapted from McQuillan (1973a).

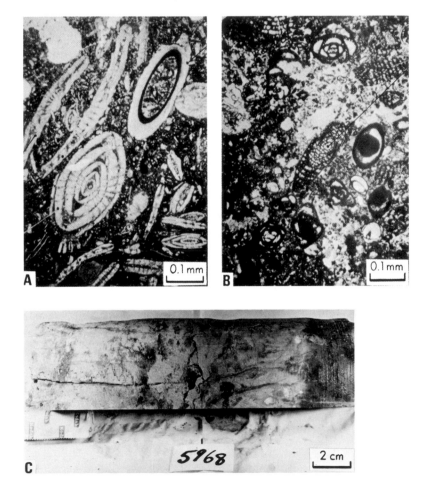

Fig. 32-4. A. Photomicrograph of a core sample from Lower Asmari Formation in Gachsaran well 25, depth 9853 feet (3234 m). Lime wackestone with a faunal assemblage that includes *Nummulites, Operculina*, and *Ditrupa. B.* Photomicrograph of a core sample from Middle Asmari Formation in Gachsaran well 25, depth 9496 feet (3117 m). Lime wackestone with a faunal assemblage that includes *Archaias* and *Miliola. C.* Photograph of core from the Asmari Formation in Bibi Hakimeh well 18, depth 5968 feet (1959 m). Open, near-vertical fractures occur, oozing crude oil. A shale-coated stylolite is oriented normal to the open fractures.

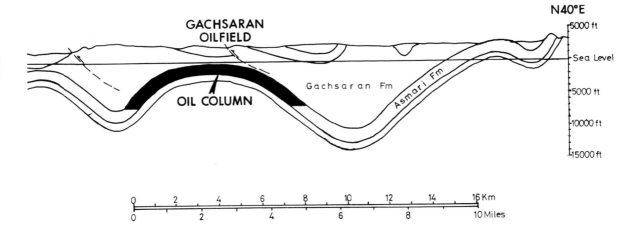

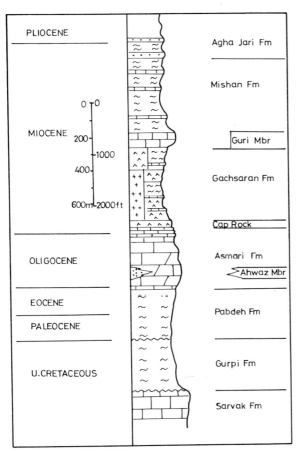

Fig. 32-5. Generalized stratigraphic column showing the Cretaceous-Tertiary succession in the Zagros foothills of southwest Iran.

ship to theoretical concepts of fracture distribution on anticlinal folds. A solution to the problem was sought by making detailed observations of small-scale fractures on surface structures and, later, by examining larger-scale fractures at airphoto scale.

Small-Scale Fractures

Direct field observations of these surface-exposed features indicate that: (1) an inverse semi-logarithmic relationship exists between bed thickness and fracture spacing; (2) in general, two major fracture sets normal to one another are present at a given station; (3) the orientations of fracture planes at this scale bear no relation to tectonic structure, and in fact in a given area all azimuth classes of orientation are present; and (4) for a given bed thickness in rocks of similar lithology, fracture spacing can be predicted and is constant regardless of structural position (Fig. 32-6). These relationships are directly applicable to reservoir engineering studies where block heights in the reservoir rocks are pertinent.

It has been suggested (McQuillan, 1973a) that fractures at this scale were initiated prior to folding, as a partial response to the diagenetic process. Then fracture-plane orientations were related to the morphology of the depositional surface. Such results shed little light on the problem of production variations.

Photoscale Fractures on Surface Asmari Anticlines

By mapping fractures on the surface anticline of Kuh-e Asmari on 1:10,000 scale air photographs,

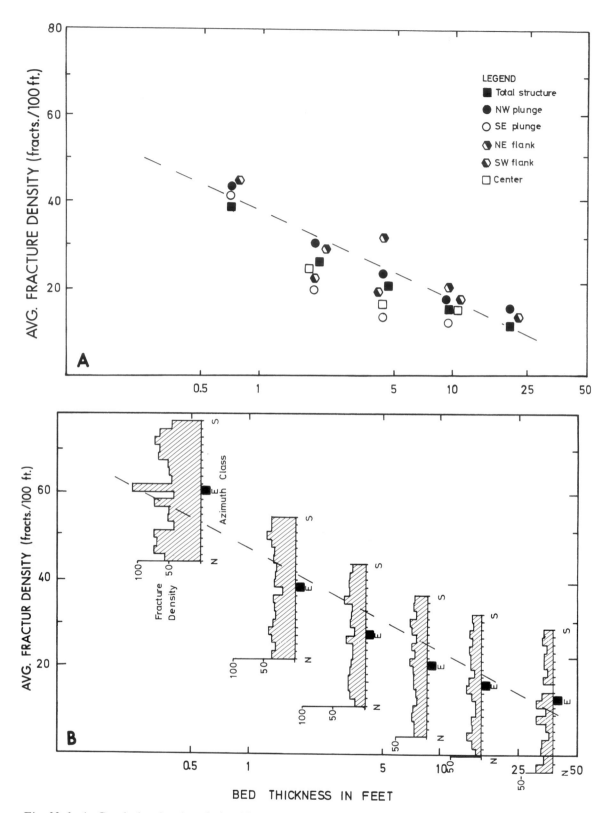

Fig. 32-6. A. Graph showing the relationship between bed thickness and density of small-scale fractures in various structural positions and for the overall structure of Kuh-e Asmari anticline. *B.* Graph showing the rela-

tion between bed thickness and density of small-scale fractures from anticlines selected from a large area of the Zagros foothills. Histograms show fracture-density distributions by azimuth classes.

the small-scale fractures studied earlier were effectively filtered out to leave only larger features (McQuillan, 1974). The results of this investigation showed that larger-scale features are present as two well-defined tension sets aligned normal and parallel to the anticlinal axis. Shear fractures are not apparent. Fracture densities are greater on the plunging noses and in the axial zone, but otherwise no real pattern of structural control is evident.

The photo-scale fractures contrast sharply with the smaller-scale features in both orientation patterns and density distributions. However, the superficial nature of the larger-scale features on structures unloaded of their overburden of sediments seems probable (McQuillan, 1974). This casts some doubt on the feasibility of drawing close analogies between surface structures and their buried oil field counterparts.

Surface Photoscale Fractures and Lineaments Above Producing Oil Fields

Results of research indicating a relationship between surface fracturing and natural steam production in Larderello, Italy (Marchesini et al., 1963) prompted the adoption of a similar investigative method over producing oil fields in Khuzestan. At the same time, a communication with R. A. Hodgson regarding his findings of genetic and geometric relationships between basement and overlying rocks in Colorado and Wyoming gave further impetus to the study (Hodgson, 1965).

The area studied lies between longitude 50° and 51°E and latitude 29° 30′ and 30° 30′N and includes two giant oil fields, Gachsaran and Bibi Hakimeh (Fig. 32-7). For the most part, Gachsaran oil field has surface exposures of Gachsaran Formation evaporites, while siliciclastic rocks of the Agha Jari Formation, together with inliers of Mishan Formation marls and limestones, form a cover over the Bibi Hakimeh oil field. The Gachsaran evaporites are deformed in a multitude of minor structures consequent on their mobile properties and form an interval of disharmony between the Asmari reservoir rock and the overlying rocks of the Mishan and Agha Jari formations.

In addition to recording actual breaks in the surface rocks, the observations were extended to include linear features produced by tonal variations, topographic features, and stream and vegetation alignments. These features were mapped at a scale of 1:60,000 on air-photo stereo pairs. The time of search per unit area was kept constant in order to assure consistent results across the whole area. The mapped traces were classified into 18 azimuth classes, each of 10 degrees range, and were processed in terms of length and number percent (McQuillan, 1974). The 8611 linear features that were mapped showed a distinct three-set pattern and marked variations in density. There were no distribution differences associated with different surface formations. Length and orientation diagrams were drawn on a grid pattern for areas of 225 square kilometers. These indicate a constantly oriented three-set pattern with trends of N10°–20°W, N20°–30°E, and N80°E (Fig. 32-8). The absence of tension sets related to surface Asmari structures is significant and supports the suggestion (McQuillan, 1974) of the superficial nature of some photo-scale fractures on surface anticlines.

A fracture density map was plotted on the basis of the number of linear features intersecting grid areas of 1 square kilometer. This showed an increase in density in zones trending parallel to the prominent linear set at N20°E where they crossed the Bibi Hakimeh and Gachsaran oil fields. The three-set orientation pattern of linear features bears no relation to folds resulting from the late Tertiary Zagros orogeny. The author assumes that the pattern is related to features reflecting basement discontinuities, which are manifest in southeast Iran as abrupt strike changes along large-scale features such as the Kazerun and Oman lines (Falcon, 1969; McQuillan, 1973a). These findings were then compared to the pattern of production trends in the oil fields (Figs. 32-8, 32-9).

The Oil Fields

No recent records of production from the Gachsaran and Bibi Hakimeh oil fields are available. The data used here are now more than 10 years old, but since both oil fields were then well

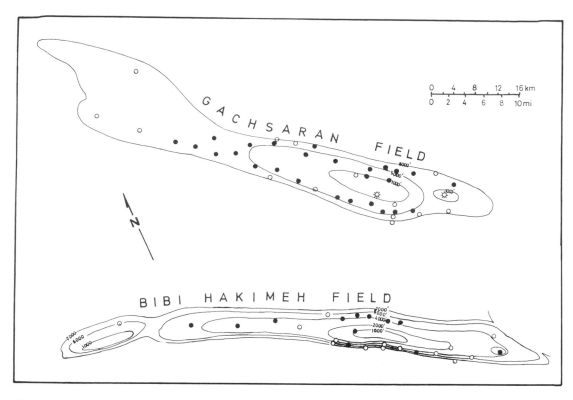

Fig. 32-7. Structure-contour map on top of Asmari Formation in Gachsaran and Bibi Hakimeh oil fields. Solid circles are producing wells, open circles are non-economic or non-producing wells.

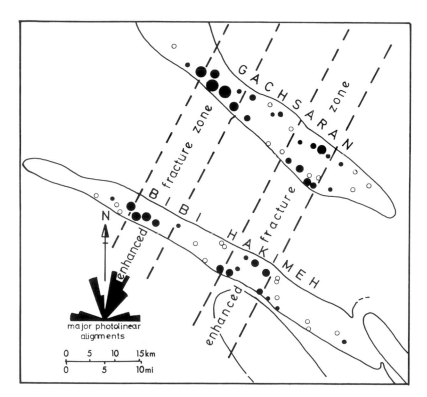

Fig. 32-8. Form-line map of Gachsaran and Bibi Hakimeh oil fields. Areas of solid circles are proportional to maximum allowed production rates. Open circles are non-commercial or non-producing wells. High-production wells lie in zones of enhanced fracturing related to trends of basement features.

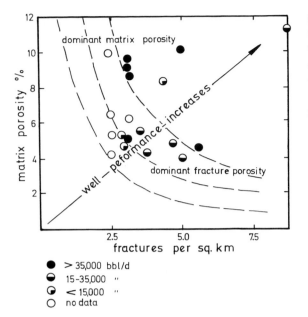

Fig. 32-9. Graph showing well production rates in Bibi Hakimeh Field as a function of matrix porosity and fracture density.

(non-fracture) porosity of 9 percent. The field has two gas caps, and reservoir engineers have noted that only a limited connection exists between production sectors along the axis of the field. The oil-water contact is tilted from 7400 feet (2255 m) subsea on the northeast flank to 8100 feet (2468 m) subsea on the southwest flank. The oil column is 6835 feet (2082 m) thick. Maximum allowable production rates emphasize the marked variations in offtake along the length of the field. A northwest zone of very high production has wells with production potentials of more than 80,000 barrels of oil per day. To the southeast, this is separated by a low-productivity zone from a section of the field in which wells produce up to 40,000 barrels of oil per day. The southeast plunge is again an area of low-productivity wells.

developed, the data are pertinent to the fracture study. More recently, the political unrest associated with the 1979 Islamic revolution and the Iran-Iraq conflict of 1980 have resulted in reduced production from these oil fields to a small fraction of their potential.

In the late 1960s, the Gachsaran and Bibi Hakimeh oil fields together provided upwards of one quarter of Iran's production. They are typical elongate asymmetrical folds of the Asmari Formation, which exhibit production variations similar to those in most of the oil fields of Khuzestan Province (Fig. 32-7).

Gachsaran Oil Field

The Gachsaran oil field, located in the southeastern part of Khuzestan, is 50 miles long by 6 miles wide (80 km × 9.6 km). Its estimated oil-in-place is 32 billion barrels. Since production began in 1940, the average daily production has increased to close to 1 million barrels from 22 wells. By the late 1960s, the cumulative production was 1.3 billion barrels.

The Asmari Formation at Gachsaran is 1600 feet (487 m) thick and has an average matrix

Bibi Hakimeh Oil Field

The discovery well was drilled on Bibi Hakimeh in 1961. The structure lies 15 miles (24 km) southwest of and parallel to Gachsaran oil field. With oil-in-place reserves estimated at 31 billion barrels, Bibi Hakimeh is second only to Gachsaran as the largest source of Iranian low-gravity crude (average 30° API). The field is 54 miles (86 km) long and has an average width of 3 miles (4.8 km). The original gas-oil contact is at 3035 feet (925 m) subsea, while the oil-water contact is tilted from 6200 feet (1890 m) subsea on the northeast flank to 6450 feet (1965 m) subsea on the southwest flank. This gives an oil column of between 3165 and 3415 feet (964 and 1040 m).

Reservoir data show different static gas-column pressures for each of three structurally controlled gas caps along the length of the field, which indicates zones of variable reservoir characteristics. The Asmari Formation has a thickness similar to that of Gachsaran, 1600 feet (487 m), and a range in matrix porosity from 4 to 11 percent. In 1966, the field had a daily production capacity of 320,000 barrels from 15 of the 23 drilled wells. At that time, the cumulative production was in excess of 40 million barrels. Maximum allowable production figures show a remarkable similarity in distribution to those for Gachsaran oil field, and a central pressure sink separates the very highly productive northwest area from the less productive southeast area.

Fracture Data and Production Variations

By plotting circles of areas proportional to maximum allowable production rates of the Gachsaran and Bibi Hakimeh fields, similar broad patterns of production trends are clear (Fig. 32-8). Examination of surface linear densities along the fields reveals an obvious relationship between fracture density and production potential. Moreover, areas of comparable productivities on the two fields lie along the N20°E major fracture trend. However, a simple relationship between maximum allowable production and fracture density cannot be established.

A study of matrix porosity from core plugs and thin sections of the Asmari Formation at Bibi Hakimeh indicates significant variations. Though these variations are not great enough to account for the large productivity variations, a plot of the combining matrix porosities and surface linear densities permits the maximum allowable production rates of the wells to fall into more meaningful classes (Fig. 32-9). It is clear that although surface linear density provides some indication of well production potential, the variations in reservoir matrix porosity also influence production. No detailed porosity data are available from Gachsaran Field, but a similar situation of production from both fracture and matrix porosity is expected there.

The alignment of trends exhibiting high surface linear fracture densities with zones of high production potential along a N20°E trend is most significant. It suggests that over such zones the small-scale fractures, which have been present in the rock since early in its diagenetic history, have been subjected to renewed movements. These movements have been propagated upward from deep-seated adjustments along basement features. These disturbances would have subjected the reservoir rocks to differential movements between limestone blocks bounded by bedding and small-scale fracture planes. The result would have been enhancement of fracture porosity and permeability over preferred zones of the oil fields.

During the course of the investigation, it was learned that an independent hydrological study to the south in Saudi Arabia defined the same three trends of surface linears. Even more significant was the fact that zones of high water production were aligned with the same N20°E trend associated with high oil production zones over the Gachsaran and Bibi Hakimeh oil fields (A. Pistolesi, personal communication, 1970).

Conclusions

Detailed studies of the Asmari Formation in surface anticlines adjacent to their buried oil-productive counterparts in the foothills of the Zagros Mountains, Iran, have shown that production trends cannot be simply related to fracture patterns. For a given lithology and bed thickness, the densities of small-scale fractures are constant regardless of structural position. Instead, production trends in an area that includes the Gachsaran and Bibi Hakimeh oil fields are coincident with lineaments observed on air photographs. These features are related to basement structure.

Reactivation movements along large-scale basement features have, by propagation to higher levels in the sedimentary cover, resulted in the development of preferred zones of enhanced small-scale fracture porosity and permeability.

References

DUNNINGTON, H.V., 1958, Oil in northern Iraq: *in* Weeks, L.G., ed., Habitat of Oil: Amer. Assoc. Petroleum Geologists, Tulsa, OK, p. 1194–1251.

FALCON, N.L., 1967, The geology of the northeast margin of the Arabian basement shield: Brit. Assoc. Advancement Sci., v. 24, p. 31–42.

FALCON, N.L., 1969, Problems of the relationship between surface structure and deep displacements illustrated by the Zagros Range: *in* Time and Place in Orogeny: Geol. Soc. London Spec. Publ. 3, p. 9–22.

GILL, W.D., and M.A. ALA, 1972, Sedimentology of Gachsaran Formation (Lower Fars Series), southwest Iran: Amer. Assoc. Petroleum Geologists Bull., v. 56, no. 10, p. 1965–1974.

HAYNES, S.J., and H. MCQUILLAN, 1974, Evolution of the Zagros suture zone, southern Iran: Geol. Soc. Amer. Bull., v. 85, p. 739–744.

HODGSON, R.A., 1965, Genetic and geometric relationships between structures in basement and overlying sedimentary rocks with examples from the Colorado plateau and Wyoming: Amer. Assoc. Petroleum Geologists Bull., v. 49, no. 7, p. 935–965.

JAMES, G.A., and J.G. WYND, 1965, Stratigraphic nomenclature of Iranian oil consortium agreement area: Amer. Assoc. Petroleum Geologists Bull., v. 49, no. 12, p. 2182–2245.

LEES, G.W., 1948, Some structural and stratigraphical aspects of the oil fields in the Middle East: Internat. Geol. Congress, London, 1948, pt. VI.

MARCHESINI, E., A. PISTOLESI, and M. BOLOGNINI, 1963, Fracture patterns of the natural steam area of Larderello, Italy, from air photographs: in Symposium on Photo Interpretation, Delft, 1962, Trans. Internat. Archives Photogrammetry, v. 14, p. 524–532.

MCQUILLAN, H., 1973a, Small-scale fracture density in Asmari Formation of southwest Iran and its relation to bed thickness and structural setting: Amer. Assoc. Petroleum Geologists Bull., v. 57, no. 12, p. 2367–2385.

MCQUILLAN, H., 1973b, A geological note on the Qir earthquake, SW Iran, April, 1972: Geol. Mag., v. 110, no. 3, p. 243–248.

MCQUILLAN, H., 1973c, Factors influencing the development of collapse structures with examples from Fars province: Iranian Petroleum Inst. Bull., no. 51, p. 1–8.

MCQUILLAN, H., 1974, Fracture patterns on Kuh-e Asmari anticline southwest Iran: Amer. Assoc. Petroleum Geologists Bull., v. 58, no. 2, p. 236–246.

O'BRIEN, C.A.E., 1953, Discussion of fractured reservoir subjects, Amer. Assoc. Petroleum Geologists Bull., v. 37, no. 2, p. 325.

STOCKLIN, J., 1968, Structural history and tectonics of Iran—a review: Amer. Assoc. Petroleum Geologists Bull., v. 52, no. 7, p. 1229–1258.

West Cat Canyon Field
Perry O. Roehl and R. M. Weinbrandt

RESERVOIR SUMMARY

Location & Geologic Setting	Santa Barbara Co., T9N, R33W, Santa Maria Valley, California, USA
Tectonics	At plate margin
Regional Paleosetting	Open ocean and closed marginal basin with oxygen minimum zone
Nature of Trap	Structural with local fracturing in impermeable shale
Reservoir Rocks	
Age	Miocene (Luisian, Mohnian)
Stratigraphic Units(s)	Monterey Fm (Dark Brown, V, and lower Buff and Brown, IV zones)
Lithology(s)	Dolomite, porcellanites, and opal CT cherts, diatomaceous organic shales and siltstones
Dep. Environment(s)	Marginal bathyal ocean basin, open to restricted
Productive Facies	Diagenetic facies of dolomites predominant, associated with fine clastics and volcanics
Entrapping Facies	(see Trap above)
Diagenesis/Porosity	Dolomitization, silicification
Petrophysics	
Pore Type(s)	Fractures (with unknown volumes of finer matrix pores)
Porosity	12% avg
Permeability	186 md avg
Fractures	(see Pore Types)
Source Rocks	
Age	Miocene (Monterey Fm)
Lithology(s)	Organic shales, phosphorite
Migration Time	Miocene
Reservoir Dimensions	
Depth	3000–4800 ft (915–1460 m)
Thickness	590 ft (175 m) avg gross pay
Areal Dimensions	NA
Productive Area	2430 acres (9.72 km²)
Fluid Data	
Saturations	$S_o = 49\%$ avg
API Gravity	12–22°
Gas-Oil Ratio	NA
Other	—
Production Data	
Oil and/or Gas in Place	563 million STBO and 128 billion SCFG (3.61 billion m³); 584 million equivalent BO total
Ultimate Recovery	88 million STBO; URE = 15.6%
Cumulative Production	NA

Remarks: 350 BOPD per well current (1983) production. Initially unsaturated, vertical gravity segregation. Gas cap followed initial production. Monterey production discovered 1938.

33
Geology and Production Characteristics of Fractured Reservoirs in the Miocene Monterey Formation, West Cat Canyon Oilfield, Santa Maria Valley, California

Perry O. Roehl and R. M. Weinbrandt

Introduction

West Cat Canyon Field is a faulted anticlinal structure that is oil-productive from fractured diagenetic dolomites of the Miocene Monterey Formation (Luisian-Mohnian stages). The field is one of several comprising the productive region of central coastal California known as Santa Maria Valley (Figs. 33-1, 33-2). This is a fault-bounded geomorphic and tectonic province within which Miocene dolomite is a prominent surface and subsurface rock type (Roehl, 1972). At least six oil fields produce from Monterey Formation dolomite and dolomite-cemented shales and sands. They are: Cat Canyon (group of fields), Lompoc, Jesus Maria, Santa Maria Valley, Casmalia, and Orcutt.

Reservoir rocks in the Monterey are principally fractured, thin to thick-bedded, siliceous dolomites interpreted as early diagenetic in origin which are interbedded with fine-grained, organic-rich, siliceous, dolomitic shales. Both deposition and dolomitization of the Monterey sediments in this tectonic province are inferred to have taken place on the bathyal slope of an anoxic basin, similar to basins in modern active oblique rifting zones such as the Guaymas Basin, Mexico.

West Cat Canyon Field was discovered in 1908 and is the principal producing field of the Cat Canyon field area. Initial production was from stratigraphically younger sands of the Pliocene Sisquoc Formation. Not until 1938 was the underlying Monterey Formation (Fig. 33-3) discovered productive by Standard Oil of California in Section 27, T9N, R33W following a directional redrill across a fault to a depth of 6656 feet (2029 m). The initial daily production was 716 barrels of 14.9° API gravity oil cutting 0.8 percent water (Huey, 1954), from fractured shales of the Monterey.

West Cat Canyon Field has been produced by solution-gas drive augmented by gravity drainage. Water influx has not been active. Initial oil gravity varies from 22° API at the crest of the structure to 12° API at the base. Other oil properties vary accordingly. High water production from down-dip wells reflects high initial water saturations. Water movement has been down dip in response to solution-gas drive, liquid expansion, rock compressibility, and gravity segregation. Transmissibility is high, with some wells having initial rates of several thousand barrels per day.

General Features of Field and Reservoir Rock

West Cat Canyon Field encompasses 3.8 square miles (9.7 km²) and is essentially a multiple-faulted, doubly crested anticline (Figs. 33-4, 33-5) elongated northwest to southeast over a distance of approximately 5 miles (8 km). Plunge is northwest. Faulting is extensive but imperfectly understood except that the eastern field limit is fault-

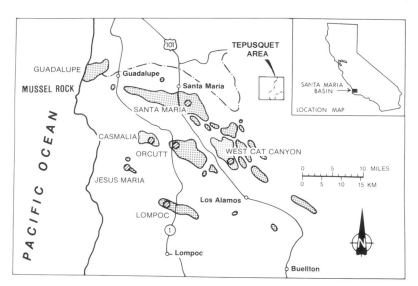

Fig. 33-1. Index map of Monterey Formation oil fields (stippled pattern) and sample localities (hatchured pattern), Santa Maria Valley area, California.*

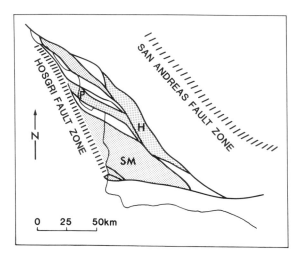

Fig. 33-2. Tectonic setting of the Santa Maria Basin. Stippled areas are depositional basins: SM = Santa Maria, P = Pismo, and H = Huasna. From Surdam and Stanley, 1981, Diagenesis and migration of hydrocarbons in the Monterey Formation, Pismo syncline, California: *in* Garrison and Douglas, eds., The Monterey Formation and Related Siliceous Rocks of California, a Special Publication of the Pacific Section of the Society of Economic Paleontologists and Mineralogists, and published by their permission. Figure also adapted from Hall, 1981, and published by permission of the American Geophysical Union.

*Figures 33-1, 33-6, 33-8, 33-9, 33-10A,B, 33-12A,B,C, 33-13A,B, 33-14, 33-15A,B,C, 33-16 are from Roehl, 1981. Published by permission, Pacific Section of Society of Economic Paleontologist and Mineralogists.

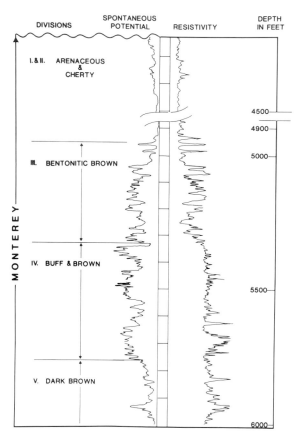

Fig. 33-3. Example of induction electrical log of Monterey Formation; Humble-Union Bell Fee (HUBF) Well #156 West Cat Canyon Field, California. See Figure 33-6 for typical stratigraphic distribution of dolomite.

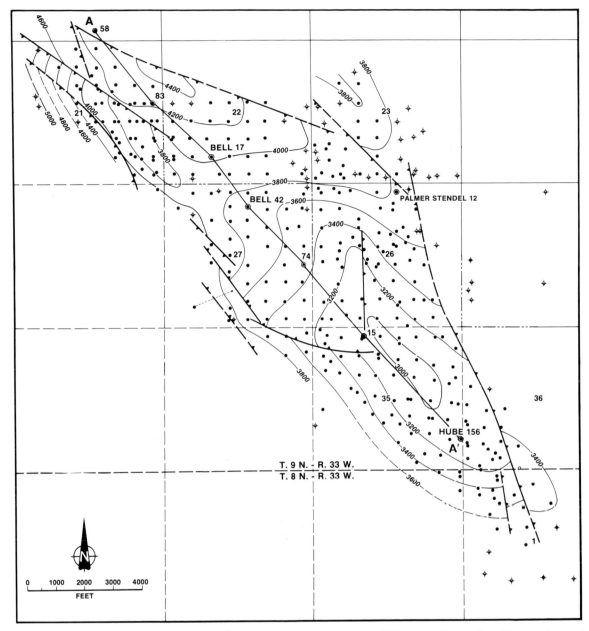

Fig. 33-4. Map of West Cat Canyon Field, structure contours on top of Monterey Formation. Contour interval 200 feet. Modified from Huey (1954).

bounded. The top of the Monterey Formation lies generally above 6000 feet (1830 m) subsea. Thickness of the productive interval within the Monterey Formation ranges from 400 to over 800 feet (120–240+ m). Initial oil in place varies from 350 to 500 stock-tank barrels per acre-foot.

The subsurface distribution and character of dolomite at West Cat Canyon Field are well represented by drill cuttings and a natural potential log obtained from Union Bell Oil Well #42 (Fig. 33-6). The interval 4800–6000 feet (1463–1829 m) was studied petrographically and by X-ray diffraction analysis. Rocks of the lowest three Miocene stages are delineated by the natural potential (SP) log curve, which appears to reflect,

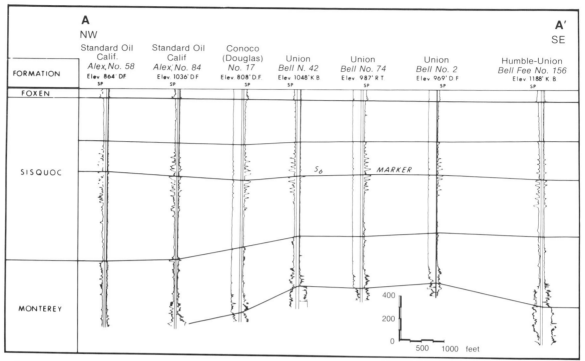

Fig. 33-5. Northwest–southeast cross-section through West Cat Canyon Field showing correlation of Sisquoc and selected Monterey stages and stratigraphic units.

in part, the vertical variations in percent of stratigraphic dolomite. Although traces of dolomite occur all through the sampled interval, it is quite clear that matrix stoichiometric dolomite is prevalent chiefly below a depth of about 5290 feet (1612 m), roughly the middle of the lower Mohnian Bentonitic Brown unit, and is coincident with the general occurrence of fracture-filling dolomites.

The matrix dolomites are of particular interest because they (1) constitute the dominant oil-productive rock type, (2) are of post-depositional origin, (3) have important rock-mechanical traits distinct from those of shale, and (4) have stable-carbon isotopic compositions that are distinctive and are considered to be uniquely characteristic of the stratigraphic and tectonic setting associated with a major plate boundary. The fracture-filling dolomites are of interest not only from a reservoir standpoint, but also because they comprise a part of a paragenetic sequence directly related to a rock mechanical origin, and appear to have been responsible for reinitiating embrittled conditions for additional subsequent fracturing episodes.

Miocene Stratigraphy and Sedimentation

The Monterey Formation within the Santa Maria Basin immediately overlies an Oligo-Miocene sequence of shallow-marine and/or continental clastic rocks. Overall, the Monterey is generally a thick, highly stratified, very thin-bedded succession of fine-grained clastics that are dominantly organic and phosphatic shale containing an abundance of calcareous nannofossils, diatom frustules, coccoliths, and foraminifers. The intermediate and basal strata are commonly interbedded with dolomitized carbonates (Roehl, 1979, 1981). A greater proportion of siliceous sediment occurs in the upper part of the formation. This sediment is variously porcelaneous and/or cherty, depending on the state of diagenetic conversion of diatomaceous sediments from opal-A to opal-CT and quartz (Isaacs, 1981).

The microfauna, lithologies, sedimentary structures, and rhythmic bedding suggest that these

Fig. 33-6. Diagram showing stratigraphy of selected petrological and X-ray data compared to natural potential log curve, Bell Union Well #42, West Cat Canyon Field. A = abundant, C = common, R = rare.

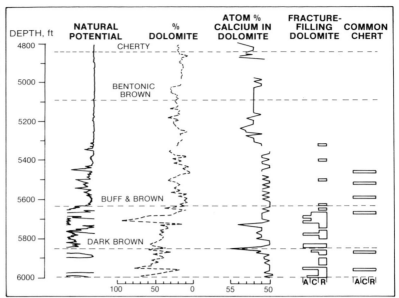

sediments are of deep-marine origin and were emplaced following rapid subsidence in association with the appearance of silled basins within the continental margin.

Five major subdivisions, largely correlative with faunal stages, are adopted from general oil field usage (Table 33-1). They are based on a combination of rock types, stratification, and faunal content. Thickness varies greatly, primary lithic types are repetitious, and fauna may be absent or rendered unrecognizable by diagenesis; therefore, subdivisions are not always identifiable in the subsurface and correlate with difficulty to surface outcrops. The dolomites apparently have several habits, occurring as discrete continuous beds; giant nodules, concretions, and lenses; partial to pervasive cements in sandstones and siltstones; vein or dike-like fillings in fractures; and pore fillings in tuff and other volcanic deposits. Some examples are discussed later in the section on *Fractured Dolomites.*

Early (Saucesian) to Middle (Relizian, Luisian) Miocene sedimentation in the Santa Maria Basin began with deposits of deep-water, hemipelagic origin composed of calcareous nannofossil, foraminifer-rich muds, and silts (Table 33-1). The hemipelagic deposits are overlain by calcareous shales and thin carbonates, together with phosphate-rich shale and foraminiferal or diatomaceous mudstones comprising the lower

Mohnian part of the Monterey Formation. These are highly parallel-laminated, organic-rich sediments. They probably represent anoxic deposition below an oxygen-minimum zone (Pisciotto and Garrison, 1981). This is an important feature bearing on the preservation of probable source beds for Monterey oil. The upper Mohnian records a fundamental change with the onset of microfaunas and floras that were transported by upwelling waters, probably as a result of expanding Antarctic ice caps (Ingle, 1981). The result is a gradual change upward in the Monterey from phosphatic rock to diatom-rich rock of the so-called "siliceous facies."

The Monterey Formation dolomite is early diagenetic in origin, and its development and extent depended on the post-depositional burial history of many primary rock types. The depositional setting was established by plate-margin relationships involving shear orientations of contesting blocks and segments, whose positions appear to have resulted from attempts to resolve collision subduction and by-passing movements about 22–25 million years B.P. (Oligo-Miocene age). A general subsidence of the central California continental margin ensued about this time, setting up a number of extensional, interrelated silled basins. One "set" of these includes the Santa Maria, Pismo, and Huasna basins (Fig. 33-2) (Hall 1981; Surdam and

Table 33-1. Oil Field Stratigraphy of Mio-Pliocene Age, Santa Maria Valley Area, California[1]

Stage and (Stratigraphic Unit)	Formation and Oil Field Subdivisions	Lithology
Pliocene Upper Delmontian (Sisquoc)	Sisquoc	Sand, gray brown, hard
Lower Delmontian (Monterey)	Montery Arenaceous (1)	Siltstones, medium brown, hard siliceous
Miocene Upper Mohnian	Cherty (II)	Shale, light buff gray, diatomaceous and hard siliceous banded, platy. Rare streaks of opaline chert and limestone
Lower Mohnian	Bentonitic Brown (III)	Shale, gray to dark-brown, brittle, thin-bedded, with thin betonitic streaks, light-gray porcelaneous shale and white, gray, and brown chert
	Buff and Brown (IV)	Shale, dark-brown, very hard siliceous, interbedded with light- and dark-brown opaline chert and brown "limestone"
Upper Luisian	Dark Brown (V)	Shale, dark-brown, hard, calcareous, siliceous; banded with claystone and argillaceous "limestone", phosphatic streaks and nodules
Lower Luisian and Relizian	Point Sal	Siltstone and shale, gray-stone, laminated; and sand, dark-brown, lenticular. Phosphatic. Pyroclastic. Sills and dikes
Saucesian and Upper Zemorrian	Rincon	Siltstone and clay shale, poorly bedded, with yellowish brown dolomitic lenses

[1]Based in part on Canfield (1939), Kleinpell (1938), Dibblee (1966) and Hall (1967).
[2]These descriptions are quite generalized and subject to abrupt vertical and lateral variations.
[3]Dolomite occurs variably throughout the upper Luisian-lower Delmontian interval, with highest concentrations in the lower divisions.

Stanley 1981). Hall refers to the area of these basins as the San Luis Obispo Transform.

Analogs in Modern Sedimentology

Confirming criteria for the typical sequence, environment of deposition, and diagenesis inferred for petroleum reservoir strata in the Monterey are provided by the results of the Deep Sea Drilling Project (DSDP), Leg 64, (Curray and Moore, 1982; Einsele and Kelts, 1982). Of particular interest is Well Site 479 located on the deep-marine slope of the east flank of the Gulf of California adjacent to Guaymas, Mexico (Fig. 33-7A). This is just east of and marginal to the zone of translation and oblique rifting (Kelts and McKenzie, 1982). The well at Site 479 was drilled in 2450 feet (747 m) of open marine water and penetrated 1437 feet (438 m) of a Plio-Pleistocene sequence of soft, muddy, hemipelagic diatomaceous ooze

with numerous thin dolomite beds which increase in abundance and become harder and thicker with increased depth of burial (Fig. 33-7B). In general, this sequence is analogous in origin to the lower Monterey Formation in West Cat Canyon Field discussed below. The sequence at DSDP Site 479 is not, however, completely analogous to the Monterey Formation in either total thickness, duration, depth of burial, or geothermal history. For example, the modern site represents about one million years, for which Kelts and McKenzie note substantial enrichment of early diagenetic dolomites in ^{13}C. In contrast, subsurface dolomites of the Monterey are generally enriched in ^{12}C. Kelts and McKenzie (1982) did observe a trend toward isotopically lighter carbon with increased depth below 100 meters (330 ft) as diagenetic dolomite presumably continues to form. We propose that, with increasing depth of burial (time and temperature) in a closed system, this trend will continue, resulting in greater quan-

Fig. 33-7A. Location map of DSDP drill sites 477-480, Leg 64, Gulf of California. Bathymetry in meters. Adapted from Kelts and McKenzie (1982). *B.* Simplified lithologic results from drilling, and a line drawing of a seismic section through Sites 479 and 480, DSDP Leg 64, Gulf of California. TF = transform fault location. Modified from Curray and Moore (1982).

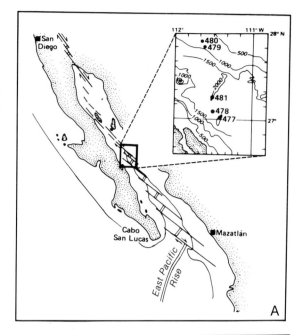

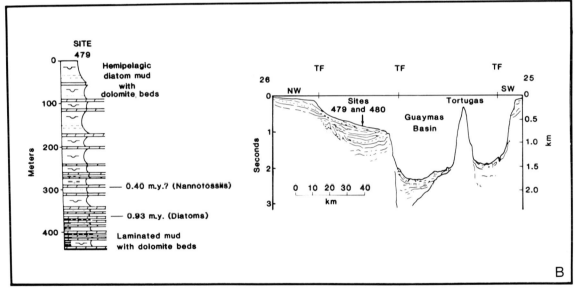

tities of dolomite that show progressive depletion in ^{13}C. Substantial negative shifts of $\delta^{13}C$ in some Monterey dolomites probably result from incorporation of light carbon acquired from newly generated petroleum during progressively deeper diagenesis (Fig. 33-8).

The principal conclusions of note from study of Plio-Pleistocene sediments at Site 479 relative to the Monterey Formation are the following:

1. Dolomite beds develop early by diagenetic replacement and are hard, resistant beds.

2. The carbon-isotopic compositions of the dolomite reflect both burial and diagenetic history. Initial $\delta^{13}C$ values will be retained by early dolomite through episodes of deeper, longer burial; dolomites of somewhat later (deeper) origin will acquire initial values enriched in ^{12}C.

3. Frequency and distribution of stratified dolomite depend in part on the specific bathymetric location of burial and thus the lithologic type and burial history of sediment on the depositional slope.

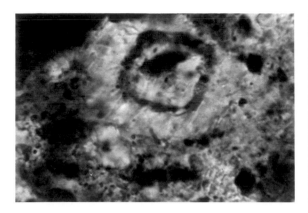

Fig. 33-8. Photomicrograph of dolomite euhedra with hydrocarbon nuclei and rhombic zoning. Euhedra average 25 microns in size. HUBF Well #156, depth 5485 feet (1672 m). Plane light.

4. Dolomites occur first as replacements of foraminifers, later as replacements of both calcareous and siliceous nannofossils, and finally as precipitated, accretionary infilling dolomite.

Miocene Analog: Mussel Rock Surface Exposure

The middle Miocene is extensively exposed along the shoreline at Mussel Rock, west of Santa Maria (Fig. 33-1). Monterey strata there are directly analogous to subsurface counterparts productive of petroleum. The similarity includes the stratigraphic succession, lithology, interstratal deformation, fracturing, and brecciation. All of these factors have relevance to reservoir origin and petroleum migration.

The lithology is generally consistent with the descriptions given for divisions IV and V of Table 33-1. However, both divisions have large amounts of unreported dolomite, the preponderance occurring in the "Dark Brown" division V (Fig. 33-9). The entire sequence displays multiple fracturing of all the competent layers. Both fractures and bedding planes exude viscous oil.

The Dark Brown division is largely undeformed except for faulting; however, some horizons high in the section show zones of flow folding and bed swelling, which appear to have occurred during or

Fig. 33-9. Thick-bedded, undeformed, lower Middle Miocene dolomite forming resistant cliff face at Mussel Rock, Santa Barbara County, California.

shortly after initial deposition. The dolomite has replaced both limestone and siliceous beds, many of which show small-scale turbidite graded bedding. Although the overall aspect is one of structural competence, large-scale fractures occur in the unit and exude very large quantities of viscous petroleum.

The overlying Buff-and-Brown division is very thinly bedded and composed of siliceous shales, organic-rich shales, claystones, and cherts, along with many dolomitized intervals. This sequence displays a gross geometry of large-scale, soft-sediment deformation, which perhaps developed on the initial depositional slope. At closer range, intrastratal flexure and flow-folding are evident (Figs. 33-10A, B). These small folds are generally intact and show little or no internal fracturing.

A

B

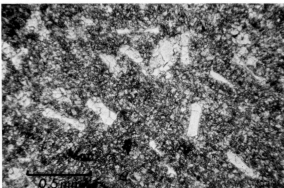

Fig. 33-11. Photomicrograph of diatom wackestone, Buff-and-Brown division, Well HUBF #156. Both diatoms and matrix replaced by dolomite. Plane light.

Fig. 33-10A. Photograph of dolomitized thin beds at Mussel Rock equivalent to the Buff-and-Brown division (IV), showing crenulated flow-folding of pre-dolomitization origin. *B*. Photograph of intrastratal flex folding of Buff-and-Brown division equivalent, thin organic shale and dolomitic interbeds. Dolomite (light color) has preferentially replaced original mineralogy along the fold axis and is post-deformational. Location: Mussel Rock.

Certainly no concentration of fractures occurs at these sites compared to adjacent and overlying sequences that display intense conjugate and nonconjugate shattering believed to have resulted from rock-mechanical stresses applied after lithification.

Brittle fracturing and brecciation are confined to specific silicified and dolomitized beds and do not occur in the more abundant soft, ductile beds. To this extent, the sequence is similar to that at DSDP Site 479 off Guaymas, Mexico. The processes of silicification and dolomitization are therefore necessary precursors, both to fracturing and brecciation, and to initial expulsion of petroleum from interbeds of great organic richness.

Fractured Monterey Dolomites

Subsurface Reservoir Rocks

Until recent years, representative cores of productive fractured intervals in the West Cat Canyon Field have been extremely rare. The first of a new generation of core suites of the subsurface Monterey Formation was recently obtained from Humble-Union Bell Fee (HUBF) Well #156 between depths 4952 and 5783 feet (1509–1763 m). Except for dolomite, the mineralogy in that well is generally the same as reported earlier. However, many of the dolomitized intervals show very little vestige of earlier lithology, and diatoms are replaced and infilled with dolomite (Fig. 33-11). X-ray diffraction analyses indicate only traces of illite, kaolinite, hydroxyapatite, and plagioclase together with large variations in amounts of silica-mineral species.

Of all recovered core, 88 percent is dolomite, about 65 percent of which displays evidence of fractures incompletely healed by dolomite cement (Roehl, 1981). The fracture-filling dolomites show evidence of multiple generations (Fig. 33-12A) and a predominant habit of incomplete cementation, which means that reservoir void space is almost always retained (Fig. 33-12B). Highly significant is the fact that petroleum commonly occurs within these cavities as a product of a pre-dolomite or intra-dolomite interlude during the course of paragenesis. Figure 33-12C shows a

A

B

C

Fig. 33-12. Photographs of three core pieces from Well HUBF #156 that display representative fracture development of Monterey "shale." Bar scales are 1 inch (2.54 cm). *A*. Multiple fracture sets healed with dolomite. Residual fracture porosity at left. Depth 5697–5699 feet (1738 m). *B*. Fracture porosity lined with dolomite euhedra. Depth 5773–5775 feet (1760 m). *C*. Partially dolomitized fracture with oil-stained zone (arrow), evincing petroleum involvement in paragenesis. Depth 5694 feet (1735 m).

fracture filled with secondary dolomite except for a 1-inch (2.54-cm) segment formerly occupied by petroleum as attested by live-oil staining and an absence of void-filling dolomite. This suggests that the petroleum was in place prior to, or more likely entrained with, the dolomitizing waters.

As mentioned in the discussion on sedimentology, the Plio-Pleistocene sequence at DSDP Site 479 shows a progressive increase in frequency of dolomite beds and amounts of matrix dolomite with depth. Figure 33-6 reveals a similar trend of increasing matrix dolomite with depth in the lower two stages of the Monterey Formation in

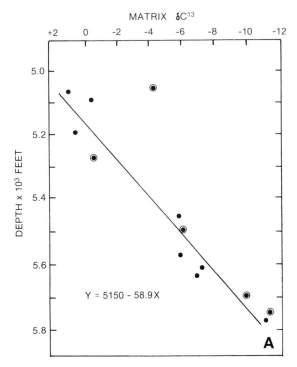

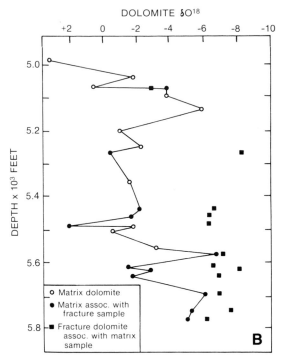

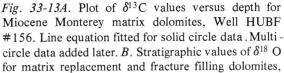

Fig. 33-13A. Plot of $\delta^{13}C$ values versus depth for Miocene Monterey matrix dolomites, Well HUBF #156. Line equation fitted for solid circle data. Multi-circle data added later. *B.* Stratigraphic values of $\delta^{18}O$ for matrix replacement and fracture filling dolomites,

Well HUBF #156. General uniformity of values for fracture-filling dolomites below 5200 feet suggests a common source water and/or temperature and time of precipitation.

Isotopic Composition

In addition to data reported for Union Bell Well #42, carbon and oxygen stable isotope compositions were obtained for 35 samples of dolomite from core material of the HUBF Well #156 (Fig. 33-13A,B). Eleven of these were from non-fracture associated material. The remaining 24 were paired data where samples were taken of matrix and closely adjacent fracture-filling dolomites.

The data indicate that the distributions of $\delta^{13}C$ and $\delta^{18}O$ in per mil values are crudely similar, each displaying general depletion in light isotopes

Union Bell Well #42. The data in this figure also suggest that the occurrence of most intense and frequent fracturing and fracture-filling dolomite depends on the presence of matrix dolomite in amounts exceeding about 25 percent. Thus, apparently not all dolomites, either by type or amount, will necessarily fracture.

with increasing depth. This same trend was observed at DSDP Site 479, where it was attributed to " . . . a closed system characterized by low sulfate concentration, high alkalinity, and high ammonia contents commonly in conjunction with zones of methanogenesis" (Kelts and McKenzie, 1982).

From a reservoir viewpoint, the $\delta^{18}O$ values of fracture-filling dolomites are of greater interest because (1) they demonstrate a striking uniformity, averaging about -7 ± 1 per mil, and (2) except for the stratigraphically highest sample, they are all much lighter than their matrix counterparts. The data imply a common origin for the fracture-filling dolomites below 5200 feet (1585 m) and no apparent relationship to their matrix counterparts. They are, however, dependent on the embrittlement and rupture of the earlier matrix dolomite, and the stratigraphic interval in the HUBF Well #156 of over 550 feet (168 m) is inferred to have been in fracture communication throughout.

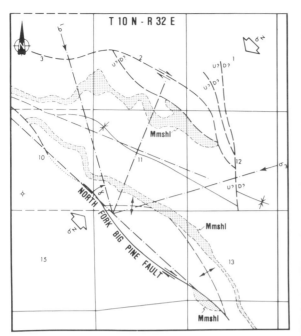

Fig. 33-14. Geologic map of the lower Monterey "shale" (Mmshl) exposed in the Tepusquet Canyon Area. A hypothetical stress field has been superimposed. It is assumed that the Big Pine Fault is a principal right-lateral shear and that it lies at an angle of approximately 30 to the north–northwest-trending principal stress axis. Base map and geology courtesy of L. E. Redwine.

Miocene Analog: Tepusquet-Colson Canyon Outcrop

It is necessary now to review another surface outcrop analog. The location is the area of Tepusquet and Colson Canyons in the San Rafael Mountains east of Santa Maria (Fig. 33-1 and 33-14), where a dolomitized sequence of the Monterey Formation contains breccias and fractures quite analogous to the reservoir rock at West Cat Canyon Field (Roehl, 1979, 1981; Redwine, 1981). Many of these surface exposures retain fracture porosity and commonly are filled with viscous to solid petroleum residues (Fig. 33-15, compare to Fig. 33-12A). Detailed accounts of Tepusquet area geology have been provided by Redwine (1958).

The outcrop area consists of a syncline and an anticline, the latter truncated roughly along strike by a major right-lateral fault (Fig. 33-14). Fault-truncated units of the Monterey have been converted to dolomite by replacement and pore-filling processes, resulting in good preservation of the

Fig. 33-15. Photographs of brecciation development in Monterey Formation outcrops in Tepusquet Canyon area, San Rafael Mountains. *A.* First stage dilation brecciation and cementation. Echelon orientation and quasi-sigmoidal form of vugs shown by arrows. Clasts show typical matching boundaries of early intrastratal rupturing. Clast ratio (clast volume: void volume) $\gg$ 1. *B.* Advanced dilation and probable hydraulic fracture brecciation, with angular and rounded clasts floating in a suspension of granular and crystalline cements. Dark areas are macrovugs. Clast ratio approaching 1. *C.* Well-advanced dilation brecciation showing near absence of original breccia clasts. Rock has pervasive development of granular and reticulate dolomite crystal interlayering with large vugs. Clast ratio $\ll$ 1.

fine sedimentary laminae. Along the faulted anticlinal limb, the normal stratal configuration deteriorates abruptly in several differing modes. Initially, ruptured clasts show closely fitted, concordant boundaries cemented by one or more layers of white precipitated dolomite crystals (Fig. 33-16A). More advanced forms are composed of unsorted oligomictic fragments that appear to be suspended in the precipitated dolomite on the scale of several inches to a foot or more (Fig. 33-16B). Fragments range from angular to subrounded, and the mass resembles a slurry. Finally, there are unique exposures of crystalline, secondary pore-filling dolomite in rock masses that defy description. These masses show relicts of earlier brecciation stages but are composed primarily of secondary layers of precipitated dolomite. Many of these layers occur in pre-existing voids and fail to fill such voids completely (Fig. 33-16C).

Development of Fracturing and Brecciation

Dilatancy

Before mechanical failure of a rock under triaxial compression, there is a brief interval of time during which actual physical expansion occurs. The resulting increase in unit volume has been referred to as *dilatancy*. Dilatancy is generally attributed to the formation of a myriad of sub-micron-sized microcracks and other matrix defects not to be confused with larger interparticle or cementational porosity. This subject is considered at greater length by Secor (1965), Brace *et al.* (1966), Bieniawski (1967), Bruhn (1972), and others (see Roehl, 1981).

During dilatancy, pore-fluid pressure drops relative to increasing rock stress until such time as additional fluid can expand into and fill the new voids. Under increasing compression, the rock stress buildup during dilatancy is gradually offset by the returning pore fluids at confining pressures higher than those at the start of dilatancy. The redistribution of pore fluids and the net increase in unit rock volume due to increasing porosity lead to an overall reduction in the strength of the rock body. At constant differential stress, reapplication of the pore-fluid pressure could result in one or a combination of regional shear failure (fault movement), large-scale deformation (folding), and local or small-scale shearing and/or exten-

sion separation. This last may be manifested as expansion fabric composed of detached ("floating") but closely fitted clasts dissociated from the original brittle matrix (Fig. 33-16A).

Each clast has roughly concordant boundaries with its neighbors. The separation and configuration will be dependent on (1) the orientation and abundance of original microcracks and any subsequent microfractures, (2) the physical properties of the original rock-matrix mineralogy comprising the bulk clasts, and (3) the interstitial water and/or hydrocarbons that comprise the fracturing fluid under excess pressures.

Advanced states of expansion packing under conditions of hydraulic fracturing, hydroplastic flow, and injection result in a non-rigid framework that requires maintenance of the directed fluid flow pressure, lest collapse of the unsupported clasts results. In other cases, rapid cementation prevents further adjustments.

Post-Expansion Dolomite

Any fracture-filling dolomite precipitated under the aforementioned process is volumetrically small compared to matrix dolomite and is usually attributed to unusual conditions of water alkalinity, temperature, or pressure. The water need not have a very high Mg:Ca concentration ratio. If mobilized during active fracturing, the water would be quite susceptible to changes in temperature and pressure (including P_{CO_2}) during its passage up extension fractures and faults, thus leading to supersaturation and emplacement of precipitated dolomite in all accessible ruptured beds. If this kind of dolomitization is real, it is both distinctive and extremely important to the problem at hand:

1. It is genetically dependent on dilatancy and subsequent hydraulic fracturing phenomena.
2. Its occurrence results in instantaneous preservation of early rupture geometry.
3. It "sets" the rock for future recurrence of dilation-induced failure and hydraulic fracturing.
4. Almost all dolomitization events are self-limiting. This is especially true of precipitated dolomite formation because the finite volume of contributing solute becomes progressively depleted to the point of undersaturation. The incomplete dolomitization leaves behind a diminished but ever-present residual, interfragmental fracture or breccia porosity.

The periodicity and economic importance of this paragenesis is evident where oil occurs in fractures of different ages and in macrovugs, which postdate all of the earlier paragenetic dolomitization events. Several cored intervals from the subsurface at West Cat Canyon Field display similar paragenetic features, and drill cuttings also show signs of the same diagenetic attributes (Roehl and Weinbrandt, 1983). One of the major observations is that these dolomite-cemented fractures are commonly porous and petroleum has filled many of them, some several times as a non-wetting, paragenetic participant.

Possible Mechanism of Petroleum Migration and Accumulation

The Monterey Formation is considered by most workers to be both a source and a reservoir for petroleum. A problem of long standing has been the apparent lack of physicochemical conditions to expel petroleum from fine-grained, shaly rocks such as the Monterey Formation. Conventional fluid-displacement pressures are considered to be insufficient to overcome the several hundred atmospheres of capillary pressure required to expel initial oil droplets. However, the compressive stresses of very large magnitude localized in zones of dilation and natural hydraulic fracturing are certainly capable of expelling oil from the rock matrix.

In dilatancy, an extremely large number of microcracks and imperfections are enlarged, thus providing potential storage space for petroleum derived from the adjacent matrix. At the same time, *in situ* water would sustain a pressure decrease of perhaps several kilobars. At this pressure differential, the pore water would increase in volume by about 20 percent, and much of this may be transformed temporarily into a vapor phase (Brace *et al.*, 1966). Should the microcrack pore space double under dilatancy, pore pressure would reduce to zero. The effect would be a marked increase of effective stress on those portions of the expanding microcrack interfaces in contact with each other prior to a re-entry of water from adjacent unstressed rock. The result would be an increase in lithostatic pressure on the matrix, and any indigenous oil should therefore be expressed from the walls of the microcracks dur-

Fig. 33-16. Photograph of polished rock slab from an oil-bearing, tightly folded sequence of fractured Monterey Formation cropping out in Colson Fork of Tepusquet Canyon. Specimen shows a dramatic paragenetic sequence beginning with dolomitization of original laminated Monterey "shale": 1. First-stage crystalline dolomite cement that binds larger clasts. 2. Solid bitumen filling fracture porosity. 3. Second-stage crystalline dolomite cement flowing directly on first-stage cement. 4. Third-stage dolomite cement lining a fracture that breaches both the original matrix and the first two stages of cementation. 5. Solid bitumen filling third-stage fracture porosity. This is a distinct, later occurrence than 2 above. Scale 1 inch (2.54 cm).

ing the transient period of water undersaturation. This would be especially true if, as is commonly the case with rocks such as the Monterey Shale, the matrix should be substantially less than 100 percent water wet. Upon water re-entry and elongation of the microcracks, the oil would be free to migrate with the water and, as a non-wetting phase, would behave like a mobile zone of mineralization and be caught up in each period of dilatant crack growth, hydroplasticity, and/or hydraulic injection, and would be retained as part of the paragenetic sequence that involves the formation of dolomite from the associated saturated waters.

A rock specimen from Colson Canyon, SW Baseline, Section 13 outcrop (Figs. 33-14, 33-15) shows two distinct events of petroleum migration that occurred between three episodes of dolomite cementation. At this stage of research, it cannot be presently shown whether this oil occurrence came about as a direct result of the individual dilation of microcracks just described. What is clear, however, is that during several of these brittle-failure episodes, oil was available for movement and possible relocation from fracture to fracture.

The constancy of the stable-oxygen isotopic populations in the fracture zones at West Cat Canyon (Fig. 33-13B) is evidence for the uniformity and singular origin of the waters that permeated the entire stratal rupture system during a single episode of great hydraulic fluid displacement. If true, the geochemical history is consistent with the proposed dilation origin for fractures and breccias present in outcrop exposures of the Monterey Formation at Tepusquet Canyon, and with the concept of mobilized formation water that provides a source for dolomite precipitated during a single event throughout several hundred feet of ruptured strata.

Assuming, then, that oil expulsion is possible under these conditions, an important possibility arises. Any petroleum contained in and around a zone of dilatancy and hydraulic rupture will be set in motion together with the interstitial water, and will accumulate in the newly created expansion pore space and open fracturing that is attendant to brittle failure. In effect, the source bed both creates and fills its own reservoir pore space!

Reservoir Engineering

The Monterey reservoir in the West Cat Canyon Field is difficult to describe from an engineering viewpoint because of its unusual geologic origin (Roehl and Weinbrandt, 1983). A reservoir's geometry can generally be described by structure and isopach maps. In the case of the Monterey Formation in West Cat Canyon, the productive interval did not obey stratigraphic constraints as is normally the case in a reservoir that produces

from non-fracture "matrix" porosity. The unique origin of the reservoir caused the fractures that control fluid conductivity to transgress lines of stratigraphy. Precise identification of the productive intervals is also a significant problem. The older logs are difficult to evaluate quantitatively, and quantitative core analysis is compromised by poor core recovery in the more fractured sections of the reservoir. Uncertainty in defining the productive interval necessitated well completions over large intervals of the formation using slotted liners, some as long as 1600 feet (488 m). Although completion practice has varied from operator to operator, completions in the north end of the field are about 500 feet (152 m) higher in the Monterey section than those in the south end of the field.

Structure

Due to the unusual origin of the Monterey reservoir, the top of the productive interval is difficult to define. Contours on the top of the Buff-and-Brown division provided a reasonable approximation to the top of the productive reservoir. The contours show that the field is a faulted anticlinal-stratigraphic trap with an oil column of approximately 1000 feet (~300 m). The areal limits of the structure have been established by drilling. Wells outside the productive limits of the field either would produce no fluid if they were completed in unfractured rock, or would produce water if they encountered fractures low on the structure. On the east side of the field, the West Cat Canyon fault forms a boundary. Some of the best wells in the field lie just to the west of this fault, indicating that the fault may have some bearing on the degree of fracturing.

The current low value of reservoir pressure suggests that water influx is not active. Water production in the down-dip wells appears to be adequately explained by the high initial water saturation coupled with high transmissibility and gravity drainage.

Rock Properties

Obtaining information on rock properties for the West Cat Canyon Field is difficult because core recovery is poor (approximately 50%) and a quantitative method of interpreting the oil well logs is lacking. On the one full-interval core obtained from the field (HUBF Well #156), the average porosity from core analysis was 8.6 percent. Based on performance data, porosity was adjusted upward, resulting in an average overall porosity of 12 percent, with a range of 2 to 24 percent. The upward adjustment in porosity from that obtained in core analysis indicates that some of the better parts of the reservoir were not recovered in the core. Average permeabilities from core analyses of less than 1 millidarcy had to be adjusted upward to match field performance. The core analysis permeability and porosity values are reasonable considering that only the less fractured rock was obtained in the coring process. Based on performance, the average permeability-thickness product for the field was 11 darcy-feet (590 feet average thickness × average permeability of 186 md). Permeability-thickness products as high as 300 darcy-feet were required to match performance at some wells. Drillers' logs make frequent mention of bit drops and lost-circulation problems when the initial development of the West Cat Canyon Field was in progress. Also, drilling mud was noted in producing wells several miles distant from development drilling.

Relative Permeability

Due to the high vertical and horizontal transmissibility via fractures, straight-line relative permeability curves were used, indicating that the field is at vertical equilibrium. Relative permeability curves representing only the matrix would be concave upward and invalid for describing inter-well flow dominated by fractures.

Initial Water Saturation

Water saturations were obtained from analyses of the HUBF Well #156 cores. Since the well was drilled with water-base mud, considerable mud filtrate invasion occurred. The average initial water saturation from this well was 66 percent. This value was reduced to 51 percent, based on performance data. The water saturation developed from performance data is an average of high values of initial water saturation in the tight matrix together with low values in the fractures.

Capillary Pressure

Capillary pressure will be high in the tight matrix, while capillary pressure will be essentially zero in the fractures. Since both macrovugs and fractures dominate flow, the assumption of zero capillary pressure was used.

Fluid Properties

Oil properties in the West Cat Canyon Field vary significantly with depth. Oil gravity decreases with increasing depth (Fig. 33-17A), and initial gas-oil ratio decreases with increasing depth (Fig. 33-17B). Bubble-point pressure also decreases with increasing depth following Standing's (1952) correlations for California crude oil. Oil at the top of the structure was initially saturated, while oil at the base of the structure was 500 psi (35×10^4 kg/m^2) undersaturated.

The average reservoir temperature is 195°F (90.5°C). Temperature increases with depth following the normal geothermal gradient for the area. The oil viscosity also increases with depth due to the decreases in oil gravity, but the effect is mitigated by temperature increase.

Production Performance

Several producing mechanisms have operated simultaneously in the West Cat Canyon Field. Early in the life of the field, liquid expansion and rock compressibility were dominant. As the pressure fell, solution-gas drive created a secondary gas cap. Gravity drainage moved mobile gas updip and mobile water downdip. This left a band of mid-structure wells producing relatively dry oil at solution gas-oil ratio. Most of the field's production has come from wells in a mid-structure position.

Conclusions

The West Cat Canyon oil field produces from fractured diagenetic dolomites of the Monterey Formation. Original Monterey sediments have depositional and stratigraphic similarities to modern deep-marine, anoxic deposits of the Gulf of California. Applying this analog to diagenetic alterations suggests that Monterey dolomites

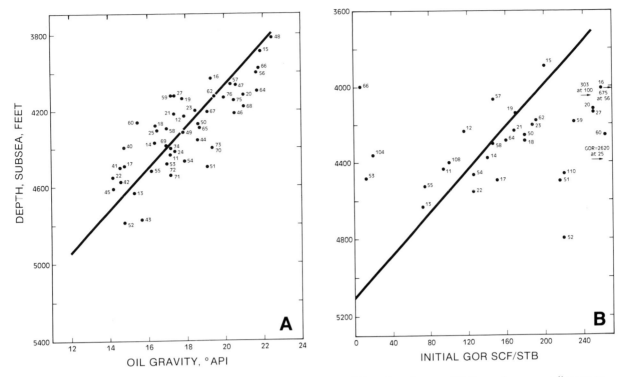

Fig. 33-17A. Initial API gravity versus depth for Bell Lease, West Cat Canyon. Well numbers are adjacent to data points. *B.* Initial gas-oil ratio versus depth for Bell Lease, West Cat Canyon. Well numbers are adjacent to data points.

formed initially within the zone of methanogenesis and continued with burial whereby ^{12}C became more prevalent via thermocatalytic reactions. Nucleation of Monterey matrix dolomites commonly involved liquid or gaseous hydrocarbons.

Fracturing of Monterey dolomite appears to have been initiated by dilatancy under tectonic compression together with elevated pore pressures. The post-dilation mechanism of rupture and brecciation appears to have been natural hydraulic or petroleum fluid fracturing. The latter requires that adequate depth of burial and petroleum maturation temperature be attained.

Based on reservoir performance, the following is concluded:

1. The productive interval transgresses stratigraphy.
2. Actual flow rates far exceed theoretical for rock matrix properties.
3. Porosity exceeds those values measured on the rock matrix of cores.

4. Quantitative evaluation of older well logs is difficult.
5. Early in the field's history, liquid expansion and rock compressibility were the dominant sources of reservoir energy.
6. Solution-gas drive, formation of a secondary gas cap, and gravity segregation came into play as the pressure declined.
7. Water influx is not evident. Downdip water production can be explained by gravity segregation.

Possible clues to reservoirs with fracture-dominated properties of similar origin may include the following:

1. Mature oil-generating rocks that comprise part or all of a prospective reservoir zone.
2. Original fine-grained sediment that has been diagenetically embrittled by dolomitization and retains zero-to-low matrix porosity and zero-to-low permeability.
3. Abnormal fluid pressures or history of same.

4. Basement terrane variations and/or a tectonic setting subject to episodic structural adjustment that enhances anisotropic responses between zones of embrittlement and less competent zones of primary sedimentary strata.

5. Little or no formation water recovery on drill stem testing or initial production.

Acknowledgments The authors are grateful for the support and encouragement of many professional associates at Union Oil Company of California. Special thanks are extended to J. P. Chauvel, R. A. Fellows, M. Metz, J. W. Randall, R. C. Ransom, and W. W. Wornardt; E. Goldish and D. W. Brink provided analytical support. Exxon Company, U.S.A., graciously provided core material from the HUBF Well #156.

References

BIENIAWSKI, Z.T., 1967, Mechanism of brittle fracture of rock; Rept. MEG, 580, Council for Scient. and Indust. Research, S. Africa, 226 p.

BRACE, W.F., B.W. PAULDING, JR., and C.H. SCHOLZ, 1966, Dilatancy in the fracture of crystalline rocks: Jour. Geophys. Research, v. 71, p. 3939–3953.

BRUHN, R.W., 1972, A study of the effects of pore pressure on the strength and deformability of Brea Sandstone in triaxial compression: Tech. Rept. MRDL 1-72, Engr. Study No. 552, MRD Lab. No. 64/493, Dept. Army Corps of Engrs., Div. Lab., Omaha.

CANFIELD, C.R., 1939, Surface stratigraphy of Santa Maria Valley Oil Field and adjacent parts of Santa Maria Valley, Calif.: Amer. Assoc. Petroleum Geologists Bull., v. 23, no. 1, p. 45–81.

CURRAY, J.R., D.G. MOORE, *et al.*, 1982, Initial Reports, Deep Sea Drilling Project, Leg 64: U.S. Govt. Printing Office, Washington, D.C., p. 417–445.

DIBBLEE, T.W., JR., 1966, Geology of the central Santa Ynez Mountains, Santa Barbara County, Calif.: Calif. Div. Mines and Geol., Bull. 186.

EINSELE, G., and K. KELTS, 1982, Pliocene and Quaternary mud turbidites in the Gulf of California: Sedimentology, mass physical properties and significance: *in* Curray, J.R., D.G. Moore, *et al.*, eds., Initial Reports, Deep Sea Drilling Project, Leg 64, Pt. 2: U.S. Govt. Printing Office, Washington, D.C., p. 511–528.

HALL, A.H., JR., 1967, Stratigraphy and structure of Mesozoic and Cenozoic rocks, Nipomo Quadrangle, Southern Coast Ranges, California; Geol. Soc. America Bull., v. 78, p. 559–582.

HALL, C.A., 1981, San Luis Obispo Transform fault and middle Miocene rotation of the western Transverse Ranges, California: Jour. Geophys. Res., v. 86, no. B2, p. 1015–1031.

HUEY, W.F., 1954, West Cat Canyon area of Cat Canyon oil field: Calif. Div. of Oil and Gas Summary of Oper. Calif. Oil Fields, v. 40, no. 1, p. 14–21.

INGLE, J.C., JR., 1981, Origin of Neogene diatomites around the north Pacific rim: *in* Garrison, R.E., R.G. Douglas, *et al.*, The Monterey Formation and Related Siliceous Rocks of California: Pac. Sect., Soc. Econ. Paleontologists and Mineralogists, Spec. Publ., p. 159–179.

ISAACS, C.M., 1981, Porosity reduction during diagenesis of the Monterey Formation: *in* Garrison, R.E., R.G. Douglas, *et al.*, The Monterey Formation and Related Siliceous Rocks of California: Pac. Sect., Soc. Econ. Paleontologists and Mineralogists, Spec. Publ., p. 257–271.

KELTS, K., and J.A. McKENZIE, 1982, Diagenetic dolomite formation in Quaternary anoxic diatomaceous muds of Deep Sea Drilling Project, Leg 64, Gulf of California: *in* Curray, J.R., D.G. Moore, *et al.*, eds., Initial Reports, Deep Sea Drilling Project Leg 64: U.S. Govt. Printing Office, Washington, D.C., p. 553–569.

KLEINPELL, R.M., 1938, Miocene stratigraphy of California: Amer. Assoc. Petroleum Geologists Bull., 450.

PISCIOTTO, K.A., and R.E. GARRISON, 1981, Lithofacies and depositional environments of the Monterey Formation, California: *in* Garrison, R.E., R.G. Douglas *et al.*, eds., The Monterey Formation and Related Siliceous Rocks of California: Pac. Sect., Soc. Econ. Paleontologists and Mineralogists, Spec. Publ., p. 97–122.

REDWINE, L.E., 1958, Geology and oil exploration of the Tepusquet area, Santa Barbara Co., Calif. (abst.): Pac. Petroleum Geol. (Amer. Assoc. Petroleum Geologists, Newsletter), v. 12, no. 11, p. 1–2.

REDWINE, L.E., 1981, Hypothesis combining dilation, natural hydraulic fracturing, and dolomitization to explain petroleum reservoirs in Monterey shale, Santa Maria area, California: *in* Garrison, R.E., R.G. Douglas, *et al.*, eds., The Monterey Formation and Related Siliceous Rocks of California: Pac. Sect., Soc. Econ. Paleontologists and Mineralogists, Spec. Publ., p. 221–248.

ROEHL, P.O., 1972, The isotopic origin and rock mechanical properties of matrix dolomite in the California Miocene may be a key to petroleum exploitation in heterogeneous mobile belts: Unpubl. Report, Union Oil Co. of Calif., E & PE 72-22M.

ROEHL, P.O., 1979, Dilation brecciation—a proposed mechanism of fracturing, petroleum expulsion and dolomitization in the Monterey Formation, California (abst.); Amer. Assoc. Petroleum Geologists Bull., v. 63, p. 1856.

ROEHL, P.O., 1981, Dilation brecciation—a proposed mechanism of fracturing, petroleum expulsion and dolomitization in the Monterey Formation, California: in Garrison, R.E., R.G. Douglas et al., eds., The Monterey Formation and Related Siliceous Rocks of California: Pac. Sect., Soc. Econ. Paleontologists and Mineralogists, Spec. Publ., p. 285–315.

ROEHL, P.O., and R.M. WEINBRANDT, 1983, Geology and production characteristics of the Monterey Formation fractured reservoir, West Cat Canyon oil field, Santa Maria Valley, California (abst.): in Isaacs, C.M., R.E. Garrison, et al., eds., Petroleum Generation and Occurrence in the Miocene Monterey Formation, Calif.: Pac. Sect. Soc. Econ. Paleontologists and Mineralogists, Spec. Publ., p. 226.

SECOR, D.T., JR., 1965, Role of fluid pressure in jointing, Amer. Jour. Sci., v. 263, p. 633–646.

STANDING, M.B., 1952, Volumetric and Phase Behavior of Oil Field Hydrocarbon Systems: Reinhold Publ. Corp.

SURDAM, R.C., and K.O. STANLEY, 1981, Diagenesis and migration of hydrocarbons in the Monterey Formation, Pismo syncline, California: in Garrison, R.E., R.G. Douglas et al., eds., The Monterey Formation and Related Siliceous Rocks of California: Pac. Sect., Soc. Econ. Paleontologists and Mineralogists, Spec. Publ., p. 317–327.

34
Nido A & B Fields
Mark W. Longman

RESERVOIR SUMMARY

Location & Geologic Setting	South China Sea 30 mi (50 km) NW of Palawan, The Philippines
Tectonics	Stable subsiding plate near spreading center
Regional Paleosetting	Reefs developed on a subsiding Oligo-Miocene carbonate platform
Nature of Trap	Stratigraphic; "pinnacle" reef and reef talus enclosed in pelagic shales
Reservoir Rocks	
Age	Early Miocene
Stratigraphic Units(s)	St. Paul Limestone
Lithology(s)	Limestone
Dep. Environment(s)	Reef and associated talus
Productive Facies	Skeletal packstone and grainstone, lithoclast packstone
Entrapping Facies	Calcareous shale
Diagenesis Porosity	Extensive marine cementation, minor dissolution and freshwater phreatic cementation, extensive fracturing
Petrophysics	
Pore Type(s)	Fracture, moldic and vuggy
Porosity	1–9%, avg 3% (matrix)
Permeability	<0.01–3.3 md, avg 1 md (matrix)
Fractures	Extensive, vertical, open; major contribution to productivity
Source Rocks	
Age	NA
Lithology(s)	NA
Migration Time	NA
Reservoir Dimensions	
Depth	6800 ft (2070 m)
Thickness	660 ft (200 m)
Areal Dimensions	0.9 × 0.6 mi (1.5 × 1 km)
Productive Area	NA
Fluid Data	
Saturations	S_w − 27% avg
API Gravity	27°
Gas-Oil Ratio	NA
Other	—
Production Data	
Oil and/or Gas in Place	NA
Ultimate Recovery	NA
Cumulative Production	·8.5 million BO through 1979

Remarks: IP 800 BOPD per well, IP mechanism water drive. Discovered 1977 and 1978. Nido A and B buildup produced 40,000 BOPD from 5 wells in 1979.

34

Fracture Porosity in Reef Talus of a Miocene Pinnacle-Reef Reservoir, Nido B Field, The Philippines

Mark W. Longman

Location and Regional Setting

The first commercial oil production in the Philippines was from the Nido Field, located in the South China Sea about 30 miles (50 km) northwest of the island of Palawan in the northwest part of the Palawan Basin (Fig. 34-1). As of mid-1980, all production in the Nido Field was from two Lower Miocene pinnacle reef complexes at a depth of approximately 6800 feet (2070 m). Some other reefs in the area have been drilled and found to be non-commercial.

Information on the regional setting has been presented by Saldivar-Sali (1979) and Hatley (1980). Underlying the reefs is an extensive Oligocene limestone that was deposited on a broad carbonate platform. During regional transgression in the Late Oligocene, local variations in the rate of carbonate deposition resulted in the formation of carbonate buildups. Terrigenous shale containing planktonic foraminifers was deposited in the deeper water between the buildups. Up to about 980 feet (300 m) of vertical reef growth then ensued, after which the reefs were drowned by a late Early Miocene transgression and covered with terrigenous shales similar to those previously deposited between the reefs. From the Middle Miocene through the Pleistocene, over 6600 feet (2000 m) of shales, limestones, sandstones, and conglomerates accumulated in the Northwest Palawan Basin over the Lower Miocene reefs of the Nido oil field. An idealized stratigraphic column for the Nido B Field is shown in Figure 34-2.

Discovery

Exploration leading to the discovery of the Nido Field began, at least in terms of involvement by Cities Service Company, in 1975 with a geophysical and geological evaluation of the Palawan Basin. Review of seismic sections revealed that several "reef-like" structures were present in the area, and in early 1975, the Nido-1 well was spudded to test one of the carbonate buildups. The well was drilled to 9026 feet (2751 m) and then plugged back to 6040 feet (1841 m). During testing, the well flowed at a rate of 1440 barrels of oil per day from the top of a Lower Miocene carbonate buildup, but this eventually proved to be non-commercial because there were only 17 feet (5 m) of net pay (Hatley, 1978).

After a second structure was drilled updip from the Nido-1 well and found to be dry, the decision was made to move downdip from the original Nido well in the hope that the buildups there would be filled with oil to the spillpoint. In 1977 the Nido A-1 well (originally called the South Nido-1) was drilled and encountered the top of a Lower Miocene carbonate buildup at 6796 feet (2071 m), some 837 feet (255 m) lower than encountered by the Nido-1 well. During testing, the

Fig. 34-1. Location map of the Nido Field off the coast of Palawan in the Philippines.

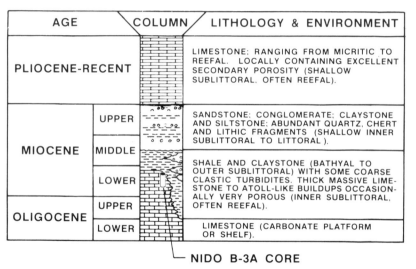

Fig. 34-2. Idealized stratigraphic section for the Nido B Field. Depth to the top of the Lower Miocene carbonate buildup is about 6800 feet (2070 m). Adapted from Hatley (1978).

Nido A-1 well flowed at a maximum rate of 7340 barrels of oil per day.

Additional drilling was needed to determine if the Nido A-1 well was commercial, and the decision was made to conduct a major 3-D seismic survey in the Nido area to gain a better understanding of the distribution of carbonate buildups. Based on both new and original seismic lines (one of which is shown in Fig. 34-3), a well was drilled into a carbonate buildup located 2-1/2 miles (4 km) southwest of the Nido A buildup. This well, the Nido B-1 (formerly called South Nido West #1) encountered an oil-filled carbonate buildup and tested at production rates of about 9800 barrels of oil per day.

To date, three development wells have been drilled. A fourth well, the Nido B-3, was drilled to test a possible lobe of the reef seen on seismic record sections but missed the reefal buildup. The B-3 well was then sidetracked and redesignated the B-3A. It penetrated the marginal part of the buildup.

Reservoir Trap Characteristics

The reservoir of the Nido B oil field is a pinnacle reef complex about 0.9 mile (1.4 km) in diameter at a depth of about 6800 feet (2070 m) (Fig. 34-4). The buildup is slightly elliptical in a north-

Fig. 34-3A. Seismic section across the Nido carbonate buildups. *B.* Interpretation of seismic section combined with well data. Adapted from Hatley (1980).

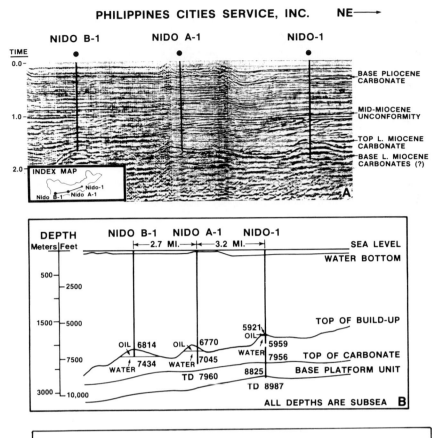

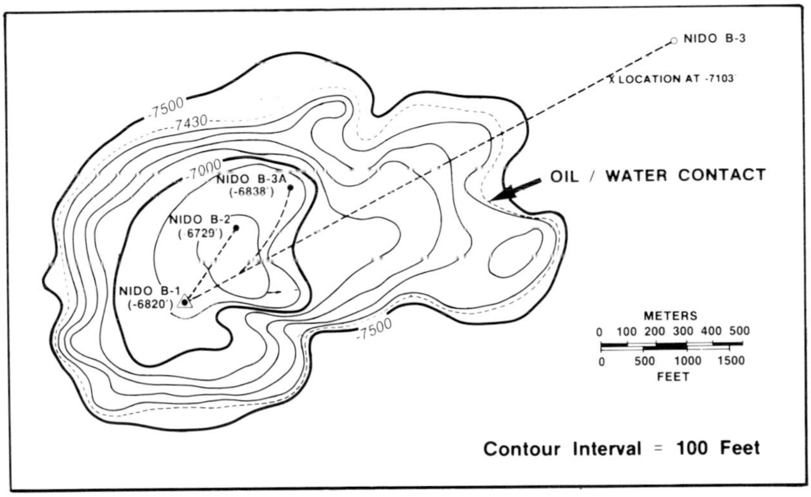

Fig. 34-4. Map showing depth to the top of the Lower Miocene carbonate buildup in the Nido B Field and the locations of wells. Based on seismic sections from a 3-D program and modified to fit well data. Prepared by B. Ayme.

east–southwest direction, about 900 feet (275 m) thick, and is surrounded and covered by a calcareous shale that is rich in illite, mixed layer illite-smectite, kaolinite, and quartz. Planktonic foraminifers are common in the shale and indicate that the shale was deposited in an open-marine environment (Fig. 34-5). Major faulting and

regional folding are not apparent on seismic sections, but marked differences in depths to the top of the pinnacle reefs in the region suggests that faulting has occurred, that the reefs were drowned at significantly different times, or both. Each of the three producing wells in the Nido B Field encountered a different facies of the reef complex,

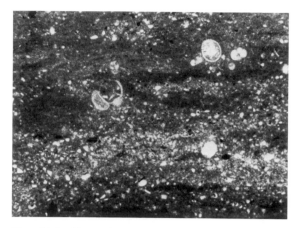

Fig. 34-5. Photomicrograph of shale that forms the seal over the Nido B carbonate buildup. Planktonic foraminifers and pyrite are visible in the silty calcareous shale. Plane light.

and each has different porosities and production characteristics, although all tested at rates in excess of 7000 barrels of oil per day.

Production

The Nido complex was put on production in early 1979 within a year of the discovery of the Nido B reservoir. Information on total reserves of the Nido oil field is not available, but the field was put on production at 40,000 barrels of oil per day from five wells, two in the Nido A buildup and three in the Nido B buildup. By the end of 1979, the field had already produced 8.5 million barrels of oil (Wood *et al.*, 1980). Details of production have been discussed by Harry (1979) and Hatley (1980). Nido crude oil is typically black-brown in color with an average API gravity of 27°. The reservoirs are undersaturated, with solution-gas ratios of only 7–10 standard cubic feet per stock-tank barrel (Hatley, 1980). The oil contains about 1500 parts per million of H_2S.

Geology of the Reservoir

The Nido B oil field produces from limestones of a Lower Miocene reef complex with marked lateral facies variations. Although each of the three wells in the buildup encountered a different facies, only the rocks of the B-3A core are described here. This core is particularly interesting for three reasons: (1) it is an excellent example of proximal forereef talus, (2) there has been extensive cementation and minor leaching of the limestones, and (3) extensive fracturing has occurred, which provides the permeability that allowed this well to produce over 10,000 barrels of oil per day.

The 30 feet (9 m) of core recovered from the Nido B-3A well shows a gradational change upward from a lithoclastic packstone containing clasts of marine-cemented reef limestone in a shaly matrix (Fig. 34-6A) into a massive, highly fractured packstone/grainstone dominated by fragments of red algae, corals, and intraclasts (Longman, 1980a). Although these two facies are gradational through the cored interval, only the end-member facies are described below for the sake of brevity.

The lithoclastic packstones consist of pieces of hard limestone within a matrix of greenish-gray calcareous shale rich in planktonic foraminifers (Fig. 34-6B). The shale consists dominantly of illite, mixed-layer illite-smectite, and kaolinite clays and forms between 5 and 30 percent of the rock. Some pressure solution occurred between the clasts where they are in contact. The clasts in the lithoclastic packstone consist of fragments of corals, red algae, and foraminifers (Figs. 34-6C, D). Fragments of large coral heads up to 4 inches (10 cm) in diameter are present, but most of the skeletal debris in the clasts is less than 0.4 inch (1 cm) in diameter. Extensive cementation of all grains in the clasts occurred prior to their becoming clasts because edges of the clasts cut across the cements (Fig. 34-7A). Most of this cement is fibrous and isopachous, which suggests a marine origin (Fig. 34-7B). The presence of geopetal micrite, which has preserved the square-ended shapes of some of the fibers, indicates that they were originally aragonite (Assereto and Folk, 1976; Longman, 1980b) and supports the interpretation of a marine origin for the fibrous cements.

Sorting of the clasts is poor, with size range up to about 6 inches (15 cm) in diameter. No graded bedding is apparent. Most of the clasts are moderately to well rounded, which is surprising considering the poor sorting, abundance of mud, and short transport distance. Based on present relief and the assumption that the reef crest was near sea level, the clasts could not have moved

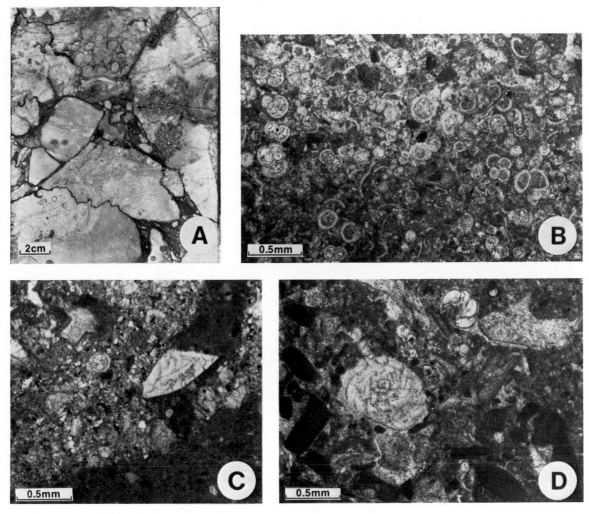

Fig. 34-6A. Photograph of lithoclasts of reef debris in calcareous shale from the lower part of the Nido B-3A core. Fragments of coral, coralline algal packstones, and other reef-derived debris are visible. Stylolites are well developed between lithoclasts, and some fractures are visible within clasts. Sample depth 7483 feet (2281 m). Plane light. *B.* Photomicrograph of shale matrix between clasts showing abundant planktonic foraminifers. Sample depth 7468 feet (2276 m). Plane light. *C.* Photomicrograph of clast margin containing micrite, large foraminifers, and coralline algae and adjacent shaly matrix. Sample depth 7485 feet (2281 m). Plane light. *D.* Photomicrograph of typical grains found within the clasts and in the upper part of the core. Coralline algae, benthonic foraminifers, and an echinoderm fragment are visible here. Sample depth 7488 feet (2282 m).

downslope more than the 165 feet or so (~50 m) vertical distance considered the probable water depth that existed during deposition. There are no borings or encrusting organisms on the surfaces of any of the clasts, perhaps indicating that the clasts moved into an anoxic environment soon after breaking away from the reef.

The limestone at the top of the core appears more massive because it contains less argillaceous material than limestone in the lower part of the core. However, it contains similar clasts packed together with fragments of corals, red algae, and mollusks. Cements, most of which formed in a marine phreatic environment, produced the apparent massive texture. Extensive networks of hairline fractures have formed subsequently. Geopetal micrite occurs within some intraparticle cavities in the debris.

On the basis of the depositional textures and fossils present, it appears that the rocks of the

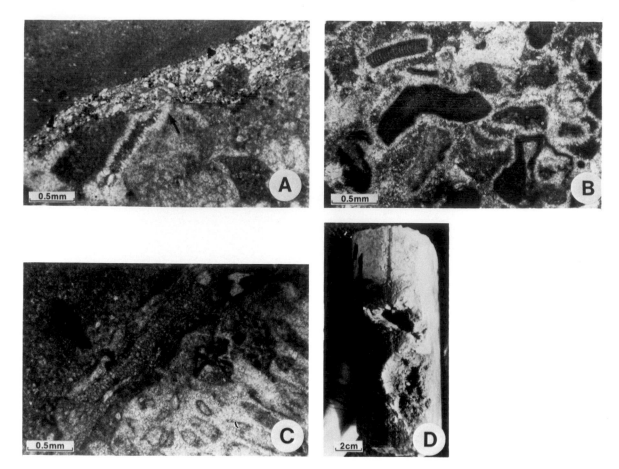

Fig. 34-7A. Photomicrograph showing the border of lithoclast with truncated cement (arrow) suggesting that lithification of the clast preceded transport. Sample depth 7485 feet (2281 m). Plane light. *B.* Photomicrograph of isopachous fibrous cement coating grains (mostly coralline algae) and covered with micrite. Shape of the fibers suggests they were originally micrite. Sample depth 7459 feet (2274 m). Plane light. *C.* Photomicrograph of a portion of a clast (lower right), composed of crystallized coral, that is encrusted with coralline algae. Sample depth 7475 feet (2278 m). Texture of the calcite in coral suggests recrystallization occurred in a fresh-water phreatic environment. Plane light. *D.* Photograph of large vugs in a core from depth 7483 feet (2281 m). Such large vugs are uncommon.

Nido B-3A core were deposited as proximal forereef talus. The abundance of planktonic foraminifers between the clasts, the poor sorting and moderate abrasion of most grains, the absence of *in situ* shallow-water organisms, the abundance of fragments of reef-dwelling organisms, and the abundance of marine cement all support this interpretation; but perhaps the strongest evidence is the position of this well between the "reefal" wells near the center of the buildup and the "basinal" shale penetrated by the Nido B-3 well (Fig. 34-4). The decrease in amount of argillaceous matrix and planktonic foraminifers toward the top of the core probably indicates a progradational sequence that formed as the reef built out over its own talus. If deposition had not been interrupted by subaerial exposure and/or the influx of clastics, the reef itself might have prograded out across the site of the B-3A well.

Diagenesis

Diagenesis within the core can be divided into several major events separated in time by long periods during which little or no apparent record of diagenesis was formed (Fig. 34-8). The first

Fig. 34-8. Paragenetic sequence of events observed in the Nido B-3A core. Equivalent changes in mineralogy and porosity are proposed in the lower diagram.

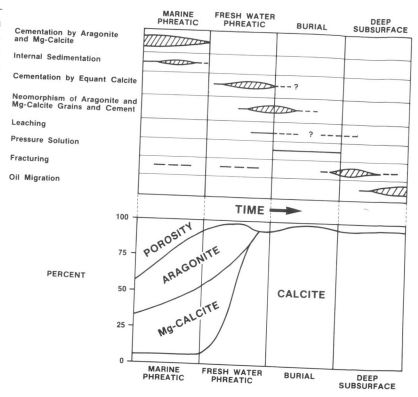

event was the extensive cementation by fibrous aragonite and micrite-sized Mg-calcite in the marine-phreatic environment. Estimated original porosity in the rocks was between 30 and 50 percent, but marine cementation reduced this porosity to less than 10 percent. Marine cements formed only within the relatively pure limestones; where terrigenous shales surround lithoclasts, no marine cementation occurred.

Following marine cementation, non-ferroan equant calcite filled the remaining primary pore spaces in the limestones. Textures of these cements, such as coarsening of crystals toward pore centers, interlocking mosaics of small crystals, and clarity suggest that cementation occurred in a freshwater phreatic environment. It is probable that the aragonite and Mg-calcite of the skeletal grains and marine cements were recrystallized to calcite at this time in the presence of the fresh water, resulting in the preservation of many original textures (Fig. 34-7C).

Still later, minor solution and the formation of secondary vuggy porosity occurred. Most of the vugs are only a millimeter or two in diameter, but some are up to several centimeters across (Fig. 34-7D). The boundaries of the vugs cut across grains and cements, indicating that they formed

after the rock recrystallized to calcite; otherwise, solution would have been at least partly controlled by original mineralogy. Solution was almost certainly caused by fresh water undersaturated with respect to calcium carbonate, but whether it occurred in the vadose zone or phreatic zone could not be determined. No evidence of vadose cements was observed in this core.

Some pressure solution of calcite occurred in the rocks after significant burial, particularly where clasts surrounded by terrigenous clays were in contact (Fig. 34-6A). Solution probably occurred when water was squeezed out of adjacent or downdip shales by compaction. There is no evidence of cementation post-dating the stylolites, and the observation that no vuggy porosity exists within the clasts near the major sites of pressure solution suggests that dissolution at depth by most, if not all, undersaturated fluids produced stylolites rather than vuggy porosity.

Porosity and Permeability

Average porosity of 17 pieces of whole core from the B-3A well is 3.9 percent, and all but four of the corresponding permeability values are less

Table 34-1. Porosity and permeability data based on whole core analyses of selected samples from the Nido B-3A well.

Measured Depth (ft)	Permeability (md)	Porosity (%)
7459	1.2	2.3
7460	0.4	1.7
7462	0.1	3.3
7463	0.2	1.4
7465	1.9	5.9
7466	0.4	4.0
7469	<0.01	9.1
7471	0.9	5.0
7473	3.3	8.3
7475	0.7	3.1
7476	0.9	2.8
7478	0.1	3.8
7480	<0.01	1.9
7481	<0.01	1.4
7483	1.1	1.9
7485	0.1	4.8
7487	<0.01	4.0

than 1 millidarcy (Table 34-1). These low values are surprising because the well was able to sustain production at over 10,000 barrels of oil per day for almost a year. However, the anomaly is easily explained by the fractured nature of the reservoir and the fact that pieces of core selected for whole core analysis were those with few fractures.

Fractures

Several stages of fracturing are present in the Nido B-3A core. The earliest fractures formed contemporaneously with deposition and are filled with sediment and marine cements (Fig. 34-9A). A second generation of fractures is filled with equant calcite probably formed during freshwater diagenesis. Because both of these early fracture sets are now filled, neither contributes to the permeability of the reservoir.

A later-formed set of fractures is characterized by: (1) complete absence of cementation (Fig. 34-9B), (2) extensive oil-staining, (3) a dominant vertical orientation with a complementary set of random and horizontal fractures that divide the core into irregular blocks ranging from a few millimeters to a few centimeters in diameter (Figs. 34-9C, D, and 34-10), (4) rare offsets, (5) sides

that tend to match perfectly, and (6) cross-cutting relationships indicating that the fractures post-date all cementation, early fractures, and stylolitization (Fig. 34-9D). These characteristics, particularly the absence of any cementation or leaching, suggest that the fractures formed after fairly deep burial, possibly nearly contemporaneously with oil migration.

Analysis of fractured pieces of whole core with few vugs yields average porosity values of 2 to 3 percent. Fracture abundance shows an inverse correlation with the amount of argillaceous material in the rock, apparently because the argillaceous material helped diffuse stress prior to fracturing.

Since this last stage of fracturing controls virtually all the production from the Nido B-3A well, determining the origin of the fractures is important to understanding the reservoir. Clues to the origin of the fractures include (1) their abundance near the margins of the pinnacle reef complex, (2) their late formation after significant burial, (3) their orientation, (4) the paucity of offsets, and (5) the closely spaced anastomosing network they form.

A tectonic origin of the fractures is considered unlikely because of the absence of any shear displacement. More likely, the fractures formed by one of the three following processes: (1) natural hydraulic fracturing related to fluid overpressuring within the shales surrounding the reef complex or the reef complex itself, (2) flexure of tongues of the reef talus due to compaction of interbedded and underlying shales, or (3) regional arching of the whole reef complex. Each of these processes is discussed briefly below.

Natural Hydraulic Fracturing

Abnormally high, natural pore fluid pressures in rocks are common (Gretener, 1976) and may form during rapid sedimentation and sediment loading[1], aquathermal pressure[2], smectite-illite transformation[3], simple osmosis[4], or several other processes. Although rocks in the Nido B oil field are not now overpressured, rapid deposition since

---

[1]Hubbert and Rubey, 1959
[2]Barker, 1972; Magara, 1975a
[3]Powers, 1967; Burst, 1969; although see Magara, 1975b, for some qualifications on this mechanism
[4]Hanshaw and Zen, 1965

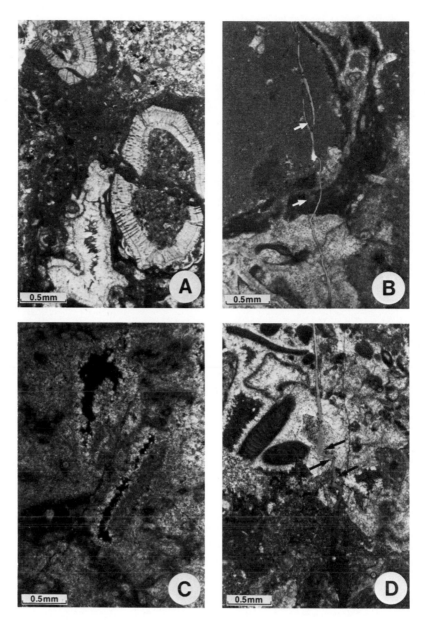

Fig. 34-9A. Photomicrograph of early fractures filled with lithified sediments. These contribute little to the reservoir permeability. Plane light. *B.* Photomicrograph of anastomosing fracture surrounding "islands" of the host rock (arrows). Sides of the fracture match up well and show little evidence of leaching and no cementation. Plane light. *C.* Photomicrograph of vugs (black) connected by hairline fractures. Such small fractures bled oil when the core was taken, indicating that they contribute to reservoir quality. Cross-polarized light. *D.* Photomicrograph of clean vertical fracture cutting a stylolite, suggesting that major fracturing may have postdated stylolitization. Sharp corners where fracture bends (arrows) indicate there has been no solution along the fracture. Plane light.

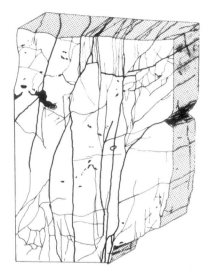

Fig. 34-10. Photograph and sketch of fractures in a slab of core from 7459 feet (2273 m) near the top of the carbonate buildup. The dominant vertical orientation and parallel nature of the fractures are clearly visible. Most of the remaining core is not so highly fractured.

initial burial of the reef complex and the presence of mixed-layer smectite-illite clay in both the seal over the reef and the inter-reef shales suggest that overpressuring could have occurred. If sufficient, such overpressuring could have produced a set of natural hydrofractures across the relatively tightly cemented proximal reef talus that separated the porous carbonates of the reef complex from the shales surrounding the reef. Thus, fracture orientation may provide a way to determine the validity of this process.

Several lines of evidence support the idea that the fractures originated as the result of a natural hydrofracture. These include the parallel nature and close vertical spacing of the fractures (Fig. 34-10) as well as the observation that the fractures formed nearly contemporaneously with oil migration and have few offsets. Furthermore, the vertical orientation of the fractures is consistent with that inferred to form from artificial hydrofracturing of oil wells where the maximum stress is due to overburden pressure (Hubbert and Willis, 1972; Gretener, 1976). Unfortunately, the orientation of the fractures relative to the reef trend cannot be determined directly from the unoriented core. However, production tests suggest that excellent permeability exists between the B-2 and B-3A wells, which are aligned perpendicular to the reef margin, and support the idea that the fractures trend across the reef.

Meissner (1980) proposed an interesting variation on the hydrofracture model that involves overpressured hydrocarbons formed during thermal maturation of kerogen in the shales. He states that this process has been active in several basins in North America; however, while such an "oil-frac" would fit the evidence for penecontemporaneous fracturing and oil migration in the B-3A well, geochemical analyses of oils and Miocene shales in the Nido area suggest that the shales were not the source of the oil.

Yet another variation of natural hydrofracturing involves localized fluid overpressuring in shales interbedded with tongues of reef talus. During compaction, water in these shales may have migrated laterally and updip until the shale pinched out into the reef limestone. Concentration of fluid pressure in this zone could have created relatively narrow bands of highly fractured rocks as the overpressured fluids escaped into the normally pressured porous limestone in the central part of the reef. The fact that most fractures in the B-3A core occur in the relatively pure limestones above the most shale-rich interval, coupled with the apparent absence of regional overpressuring, lends credence to this idea.

Reef-Margin Compaction Model

Study of the cored interval in the B-3A well and logs below the cored interval indicates that considerable shale is present within the proximal talus in certain intervals. Compaction of these shales during burial of the reef complex could have resulted in fracturing of the cleaner interbedded limestones as they bent downward around the reef. Most fractures formed by such compaction should trend parallel to the reef crest and be oriented vertically or nearly so. Such fracturing could be assisted by overpressured fluids by the process discussed earlier. Spacing of the fractures formed in this way would be affected by the composition of the limestone, especially grain size and micrite content (Hugman and Friedman, 1979), amount of compaction, geometry of the talus beds relative to the reef, etc.

Other than the presence of shale in the talus and the clear evidence of compaction in the form of pressure solution along stylolites (Fig. 34-6A), there is little direct evidence to indicate that the fractures formed during compaction. It seems probable that compaction would occur over a long time and that fractures forming early would show more evidence of cementation or leaching than those formed later. No obvious differences are observed in the fractures, which contribute to the B-3A reservoir. However, this is negative evidence and the compaction model for the fractures cannot be ruled out.

Regional Arching

Local uplift of rocks over structures such as salt domes can create fracture sets with many vertically oriented fractures (Cloos, 1968). Although it is unlikely that salt underlies the Miocene reef complex off Palawan, some other type of regional uplift might have contributed to the vertical fractures in the B-3A core.

Evidence tends to oppose this explanation, however. Not only do the fractures show a biased distribution (being most common at the margins of the reef complex), but also they are exceptionally closely spaced for fractures induced by regional tectonics and lack the shear displacement common along such fractures. Furthermore, there is no evidence on seismic lines for any significant uplift of the Nido pinnacle reefs.

Conclusions

Both natural hydrofracturing and compaction of talus tongues at the reef margin could have created the set of closely spaced vertical fractures observed in the Nido B-3A well. Additional data are required to evaluate either process at Nido B

Field. However, knowledge of the orientation of the fractures relative to the trend of the reef could aid interpretation. Naturally formed hydrofractures should be oriented perpendicular to the reef trend, whereas those formed by compaction loading should parallel the trend. Production tests indicate that excellent communication exists between the B-2 and B-3A wells, and that fractures trend perpendicular to the reef, thus supporting the concept of formation by natural hydrofracturing.

In summary, analysis of the Nido B-3A core of the Nido B oil field offshore Palawan, Philippines, shows that: (1) forereef talus is susceptible to extensive marine cementation, which can destroy its reservoir potential unless secondary porosity forms; (2) small fractures can form a significant reservoir and conduit capable of producing in excess of 10,000 barrels of oil per day; and (3) when associated with interbedded shales, forereef talus is susceptible to extensive fracturing by either overpressured fluids or flexure due to compaction. The former process seems to have been more important in the Nido B-3A core.

Acknowledgments   My thanks to V. Schmidt, R. Ginsburg, and T. Connally, who suggested the various hypotheses to explain the fracturing in the Nido B-3A core and to the many other people who contributed ideas on the core during the SEPM core workshop held in Denver (1980). I also thank L. Baie, D. Brownlee, C. Feazel, G. Hicks, J. Jamison, W. Rizer, and R. Slatt for their reviews of various drafts of the manuscript. Finally, I wish to acknowledge the support of the management of Cities Service Company, the Philippines Bureau of Energy Development, and Cities' exploration partners in the Philippines who obtained the cores from the Nido B Field and permitted publication of this paper.

References

ASSERETO, R., and R.L. FOLK, 1976, Brick-like texture and radial rays in Triassic pisolites of Lombardy, Italy: a clue to distinguish ancient aragonitic pisolites: Sedimentary Geology, v. 16, p. 205–222.

BARKER, C., 1972, Aquathermal pressuring—role of temperature in development of abnormal-pressure zones: Amer. Assoc. Petroleum Geologists Bull., v. 56, no. 10, p. 2068–2072.

BURST, J.F., 1969, Diagenesis of Gulf Coast clayey sediments and its possible relation to petroleum migration: Amer. Assoc. Petroleum Geologists Bull., v. 53, no. 1, p. 73–93.

CLOOS, E., 1968, Experimental analysis of Gulf Coast fracture patterns: Amer. Assoc. Petroleum Geologists Bull., v. 52, no. 3, p. 420–444.

GRETENER, P.E., 1976, Pore pressure: Fundamentals, general ramifications and implications for structural geology: Continuing Education Course Note Series 4: Amer. Assoc. Petroleum Geologists, 87 p.

HANSHAW, B.B., and E-AN ZEN, 1965, Osmotic equilibrium and overthrust faulting: Geol. Soc. Amer. Bull., v. 76, no. 12, p. 1379–1386.

HARRY, R.Y., 1979, Cities develops offshore Philippines: Oil and Gas Jour., v. 77, no. 18, p. 180–195.

HATLEY, A.G., 1978, Palawan oil spurs Philippine action: Oil and Gas Jour., v. 76, no. 9, p. 112–118.

HATLEY, A.G., 1980, Offshore oil field put onstream in 18 months: World Oil, v. 171, no. 7, p. 43–47.

HUBBERT, M.K., and W.W. RUBEY, 1959, Role of fluid pressure in mechanics of overthrust faulting, I. Mechanics of fluid-filled porous solids and its application to overthrust faulting: Geol. Soc. Amer. Bull., v. 70, no. 2, p. 115–166.

HUBBERT, M.K., and D.G. WILLIS, 1972, Mechanics of hydraulic fracturing: in Underground Waste Management and Environmental Implications: Amer. Assoc. Petroleum Geologists Mem. 18, p. 239–257.

HUGMAN, R.H.H., III, and M. FRIEDMAN, 1979, Effects of texture and composition on mechanical behavior of experimentally deformed carbonate rocks: Amer. Assoc. Petroleum Geologists Bull., v. 63, no. 9, p. 1478–1489.

LONGMAN, M.W., 1980a, Carbonate petrology of the Nido B-3A core, offshore Palawan, Philippines, in Halley, R.B. and R.G. Loucks, eds., Carbonate Reservoir Rocks: notes for SEPM Core Workshop No. 1, Denver, Colorado, June 1980: Soc. Econ. Paleontologists and Mineralogists, p. 161–183.

LONGMAN, M.W., 1980b, Carbonate diagenetic textures from nearsurface diagenetic environments: Amer. Assoc. Petroleum Geologists Bull., v. 64, no. 4, p. 461–487.

MAGARA, K., 1975a, Importance of aquathermal pressuring effect in Gulf Coast: Amer. Assoc. Petroleum Geologists Bull., v. 59, no. 10, p. 2037–2045.

MAGARA, K., 1975b, Reevaluation of montmorillonite dehydration as cause of abnormal pressure and hydrocarbon migration: Amer. Assoc. Petroleum Geologists Bull., v. 59, no. 2, p. 292–302.

MEISSNER, F.F., 1980, Examples of abnormal fluid pressure produced by hydrocarbon generation: Amer. Assoc. Petroleum Geologists Bull., v. 64, no. 5, p. 749.

POWERS, M.C., 1967, Fluid-release mechanisms in compacting marine mudrocks and their importance in oil exploration: Amer. Assoc. Petroleum Geologists Bull., v. 51, no. 7, p. 1240-1254.

SALDIVAR-SALI, A., 1979, Reef exploration in the Philippines: Presented at Second Circum-Pacific Energy and Mineral Resources Conference, Honolulu, Hawaii, August, 1978: Publ. by Philippines Bur. of Energy Development, 54 p.

WOOD, P.W.J., and CITIES SERVICE EXPLORATION STAFF, 1980, Hydrocarbon plays in Tertiary SE Asia Basins: Oil and Gas Jour., v. 78, no. 29, p. 90–96.

35
Fukubezawa Field
Koichi Aoyagi

RESERVOIR SUMMARY

Location & Geologic Setting	Akita, NE Honshu, Japan, in Katanishi Basin
Tectonics	Faulted north-trending asymmetrical anticline; Mio-Pliocene folding
Regional Paleosetting	Central, deep-water part of Katanishi Basin
Nature of Trap	Structural
Reservoir Rocks	
Age	Miocene
Stratigraphic Units(s)	Onnagawa Formation
Lithology(s)	Dolomite, some tuff
Dep. Environment(s)	Deep water (bathyal) and stagnant
Productive Facies	Crystalline dolomite and dolomitic mudstone and wackestone
Entrapping Facies	Siliceous shale (capping facies)
Diagenesis/Porosity	Alteration of tuffs to chert and limestone; then dolomitization of limestone
Petrophysics	
Pore Type(s)	Interparticle and intercrystal
Porosity	0.6–30.9%, avg 9.7%
Permeability	0.3–8.5 md, avg 3.2 md
Fractures	Negligible
Source Rocks	
Age	Miocene
Lithology(s)	Shales
Migration Time	Mio-Pliocene
Reservoir Dimensions	
Depth	3940–4590 ft (1200–1400 m)
Thickness	50 ft (15 m) avg 230 ft (70 m) maximum
Areal Dimensions	NA
Productive Area	290–1370 acres (1.4–6.6 km)*
Fluid Data	
Saturations	S_o = 2–29%, S_w = 49–93%
API Gravity	35° avg
Gas-Oil Ratio	80–110:1
Other	Paraffinic oil
Production Data	
Oil and/or Gas in Place	13.96 million BO
Ultimate Recovery	3.30 million BO; URE = 23.6%
Cumulative Production	NA

Remarks: *Different producing zones have different thicknesses and area extents. Discovered 1964.

35
Origin of the Miocene Carbonate Reservoir Rocks, Fukubezawa Oil Field, Akita Province, Northeast Honshu, Japan

Koichi Aoyagi

Exploration History

Fukubezawa oil field is located in the western area of Hachiro-gata Lagoon, Akita, northeast Honshu, Japan (Fig. 35-1). The area is also the site of the Sarukawa and Hashimoto oil fields and Katanishi gas field, and is one of the most productive areas in Japan. The first wildcat in Fukubezawa Field, Well No. AK-1, was drilled in September 1964. Subsequently, 25 wells were drilled. The deepest well (SK-25D) was drilled in 1978 and reached a total depth of 7095 feet (2163 m). Eighteen wells were successful, and the cumulative production through the end of 1978 was 2.64 million barrels of crude oil and 1836 million cubic feet (52 million m^3) of natural gas. Fukubezawa SK-6, which has been the most productive well in the field, produced 0.76 million barrels of oil and 388 million cubic feet (11 million m^3) of gas through 1978.

Geology

The standard stratigraphic section in the Fukubezawa area consists of the Miocene Nishikurosawa, Onnagawa, and Funakawa formations, and the Pliocene lower and upper Tentokuji, Sasaoka, and Shibikawa formations. A generalized columnar section showing the thickness and foraminiferal assemblage in each of the Miocene and Pliocene formations is indicated in Figure 35-2.

The Nishikurosawa Formation (Miocene) is generally composed of pyroclastic detritus and volcanic rock fragments. Foraminifers are generally absent, but sedimentary facies include abundant plant fossils, such as *Picea kanoi*, *Tsuga aburaensis*, and *Metasequoia occidentalis*, which indicate that the depositional environment was littoral to inner neritic. The Onnagawa Formation consists of alternating layers of acidic tuff and siliceous shale. The lower part of the formation is intruded locally by basalt. Carbonate reservoirs in Fukubezawa Field are found in the middle part of the Onnagawa Formation (Miocene). Foraminiferal assemblages indicate that the depositional environment for these reservoir rocks was bathyal. The Funakawa Formation is mainly composed of black shale with several interbeds of acidic tuffs. Its depositional environment was also bathyal.

The lower part of the Lower Tentokuji Formation (Pliocene) is composed chiefly of tuffaceous coarse-grained sandstone, which is the main reservoir in Sarukawa oil field and Katanishi gas field. The upper part of the Tentokuji Formation generally consists of dark gray siltstone interbedded with mudstone and tuff. Foraminiferal assemblages indicate that the general paleoenvironment was outer to inner neritic and was affected by warm currents. The upper Tentokuji Formation is composed principally of greenish-gray siltstone, whereas the lower part is more muddy and in places contains several tuffs and sandstones. Depositional environment was inner

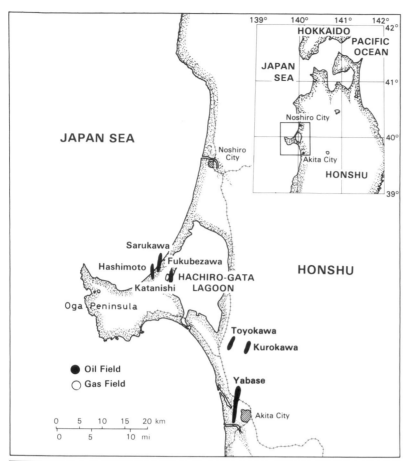

Fig. 35-1. Locality map of area studied. Fields occur on north-trending asymmetrical anticlines.

AGE	FORMATION	THICK-NESS (m)	COLUMNAR SECTION	FORAMINIFERA
PLIOCENE	Shibikawa	150+		
	Sasaoka	150+		Cribrononion clavatum Buccella frigida
	Upper Tentokuji	400–500		Epistominella pulchella Uvigerina akitaensis Globorotalia inflata
	Lower Tentokuji	150–200		Praeglobobulimina pupoides Globorotalia orientalis
MIOCENE	Funakawa	300		Martinottiella nodelosa Miliamina echigoensis Cyclammina japonica
	Onnagawa	500		Cribrostomoides rengi Cyclammina egoensis Bulimina kamedaensis
	Nishikurosawa	500+		

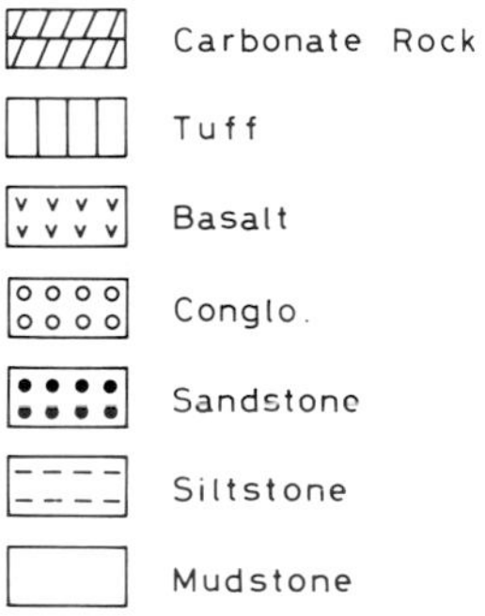

Legend:
- Carbonate Rock
- Tuff
- Basalt
- Conglo.
- Sandstone
- Siltstone
- Mudstone

Fig. 35-2. Generalized columnar section showing the thickness and foraminiferal assemblage in each formation of the Miocene and Pliocene in the Fukubezawa area.

Fig. 35-3. Geologic cross section showing the structure between Katanishi gas field and Fukubezawa oil field. Modified after JAPT, 1973, and published with their permission.

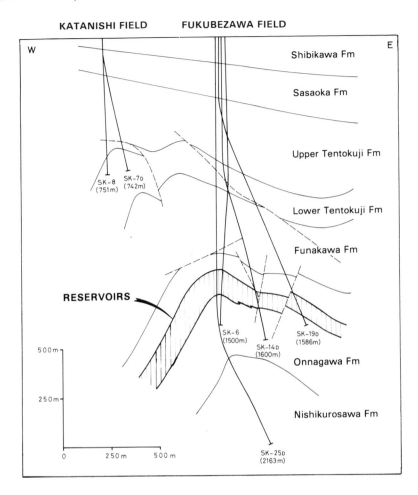

to outer neritic. The Sasaoka Formation consists mainly of bluish-gray siltstone; the upper part is more sandy and contains abundant molluscan fossils. Paleo-environment was littoral to inner neritic. The Shibikawa Formation overlies the lower formations and is composed of coarse-grained sand, gravel, and lignite. Depositional environment was littoral.

Figure 35-3 is a geologic cross-section of Fukubezawa and Katanishi fields (JAPT, 1973). As it suggests, the traps in these fields are part of an asymmetrical anticline that is elongated parallel with the regional north–south strike of fields and fold axes. The shallower trap of the anticline constitutes the Katanishi gas field, and the deeper one forms the Fukubezawa oil field. The structure is about 2.4 miles long (4 km) north–south and 0.6 mile wide (1 km) east–west.

Reservoirs

The main reservoirs in Fukubezawa Field occur in the middle part of the Miocene Onnagawa Formation at depths of 3940 to 4590 feet (1200–1400 m). Reservoirs have an average thickness of 50 feet (15 m) and a maximum of 230 feet (70 m). A representative lithologic column and various mechanical logs of the reservoirs in Fukubezawa SK-6 well are shown in Figure 35-4 (after Takuma, 1969), which illustrates the high spontaneous potential and resistivity of the reservoir intervals. In the center of the field, these reservoirs are generally composed of dolostone, calcitic dolostone, and dolomitic limestone. They gradually change in lithologic character to acidic tuff and siliceous shale at the periphery of the

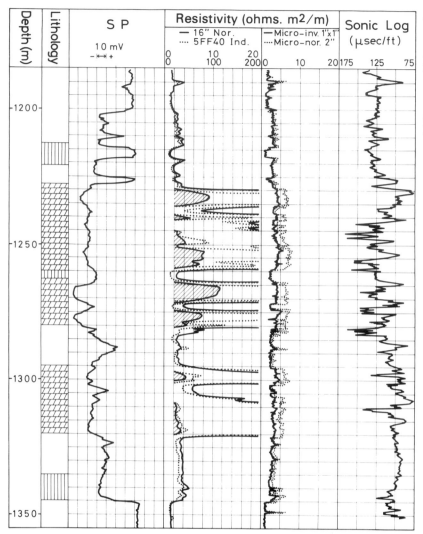

Fig. 35-4. Lithologic columnar section and various electric logs showing the petrophysical properties of carbonate reservoir rocks in Fukubezawa SK-6 well. Modified after Takuma, 1969, and published by permission, Japanese Association of Petroleum Technologists.

reservoirs. Oil and gas production is much higher in the carbonate than in the tuff reservoirs.

Porosity of the dolostones and dolomitic limestones measured by core analysis ranges from 5 to 30 percent, and permeability ranges from 0.1 to 12.5 millidarcys. Compared with porosity and permeability data of carbonate reservoirs in the USA and elsewhere summarized by Chilingar *et al.* (1972), the carbonate reservoirs in Fukubezawa show fair to good porosity and poor permeability.

Petrographic thin-section and X-ray diffraction studies of the carbonate reservoirs in Fukubezawa suggest that kaolinite, quartz, plagioclase, calcite, and dolomite are generally common in these carbonates and that dolomite content is inversely proportional to quartz content. Kinoshita

(1967) reported that some of these carbonate reservoir rocks also contain abundant tests of diatoms.

Carbonate Rocks of the Onnagawa Formation

Dolostone, calcitic dolostone, and dolomitic limestone occur discontinuously in the Onnagawa Formation in areas surrounding Fukubezawa Field. Detailed sedimentological and micropaleontological studies on these carbonates were done by the writer and his colleagues (Aoyagi *et al.*, 1970; Aoyagi and Kazama, 1971; Aoyagi, 1972; Aoyagi and Chilingarian, 1972; and Aoyagi *et al.*, 1972).

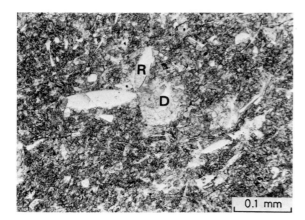

Fig. 35-5. Photomicrograph of biogenic dolostone in the Miocene Onnagawa Formation on the southern shore of Oga Peninsula, Akita, Japan (sample 68-1132). D diatom, R radiolaria.

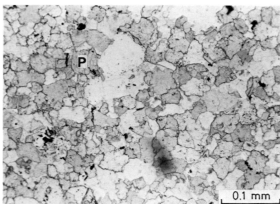

Fig. 35-6. Photomicrograph of finely crystallized dolomitic limestone in the Miocene Onnagawa Formation at the southern shore of Oga Peninsula, Akita, Japan (sample 68-1343). P plagioclase.

Typical outcrops of similar carbonate rocks are recognized along the southern shore of Oga Peninsula southwest of the oil field. These carbonates generally occur in units 1.6 to 3.3 feet (0.5–1.0 m) thick that are lenticular and grade laterally to acidic tuff and siliceous shale, just as they do in the Miocene carbonate reservoirs of Fukubezawa Field.

These carbonates in the Onnagawa are grouped into two types: (1) biogenic dolostone and dolomitic lime wackestone or packstone (Dunham, 1962) with abundant foraminifers, diatoms, radiolarians, sponge spicules, and calcareous nannoplankton; and (2) crystalline dolostone and dolomitic lime mudstone. These two types of carbonate rocks are shown in Figures 35-5 and 35-6, respectively. Based on the classification of Honjo (1969), the latter type was further classified into two subtypes from electron-microscopic observation: nannoagorite with abundant nannoplanktons, and dolomitized orthomicrite. The former subtype is more common than the latter.

Mineralogically, these rocks consist mainly of dolomite, calcite, quartz, plagioclase, and clay minerals. As reported by Aoyagi et al. (1970), dolomite content in the carbonates is directly proportional to plagioclase content and inversely proportional to quartz content. On the basis of clay mineral content, these carbonate rocks can be grouped into two types containing predominantly kaolinite and montmorillonite.

The foraminiferal assemblage in argillaceous rocks of the Onnagawa Formation is mainly composed of Martinottiella communis, Cyclammina japonica, Spirosigmoilinella compressa, Cribrostomoides renzi, and Globorotalia pseudopachyderma (Maiya, 1978). The upper part of the formation generally includes abundant radiolarians and diatoms, indicating that the formation was deposited in bathyal and stagnant water and was affected by cold ocean currents in the later stages of deposition.

Total porosity of these carbonate rocks ranges from 0.6 to 30.9 percent and averages 9.7 percent. According to the classification of Choquette and Pray (1970), porosity in these carbonates is mainly interparticle and intercrystalline. Aoyagi (1973) discussed the effects of physical and chemical processes on the total porosity of middle Carboniferous carbonate rocks in Canada. Based on this discussion, Aoyagi (1972) considered that the most fundamental and significant porosity-reducing process in these Miocene carbonates of Japan was compaction, and the next most important process tending to reduce porosity was cementation. Modifications effective in increasing pore space were dolomitization and conversion of opal to cristobalite, quartz and calcite. In addition, Aoyagi and Chilingarian (1972) recognized that total porosity of the montmorillonitic carbonate rocks is much higher than that of the kaolinitic rocks.

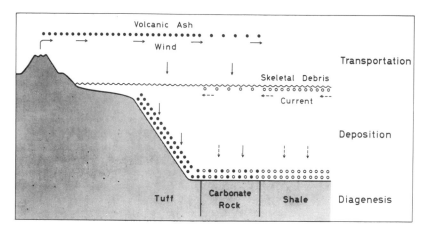

Fig. 35-7. Schematic diagram showing the location and source of sediments diagenetically converted to carbonate rocks in the Miocene Onnagawa Formation of studied area.

Origin of Carbonate Reservoir Rocks

The above-mentioned field and laboratory work led to an interpretation of the origin of the carbonate rocks in the Onnagawa Formation that are exposed on the Oga Peninsula southwest of Fukubezawa Field (Aoyagi *et al.*, 1970; Aoyagi and Kazama, 1971). A schematic diagram explaining the origin is provided by Figure 35-7.

During the Middle Miocene, bioclastic and siliciclastic sediment was separately supplied to the basin and deposited in a stagnant bathyal environment. The bioclastic sediments, transported by cold sea currents, consisted mainly of small skeletal debris derived from diatoms, foraminifers, radiolarians, sponge spicules, and calcareous nannoplankton. Diatoms are the most abundant bioclastics. Siliciclastic sediments, transported by the wind, are chiefly composed of volcanic detritus including much acidic glass and plagioclase. These two sediment types were intermixed at special sites of deposition within the basin.

Almost pure siliciclastic sediment, rich in volcanic detritus, was altered to tuff and tuff breccia. Bioclastic sediment, rich with the debris of small skeletons, was changed to siliceous shale by various geologic processes such as compaction, cementation, and the inversion of various minerals during diagenesis. Source materials for silica minerals in the shale, such as cristobalite and quartz, were mainly amorphous silica from diatoms and partly inorganic amorphous silica transformed by solution and dehydration of volcanic glass.

Mixtures of bioclastic and siliciclastic sediments deposited at special sites in the basin gradually became altered to carbonate rocks by various diagenetic reactions. During early diagenesis, calcium and carbon dioxide derived from sea water, as well as from volcanic detritus within the basin, changed the originally siliceous sediments to fine-grained limestones. During the late stage of diagenesis, these limestones were gradually altered to dolostones. A source of magnesium is inferred to have been buried sea water mixed with volcanic detritus. This interpretation is supported by the directly proportional relationship between dolomite content and plagioclase content and by the inversely proportional relationship between dolomite content and quartz content in carbonate rocks. The primary textures of originally siliceous sediments were well preserved during the diagenetic process of inversion from silica to calcite, but were gradually destroyed by dolomitization and by recrystallization of calcite during a later stage of diagenesis.

Many stratigraphic and sedimentological similarities are recognized between the Miocene carbonate reservoirs in Fukubezawa Field and carbonate rocks along the southern shore of the Oga Peninsula. This suggests that the origins of both carbonates were similar. Therefore, it is reasonable to infer that the carbonate reservoirs in Fukubezawa also originated from biogenic siliceous sediments mixed with volcanic detritus, which were deposited in bathyal environments and later altered to carbonate rocks by transformation of amorphous silica to calcite and dolomite during diagenesis.

Acknowledgments The writer would like to thank Japan Petroleum Exploration Company (JAPEX) and Japan National Oil Corporation (JNOC) for permission to publish the paper. Ms. Toshie Takamasa of JAPEX and Ms. Noriko Ishikawa of JNOC drafted the figures and typed the manuscript.

References

AOYAGI, K., 1972, Origin of porosity in Neogene carbonate rocks of Akita, Japan: Jour. Jap. Assoc. Petrol. Technol., v. 37, p. 169–176 (in Japanese).

AOYAGI, K., 1973, Petrophysical approach to origin of porosity of carbonate rocks in middle Carboniferous Windsor group, Nova Scotia, Canada: Amer. Assoc. Petroleum Geologists Bull., v. 57, p. 1692–1702.

AOYAGI, K., and G.V. CHILINGARIAN, 1972, Clay minerals in carbonate reservoir rocks and their significance in porosity studies: Sedimentary Geol., v. 8, p. 241–249.

AOYAGI, K., and T. KAZAMA, 1971, Electron-microscopic observation of Miocene carbonate rocks from Akita area, Japan, and discussion of their origin: Jour. Jap. Assoc. Petrol. Technol., v. 36, p. 357–362 (in Japanese).

AOYAGI, K., T. KAZAMA, and G.V. CHILINGAR, 1972, Clay minerals and organic matter in Neogene carbonate rocks of Akita oil field of Japan: Jour. Clay Sci. Soc. Jap., v. 12, p. 1–10 (in Japanese).

AOYAGI, K., T. SATO, and T. KAZAMA, 1970, Distribution and origin of the Neogene carbonate rocks in the Akita oil fields, Japan: Jour. Jap. Assoc. Petrol. Technol., v. 35, p. 67–76 (in Japanese).

CHILINGAR, G.V., R.W. MANNON, and H.H. RIEKE III, 1972, Oil and Gas Production from Carbonate Rocks: Elsevier Scientific Publ. Co., New York, 408p.

CHOQUETTE, P.W., and L.C. PRAY, 1970. Geologic nomenclature and classification of porosity in sedimentary carbonates: Amer. Assoc. Petroleum Geologists Bull., v. 54, p. 207–250.

DUNHAM, R.J., 1962, Classification of carbonate rocks according to depositional texture: *in* Ham, W.E., ed., Classification of Carbonate Rocks—a symposium: Amer. Assoc. Petroleum Geologists Mem. 1, p. 108–121.

HONJO, S., 1969, An electron-microscopic study of fine-grained carbonate matrix: Jour. Geol. Soc. Jap., v. 75, p. 349–364 (in Japanese).

JAPANESE ASSOCIATION OF PETROLEUM TECHNOLOGISTS (JAPT), 1973, Oil-Mining Industry in Japan and Their Technology: JAPT, Tokyo, 430 p. (in Japanese).

KINOSHITA, K., 1967, Dolomite reservoirs, with special references to the reservoirs in Fukubezawa oil field in Akita Prefecture, Japan: Jubilee publ. in commemoration of Prof. Yasuo Sasa, p. 595–612 (in Japanese).

MAIYA, S., 1978, Late Cenozoic planktonic foraminiferal biostratigraphy of the oil field region of northeast Japan: *in* Fujita, K., ed., Cenozoic Geology of Japan, p. 35–60.

TAKUMA, T., 1969, Exploration of Fukubezawa oil field: Jour. Jap. Petrol. Inst., v. 12, p. 43–48 (in Japanese).

Appendix
Classifications Used in Case Studies

Rationale

The authors contributing to this volume were asked to use specific classifications of carbonate rocks and associated pore systems. Classifications have at least two vital functions. One is to provide investigators with the intellectual benefits of organizing data and interpretations into clear, concise language (Ham and Pray, 1962). The other is to supply a means of communicating that information in a "normalized" form so that it can be compared readily with similar kinds of information about other occurrences. Both of these benefits apply, of course, to any classification of natural systems. For carbonate rocks and their pore systems, the rationale and historical perspectives for classifications have been ably summarized by others and need no elaboration here. Among the more useful treatments of the subject are the papers on classification of carbonate rocks in Memoir 1 of the American Association of Petroleum Geologists including the lead article by Ham and Pray (1962); a discussion of porosity types and their origins by Choquette and Pray (1970); and a concise summary and comparison of many pore-system and carbonate-rock classifications by R. Cussey and his associates in Elf Aquitaine, translated by Reeckman and Friedman (1982).

Outline

Four main systems for describing or characterizing carbonates and their pore systems are used in this book:

1. A textural classification by Dunham (1962), supplemented for coarse biogenic limestones by Embry and Klovan (1971).
2. A petrographic-genetic classification of porosity by Choquette and Pray (1970).
3. A classification relating "matrix" texture to pore size and pore frequency, proposed by Archie (1952) and modified by Roehl (unpublished).
4. A detailed system of symbols and descriptive procedures developed in the Shell companies for the description and logging of sedimentary rocks and their constituents, and published recently by Swanson (1981).

Some authors in this book chose to employ the limestone classification of Folk (1959, 1962), in lieu of or in addition to that of Dunham. All of the systems just cited are in relatively widespread use by sedimentary geologists in the petroleum industry and elsewhere. The brief summaries that follow are intended for readers not closely conversant in carbonate sedimentology; more extended information can be found in the original publications (see References), and key terms are defined in the GLOSSARY.

Carbonate Sediment and Rock Classifications

Both of the classifications currently in widespread use, by Dunham (1962) and Folk (1959, 1962), recognize that limestones consist in general of three kinds of components: "grains", "matrix" or lime mud (the "micrite" of Folk when indurated), and cement. Both classifications also make a basic distinction between limestones made up dominantly of components bound together biogenically during deposition, and limestones with mostly unbound components. Neither classification provides for clear distinctions between different dominant grain-size characteristics, a feature partly responsible for the system proposed by Embry and Klovan (1971).

The Dunham Classification

This system, elegant in its terminological simplicity, can be used equally well, despite the suffix "-stone" in all of its terms, to name and characterize lime sediments as well as limestones. It is illustrated in Figures A-1 and A-2. Dunham suggested two basic textural features, in addition to the presence or absence of organic binding of sedimentary particles, as a basis for classifying carbonates: (1) the presence or absence of carbonate mud, which differentiates muddy carbonates from essentially mud-free *grainstones*; and (2) the relative abundance of grains and lime mud, which determines whether the load-bearing components are grains or mud (grain support" *vs.* "mud support"). "Grains" mean any sedimentary particles larger than 0.02 mm for Dunham (or 0.03 mm for Embry and Klovan), *i.e.*, particles of coarse silt size and larger. Muddy carbonates are called *packstone, wackestone,* or *mudstone* depending on their lime-mud content; packstone is grain-supported, whereas wackestone and mudstone are mud-supported. Mud content is one indication of whether and how effectively winnowing was operative in the depositional environment, and can be a critical influence in the course of later diagenesis and pore-space evolution. Biogenically bound sediments are called *boundstones*. Where deposi-

tional texture is no longer recognizable, the term *crystalline carbonate* is used.

The Embry-Klovan Classification

Embry and Klovan (1971) proposed important modifications to the Dunham system, specifically to accommodate the fact that biogenically bound carbonates, such as many biogenic reefs and reef-mounds (see James, 1983, for definitions) commonly contain components much coarser than sand size (>2 mm) and may be bound together in different ways. As shown in Figure A-3, this classification retains the basic grain-support/ mud-support terms of Dunham, adding categories as follows: For carbonates not biogenically bound ("organically" in the usage of Embry and Klovan), but containing significant amounts of coarse components, the terms *floatstone* and *rudstone* were proposed, depending on whether the matrix or coarser fraction is load-bearing. For biogenically bound carbonates, the terms *bafflestone, framestone,* and *bindstone* were proposed. Examples of the kinds of reefal organisms that can occur in these carbonates also are illustrated by James (1983). They might include platy or "potato-chip" calcareous algae that collect and shelter sediment, in a *bafflestone*; robust framework builders such as corals or rudistids in life positions, in a framestone; or encrusting coralline algae, serving to bind corals together in a bindstone. The relationships between these terms of Embry and Klovan and those of Dunham are shown in Figure A-3.

The Folk Classification

The main elements of this classification are shown in Figures A-4 and A-5 from Folk's two articles on the subject (1959, 1962). The basis for this system is threefold: (1) relative proportions of carbonate grains or allochems versus carbonate mud or micrite; (2) sorting of allochems, and (3) rounding of allochems. Allochems include ooids (oolites in Figs. A-4 and A-5A), pellets, intraclasts, and bioclasts. These four types supply the prefixes for rock-textural categories that are dependent mainly on the proportiions of allochems and micrite, and on whether the allochems are cemented by sparry calcite (-sparite) or lime mud (-micrite).

As an example, using the chart in Figure A-4, a

DEPOSITIONAL TEXTURE RECOGNIZABLE					DEPOSITIONAL TEXTURE NOT RECOGNIZABLE
Original Components Not Bound Together During Deposition				Original components were bound together during deposition . . . as shown by intergrown skeletal matter, lamination contrary to gravity, or sediment-floored cavities that are roofed over by organic or questionably organic matter and are too large to be interstices.	**CRYSTALLINE CARBONATE** (Subdivide according to classifications designed to bear on physical texture or diagenesis.)
Contains mud (particles of clay and fine silt size)			Lacks mud and is grain-supported		
Mud-supported		Grain-supported			
Less than 10 percent grains	More than 10 percent grains				
MUDSTONE	**WACKESTONE**	**PACKSTONE**	**GRAINSTONE**	**BOUNDSTONE**	

Fig. A-1. The textural classification of carbonates by Dunham (1962). Published by permission, American Association of Petroleum Geologists.

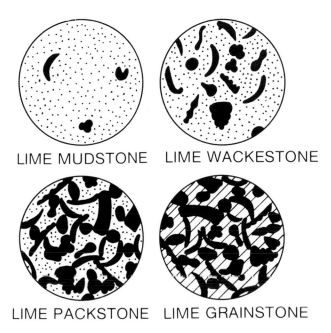

LIME MUDSTONE   LIME WACKESTONE

LIME PACKSTONE   LIME GRAINSTONE

Fig. A-2. Diagrams illustrating terms proposed by Dunham (1962) for those carbonates in which components are not organically bound together. Published by permission, American Association of Petroleum Geologists.

CARBONATE GRAIN

CALCITE MUD

SECONDARY SPARRY CALCITE

limestone with an allochem content of more than 25 percent ooids (oolites) would be either an *oosparite*, an *oomicrite*, or an *oolite-bearing micrite* depending on whether allochems made up more than 10 percent and were calcite-cemented or had a micrite matrix, or only 1 to 10 percent, with a predominance of micrite. Limestones with less than 1 percent allochems would be called micrite

regardless of which allochems were dominant. In Folk's more recent (1962) version of his system, particular recognition was accorded to textural maturity in terms of sorting and abrasion of allochems as well as winnowing or bypassing of micrite, as illustrated in Figure A-5B. Several of those terms are more fully defined in the Glossary.

<table>
<tr><td colspan="6">ALLOCHTHONOUS LIMESTONES
ORIGINAL COMPONENTS NOT ORGANICALLY
BOUND DURING DEPOSITION</td><td colspan="3">AUTOCHTHONOUS LIMESTONES
ORIGINAL COMPONENTS ORGANICALLY
BOUND DURING DEPOSITION</td></tr>
<tr><td colspan="4">LESS THAN 10% > 2mm COMPONENTS</td><td colspan="2">GREATER THAN
10% > 2mm
COMPONENTS</td><td rowspan="4">BY
ORGANISMS
WHICH
ACT
AS
BAFFLES</td><td rowspan="4">BY
ORGANISMS
WHICH
ENCRUST
AND
BIND</td><td rowspan="4">BY
ORGANISMS
WHICH
BUILD
A RIGID
FRAME-
WORK</td></tr>
<tr><td colspan="3">CONTAINS LIME MUD(< .03mm)</td><td>NO
LIME
MUD</td><td rowspan="3">MATRIX
SUPPORTED</td><td rowspan="3">COMPONENT
SUPPORTED</td></tr>
<tr><td colspan="2">MUD SUPPORTED</td><td colspan="2" rowspan="2">GRAIN
SUPPORTED</td></tr>
<tr><td>LESS THAN
10%GRAINS
(>.03mm<2mm)</td><td>GREATER
THAN 10%
GRAIN</td></tr>
<tr><td>MUD-
STONE</td><td>WACKE-
STONE</td><td>PACK-
STONE</td><td>GRAIN-
STONE</td><td>FLOAT-
STONE</td><td>RUD-
STONE</td><td>BAFFLE-
STONE</td><td>BIND-
STONE</td><td>FRAME-
STONE</td></tr>
</table>

Fig. A-3. The textural classification of carbonates proposed by Embry and Klovan (1971) using elements of the Dunham classification (left side). This expanded scheme allows the differentiation between types of coarse-grained (>2 mm) carbonates as well as between different types of organic binding. Published by permission, Canadian Society of Petroleum Geologists.

<table>
<tr><td colspan="5" rowspan="3"></td><td colspan="6">LIMESTONES AND PARTLY DOLOMITIZED LIMESTONES</td><td colspan="3">DOLOMITES</td></tr>
<tr><td colspan="2">> 10% ALLOCHEMS</td><td colspan="3">< 10% ALLOCHEMS</td><td rowspan="2">UNDIS
TURBED
BIOHERM
ROCKS</td><td colspan="2" rowspan="2">ALLOCHEM
GHOSTS</td><td rowspan="2">NO ALLOCHEM
GHOSTS</td></tr>
<tr><td>CEMENTED BY
SPARRY CALCITE</td><td>MATRIX OF
MICROCRYSTAL-
LINE OOZE</td><td colspan="2">1-10%
ALLOCHEMS</td><td>< 1%
ALLOCHEMS</td></tr>
<tr><td rowspan="5">VOLUMETRIC ALLOCHEM COMPOSITION</td><td colspan="4">> 25%
INTRA
CLASTS</td><td>INTRASPARITE</td><td>INTRAMICRITE</td><td rowspan="5">MOST ABUNDANT ALLOCHEM</td><td>INTRACLAST-
BEARING
MICRITE</td><td rowspan="5">MICRITE</td><td rowspan="5">BIOLITHITE</td><td rowspan="5">EVIDENT ALLOCHEM</td><td>INTRACLASTIC
DOLOMITE</td><td rowspan="5">DOLOMITE</td></tr>
<tr><td rowspan="4">< 25% INTRA-CLASTS</td><td colspan="3">> 25%
OOLITES</td><td>OOSPARITE</td><td>OOMICRITE</td><td>OOLITE-
BEARING
MICRITE</td><td>OOLITIC
DOLOMITE</td></tr>
<tr><td rowspan="3">< 25% OOLITES</td><td rowspan="3">VOLUME RATIO OF FOSSILS TO PELLETS</td><td>> 3:1</td><td>BIOSPARITE</td><td>BIOMICRITE</td><td rowspan="2">FOSSILI-
FEROUS
MICRITE</td><td rowspan="2">BIOGENIC
DOLOMITE</td></tr>
<tr><td>3:1 - 1:3</td><td>BIOPELSPARITE</td><td>BIOPELMICRITE</td></tr>
<tr><td>< 1:3</td><td>PELSPARITE</td><td>PELMICRITE</td><td>PELLETI-
FEROUS
MICRITE</td><td>PELLET
DOLOMITE</td></tr>
</table>

Fig. A-4. The textural classification of carbonates by Folk (1959, 1962). Published by permission, of American Associates of Petroleum Geologists.

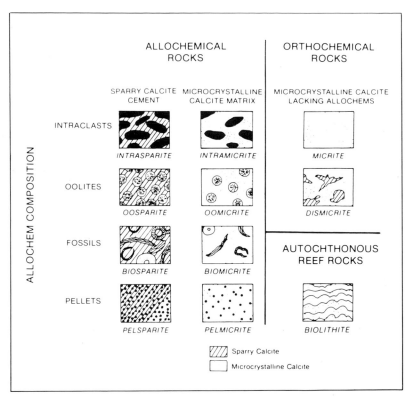

The following is the lower portion (B) diagram table:

	OVER 2/3 LIME MUD MATRIX				SUBEQUAL SPAR & LIME MUD	OVER 2/3 SPAR CEMENT		
Percent Allochems	0-1 %	1-10 %	10-50%	OVER 50%		SORTING POOR	SORTING GOOD	ROUNDED & ABRADED
Representative Rock Terms	MICRITE & DISMICRITE	FOSSILI-FEROUS MICRITE	SPARSE BIOMICRITE	PACKED BIOMICRITE	POORLY WASHED BIOSPARITE	UNSORTED BIOSPARITE	SORTED BIOSPARITE	ROUNDED BIOSPARITE
1959 Terminology	Micrite & Dismicrite	Fossiliferous Micrite	Biomicrite			Biosparite		
Terrigenous Analogues	Claystone		Sandy Claystone	Clayey or Immature Sandstone		Submature Sandstone	Mature Sandstone	Supermature Sandstone

■ LIME MUD MATRIX
▨ SPARRY CALCITE CEMENT

Fig. A-5. Diagrams illustrating aspects of the Folk classification. Published by permission, American Association of Petroleum Geologists. *A.* Main textural types based on allochem composition and presence/abundance of cement vs. micrite. From Folk (1962). *B.* A textural maturity spectrum for carbonates show-ing eight stages from low-energy sediments at left to higher-energy sediments toward the right. Bioclast-dominated allochem composition is used in this example, but parallel stages and terms could be constructed using other allochem types. From Folk (1962).

Porosity and Pore/Matrix-Type Classifications

Although many classifications have been proposed[1], the authors contributing to this volume were asked to employ either one or both of two published by Choquette and Pray (1970), and by Archie (1952; modified by Roehl, unpublished).

The Choquette-Pray Classification

This classification relates porosity types to their origins. Some 15 basic pore types, shown in Figure A-6A, are distinguished depending on whether primary or diagenetic constituents and textures determine their distribution ("fabric-selective") or such a relationship is not demonstrable ("not fabric-selective"). The general time and realm or zone in which a given pore type is believed to have been created or modified can be specified, as can various processes and effects of pore modifications, general pore-size characteristics, and volumetric abundance (Fig. A-6B). Porosity terms, or code designations of them, can be constructed using a succinct format (Fig. A-7A). Common stages in the evolution of a type of pore, a mold in this case, are illustrated in Fig. A-7B. The classification also distinguishes three main post-depositional time-space zones in which pore systems may be created or modified in carbonates, illustrated in Figure A-8. These three zones—*eogenetic, mesogenetic,* and *telogenetic*—are defined in the GLOSSARY.

The Archie Classification

This widely used classification is based on the relationships observed between pore size and rock-matrix texture, as summarized in Table A-1. It was designed for use in well-site or laboratory studies of well cores, cuttings, and core-analysis samples, whether "raw" and in untreated or on sawed surfaces, with the unaided eye or at low magnifications generally around 10 to 15X. Archie distinguished three basic types of rock-matrix textures and provided simple criteria for their recognition, as well as providing a simple classification of visible pore sizes and pore frequencies. In this form, the classification can be used by individuals with little or no training in sedimentology, and thus provides a basis for organizing and interpreting petrophysical data by either geologists or petroleum engineers.

Roehl (unpublished reports, 1959, 1961), recognizing that many carbonates have matrix textures that represent combinations and variations of the types defined by Archie, proposed a more elaborate classification, which he specifically related to parameters that can be determined from mercury-injection capillary pressure curves. The Roehl classification is outlined in Table A-2. In that table, the parameter r is the average largest pore radius in microns based on minimum entry pressure, P_d is the displacement pressure in psia, the G-factor is a dimensionless hyperbolic measure of capillary-pressure curve shape defined by Thomeer (1960), and ϕri is relative ineffective porosity not occupied by mercury at high capillary pressure. Figure A-9 shows diagramatically the relationships observed between grain size or crystal size of carbonates, and the matrix-texture types defined originally by Archie (1952) and subsequently by Roehl (unpublished).

Shell System for Logging Descriptive Rock Attributes

The contributors to this book were asked to portray lithologic columns using a detailed system of symbology developed in the Shell companies, and available now to the general profession as recently published by Swanson (1981). Figures A-10A–E portray those parts of the system from which most authors in this volume drew symbols utilized in the case studies. Provision is made for portrayal of the Dunham textural types (Fig. A-10A) and Archie matrix-texture types ("chalky" etc., Fig. A-10A); certain types of cements and other subordinate components (Fig. A-10B); most of the skeletal and other components generally found in carbonate rocks (Figs. A-10C,D); and the more common sedimentary structures (Fig. A-10E).

---

[1] See Reeckman and Friedman (1982, p. 82–94) for a good sampling.

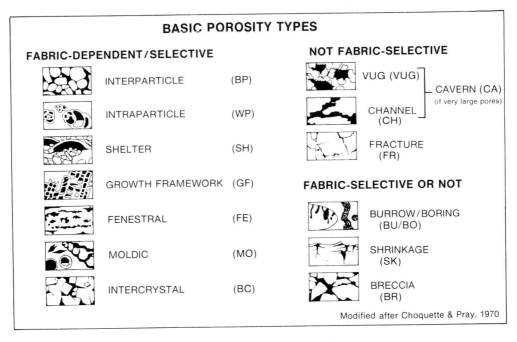

Fig. A-6. The porosity classification proposed by Choquette and Pray (1970). Published by permission, American Association of Petroleum Geologists. A. Basic porosity types. Diagrams representing porosity types from Gagliardi et al. (1980). Published by permission, Consiglio Nazionale Delle Richerche, Rome, Italy. B. Genetic, size, and abundance modifiers.

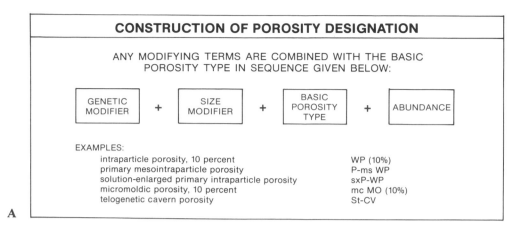

A

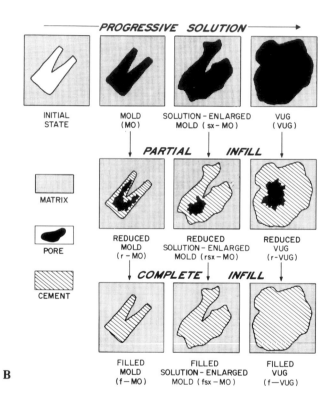

B

Fig. A-7. The porosity classification of Choquette and Pray (1970). Published by permission, American Association of Petroleum Geologists. *A*. Format for construction of porosity name and code designations. *B*. Common stages in evolution of one basic type of pore, a mold, showing application of genetic modifiers and classification code.

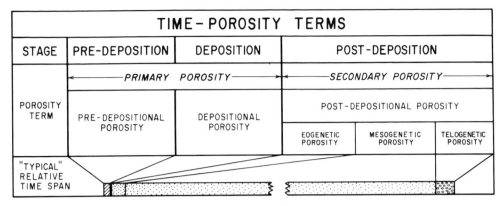

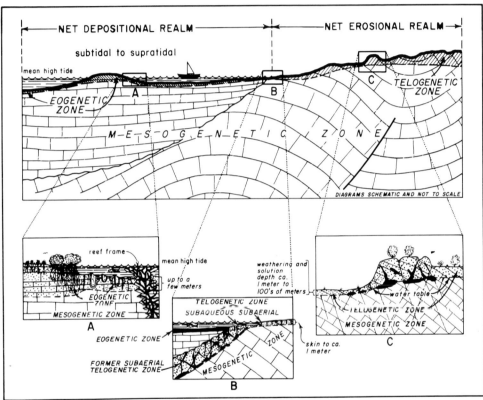

Fig. A-8. Time-porosity terms and zones of creation and modification of porosity in sedimentary carbonates, from Choquette and Pray (1970). Published by permission, American Association of Petroleum Geologists. *Upper:* Interrelation of major time-porosity zones. *Lower:* Diagrammatic representation of major surface and burial zones where porosity is created or modified.

Table A-1. Classification of Carbonate Reservoir Rock Types According to Matrix Texture, Visible Pore Size, and Pore Frequency (from Archie, 1952)

Texture of Matrix	Macroscopic Appearance	Microscopic Appearance 10X to 15X
TYPE I Compact Crystalline	Crystalline, hard, dense, sharp edges and smooth faces on breaking Resinous	Matrix made up of tightly interlocking crystals, no visible pore space between crystals, commonly produced "feather edge" and thin flakes on breaking
TYPE II Chalky	Dull, earthy, siliceous or argillaceous. Crystalline appearance absent because small crystals are less tightly interlocked, thus reflecting light in different directions, or made up of extremely fine particles (skeletal or other)	Crystals joined at different angles. Extremely fine texture may still appear "chalky" under this power, but others may begin to appear crystalline. Grain size <0.05 mm generally. Coarser textures classed as Type III
TYPE III Granular or Saccharoidal	Sandy or sugary appearance (sucrose) Size of crystals or particles classes as: Very fine = 0.05 mm Fine = 0.1 mm Medium = 0.2 mm Coarse = 0.4 mm	Crystals interlock at different angles, but considerable porosity between crystals. Oolitic and other granular textures fall in this class

Classification of Visible Pores

Class A: No visible porosity under 10X microscope or where pore size is <0.01 mm
Class B: Visible porosity, 0.01 but <0.1 mm
Class C: Visible porosity, 0.1 mm but size of cuttings
Class D: Visible porosity as shown by secondary crystal growth on faces of cuttings or "weathered-appearing" faces showing evidence of fracturing or solution channels; where pore size > size of cutting

Classification of Visible-Pore Frequency

Description	Frequency—Percentage of Surface Covered by Pores
Excellent	20
Good	15
Fair	10
Poor	5

Table A-2. Gross Petrographic-Petrophysical Relationships of Modified Archie Rock Types (from Company Reports by Roehl, 1959, 1961)

Rock Type	Visual and Petrographic Character	Matrix Particle Size General Range (μ)	Petrophysical Properties $\bar{r}$ Matrix[3]	P_d(psia)	G Factor	ϕri (%)[4]
I	Dense or compact; vitreous or dull	1–100	—	>20 variable	>0.4	—
II	Chalky, earthy or dull	5–20	~1.1	>100	<0.4	24
II–III	Microsucrosic or extra-fine, uniform particulate matrix	20–50	~5.5	20–100	<0.5	18
III	Sucrosic or very fine to medium heterogeneous particulate matrix	50–500	>5.5	<20	<0.7	<18
[1](III/1)$_L$	Hetero-particulate or crystalline with complex fabrics, variable mineralogy	Variable	>5.5	<20	≤0.7	15.5
[2](III/1)$_H$	Hetero-particulate or crystalline with complex fabrics, variable mineralogy	Variable	>3.5	<30	>0.7	48

[1]Highly variable mixtures of Archie Types I and III with low G factor.
[2]Highly variable mixtures of Archie Types I and III with high G factor.
[3]$\bar{r}$-average largest pore radius based on minimum entry pressure, in microns.
[4]Based on relative ineffective porosity at high P_c and extrapolated bulk volume occupied (Thomeer, 1960) of 10%.

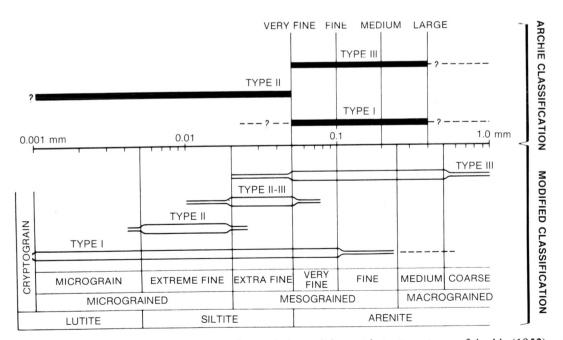

Fig. A-9. Relationships between grain size and crystal size and the matrix-textures types of Archie (1952) and Roehl. From unpublished company reports by Roehl (1959, 1961).

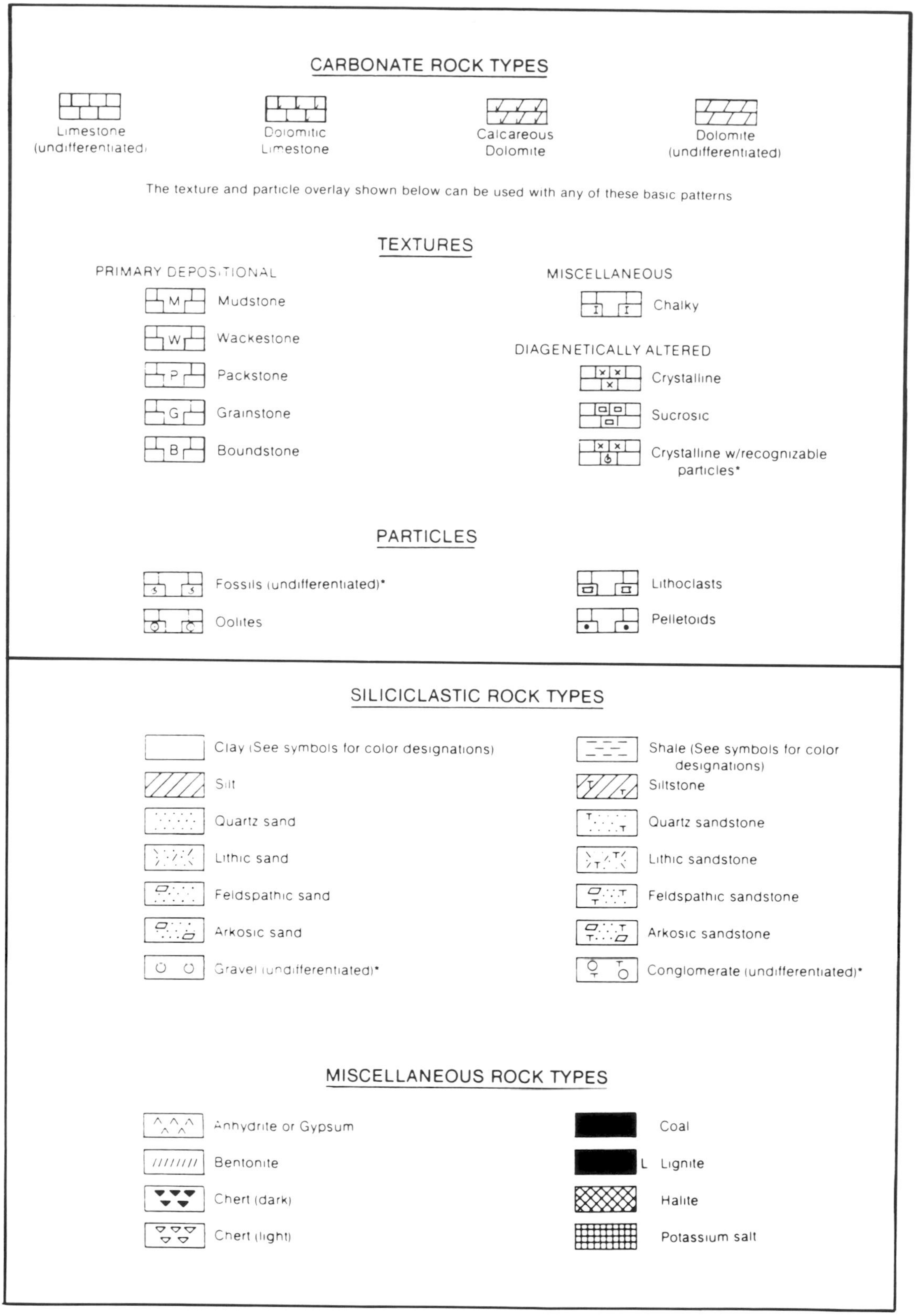

CARBONATE ROCK TYPES

Limestone (undifferentiated) Dolomitic Limestone Calcareous Dolomite Dolomite (undifferentiated)

The texture and particle overlay shown below can be used with any of these basic patterns

TEXTURES

PRIMARY DEPOSITIONAL

M Mudstone
W Wackestone
P Packstone
G Grainstone
B Boundstone

MISCELLANEOUS

I I Chalky

DIAGENETICALLY ALTERED

x x x Crystalline
Sucrosic
Crystalline w/recognizable particles*

PARTICLES

Fossils (undifferentiated)*   Lithoclasts
Oolites   Pelletoids

SILICICLASTIC ROCK TYPES

Clay (See symbols for color designations)   Shale (See symbols for color designations)
Silt   Siltstone
Quartz sand   Quartz sandstone
Lithic sand   Lithic sandstone
Feldspathic sand   Feldspathic sandstone
Arkosic sand   Arkosic sandstone
Gravel (undifferentiated)*   Conglomerate (undifferentiated)*

MISCELLANEOUS ROCK TYPES

Anhydrite or Gypsum   Coal
Bentonite   L   Lignite
Chert (dark)   Halite
Chert (light)   Potassium salt

Fig. A-10A

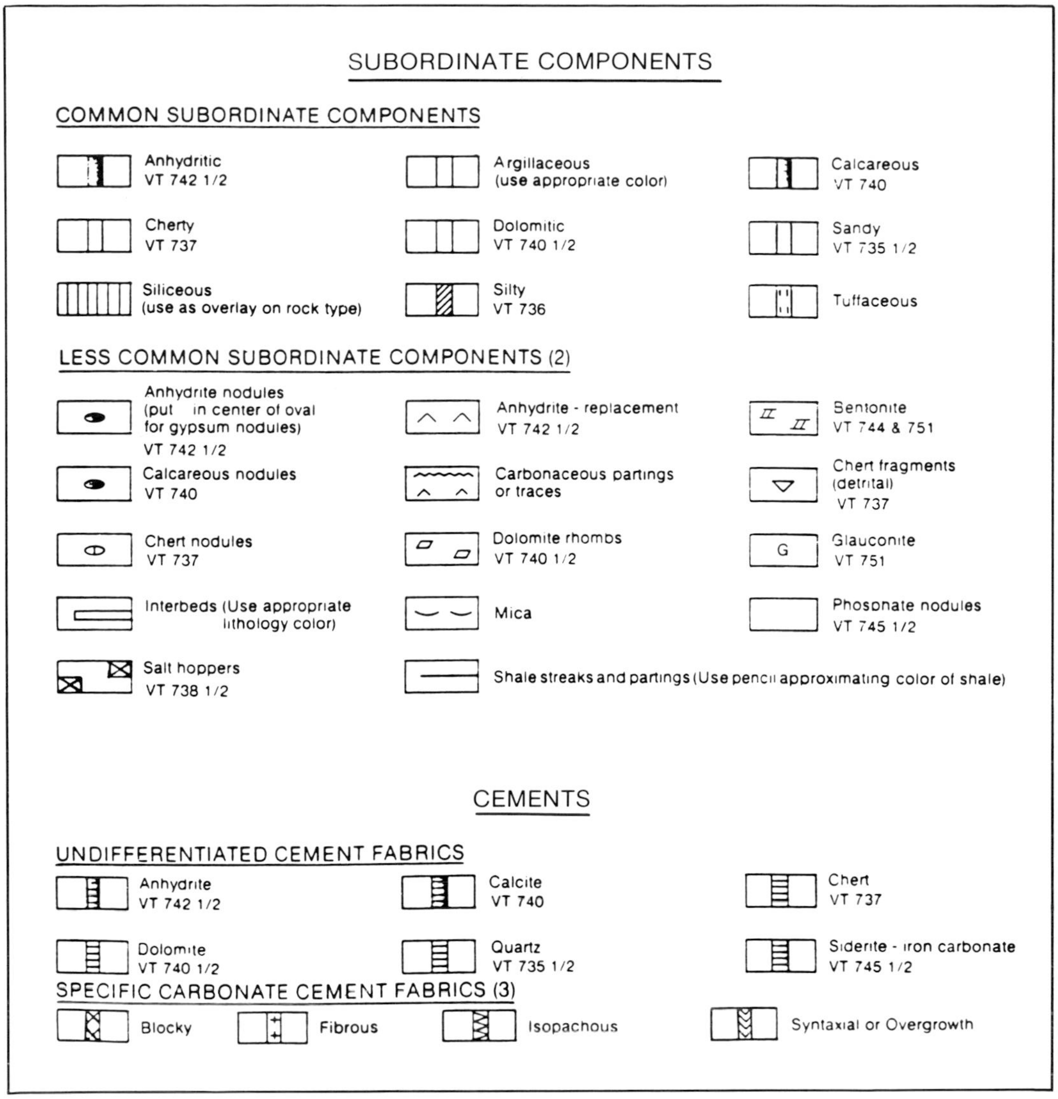

SUBORDINATE COMPONENTS

COMMON SUBORDINATE COMPONENTS

Anhydritic VT 742 1/2

Argillaceous (use appropriate color)

Calcareous VT 740

Cherty VT 737

Dolomitic VT 740 1/2

Sandy VT 735 1/2

Siliceous (use as overlay on rock type)

Silty VT 736

Tuffaceous

LESS COMMON SUBORDINATE COMPONENTS (2)

Anhydrite nodules (put in center of oval for gypsum nodules) VT 742 1/2

Anhydrite - replacement VT 742 1/2

Bentonite VT 744 & 751

Calcareous nodules VT 740

Carbonaceous partings or traces

Chert fragments (detrital) VT 737

Chert nodules VT 737

Dolomite rhombs VT 740 1/2

Glauconite VT 751

Interbeds (Use appropriate lithology color)

Mica

Phosphate nodules VT 745 1/2

Salt hoppers VT 738 1/2

Shale streaks and partings (Use pencil approximating color of shale)

CEMENTS

UNDIFFERENTIATED CEMENT FABRICS

Anhydrite VT 742 1/2

Calcite VT 740

Chert VT 737

Dolomite VT 740 1/2

Quartz VT 735 1/2

Siderite - iron carbonate VT 745 1/2

SPECIFIC CARBONATE CEMENT FABRICS (3)

Blocky

Fibrous

Isopachous

Syntaxial or Overgrowth

Fig. A-10B

Fig. A-10. Parts of the Shell system for logging geologic attributes of carbonate rocks in cores, cuttings, and outcrop samples. From Swanson (1981). Published by permission, American Association of Petroleum Geologists. *A.* Major sedimentary rock types, Dunham (1962) and other textural carbonate types, and major particle types. *B.* Subordinate components, and cements. *C.* Fossil components. *D.* Nonskeletal and other carbonate particles, lithoclasts, mineralization occurrences, and chert varieties. *E.* Common sedimentary (syngenetic) and burial-diagenetic to tectonic (epigenetic) structures found in carbonates.

FOSSILS

General Symbols

Macrofossil fragments, undifferentiated

* { Macrofossil fragments, rounded
     Macrofossils, whole

Microfossils, undifferentiated

** Fossils, encrusting

Specific Symbols

Algae, undifferentiated
Algae, Red
Algae, Green
Algal plates
Algal balls, oncolites, rhodoliths
Algal stromatolites
Brachiopods, undifferentiated
Brachiopods, phosphatic
Brachiopods, productid
Bryozoa, tube-like forms
Bryozoa, fenestellid forms
Calcispheres
Chara
Conodonts and scolecodonts
Corals, colonial
Corals, Chaetetes
Corals, Syringopora
Corals, solitary
Crinoids
Diatoms
Echinoderms

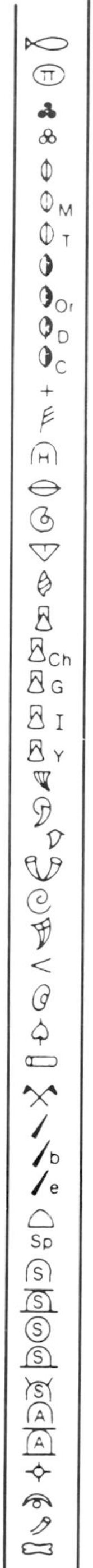

* Use appropriate fossil symbol
  within circle or square if
  fossil identifiable

** Use as underline under
   appropriate fossil symbol.
   Ex: ♣ = encrusting
   foraminifera

Fish remains
Fish scales
Foraminifera, undifferentiated
Foraminifera, pelagic
Foraminifera, small benthonic
        Miliolids
        Tubular forams
Foraminifera, large benthonic
        Orbitolina
        Dictyoconus
        Coskinolina and Coskinolinoides
        Fusulinids
Graptolites
Hydrozoa
Molluscs, undifferentiated
        Cephalopods
        Belemnites
    Gastropods
    Pelecypods (clams)
        Chondrodonta
        Gryphaea
        Inoceramus
        Oysters
Rudists, undifferentiated
        Caprinids
        Caprotinids
        Monopleurids
        Requieniids
        Radiolitids
    Tentaculites
Ostracods
Plant remains
Silicified wood
Spicules

Spines
Spines, brachiopod
Spines, echinoid
Sponges
Spores and/or pollen
Stromatoporoids, undifferentiated
Stromatoporoids, lamillar
Stromatoporoids, spherical
Stromatoporoids, hemispherical
Stromatoporoids, branching
Stromatoporoids, Amphipora, undifferentiated
Stromatoporoids, Amphipora, lamillar
Radiolarians
Trilobites
Worm tubes
Vertebrates

Fig. A-10C

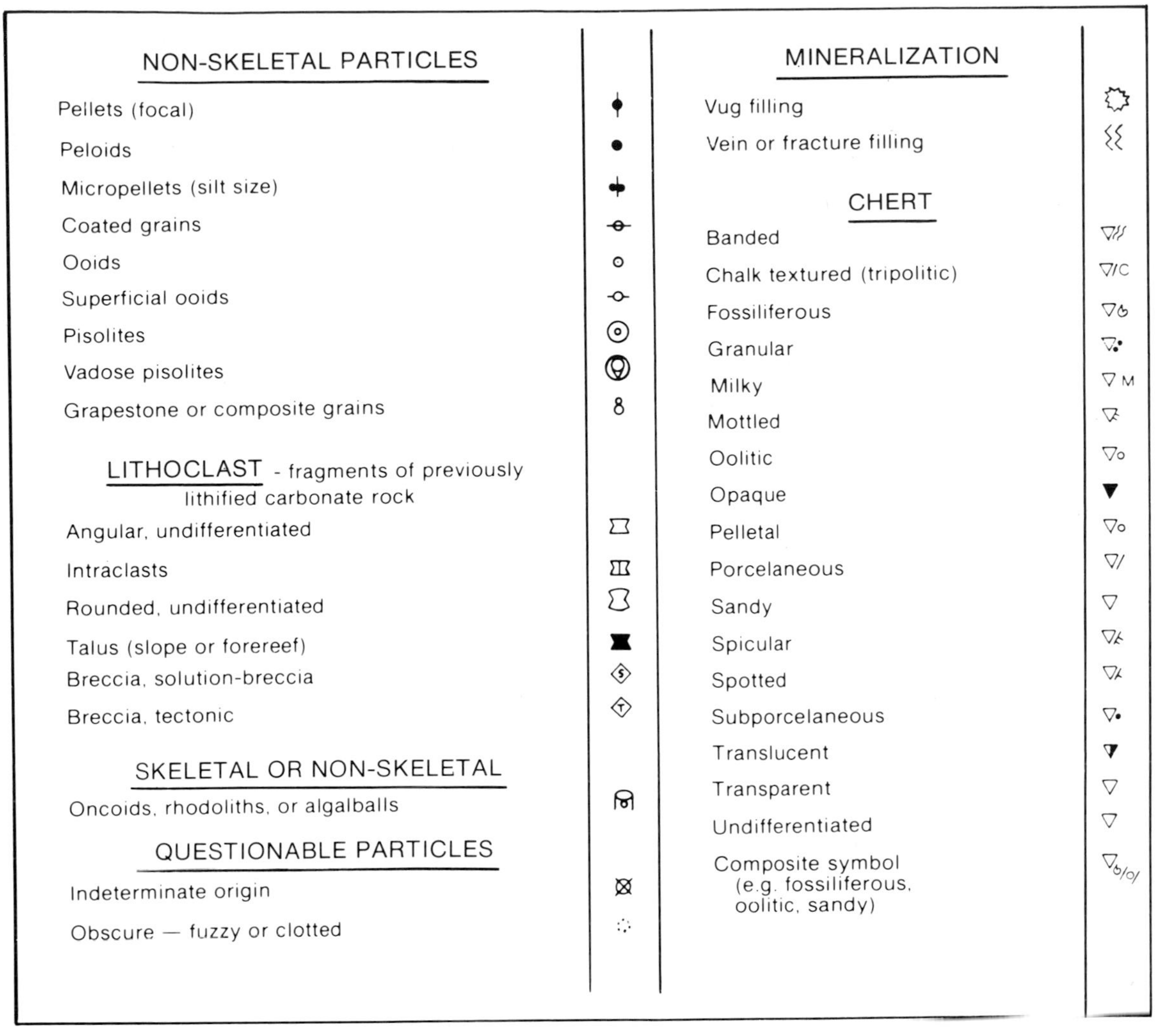

NON-SKELETAL PARTICLES

Pellets (focal)	
Peloids	
Micropellets (silt size)	
Coated grains	
Ooids	
Superficial ooids	
Pisolites	
Vadose pisolites	
Grapestone or composite grains	

LITHOCLAST - fragments of previously lithified carbonate rock

Angular, undifferentiated	
Intraclasts	
Rounded, undifferentiated	
Talus (slope or forereef)	
Breccia, solution-breccia	
Breccia, tectonic	

SKELETAL OR NON-SKELETAL

Oncoids, rhodoliths, or algalballs

QUESTIONABLE PARTICLES

Indeterminate origin	
Obscure — fuzzy or clotted	

MINERALIZATION

Vug filling	
Vein or fracture filling	

CHERT

Banded	
Chalk textured (tripolitic)	
Fossiliferous	
Granular	
Milky	
Mottled	
Oolitic	
Opaque	
Pelletal	
Porcelaneous	
Sandy	
Spicular	
Spotted	
Subporcelaneous	
Translucent	
Transparent	
Undifferentiated	
Composite symbol (e.g. fossiliferous, oolitic, sandy)	

Fig. A-10D

SYNGENETIC STRUCTURES

A.   Stratification

   1.   Parallel type

      Thickness of Bedding
        Metric System

millimeter bed	1 mm - 10 mm	mm
centimeter bed	1 cm - 10 cm	cm
decimeter bed	1 dm - 10 dm	dm
meter bed	1 m - 10 m	m

When greater precision is desired,
the modal thickness can be indicated,
e.g., beds with modal thickness 3 meters.   3m

      British System

thin lamina	< 1/10 inch	thlam
lamina	1/10 - 1/2 inches	lam
very thin bed	1/2 - 2 inches	(tn)
thin bed	2 - 6 inches	tn
medium bed	1/2-1 1/2 feet	med
thick bed	1 1/2 - 5 feet	tk
very thick bed	> 5 feet	tk

   2.   Cross-bedding
      in general
      with angle indicated
      chevron
      climbing
      festoon
      planar

   3.   Irregular bedding

   4.   Graded bedding

   5.   No apparent bedding

   6.   Nodular bedding

B.   Current - produced markings

   1.   Ripple marks
      asymmetrical
      interference
      symmetrical

   2.   Pull-over flame structure

   3.   Scour and fill

   4.   Flute cast

   5.   Groove cast

   6.   Striation

   7.   Parting lineation

C.   Organism - produced markings

   1.   Burrowed
      slightly burrowed
      moderately burrowed
      well burrowed

   2.   Churned

   3.   Bored

   4.   Bored surface

   5.   Organism tracks and trails

   6.   Plant root tubes

   7.   Vertebrate tracks

D.   Penecontemporaneous deformation structures

   1.   Mud cracks

   2.   Rain or hail prints

   3.   Pull-apart

   4.   Slump structures and contorted bedding

   5.   Convolute bedding

   6.   Load cast

   7.   Tepee structure

   8.   Birdseye, fenestral fabric

EPIGENETIC STRUCTURES

A.   Solution structures

   1.   Breccia, solution, collapse

   2.   Dissolution - compaction (horse tail)

   3.   Stylolite

   4.   Vadose pisolite

   5.   Vadose silt

   6.   Boxwork

   7.   Salt hoppers or casts

B.   Tectonic structures

   1.   Fractures

   2.   Slickensides

   3.   Breccia, tectonic

C.   Miscellaneous

   1.   Geopetal fabric

   2.   Cone-in-cone

   3.   Stromatactis

   4.   Boudinage, ball and flow structure

Fig. A-10E

References

ARCHIE, G.E., 1952, Classification of carbonate reservoir rocks and petrophysical considerations: Amer. Assoc. Petroleum Geologists Bull., v. 36, no. 2, p. 278–298.

CHOQUETTE, P.W., and L.C. PRAY, 1970, Geologic nomenclature and classification of porosity in sedimentary carbonates: Amer. Assoc. Petroleum Geologists Bull., v. 54, p. 207–250.

DUNHAM, R.J., 1962, Classification of carbonate rocks according to depositional texture: *in* Ham, W.E., ed., Classification of Carbonate Rocks—a symposium: Amer. Assoc. Petroleum Geologists Mem. 1, p. 108–121.

EMBRY, A.F., and J.E. KLOVAN, 1971, A Late Devonian reef tract on northeastern Banks Island, N.W.T.: Bull. Canadian Petroleum Geology, v. 19, p. 730–781.

FOLK, R.L., 1959, Practical petrographic classification of limestones: Amer. Assoc. Petroleum Geologists Bull., v. 43, p. 1–38.

FOLK, R.L., 1962, Spectral subdivision of limestone types: in Ham, W.E., ed., Classification of Carbonate Rocks—a symposium: Amer. Assoc. Petroleum Geologists Mem. 1, p. 62–84.

GAGLIARDI, R., O. KALIN, E. MORREALE, E. PATACCA, P. SCANDONE, and R. SPRUGNOLI, 1980, Una Banca dai Geologici del Progetto Finalizzato Geodinamica: Consiglio Nazionale Delle Richerche, Rome, Italy.

HAM W.E., and L.C. PRAY, 1962, Modern concepts and classifications of carbonate rocks: *in* Ham, W.E., ed., Classification of Carbonate Rocks—a symposium: Amer. Assoc. Petroleum Geologists Mem. 1, p. 2–19.

JAMES, N.P., 1983, Reefs: *in* Scholle, P.A., D.G. Bebout, and C.H. Moore, Carbonate Depositional Environments: Amer. Asoc. Petroleum Geologists Mem. 33, p. 345–440.

REECKMAN, A., and G.M. FRIEDMAN, 1982, Exploration for Carbonate Petroleum Reservoirs: John Wiley and Sons, New York, 213 p.

SWANSON, R.G., 1981, Sample Examination Manual; Methods in Exploration Series: Amer. Assoc. Petroleum Geologists, Tulsa, OK.

THOMEER, J.H.M., 1960, Introduction of a pore geometrical factor defined by the capillary pressure curve: Jour. Petroleum Technology, Tech. Note 2057.

Glossary of Terms*

Algae Photosynthetic, almost exclusively aquatic plants of a large and diverse division of the thallophytes.

Allochem Allochemical constituent (Folk, 1959); chemically and biochemically precipitated carbonate aggregates occurring in mechanically deposited limestones.

Allochthonous Said of material formed in an environment other than that in which it is found, and transported therefrom.

Alluvial fan A fan-shaped deposit of stream alluvium laid down where a change in stream gradient occurs.

Anhedral A descriptor for a mineral that has no planar crystal faces.

Anhydrite A mineral consisting of anhydrous calcium sulfate, $CaSO_4$.

Aquiclude A body of relatively impermeable rock that acts wholly or in part to confine an aquifer.

Aquathermal A term pertaining to abnormally high pore pressures produced by heating of confined connate water in sediments being compacted during burial.

Aragonite The orthorhombic mineral consisting of calcium carbonate, $CaCO_3$.

Arenaceous A textural term for clastic sediments or sedimentary rocks of average grain size ranging from 1/16 to 2 mm (*n*: arenite).

Argillaceous Pertaining to a clastic sediment or sedimentary rock containing clay-size particles; claycy or shaly.

Arthropod Any one of a group of invertebrates belonging to the phylum Arthropoda characterized chiefly by jointed appendages and segmented bodies; includes trilobites and crustaceans.

Atoll reef A coral reef complex encircling a lagoon, devoid of pre-existing land of non-coral origin and surrounded by deep open sea.

Aulacogen A fault-bounded intracratonic or interplate trough or graben.

Authigenic Said of a mineral that has grown in place subsequent to the formation of the sediment or rock in which it occurs.

Autochthonous Formed or occurring in the place where found.

Bafflestone A textural type of biohermal limestone having a framework of calcareous organisms that have acted to "baffle" and entrap sediment.

*Except as noted, terms are from, or modified from: Amer. Geological Inst. *Glossary of Geology*, R. L. Bates and J. A. Jackson, eds., 2nd edition (1980); The Encyclopedia of Sedimentology, Earth Sciences, Vol. VI, R. W. Fairbridge and J. Bourgeois, eds. (1978), Dowden, Hutchinson & Ross, Inc.; Glossary of Terms and Expressions Used in Well Logging, Soc. Prof. Well Log Analysts (1975); A Dictionary of Petroleum Terms, 2nd edition, Petroleum Extension Service, Univ. Texas (1979); Principles of Sedimentology, G. M. Friedman and J. E. Sanders (1978), John Wiley and Sons, Publ.

Baroque dolomite Coarse dolomite crystals, generally a millimeter or larger in size, milky in color, and having curved saddle-like crystal faces due to rotating c-axes, and consequently displaying sweeping extinction in cross-polarized light (*syn*: saddle dolomite).

Bathyal Pertaining to the ocean environment or depth zone between 100 and 500 fathoms (200–1000 m); also pertaining to the organisms of that environment.

Bathymetric Pertaining to the morphology or bottom configuration of a water body.

Beachrock A friable to well-cemented sedimentary rock, formed in intertidal zones, and consisting of calcareous debris cemented with calcium carbonate; best known in tropical or subtropical regions.

Bellerophon(tid) A group (superfamily) of gastropods roughly of thumb size that have planispiral shells; found mostly in Ordovician through Devonian carbonates, but known range is Cambrian–Triassic.

Benthic *see* Benthonic

Benthonic Pertaining to the benthos, the ocean bottom environment, and associated bottom-dwelling marine life.

Bentonite A soft, plastic claystone composed principally of clays of the montmorillonite (smectite) group, which result principally from chemical alterations of volcanic rocks.

Bioclastic Aspect of a material (sediment, rock, particle) alluding to its composition of broken remains of organisms.

Biofacies A lateral subdivision of a stratigraphic unit, distinguished from other adjacent subdivisions on the basis of its biologic constituents, without respect to the non-biological features of its lithology.

Biogenesis Formation by the action of organisms (*adj*. biogenic, biogenetic).

Bioherm A circumscribed mass of rock, usually limestone, of local extent, and composed of the remains of sedentary $CaCO_3$-secreting organisms; enclosed by rock of different compositional makeup, not necessarily noncarbonate.

Biolithite A rigid framework of skeletal organic remains that grew and remained in place to form a limestone (Folk, 1959).

Biomicrite A limestone containing less than 25% intraclasts and less than 25% ooids, with a volume ratio of fossils and fossil fragments to nonskeletal allochems of more than 3 to 1, and a carbonate mud matrix more abundant than the sparry calcite cement (Folk, 1959).

Biostratigraphic Pertaining to a unit or stratum that is defined and identified by one or more distinctive fossil species, without regard to other physical features; adjective denoting the basis for an interpretation of geologic age.

Biostrome A layer of skeletal, sedentary organic fragments comprising a rock of uniform thickness, distinctly bedded and of broad areal extent.

Bioturbation The stirring and churning (*in situ* reworking) of a sediment by organisms.

Bitumen In general, the spectrum of natural flammable hydrocarbons (petroleum, asphalts, mineral waxes etc.), including semisolid and solid admixtures with mineral matter.

Boudinage A structure common in deformed sedimentary and metamorphic rocks, resulting from the stretching, thinning and breaking of a competent bed within less competent strata resembling boudins (sausage) in cross-section.

Boundstone A carbonate rock texture defined by Dunham (1962) and formed by the binding action of skeletal organic components substantially preserved in position of growth.

Brachiopod A solitary marine invertebrate characterized by two bilaterally symmetrical valves (shells) and a lophophore (circular or horseshoe-shaped feeding organ).

Brachyanticine An anticline that is wider than it is long.

Bryozoan A calcareous or chitinous skeletal, colonial invertebrate marine organism belonging to the phylum Bryozoa.

Calcarenite A limestone consisting of predominantly (>50%) calcite grains of sand size.

Calcisphere Calcified spherical grains on the order of 100 microns in diameter, believed to be reproductive cysts or gametangia of green algae.

Calcite The principal mineral, $CaCO_3$, of the hexagonal carbonate mineral group. A primary biogenic precipitate, authigenic cement, or secondary mineral after aragonite.

Calcium carbonate compensation depth That level within the ocean below which the rate of dissolution of calcium carbonate exceeds the rate of deposition (*abbrev*: CCD).

Calcspar Coarse crystalline calcite cement (*syn*: sparry calcite).

Caliche A broad term variously applied to calcareous ($CaCO_3$) layers, crusts, and porous, friable horizons, or to thick impermeable, indurated layers, but also applied to similar features composed of silica, gypsum, etc. Formed by a variety of processes, chiefly capillary action and evaporation (*syn:* calcrete).

Capillarity The action by which a fluid such as water is drawn up into small interstices such as tubes and pores as a result of surface tension.

Cathodoluminescence The emission of colored light by minerals containing certain activator elements upon bombardment at high voltage by a cathode ray (electron) beam.

Chalcedony A cryptocrystalline or variety of silica, commonly microscopically fibrous, with lower indices of refraction and mineral density than quartz.

Chert A hard, dense, dull to semi-vitreous, cryptocrystalline sedimentary rock, composed of variable amounts of silica in the form of fibrous chalcedony and microcrystalline quartz; may contain minor carbonate, iron oxide, or other impurities.

Chronostratigraphic Characterization of geologic history (or a portion thereof) based on the age and time sequence of strata.

Clastic Pertaining to a rock or sediment composed principally of broken fragments derived from pre-existing rocks or minerals that have been transported from their place of origin.

Coelenterate Any multicelled invertebrate belonging to the phylum Coelenterata, characterized by a body wall composed of two layers of cells connected by a structureless mesogloea, a single body cavity, and radial or biradial symmetry.

Coeval Coexisting during a specified period of geologic time.

Coccolith Contraction of coccolithophore, a golden-brown unicellular alga and calcareous protistid of the division Chrysophyta; microscopic remains of coccoliths are the principal sedimentary constituents of chalk deposits.

Connate water Interstitial water, originating at the same time as the enclosing sediments.

Craton A part of a continent that has had a prolonged geologic history of tectonic stability.

Crinoid(ea) A large class of echinoderms characterized by five-fold body symmetry and a globular body enclosed by calcareous plates, from which branching appendages entered in radial arrays; also characterized by a segmented stem of calcite ossicles.

Cristobalite A silica (SiO_2) polymorph of quartz that is stable only above 1470°C.

Cyclothem An informal sedimentary unit equivalent to formation which contains a sequence of beds deposited during a cycle of non-marine (regressive) and overlying marine (transgressive) sedimentation.

Dasyclad(acae) A type of green alga belonging to the family Dasycladaceae, composed of filaments whorled about a central axis that are often preserved by a layer or rind of precipitated calcium carbonate.

Dedolomitization The replacement of dolomite by calcite during diagenesis or chemical weathering.

Detritus Loose, fragmented rock and mineral debris mechanically transported from its place of origin.

Depletion drive Production of oil as a result of the expansion of gas following its release from solution below the saturation pressure.

Depocenter An area of maximum deposition, which thus comprises the thickest part of any stratigraphic unit in a depositional basin.

Diagenesis Term encompassing any post-depositional changes resulting from physical, chemical, or biological processes short of metamorphism.

Diapir A dome or anticlinal fold whose enclosing strata are usually ruptured by the plastic vertical displacement of core material, usually salt or clay shale.

Diastem A relatively short interruption in sedimentation with little or no erosion before sedimentation resumes.

Diatom A microscopic, single-celled, aquatic plant that secretes walls (frustules) of silica.

Dilatancy An increase in bulk volume during deformation, accompanied in crystalline rocks of low porosity by an increase in pore volume as a result of microfracturing.

Disconformity An unconformity above and below which the bedding planes are essentially parallel.

Dolomitization A process whereby limestone or its precursor sediment is wholly or partly converted to dolomite by replacement of the original $CaCO_3$ by magnesium carbonate, through the action of Mg-bearing water.

Dolospar A sparry dolomite crystal, generally of rather coarse size on the order of 100 microns or more.

Dolosparite A sedimentary rock composed of dolospar.

Dolostone A term employed for sedimentary rock composed of the mineral dolomite.

Druse A crust or coating of crystals lining a cavity (druse) in a rock (specif. sparry calcite lining the pores of a limestone).

Echinoderm A solitary marine invertebrate, the phylum for which (Echinodermata) is characterized by radial symmetry and an endoskeleton formed by calcite plates or ossicles and a water vascular system (*e.g.*, crinoids, sea urchins).

Echinoid An echinoderm belonging to the class Echinoidea and characterized by subspherical shape, interlocking calcareous plates, and movable appendages; *e.g.*, sea urchin and sand dollar.

Ecologic Of, or pertaining to, the environment in regard to included organisms and their interrelationships therein.

Enterolithic A sedimentary structure resembling in two-dimensional view intestinal folds and formed through physicochemical changes that involve increases and decreases in unit volumes of the rock; said of local crumpling in some thin-bedded anhydrite.

Eogenetic A term proposed by Choquette and Pray (1970) for that stage and realm of diagenesis represented by the time interval between final deposition and burial below the depth range of significant influence by processes that rely on proximity to the surface depositional interface.

Epeiric (seas) Having to do with seas on the margin of, or within, a continent (*syn*: epicontinental seas).

Epeirogeny Largely vertical diastrophic earth movement of very large scale that has formed a principal structural feature, such as a mountain range, continent, or ocean basin.

Epicontinental Pertaining to the continental shelf.

Epigenetic Pertaining to sedimentary structures, minerals, and mineral deposits formed after deposition, at low temperature and pressure changes or transformations affecting sedimentary rocks subsequent to compaction (a definable stage of diagenesis).

Euhedral A descriptive term for mineral crystals bounded wholly by crystal faces.

Eustatic Pertaining to world-wide changes of sea level (*n.* eustacy).

Evaporative pumping The third of three stages of a single hydrological cycle (others: 1. flood recharge and 2. capillary evaporation); the upward flow of groundwater in the zone of saturation to replace water lost by evaporation in the capillary zone above the groundwater table.

Facies An areally restricted part of a designated stratigraphic, paleontologic, or diagenetic unit exhibiting characteristics significantly different from those of other parts of the unit.

Fan delta A gently sloping, fan-shaped alluvial deposit produced where a mountain stream flows out onto a lowland.

Favositid Tabulate corals (family Favositidae) characterized by massive colonies of slender corallites.

Fenestral Alluding to a sediment or rock texture wherein numerous cavities occur that exceed the size of matrix grains or the associated intergranular interstices, thus apparently without framework support (*syn*: birdseye).

Ferroan Containing ferrous iron, usually in concentrations sufficient to react with potassium ferricyanide stain ($>1\%$ Fe^{++}) or to produce a rusty brown color when weathered.

Flagellate An organism, most commonly a protozoan, which bears a flagella.

Foram(inifera) Any protozoan belonging to the order Foraminiferida characterized by the presence of a test composed of agglutinated particles or of secreted calcite (rarely of silica or aragonite) and commonly found in marine to brackish environments; range: Cambrian to the present.

Framboid(s) Microscopic spheroidal clusters of pyrite grains said to be associated with organic material.

Free water level That level to which water will (or would) adjust in the subsurface when unimpeded by capillary forces and subjected only to atmospheric pressure.

Frustule The siliceous skeleton of a diatom.

Fusilinid(ae) A foraminifer (protozoa) belonging to the superfamily Fusilinacea characterized by a spindle-shaped, spheroidal or discoid test (skeleton) with complex internal structure.

Gastropod Colloquially, a snail; usually with

calcareous shell closed at apex, sometimes spiralled, without chambers and asymmetrical.

Geopetal A feature of a rock or sediment that reveals the sense of orientation relative to gravitational forces, *e.g.*, sediment-floored cavities.

Gilsonite A naturally occurring solid hydrocarbon belonging to the asphalt group.

Graben An elongate, relatively depressed crustal unit or block that is bounded by faults on its long sides.

Grainstone A mud-free, grain-supported sedimentary carbonate rock defined by Dunham (1962).

Grapestone A cluster of small calcareous pellets or grains, commonly of sand size and stuck together by incipient cementation shortly after deposition.

Gravity high An area of numerically higher gravitational attraction relative to the surrounding gravity field.

Growth fault A fault in a sedimentary rock sequence that forms contemporaneously and continuously with deposition, so that the throw increases with depth and the strata of the downthrown side are thicker than the correlative strata of the upthrown side.

Gypsum The mineral form of hydrous calcium sulfate: $CaSO_4 \cdot 2H_2O$.

Halokinesis A general term for the structure and mechanism of emplacement of salt domes and other salt-controlled structures (*syn*: salt tectonics).

Hardground A zone at the sea bottom, generally a few cm thick, the sediment of which is lithified to form a hardened surface; often encrusted, discolored, hardened and bored by organisms; implies a gap in sedimentation, and may be preserved stratigraphically as a disconformity (*syn*: hard ground).

Hemipelagic Deep-sea sediment of pelagic origin, principally composed of pelagic organisms with varying amounts of inorganic terrestrial sediment.

Homocline A general term for a rock unit within which the strata have a common uniform dip, *e.g.*, one limb of a fold, a tilted fault block, a monocline, or an isocline.

Horst An elongate, centrally uplifted crustal block that is bounded by faults on its long sides.

Hydrofrac Well stimulation by artificially induced hydraulic fracturing.

Hydrozoan *See* Coelenterate.

Hypersaline Excessively saline, with a salinity substantially greater than that of normal sea water. Often regarded specifically as having a salinity above the lowest concentration at which halite is precipitated.

Ichnofossil A sedimentary structure consisting of a fossilized non-growth activity of an animal, such as marks made by moving, creeping, crawling, feeding, browsing, running, etc., resulting in tracks, trails, burrows, tubes, tunnels, etc.

Illite A general name for a group of three-layer, mica-like clay minerals widely occurring in argillaceous marine shales.

Infratidal That water regime occurring below mean low tide (*syn*: subtidal).

Intergranular Referring to pore spaces or other characteristics existing between individual grains or particles of a sedimentary rock (*syn*: interparticle).

Intertidal The depth interval between mean high and mean low tide.

Intraclast A general term referring to sedimentary fragments derived by penecontemporaneous erosion within the initial site of sedimentation and redeposition within, or immediately adjacent to, such site.

Intracratonic Contained on or within the craton, as with an epeirogenic basin or parageosyncline. Wholly within or upon a continent or stable crustal area.

Intragranular Referring to pore spaces or other characteristics existing within individual grains, particles, or other constituents of a sedimentary (esp. carbonate) rock (*syn*: intraparticle).

Intrastratal Occurring within a stratum or strata.

Isobath In oceanography, a line on a map or chart that connects points of equal water depth.

Isopachous Of, or relating to, an isopach: that line drawn on a map that connects points of equal value.

Kaolinite A high-alumina clay mineral of the kaolin group: $Al_2Si_2O_5(OH_4)$

Karst A type of topography that is formed on limestone, dolomite, or gypsum as a result of dissolution and is characterized by sink holes, caves, and underground drainage.

Kerogen An insoluble, fossil organic complex that can be distilled to yield petroleum products.

Laminite A thinly parallel-laminated, bottom-set detrital clastic bed occurring seaward of a genetically related turbidite facies; also an evenly parallel-laminated, lime-mud-rich sediment or rock.

Laterite　A highly weathered subsoil, usually red to brown, that is rich in secondary oxides of Fe and/or Al, nearly devoid of bases and primary silicates, and sometimes containing large amounts of quartz and kaolinite.

Lenticular　Any three-dimensional body habit resembling a lens shape.

Limpid dolomite　A variety of dolomite crystal that is optically clear (essentially free of inclusions), generally <100 μ in size, and thought to have precipitated from relatively dilute pore waters.

Lithoclast　A mechanically formed and deposited fragment of a carbonate rock, normally >2mm in diameter, derived from an older limestone, dolomite, or other sedimentary rock stratum.

Lithofacies　A lateral, mappable subdivision of a stratigraphic unit distinguished on the basis of lithologic variations.

Lithographic　A sedimentary texture of some calcareous rocks, generally limestones, composed of particles of less than clay size and characterized by its extremely smooth appearance like that of stone used in lithography.

Littoral　Pertaining to the benthic ocean environment or depth zone between high and low tide levels; also, pertaining to the organisms of that environment.

Löferite　Carbonate rock containing a great abundance of shrinkage-type pores ("birdseye" or fenestral limestone or dolomite).

Lysocline　A depth or depth interval within the ocean where evidence of considerable dissolution of $CaCO_3$ is first encountered.

Marl　Soft, loose, earthy sediment or rock consisting chiefly of an intimate mixture of clay and calcium carbonate in varying proportions between 35 and 65% of each; formed under marine or freshwater conditions.

Meniscus cement　A type of calcareous cement so called because it is precipitated at grain-to-grain contacts in pores containing both air and water, in meniscus style; characteristically formed therefore in vadose groundwater zones.

Mesogenetic　A term proposed by Choquette and Pray (1970) for a period and realm between when newly buried deposits are affected mainly by

processes related to the depositional interface (eogenetic stage) and when long-buried deposits are affected by processes related to the erosion interface (telogenetic stage).

Metabentonite　Metamorphosed or altered bentonite characterized by clay minerals that no longer can absorb, or adsorb large quantities of, water.

Meteoric　Related to or associated with atmospheric manifestations; most notably wind, rain, and the resulting percolating ground waters derived therefrom.

Micrite　A term proposed by Folk (1959) for the microcrystalline calcite portion of a limestone; originally defined as <4 microns in crystal diameter, now less strictly defined and commonly understood to be <10 microns.

Micritization　Conversion of sedimentary particles partly or completely to micrite-size $CaCO_3$, possibly due to microscopic boring algae.

Microfacies　Geologically, a more restricted chronological or areal representation of characteristics generally attributable to the term facies; also, a facies defined by constituents identifiable only with a microcope (low to moderate magnifications).

Microspar　Calcite crystal mosaic of post-depositional origin in limestone up to ~50 microns in crystal diameter.

Microsucrosic　Microcrystalline texture of largely euhedral to subhedral calcite or dolomite crystal (mosaics) in the approximate size range 5 to 50 microns.

Miliolid　A foraminifer belonging to the family of Miliolidae, characterized by a test that usually has a porcelaneous and imperforate wall and has two chambers to a whorl variably arranged about a longitudinal axis.

Miogeosyncline　A geosyncline in which volcanism is not associated with sedimentation.

Moldic　Said of porosity formed by the solution removal of more soluble granular constituents of a rock.

Mudstone　Mud-supported carbonate sedimentary rock containing less than 10 percent grains with diameters greater than 20 microns.

Nannoagorite　A carbonate rock described by S. Honjo, 1969, for an assemblage of predominately nannofossil (coccolith) grains.

Nannofossils　Collective term for certain small marine fossils, such as coccoliths, near the limit of resolution of conventional light microscopes.

Nannoplankton   Plankton in the size range of 5 to 60 microns.

Natural potential   *see* Self potential

Neomorphism   Transformation of a sedimentary or diagenetic mineral that retains the gross (chemical) composition.

Occlusion   The reduction or replacement of porosity by mineral growth or internal sedimentological infilling.

Oligomictic   Said of a clastic sedimentary rock composed of a single rock type.

Oncolite(ic)   A concentrically laminated calcareous sedimentary structure, formed by the successive accretion of layered sheaths of blue-green algae.

Onlap   A regular progression of sedimentary units through time, as in a conformable sequence that transgresses shoreward, so that each succeeding younger unit terminates farther from the initial reference point toward shoaler water.

Oolite   A sedimentary rock composed of ooids (ooliths), which are single, rounded accretionary, sand-sized grains of calcium carbonate formed by precipitation around distinct nuclei.

Oomold(ic)   A spheroidal opening in a sedimentary rock resulting from the dissolution of an ooid (oolith).

Orthomicrite   Primary calcareous micrite.

Ossicles   Calcareous skeletal components of echinoderms (*e.g.*, plates).

Ostracod(e)   An aquatic crustacean generally of microscopic size, and characterized by a bivalve and generally calcified carapace with a hinge along the dorsal margin.

Packstone   A grain-supported sedimentary carbonate rock with variable amounts of intergranular calcareous mud (Dunham, 1962).

Palisade(-style)   A type of pore-lining calcium carbonate cement composed of markedly elongate crystals arranged picket-fence style on or around a grain or other substrate.

Paragenesis   A sequential order of mineral formation or transformation.

Paramorphism   The transformation of internal structure of a mineral without change of external form or of chemical composition.

Parastratigraphy   Stratigraphy based on operational units rather than the classical criteria (fossil-defined biostratigraphic zones) of orthostratigraphy.

Passive margin   A continental margin formed by the rifting apart and separation of formerly combined plates, and generally marked by a system of normal faults now mantled by younger sedimentary sequences.

Pedological   Pertaining to (the science of) soils.

Pelagic   Of, or pertaining to, the open ocean as an environment.

Pelecypod(a)   A large class of benthic aquatic Mollusca characterized by a bilaterally symmetrical bivalve shell, a hatchet-shaped foot, and sheet-like gills.

Peloid   A non-generic name for a grain composed of cryptocrystal line or microcrystalline material.

Peridotite   A coarse-grained plutonic rock, composed chiefly of olivine with or without other mafic minerals and containing little or no feldspars.

Peritidal   Pertaining to subaerial and subaqueous zones under tidal influence.

Petrophysics   Term applied to both physical and chemical properties in reference to pore frequency, pore-size distribution, and fluid properties of reservoir-related rock bodies.

Photic   Of or relating to penetration by light, specifically with regard to the zone of photosynthesis of aquatic organisms.

Phreatic   With respect to the zone of saturation below the permanent free water level (table).

Phylloid algae   Fossil calcareous algae characterized by leaf-like or curved platy forms and found principally in late Paleozoic carbonate rocks; may or may not be identifiable more precisely, but are known to include both red and green algae (Rhodophyta and Chlorophyta).

Pinnacle reef   A conical, or steep-sided, upward tapering, bioconstructed mound or reef.

Pisolite   Usually a sedimentary limestone composed of cemented pisoliths which are pin-size ($\approx$2–10 mm) grains of accretionary calcium carbonate of probable biochemical origin.

Plankton   Open-marine (pelagic) organisms that float.

Poikilitic   Rock texture in which small crystals of one mineral are irregularly scattered without common orientation in a larger crystal of another mineral.

Poikilotopic   Pertaining to the fabric of a carbonate rock in which larger crystals enclose smaller crystals or grains of another mineral.

Polyhedral   Multicrystal surfaces; multifaceted crystal form.

Porcellaneous   Having the appearance of glazed porcelain.

Pore geometrical factor A numerical notation representing pore-size distribution and interconnection of porous carbonate rocks; referenced as "G"; *see* Classifications.

Porphyroblast A crystal precipitated within, and substantially coarser than, its host sediment or sedimentary rock, and formed in part by replacement of host constituents.

Progradation The process of seaward construction of a shoreline by deposition of new beach (coastal) material.

Pseudomorph A mineral whose outward crystal form is that of another mineral.

Pyroclastic Pertaining to clastic rock material formed by volcanic explosion and rock textures resulting therefrom.

Quaquaversal Said of strata and structures that dip in all directions away from a central point, as on the flanks of domes.

Radiaxial Radially axial *e.g.*, arrays of calcite crystals lining sedimentary rock cavities and elongate normal to the cavity walls; crystals have curved cleavage concave outward and their optic axes converge outward toward the center of the cavity (*var*: radiaxial fibrous).

Radiolaria A marine pelogia, single-celled animal (protozoan of the class Achinopoda), characterized by a siliceous skeleton spheroidal shape, and codiatria protoplasmic extensions.

Radiolitid(ae) A member of the aberrant pelecypod family, Rudistid(ae).

Recrystallization The formation of new crystalline material in a rock, essentially in the solid state, without essential change in chemical composition.

Reef A community of frame-building organisms on the sea floor that comprises a natural biofacies and whose bathymetric size and configuration alter or significantly contribute to variations in adjacent marine environments.

Reflux dolomitization A theoretical process whereby dense evaporative brines, enriched in Mg^{++}, sink by seepage through carbonate sediment and initiate the diagenetic alteration of calcium carbonate to dolomite.

Regolith A general term for the layer of fragmental or unconsolidated rock material of whatever origin that nearly everywhere forms the surface of the land and overlies more coherent bedrock.

Regression (strat.) A retreat or contraction of the sea from land areas and the sedimentary evidence of such withdrawal.

Regressive (strat.) Pertaining to the sedimentary record of a regression, wherein the boundary between non-marine and marine facies or shallow- and deeper-water facies shifts seaward or toward a basin center with decreasing geologic age, in offlap arrangement.

Resistivity The resistance of a unit cross-sectional area per reciprocal unit length, expressed in ohm-meters^2/meter.

Rhodolite (sed.) A nodule of $CaCO_3$ composed largely of encrusting coralline algae arranged in more or less concentric layers about a core; generally cream to pink, spheroidal but with knobby surface, and up to several cm in diameter; forms in warm, clear, shallow sea water (*syn*: rhodolith).

Rift In tectonics, a relatively long and narrow fault-bounded trough or graben system.

Rudist(idae) A marine bivalve mollusc family, characterized by an inequivalve shell, usually attached to substrate; solitary or gregarious in reef-like masses.

Rudite A consolidated sedimentary rock composed of rounded or angular fragments coarser than sand (>2mm).

Rudstone A textural type of coarse-grained limestone grain supported by fragmented constituents mostly >2mm in diameter.

Sabkha A "salt" flat, infrequently inundated; essentially equivalent to arid supratidal; found in some inland and coastal arid settings.

Sapropel Unconsolidated, jelly-like organic ooze or sludge composed of plant remains most often algae, putrefying in an aqueous anaerobic environment, and said to be a precursor of petroleum.

Secondary porosity The porosity developed in a rock formation subsequent to its deposition or emplacement.

Seismic Pertaining to an earthquake or earth vibration, including those that are artificially produced.

Self potential The combination of electrochemical and electrokinetic potentials of fluid-saturated rock masses as measured between a movable electrode and a fixed electrode in well logging; abbrev. SP (*syn*: Natural potential).

Shelter porosity A type of primary interparticle porosity defined by Choquette and Pray (1970) as the porosity created by the sheltering effect of relatively large sedimentary particles that prevent the infilling of pore space beneath them by finer sediment.

Siliciclastic Pertaining to clastic noncarbonate rocks, or to sedimentary fragments of previous rocks, comprised dominantly of silicon-rich minerals such as quartz or feldspars.

Skeletal Pertaining to $CaCO_3$, or less commonly SiO_2 or calcium phosphate, derived from the hard parts secreted by or associated with organic tissue; may be *syn.* with bioclastic.

Smectite Group name for clay minerals with layer charge between 0.2 and 0.6 per formula unit. Principal species: Montmorillonite, beidellite, saponite, etc.

Solution breccia A collapse breccia formed where soluble material has been partly or wholly removed by solution, thereby allowing the overlying rock to settle and become fragmented.

Sparite A descriptive term for clear, transparent or translucent, relatively coarse crystalline carbonate cement (usually calcite) of post-depositional origin. May be *syn.* with calcspar or sparry limestone (*adj.* sparry).

Sparry *see* Sparite.

Spicules Minute calcareous or siliceous skeletal elements, commonly elongate or rod- or needle-shaped, which occur often in interlocking arrays and serve to stiffen and support tissues and organs of invertebrates such as sponges.

Stenohaline Referring to a narrow tolerance range of salinity by one or a group of organisms.

Stillstand A condition of stability with reference to sea level; applicable to an undisturbed sea level relative to an area of land.

Stoichiometry That measure of the ideal or theoretical chemical proportions established for a mineral compound or phase.

Strandline The ephemeral line or level at which a body of standing water such as the sea, meets the land; the shoreline.

Stromatolite An organo-sedimentary structure produced by sediment trapping, binding and/or precipitation as a result of the growth and metabolic activity of micro-organisms, principally cyanophytes (blue-green algae); vary in form from nearly horizontal to columnar, domal, or spherical.

Stromatoporoid General name for any of a group of extinct sessile benthic marine organisms characterized by a calcareous skeleton and colonial, massive, sheetlike, or dendroid growth; they have been identified as varieties of hydrozoans and corals, but their systematics are in doubt.

Stylolite A thin seam, coalescence of seams, or surface of dissolution occurring within a sedimentary or metamorphic rock, commonly characterized by concentrations of insoluble residues, most notably clay, bitumen, oxides, etc. and crudely parallel to bedding in the configuration of "tooth and socket" interpenetrations.

Subaerial Said of location, of processes, or conditions operating in open air or immediately beneath land surfaces.

Subaerial diagenetic terrane Term proposed by Roehl, 1967, for a surface of subaerial or shallow meteoric diagenesis that is related to coeval sedimentary facies, most commonly of peritidal origin.

Subhedral Descriptor for a mineral crystal that is only partly bounded by planar crystal faces.

Sucrosic A synonym of saccharoidal; commonly applied to dolomite composed of euhedral or subhedral dolomite crystals, with intercrystalline pore space.

Supratidal Pertaining to the proximal shore area above high tide.

Sylvinite A mixture of halite (sodium chlorite) and, chiefly, sylvite (potassium chloride).

Syndepositional Contemporaneous with deposition.

Syngenetic Said of a mineral or other sedimentary feature formed contemporaneously with the deposition of the sediment or during very early near-surface diagenesis.

Syntaxial The manner of crystal overgrowth that maintains and extends optical continuity of a precursor crystal or grain (*syn*: epitaxial).

Talus (reef) Downslope accumulation of poorly sorted reef debris.

Telogenetic A term proposed by Choquette and Pray (1970) for the period and realm in which long-buried carbonate rocks are affected significantly by processes related to dewatering and subaerial and subaqueous erosion.

Terra rossa A reddish-brown, residual soil found as a mantle over limestone bedrock.

Terrigenous A term commonly used for marine sediment derived by erosion of the land surface.

Test   External shell of variable mineral composition and architecture secreted by invertebrates, especially protozoans of the order Foraminiferida.

Tetrahedral   Having the symmetry or forms of a tetrahedron, which is a crystal form in cubic crystals having symmetry 4 3m or 2 3.

Thallus   The body of certain simple plants such as algae, seaweeds and liverworts, that is characterized by relatively little cellular differentiation and no true roots, stems, or leaves.

Thrombolite   A cryptalgal structure like a stromatolite but lacking lamination and characterized by macroscopic clotted fabric.

Tidalite   A sediment or facies deposited by tidal tractive currents, by an alternation of tidal currents and tidal suspension deposition, or by tidal slack-water suspension deposition.

Transgression   A spread or extension of the sea over land area and the evidence of such advance.

Transgressive   Pertaining to the sedimentary record of a transgression, wherein the boundary between non-marine and marine facies or shallow and deeper water facies shifts landward or away from a basin center with decreasing geologic age, in onlap arrangement.

Trilobite   A Paleozoic marine arthropod belonging to the class Trilobita, characterized by a three-lobed, ovoid to subelliptical exoskeleton and divisable longitudinally into axial and side regions and transversely into anterior, middle, and posterior regions.

Turbidite   A sediment or rock interpreted to have been deposited from a turbidity current and characterized by graded bedding, moderate sorting, and distinctive primary sedimentary structures that are commonly arranged in orderly vertical sequence.

Unconformity   A substantial gap in the geologic record as demonstrated where one rock unit is overlain by another that is not next in the stratigraphic succession.

Vug, vuggy   An opening or openings (voids, cavities) exceeding in size the normal grain or crystal diameter of a rock matrix.

Vadose   Pertaining to that zone of partial or complete groundwater saturation subject to aeration and lying between the land surface and the phreatic zone (above the groundwater table); may not extend up to the surface.

Vitrinite   An oxygen-rich maceral group that is characteristic of vitrain (brilliant, vitreous conchoidal coal) and composed of humic material.

Wackestone   A term defined by Dunham (1962) for a carbonate rock with mud-supported texture containing more than 10 percent of grains larger than fine silt (20–50 μ) size.

Author Index

Subject Index

The *italic* page numbers refer to figures.